THE FOUNDATIONS OF MATHEMATICS

THOMAS Q. SIBLEY

St. John's University

WILEY

JOHN WILEY & SONS, INC.

Publisher   *Laurie Rosatone*
Associate Editor   *Shannon Corliss*
Editorial Assistant   *Jeffrey Benson*
Marketing Manager   *Jaclyn Elkins*
Production Manager   *Dorothy Sinclair*
Senior Production Editor   *Sandra Dumas*
Senior Designer   *Madelyn Lesure*
Production Management Services   *Aptara, Inc.*

This book was set in *10/12 Times New Roman* by *Aptara, Inc.* and printed and bound by *Integrated Book Technology, Inc.* The cover was printed by *Integrated Book Technology, Inc.*

The book is printed on acid free paper. ∞

To order books or for customer service, please call 1-800-CALL WILEY (225-5945).

ISBN-13   978-0470-08501-1

Printed in the United States of America

10 9 8 7 6 5 4 3 2

to Jennifer
friend, companion, wife

CONTENTS

PREFACE

"Pure Mathematics is, in its way, the poetry of logical ideas."
— Albert Einstein (1879–1955)

Over the past two decades, mathematicians have realized the need to provide a transition from problem-solving courses to proof-based courses. Naturally, no one course can accomplish all the preparation that students need in order to succeed in proving mathematical statements. Even so, many mathematics departments designate some course early in the major that provides special emphasis and instruction on proofs. This book is a text for such a course and is intended for students who have successfully finished some calculus and possibly other college-level mathematics courses. It builds on the mathematical sophistication that students acquire in calculus, but, outside of Chapter 8, only occasionally draws on that content. Of course, there are many texts available now for the transitions/foundations course. I have taught from several and looked at many more. Clearly, I think there is a need for a different, improved approach, since I am writing yet another text.

In teaching the art of proving theorems, I have repeatedly observed many students overwhelmed with the need to master several aspects at once. Digesting formal definitions of new concepts while learning proof techniques and thinking of clever ways to argue from hypothesis to conclusion simply seem too much for many emerging mathematics majors. My main goal in writing this text is to help students succeed in all of these aspects of doing mathematics. For many students, such success seems to depend on separating the acquisition of these skills. In response, I introduce concepts in Part I at least one section before students need to use them in proofs. Chapter 1 introduces mathematical language, including logic. It provides experience in understanding and working with formal definitions. Chapter 2 discusses proof formats, using the definitions from Chapter 1. In addition, many of the more demanding proofs of that chapter appear in the final section, after students have had time to learn proof formats. The early sections of Chapters 3 and 4 (and 5 in Part II) introduce the key definitions, develop intuition about these concepts, and provide practice using them in the problems. Later sections in those chapters turn to proofs involving these concepts. Chapters 6, 7, and 8 in Part II introduce the themes of different areas of undergraduate mathematics. As such, they expect students to master definitions more quickly than in earlier chapters. Nevertheless, I try to fully motivate definitions and prepare students to use new definitions to find proofs.

The text includes many carefully explained examples to facilitate student learning. The symbol ◊ indicates the end of an example. Starting with Chapter 2, the examples often include proofs, along with a discussion of important aspects of the proofs. The symbol ■ announces the end of a proof. Along with model proofs, the book includes numerous questionable arguments, which end with the symbol ▲.

A number of the problems in earlier sections are used in later sections. At the end of these problems, I have noted the sections that refer to them. Problems with hints or solutions at the back of the book are noted with an * before the problem.

Problems are the heart of any textbook. I have tried to include a larger variety of problems than most textbooks for this transitions course. I have also tried to include a healthy range of difficulty in the problems. Of course, I include the standard types of problems, such as many requests for proofs and counterexamples. In addition, most sections start with a question testing reading comprehension. Once we get to proofs, most sections also end with questions about incorrect or questionable reasoning. I have included problems asking students to formulate conjectures and prove them. A number of problems ask students either to prove the statement or give a counterexample. The sections on metamathematics and philosophy of mathematics have very different exercises, especially requests for essays.

My second motivation for writing this text is to breathe into mathematics at this level some of the poetry that Einstein refers to in the quote opening this preface. Mathematicians respond to the excitement and beauty of the ideas of mathematics, as well as to the logical clarity. Students deserve to experience this excitement and beauty while they practice achieving logical clarity. This text and especially the problems will, I hope, help instructors convey the spirit of mathematics as well as the skills. Also, the final chapter on metamathematics and philosophy of mathematics provides students with a broad vision of mathematics, which is an important part of this spirit of mathematics. Unfortunately, these last topics don't seem to fit as readily into the standard curriculum as I think they should. I believe that the foundations or transitions course is the best place to include this material.

CHAPTER DESCRIPTIONS

Chapter 1 starts with the logical language needed for proofs and mathematics in general. Mathematicians have learned to turn everyday English with its hidden and misplaced quantifiers into formal mathematics. However, students often struggle to extract the precise mathematical meaning of sentences written in informal English. They first need, I believe, to spend time understanding standard mathematical wording before converting informal sentences into mathematical ones. Thus, I have tried throughout the book to give unambiguous, mathematical statements, although I have also included more natural wording in many instances. The descriptions of "working negations" make explicit how negations of mathematical sentences are formed. These descriptions appear in the first three sections of Chapter 1 and, for the negation of uniqueness, in Section 2.3.

I also discuss mathematical definitions and the difference between a definition and the intuitive concept it formalizes. Many of the problems should help students build bridges between definitions and intuitions. As a starter, the first problem of each section gives a number of true/false questions reinforcing reading comprehension. (This type of question appears in most later sections, as appropriate.) Later problems in many of the sections develop the understanding and use of definitions. Students will also develop a number of conjectures, providing propositions to prove or disprove, in Chapter 2. I encourage instructors to supplement the logical sections with some of the many entertaining logic puzzles found in many collections of recreational mathematics. (Problems 12 and 13 in Section 1.2 use two such puzzles. These can be used to motivate various proof formats in Chapter 2.)

Chapter 2 draws on the earlier introduction to definitions and logical language to present theorem proving. The first four sections introduce the basic formats for proofs. Examples of proofs in each format include discussion about how the proof is written and comments on aspects that some students find difficult. The final section discusses a variety of related issues on proofs. Most of its problems draw on the different proof formats, without any indication of which one might be successful. Also, a number of more challenging proofs appear in this last section. The next-to-last problem of each section gives several incorrect arguments for students to critique. The last problem of each section gives a mix of arguments, some correct, some incorrect, and some in between. (Later chapters include such questionable arguments, as appropriate.) I think both kinds of problems are good for students, depending on their strengths and weaknesses. It is even better for students to present their own arguments and have classmates critique them. However, such efforts require an appropriate classroom atmosphere—supportive and yet willing to challenge respectfully. In each section of Chapter 2 and the remaining chapters involving proofs, the proofs requested range from routine to challenging. Some of the problems provide outlines to focus students' development of proofs.

I should note particularly the switch in order of Chapters 3 and 4 from most texts. Relations logically precede functions, which are a special case of relations. However, my students find relations noticeably harder to treat in the abstract, since they have never before thought about general relations. Their experience in calculus makes a general function easier to approach. In turn, functions as examples of relations should make general relations more natural.

In Chapter 3, students' prior understanding of functions motivates formal definitions in Section 3.1 before they are asked to prove results. The second and third sections emphasize proofs of properties of functions. Students too often feel little connection among the topics of a foundations course. To address this frustration, the second and third sections discuss how these formal definitions involving functions serve some other areas of mathematics. Chapter 4 continues this motivational discussion.

Chapter 4, following the format of Chapter 3, introduces the general idea of a relation and its properties from numerous particular examples. Additionally, a number of problems ask students to give a variety of examples of relations in different contexts. Section 4.2 starts proofs involving relations, focusing on equivalence relations and their defining properties. The third section addresses the more involved connection between equivalence relations and partitions. The final section on partial orders is optional and can be omitted if time is a factor. There is a large variety of terms connected with partial orders. Instead of squeezing them all in the text of the first section, I ask students to develop them through examples in Problems 14 and 15 of Section 4.1. Section 4.4 on partial orders explicitly defines these terms.

Chapters in Part II introduce a variety of topics, which are essentially independent of one another. I include, as appropriate, connections among these chapters and to earlier chapters.

Chapter 5 explores infinite sets and cardinality, trying to communicate some of the wonder of the infinite as well as the careful reasoning. After the standard results on countable and uncountable sets, Section 5.4 gives an optional exposition on the Axiom of Choice and Zorn's Lemma. I don't expect students encountering proofs for the first time to be able to use these sophisticated assumptions. Still, I think students headed for graduate school benefit greatly from studying and trying their hand at proofs involving them. Since upper division texts appropriately focus on their particular subjects, it is now too easy for even willing and capable undergraduates to miss learning to employ these techniques. Section 5.4 discusses these proof techniques carefully, including problems that outline proofs using them.

Chapter 6 introduces the broad topic of discrete mathematics. While many schools require a separate course in this subject, others expect the essentials of discrete mathematics to appear in a transitions course. I have tried to limit this chapter to the material on graph theory and combinatorics that I think all mathematics majors should know. I have sought to keep the number of graph theory terms to a manageable number, while giving a flavor of the area. The section on trees emphasizes the concept of an algorithm, an essential idea for computer science. Problem solving is the central concern in the sections on elementary combinatorics, although I don't discount proofs.

Students often find the required abstract algebra and analysis courses the most demanding ones in a mathematics major. I intend for my chapters on these areas to provide a gentle introduction to understandable examples and some basic ideas underlying these areas, rather than a quick tour of the main ideas.

The abstraction of algebra, introduced in Chapter 7, too often strikes students as mysterious, intimidating, or both. I seek to provide a way into this characteristically modern approach to mathematics. Solving equations provides a fruitful theme in the first three sections to connect students' prior experience to abstract algebra. In Section 7.3, I also discuss the role of distributivity in familiar arithmetic rules. I seek to illustrate the power of structure to make the familiar transparent. The first three sections would be especially appropriate for future secondary school teachers. For instructors with the time and inclination to dig more deeply, I include some more structural aspects. These include subgroups and Lagrange's Theorem in Section 7.2 and, later, lattices and homomorphisms. Section 7.4 on lattices builds on the students' understanding of partial orders and number theory. While few abstract algebra courses have time to investigate lattices, I think it worthwhile to make these connections with earlier material in the text. In the section on homomorphisms, I try to illustrate how a homomorphism can help us understand a complicated system by connecting it with a better understood and simpler system.

Analysis, the topic of Chapter 8, depends on the sophisticated definition of a limit with its alternation of quantifiers. My experience suggests that students need significant preparation to digest this definition. The initial section explores the real numbers in some depth. It discusses the intuitive idea of a number line, but also addresses infinite decimals. Although students generally "know," for example, that $0.999\ldots = 1$, they often don't really believe it. Sections 8.1 and 8.4 attempt to provide sufficient grounding to enable students to connect formal definitions of decimals and infinite series with their previous understanding of decimals. Future high school teachers, in my view, would particularly benefit from such an understanding. Section 8.1 also introduces the formal definition of a complete ordered field. Although most analysis texts treat sequences and their limits first for good pedagogical reasons, I reverse the order here for different pedagogical reasons. I am not trying to do analysis, but rather trying to enable students to succeed in an analysis course. Since students understand more about real functions and their limits than sequences, I introduce limits of real functions in Section 8.2. Section 8.3 builds on limits to discuss continuous functions and explore counterexamples connected with continuity. I seek in that section to explain the important role that counterexamples play, something students starting analysis often don't appreciate. Students' experience in calculus makes questions about series more natural for them than questions about sequences. Hence, Section 8.4 on sequences and series introduces series more quickly than analysis texts do. Additionally, series allow a definition of decimal representation in this short introduction. The final section gives a taste of the modern area of dynamical systems, building on the elementary analysis topics already introduced. I think such an enticement is pedagogically more appropriate at this level than a quick exposition of open and closed sets or proving major analysis theorems, which are better reserved for a full analysis course.

Chapter 9 provides a short introduction to metamathematics and the philosophy of mathematics. Since few undergraduates take a mathematical logic course, this text may be the only one they use that discusses metamathematics. Yet, majors deserve to know about several metamathematical results, especially Gödel's Incompleteness Theorems. I make no attempt to prove any of these results. My students find the material on philosophy of mathematics a good way to close the course. They appreciate reflecting on the nature of mathematics and proofs after a strenuous term of doing mathematics. It also enables the introduction of a broader view of mathematics and a way to tie the many topics together.

PEDAGOGICAL ADVICE

In a semester, I have been able to cover the first four chapters, plus nine sections from Part II. I would think a quarter-long course could cover at least the first four chapters.

While it makes organizational sense in the book to put all of the proof instruction together, I don't advocate doing all of Chapter 1 before Chapter 2. The first four sections of Chapter 1 are intended to be done before starting proofs. After these sections, the last two sections can be interspersed with Chapter 2 material. Instructors deciding to do such interspersing need to judge the pace of introducing new techniques of proof versus introducing new content. In my experience, students benefit from some breaks from learning and practicing proof formats. Of course, such a nonlinear approach requires the instructor to pick among the problems in each section of Chapter 2, depending on the definitions already covered. One advantage of this alternation is the chance to revisit earlier proof techniques while using more recent content from Chapter 1. This alternating between Chapters 1 and 2 may provide a second advantage with regard to learning the content. I suspect that students at this level should not experience a gap of more than one or two classes between learning a new concept and first using it in proofs, something enabled by this alternation. To facilitate flexibility in using Chapter 1, a list of its definitions appears at the end of the chapter. I have also put the elementary algebraic and ordering properties of the fundamental number systems there. Some instructors want their students at this level to justify their algebraic manipulations, and this listing will support that focus. For equally valid pedagogical reasons, other instructors choose not to insist on such justifications. A listing of proof formats occurs at the end of Chapter 2, to facilitate student reference. Similarly, later chapters end with a list of definitions from that chapter.

As is usual, this text uses T for true and F for false in truth tables. There is a risk that students will misread these letters in a table filled with Ts and Fs, so the instructor may find it helpful to emphasize paying attention to the differences. Alternatively, instructors can adopt the computer science notation of 1 for true and 0 for false.

The sections follow the division of content rather than pedagogical pacing. Instructors should feel free to devote more than one class period to a section. In particular, I devote at least two classes to each section of Chapter 2, to give students time to digest and practice proof techniques.

ACKNOWLEDGEMENTS

I want to thank a number of people for their support in this project. First of all, I am indebted to Laurie Rosatone and Shannon Corliss, my editors, who have encouraged me and shown their confidence in me by committing John Wiley and Sons to paying the costs of the initial versions and classroom testing. My friends and colleagues who have agreed over several semesters to try out the first two chapters, and more recently the entire text, sight unseen, deserve a thank-you for their trust and willingness to give honest feedback. These professors are Dr. Ivy Knoshaug at Bemidji State University, Dr. Tom Linton and Dr. Wendy Weber at Central College, Dr. Carolyn Yackel at Mercer College, Dr. Steve Walk at St. Cloud State University, Dr. Michael Gass at St. John's University, and Dr. Philip Byrne and Dr. Gary Brown at the College of St. Benedict. In addition, their students deserve my appreciation for unwittingly performing a crucial test of these earlier versions. I appreciate the feedback I received from them, as well as Dr. Paul Campbell, Dr. Gary Brown, and Dr. Jennifer Galovich, who critiqued earlier parts of this text. I also owe a debt of gratitude to the many reviewers who carefully read and commented on the text in various stages. These include Carolyn Yackel at Mercer College, Lyn Phy at Kutztown University, Jonathan White at Coe University, Stephen Edwards at Southern Polytechnic State University, Alexander Richman at Bucknell University, Beimnet Teclezghi at New Jersey City University, and others wishing to remain anonymous. The final version has benefited from all of their many helpful suggestions, although the responsibility for any remaining errors or other weaknesses remains with me. I am grateful to St. John's University for the sabbatical during which I wrote the bulk of this text. Finally, I want to thank my wife, Jennifer Galovich, who, in addition to feedback on some of the material, has given me love, support, and encouragement throughout the planning and the many months of writing.

I welcome suggestions for improving this book and questions about it. I hope you find the book clear, helpful, and interesting.

Thomas Q. Sibley
tsibley@csbsju.edu
Department of Mathematics
St. John's University
Collegeville, MN 56321-3000

CHAPTER *1*

LANGUAGE, LOGIC, AND SETS

MATHEMATICS HAS EARNED a reputation of supplying indisputable answers to well-formed questions. From calculations to proofs to mathematical models, people depend on mathematics for a large variety of uses. Our confidence in mathematics rests on the precision of mathematical language and the explicit, careful reasoning required in mathematics as well as the verifiable exact answers it provides. This book introduces the language, concepts, and reasoning common to mathematics. In order to better enable the reader to understand both the terminology and the reasoning, we will in the earlier part of the book introduce a concept at least one section before using it in proofs. Indeed, we defer proofs until Chapter 2, focusing in this chapter on the language and concepts of logic, sets, and numbers before encountering the formats and subtleties of proofs.

1.1 LOGIC AND LANGUAGE

"Logic is the hygiene which the mathematician uses to keep his ideas healthy and strong."
—Hermann Weyl (1885–1955)

Ordinary English benefits from a rich range of nuances of words, including the logical connectives discussed in Sections 1.1, 1.2, and 1.3: "and," "or," "not," "if . . . , then," "for all," and so on. However, mathematical language and particularly proofs require precise and sometimes artificial use of language. In these sections we make explicit our interpretations of the basic logical words and how we can combine them. As in the first sentence of this paragraph, when we talk about words and sentences, we place them in quotation marks.

Mathematical Propositions

Mathematical sentences, like sentences in any language, must be grammatical to be understood. However, we need to restrict mathematical statements further if we are to have a chance of proving or disproving them. In particular, they must qualify as correct or incorrect, even if we don't know which at the start. Thus, "$2 < 4$" is true, "$4 < 2$" is false, and "The billionth prime, when written in base 10, ends

3

in a 3" is one or the other, even if we don't know which one.* However, "$x < y$" can't begin to be classified as correct or incorrect, because it has variables in it. We will address sentences with variables in Section 1.3. Even more removed from mathematical proofs are sentences that can't be either true or false. For example, questions ("Is $\pi < \sqrt{10}$?") and commands ("Factor $x^2 + 2x + 1$.") fail to be true or false. More subtle are paradoxical sentences such as "This sentence is false." Suppose the sentence in quotes were true. Since the sentence talks about itself, we see the sentence says it is false, which is impossible. Suppose, instead, the sentence were false. Then the meaning of the sentence would make it true, which is also impossible. Hence it can neither be true nor false. We call a sentence a *proposition*, provided that it is true or false, but not both. Mathematicians simply side-step all the problems associated with variables, questions, and paradoxical sentences by restricting our efforts to propositions.

Truth Tables

Elementary logic starts with building complex propositions from simpler propositions, using the logical words "and," "or," "not," "if . . . , then," and others. The fundamental nature of these words makes it seem pointless to define them, yet their mathematical use is generally quite restricted compared with English usage. Truth tables help make their mathematical use explicit.

And   We start with the least controversial logical word, "and." The following example, as all examples in the book, ends with the symbol $\Diamond$:

EXAMPLE 1

From the simple propositions "$2^3 > 3^2$" and "A minute has 60 seconds," we can build the compound proposition "$2^3 > 3^2$ and a minute has 60 seconds." Of course, "$2^3 > 3^2$" is false, making the compound proposition false. The modified proposition "$2^3 < 3^2$ and a minute has 60 seconds" is correct, since each of its parts is correct, even though they have nothing to do with one another. In English we often employ the conjunctions "but," "although," and other words to make subtle distinctions from "and." For example, we might say "$2^3 < 3^2$, but $2^5 > 5^2$." Logically, "but" functions in this sentence as "and" would. In formal mathematics we use only "and." $\Diamond$

Rather than inventing other compound propositions, let's look at the logical form more closely. If P and Q represent any propositions and $\wedge$ represents "and," then $P \wedge Q$ represents a generic compound proposition. We call P, Q, and $P \wedge Q$ *propositional forms* because they show the logical form of propositions. Logicians and computer scientists use a *truth table* to summarize the truth values a compound propositional form can have based only on the truth values of its parts. Our understanding of a proposition restricts us to two-valued logic with the truth values T for true and F for false. (See Problem 9 for another option.) There are four rows in the

---

*Proponents of the constructivist philosophy of mathematics would not consider the sentence about the billionth prime to be a proposition, since no one can currently classify it. We will briefly discuss such philosophical issues in Chapter 9.

truth table that follows, representing all of the possible options for the truth values of P and of Q. For instance, the first row lists T under P and T under Q, representing the situation when P is true and Q is true. The final column indicates the truth value of the entire propositional form, here $P \wedge Q$, based on the truth values of P and Q. The only way the compound propositional form $P \wedge Q$ can be true is when both P and Q are true. Hence, the first line of the truth table has a T under $P \wedge Q$, whereas the other rows have an F under $P \wedge Q$. For particular propositions, as in Example 1, $P \wedge Q$ has just one truth value, which depends on the values of P and Q. Thus, the proposition "$2^3 > 3^2$ and a minute has 60 seconds" is an instance of the third line of the table. Similarly, "$2^3 < 3^2$ and a minute has 60 seconds" is an example of the first line.

Truth table for "and"

P	Q	$P \wedge Q$
T	T	T
T	F	F
F	T	F
F	F	F

Or In English, context helps us distinguish different meanings of the word "or." For example, when a server in a restaurant asks "Would you like salad or soup with your meal?" the options do not include both soup and salad. But the question "Would you like cream or sugar with your coffee?" from the same server does allow the customer to request both. While we readily handle such ambiguity in restaurants, in proofs we need agreement on which meaning we will follow. Mathematicians always use "or" inclusively: Whatever P and Q represent, "P or Q" always includes the option "P and Q." The following truth table summarizes this usage, where $\vee$ represents "or":

Truth table for "or"

P	Q	$P \vee Q$
T	T	T
T	F	T
F	T	T
F	F	F

Not The word "not," represented symbolically by $\neg$, is mathematically as straight-forward as the following truth table indicates:

Truth table for "not"

P	$\neg P$
T	F
F	T

However, double negations in English and other languages allow for more variation than the next truth table would indicate:

P	$\neg P$	$\neg (\neg P)$
T	F	T
F	T	F

In Spanish, a double negative generally makes the negation more emphatic, rather than cancelling it. As the accompanying cartoon illustrates, double negations enable subtle distinctions in English. Nevertheless, in mathematics we accept the straightforward interpretation of the truth table.

Logically Equivalent

The use of symbols emphasizes the logical form. By using these symbols repeatedly, we can readily investigate complicated compound propositional forms. We use parentheses to eliminate ambiguity, an option we don't have with actual English sentences.

EXAMPLE 2

Compare the truth table of $P \wedge (Q \vee R)$ with those of $(P \wedge Q) \vee R$ and $(P \wedge Q) \vee (P \wedge R)$.

With three variables, we need more rows to list all of the possible combinations of the truth and falsity of the individual propositional forms. Since each of P, Q, and R has two options, we need $2 \times 2 \times 2 = 8$ rows. Only the first three columns and the last columns of the truth tables shown next are necessary, but the intermediate columns indicate steps helpful in determining the final column. In order to compare the truth tables of different compound propositions, it is important that the ordering of the rows match. For instance, we uniformly put the variables in alphabetical order and list all the options, with P having the truth value T first. Among these options, we list those for which Q has the truth value T before those with an F, and so on.

P	Q	R	$Q \vee R$	$P \wedge (Q \vee R)$
T	T	T	T	T
T	T	F	T	T
T	F	T	T	T
T	F	F	F	F
F	T	T	T	F
F	T	F	T	F
F	F	T	T	F
F	F	F	F	F

P	Q	R	$P \wedge Q$	$(P \wedge Q) \vee R$
T	T	T	T	T
T	T	F	T	T
T	F	T	F	T
T	F	F	F	F
F	T	T	F	T
F	T	F	F	F
F	F	T	F	T
F	F	F	F	F

P	Q	R	$P \wedge Q$	$P \wedge R$	$(P \wedge Q) \vee (P \wedge R)$
T	T	T	T	T	T
T	T	F	T	F	T
T	F	T	F	T	T
T	F	F	F	F	F
F	T	T	F	F	F
F	T	F	F	F	F
F	F	T	F	F	F
F	F	F	F	F	F

Since two of the rows of the first two truth tables differ, we see that "P and Q or R" can mean different things logically, depending on how we group the simple propositional forms. Mathematicians use parentheses to ensure clarity. However, ordinary English sentences aren't always clear, as many sentences on income tax forms and in laws demonstrate. Returning to our example, it is somewhat surprising that the final columns of the truth tables for $P \wedge (Q \vee R)$ and $(P \wedge Q) \vee (P \wedge R)$ are the same. This matching echoes the distributive law of multiplication over addition: $p \times (q + r) = (p \times q) + (p \times r)$. In terms of the following definition, $P \wedge (Q \vee R)$ and $(P \wedge Q) \vee (P \wedge R)$ are *logically equivalent*. ◇

DEFINITION. Two propositional forms are *logically equivalent* provided that their truth values are equal exactly when the truth values of their corresponding component propositional forms are equal.

EXAMPLE 3

Verify the first of De Morgan's Laws: $\neg (P \wedge Q)$ is logically equivalent to $\neg P \vee \neg Q$.

Solution

The truth tables give

P	Q	$P \wedge Q$	$\neg (P \wedge Q)$
T	T	T	F
T	F	F	T
F	T	F	T
F	F	F	T

and

P	Q	$\neg P$	$\neg Q$	$\neg P \vee \neg Q$
T	T	F	F	F
T	F	F	T	T
F	T	T	F	T
F	F	T	T	T,

verifying De Morgan's Law. ◇

EXERCISE. Verify De Morgan's other law: $\neg (P \vee Q)$ is logically equivalent to $\neg P \wedge \neg Q$.

De Morgan's Laws show a formal relationship among the logical words "and," "or," and "not" that we will need in our work. Ordinary English uses this relationship as well, although not formally. For example, a child might say, "I want cake and ice cream." Symbolically, we can satisfy the child with $C \wedge I$ (cake and ice cream). A dissatisfied child might correspond to $\neg (C \wedge I)$. And, indeed, that could happen when there is no cake ($\neg C$) or no ice cream ($\neg I$), just as De Morgan's Law indicates: $\neg C \vee \neg I$.

Working Negations

We often need to negate propositions in proofs, but simply putting a "not" in front usually doesn't help much. In general, we need to move the "not" from the front of a compound proposition to inside of the proposition, to the level of the component propositions. Example 3 and the exercise following it provide the logical means to move the "not." That is, we can rewrite $\neg (P \wedge Q)$ as $\neg P \vee \neg Q$. Because this latter form is more useful, we call it a *working negation*. Because $\neg (\neg P)$ is logically equivalent to P, we never need double negations in working negations. We'll consider working negations further in Sections 1.2 and 1.3.

WORKING NEGATION. The *working negation* of $P \wedge Q$ is $\neg P \vee \neg Q$. The *working negation* of $P \vee Q$ is $\neg P \wedge \neg Q$.

EXAMPLE 4

Write the working negation of $(A \vee B) \wedge \neg C$.

Solution The easiest way to write the negation is simply to put a "not" in front: $\neg((A \vee B) \wedge \neg C)$. However, such a form doesn't help much in proofs. We need to shift the negation inside. Moving the "not" in one layer negates the two parts of the "and" statement and switches the "and" to "or." This gives us $\neg(A \vee B) \vee \neg\neg C$. We don't need a double negation, so we can simplify this expression to $\neg(A \vee B) \vee C$. Finally, we can move the remaining negation inside further to get $(\neg A \wedge \neg B) \vee C$, which is the working negation. ◇

EXERCISE. Write the working negation of "You like chocolate or you are really odd."

Tautologies and Contradictions

Two types of propositions deserve special names: those that are always true and those that are always false.

DEFINITION. A compound propositional form is a *tautology* if and only if its truth value is always T regardless of the truth values of its simple propositional forms. A compound propositional form is a *contradiction* if and only if its truth value is always F regardless of the truth values of its simple propositional forms.

EXAMPLE 5

The "law of the excluded middle," $P \vee \neg P$, and the "law of noncontradiction," $\neg(P \wedge \neg P)$, are tautologies, as illustrated by the following truth tables:

P	$\neg P$	$P \vee \neg P$		P	$\neg P$	$P \wedge \neg P$	$\neg(P \wedge \neg P)$
T	F	T		T	F	F	T
F	T	T		F	T	F	T

The law of the excluded middle, in effect, eliminates a third truth value: Either a proposition is true or it is false (and its negation is therefore true). The law of noncontradiction effectively prohibits a proposition from being both true and false. While these two laws are logically equivalent, they highlight different aspects of two-valued logic. From the third column of the second truth table, we see that $P \wedge \neg P$ is a contradiction. ◇

REMARKS. Although many of the problems in Sections 1.1 and 1.2 involve truth tables, students should realize truth tables are not a focus of mathematics, nor do they prove theorems. They do, however, have significant roles in computer science and in mathematical logic. In general, mathematics is anything but mechanical. Indeed, much more than logic is needed to do mathematics, especially to prove theorems. We need insight into how the ideas fit together. Such insight often requires considerable effort, including trial and error. Successful mathematicians and students combine tenacity and curiosity with their logical skills.

Historical Remarks

People have used logical words and arguments for thousands of years. The ancient Greeks looked at arguments extensively, but the mathematical structure of logical words is much more recent. Gottfried Wilhelm Leibniz (1646–1716) advocated making a symbolic "calculus" for reasoning, encompassing logical words. While some people following his lead made significant progress, their work was largely forgotten.

Nineteenth-century mathematicians used a new approach to algebra to make important progress in logic. Augustus De Morgan (1806–1871) and George Boole (1815–1864) emphasized general formal properties in both areas, such as commutativity $(a + b = b + a)$ and distributivity $(a(b + c) = ab + ac)$, rather than methods of solving equations, an earlier focus of algebra. In logic, they used $+$ for "or" and $\cdot$ for "and," emphasizing the analogy of logic with algebra. Hence, the commutativity and distributivity laws hold in logic as well as algebra. Continuing this analogy, Boole used 1 for the "universal class" and $1 - x$ to represent "not x." Thus, he turned De Morgan's wording of the laws now named for him into the symbolic equations $1 - (x + y) = (1 - x)(1 - y)$ and $1 - xy = (1 - x) + (1 - y)$. Of course, these equations are not true for polynomials in algebra.

This new emphasis on formal properties and symbolism led to a separation between what symbols mean and the rules for using them. This emphasis also appears in the idea of truth tables for logical words, which proved a fruitful idea starting in the nineteenth century. Much more recently, modern computers depend explicitly on truth tables to carry out logical instructions. For more historical background see Kline, 596, 1188–1191.

PROBLEMS

(Note: An * before a problem or a part of a problem indicates that an answer or hint appears in the Selected Answers section at the back of the book.)

*1. For each statement that follows, decide whether it is true or (at least sometimes) false. If false, explain why. Cite the part of the text supporting your conclusion.

a) Mathematics and ordinary English use logical words in the same way.

b) The double negation of a proposition is logically equivalent to the original proposition.

c) When mathematicians assert "X or Y," they exclude the possibility of both X and Y being true.

d) The working negation of "A and B" is "not A and not B."

e) The working negation of "A or B" is "not A and not B."

f) We can use truth tables to prove mathematical theorems.

g) The negation of a contradiction is a tautology.

2. Determine which of the following sentences or pairs of sentences are propositions. For those that are not, explain why not. (Parts c and d are adapted from Carroll, 183–188.)

   a) $2^{101} - 1$ is prime.

   b) An insect has six legs or a spider has nine legs.

   ***c)** The sea is boiling hot.

   ***d)** Do pigs have wings?

   e) This sentence has three errors.

   f) The following sentence is true. The preceding sentence is false.

3. Write the truth table for each of the following propositional forms:

   ***a)** $A \vee (B \wedge \neg A)$.

   b) $A \wedge \neg (A \wedge B)$.

   ***c)** $(A \vee B) \wedge (A \wedge C)$.

   d) $(\neg A \vee B) \wedge (C \vee \neg B)$.

   e) $A \wedge \neg (B \wedge \neg C)$.

   f) $(A \vee \neg C) \wedge (B \vee \neg A) \wedge (C \vee \neg B)$.

   g) $(A \vee \neg B) \wedge (C \vee \neg D)$.

4. Verify that the given pairs of propositional forms are logically equivalent. The corresponding algebraic property is indicated after each (referred to in problems in Sections 1.4 and 1.5).

   a) $P \vee Q$ and $Q \vee P$ (commutativity)

   b) $P \wedge Q$ and $Q \wedge P$ (commutativity)

   c) $P \vee (Q \vee R)$ and $(P \vee Q) \vee R$ (associativity)

   d) $P \wedge (Q \wedge R)$ and $(P \wedge Q) \wedge R$ (associativity)

   e) $P \wedge (Q \vee R)$ and $(P \wedge Q) \vee (P \wedge R)$ (distributivity)

   f) $P \vee (Q \wedge R)$ and $(P \vee Q) \wedge (P \vee R)$ (distributivity; referred to in Section 2.1.)

   g) $P \vee (P \wedge Q)$ and P (absorption)

   h) $P \wedge (P \vee Q)$ and P (absorption)

5. Determine which of the given pairs of propositional forms are logically equivalent. Justify your answers.

   ***a)** $J \wedge \neg K$ and $\neg (K \wedge \neg J)$.

   b) $L \vee \neg J$ and $\neg (J \wedge \neg L)$.

   ***c)** $J \vee \neg (K \vee L)$ and $(J \vee \neg K) \wedge (J \vee \neg L)$.

   d) $K \wedge (L \vee \neg J)$ and $\neg (L \wedge (J \vee K))$.

   e) $(L \wedge \neg J) \vee K$ and $(K \vee L) \wedge \neg (J \wedge K)$.

   f) $(K \vee J) \wedge \neg L$ and $\neg (L \vee (\neg J \wedge \neg K))$.

6. Write working negations of the following propositions and propositional forms:

   ***a)** We will win the next game, or we won't win the tournament.

   b) You are my best friend, but I won't tell you my secret.

   c) $X \vee (Y \wedge Z)$.

   d) $Y \wedge (\neg Z \wedge \neg X)$.

   ***e)** $(X \vee \neg Y) \wedge \neg (Z \wedge \neg X)$.

   f) $Z \wedge ((X \vee \neg Y) \wedge W)$.

7. Determine whether each of the given propositional forms is a tautology, a contradiction, or neither. Justify your answer.

   *a) $(F \wedge \neg G) \wedge (G \vee \neg F)$.

   b) $(G \vee \neg H) \vee (H \wedge \neg G)$.

   *c) $(F \wedge \neg H) \vee (H \wedge \neg G)$.

   d) $(H \wedge \neg (G \vee \neg F)) \wedge (F \wedge \neg H)$

   *e) $(H \vee (F \wedge \neg G)) \vee ((G \wedge F) \vee \neg H)$.

   f) $\neg (G \vee H \vee \neg F) \wedge (F \wedge H \wedge \neg G)$.

   g) $(F \vee (G \wedge H)) \vee (G \vee \neg H)$.

8. ***a)** Explain why all propositional forms with just one variable are logically equivalent with one of the following four: P or $\neg P$ or $P \vee \neg P$ or $P \wedge \neg P$.

   b) Explain why all propositional forms with two variables are logically equivalent with 1 of 16 possibilities.

   c) Using only $\vee$ and $\wedge$ and $\neg$, together with the variables P and Q, give examples of all 16 possibilities in part b, using the fewest number of symbols.

   d) Find a formula for the number of logically nonequivalent types of propositional forms of three variables. Repeat with n variables. Explain why your formulas are correct.

   *e) Explain how to rewrite all of the possibilities in part c, using only $\vee$ and $\neg$. Repeat, using only $\wedge$ and $\neg$.

   f) The propositional form $(P \wedge Q) \vee (\neg P \wedge \neg Q)$ is said to be in *disjunctive normal form*, which means that $\neg$ appears only next to a simple propositional form, in each set of parentheses each variable appears exactly once joined with $\wedge$ (and), and the parenthetical parts are joined with $\vee$. ("Disjunction" is a fancy word for "or"). Explain why each parenthetical part specifies one row of the truth table. Given any propositional form with any number of variables, explain why we can write a logically equivalent one in disjunctive normal form. (In effect, this part shows that computers can mimic all propositional forms in a uniform way, with circuits built from a few components. See Epp, 43–53 for more information.)

9. Because computers deal with 1s and 0s, computer scientists use 1 for true and 0 for false in truth tables. Thus, it is an advantage for computers to use arithmetic expressions to calculate the truth values of complicated logical expressions. We explore this idea and generalize it.

   *a) Verify that $P \wedge Q = \min(P, Q) = P \cdot Q$ and $P \vee Q = \max(P, Q)$. Find expressions for $P \vee Q$ and $\neg P$, using addition, multiplication, and subtraction, as needed.

   b) Logicians have considered alternatives to two-valued logic. Suppose for now we add just one intermediate truth value, $\frac{1}{2}$, to represent sentences that are sometimes true and sometimes false. Explain why truth tables for $\wedge$ and $\vee$ need to have nine rows with three truth values. Make truth tables for $P \wedge Q$ and $P \vee Q$ and $\neg P$, with the three truth values 1, $\frac{1}{2}$, and 0, using $\min(P, Q)$ for $P \wedge Q$, $\max(P, Q)$ for $P \vee Q$, and the formula you found in part a for $\neg P$.

   *c) Which of the pairs of propositional forms in Problem 4 are logically equivalent? Use these three truth values and the truth tables you found in part b. Would your answer change if you permitted more truth values between 0 and 1? (Hint: Use properties of min and max rather than truth tables.)

   d) Try to repeat part b, using multiplication for $P \wedge Q$ and the formula you found in part a for $P \vee Q$. What goes wrong? Adjust in an appropriate way, such as adding more truth values.

   e) Repeat part c for the modified truth tables of part d.

f) Think of examples of sentences that are sometimes true and sometimes false. Describe the strengths and weaknesses of your truth tables in parts b and d for handling such sentences. For more on alternative logics, see Rescher.

REFERENCES

CARROLL, LEWIS, 1976. *Complete works*. New York: Vintage Books.

EPP, SUSANNA, 2004. *Mathematics with applications*, 3rd ed. Belmont, CA.: Brooks/Cole.

KLINE, MORRIS, 1972. *Mathematical thought from ancient to modern times*. New York: Oxford University Press.

RESCHER, NICHOLAS, 1969. *Many valued logic*. New York: McGraw-Hill.

1.2 IMPLICATION

"Mathematics is nothing more, nothing less,
than the exact part of our thinking."

—L. E. J. Brouwer (1881–1966)

All reasoning depends on the "if . . . , then . . . " format, but mathematical proofs use it more restrictively than other reasoning. As in Section 1.1 we will develop a truth table to make these differences explicit. We write $P \Rightarrow Q$ to represent the *implication* "if P, then Q." As with previous truth tables, we need to place an F or T in each row on the basis only of the values of P and Q. First, some terminology: In $P \Rightarrow Q$ we call P the *hypothesis* (or assumption) and Q the *conclusion*.

Let's start with our ordinary use of "if" in English to understand the truth table that follows. Consider, for instance, the sentence "If it is raining, then the sidewalk is wet." Whenever the hypothesis P is true (it is raining), we require the conclusion Q to be true (the sidewalk is wet) in order for the whole proposition $P \Rightarrow Q$ to be true. That is, when both P and Q have the truth value T, then $P \Rightarrow Q$ is also true. However, the combination of rain and a dry sidewalk would make the sentence false. That is, when P has truth value T and Q has truth value F, the entire implication $P \Rightarrow Q$ must be false. This explanation justifies the first two rows of the truth table, which generally don't give students problems. The last two rows, however, tend to be more confusing, because English usage of implications is so variable. For our sentence, we can readily imagine scenarios for the other two logical possibilities. The sun could be shining and the sprinkler running, so that it is not raining and the sidewalk is wet. That would give the third line of the table: P with the value F and Q the value of T. A sunny day without the sprinkler running can give the fourth line of the table: It is not raining and the sidewalk is not wet. That is, both P and Q have truth value F. In ordinary English we would consider the original sentence "If it is raining, then the sidewalk is wet" true in these situations. From a mathematical point of view, English unfortunately has many situations where the truth value of an implication is more problematical when the hypothesis is false. In mathematics we need to have a uniform interpretation of "if . . . , then. . . ." As with the logic words of Section 1.1, we choose an interpretation fitting our use of implications in proofs. Example 1 illustrates why we want the third and fourth lines of the truth table to evaluate $P \Rightarrow Q$ as true when P is false. In everyday English, our mathematical rule is as follows: The only way an implication can be false is for the hypothesis to be true and the conclusion to be false.

P	Q	$P \Rightarrow Q$
T	T	T
T	F	F
F	T	T
F	F	T

Truth table for "if, then"

EXAMPLE 1

Consider the next three propositions, which illustrate, respectively, the first, third, and fourth lines of the previous truth table. Thus, P in each case is "___ is a multiple of 4" and Q is "___^{2} is a multiple of 4."

"If 8 is a multiple of 4, then 8^2 is a multiple of 4."
"If 6 is a multiple of 4, then 6^2 is a multiple of 4."
"If 5 is a multiple of 4, then 5^2 is a multiple of 4."

Any integer substituted for 8, 6, or 5 in these propositions gives a proposition fitting the first, third, or fourth lines of the truth table, never the second line. Section 1.3 considers propositions involving variables. The preceding propositions are all particular examples of the following proposition involving a variable:

(∗) "For all integers n, if n is a multiple of 4, then n^2 is a multiple of 4."

In Chapter 2, we will prove the general proposition (∗) is always true. Thus, we need to have the first, third, and fourth lines under $P \Rightarrow Q$ in the previous truth table to each have a T. The proof in Chapter 2 will show that the second line can never occur. If it could occur, we would have a *counter-example*, an example showing the entire proposition to be incorrect. ◇

Converse and Contrapositive

The truth table for implication, while correct for mathematical purposes, does not reflect the variety of nuances we use in English. Mathematics students need to work especially hard to make a clear separation between ordinary English and mathematical language when using the word "if" and its many equivalents. Consider our earlier example, "If it is raining, then the sidewalk is wet." If you see a wet sidewalk, you might suspect it is raining, but you know other possibilities could occur, such as the sprinkler running. Thus, even when the implication "If it is raining, then the sidewalk is wet" is true, the switched implication "If the sidewalk is wet, then it is raining" can be false. In mathematics we are not permitted to interchange the hypothesis and conclusion without justification. For example, while "If 6 is a multiple of 4, then 6^2 is a multiple of 4" is true, the switched proposition "If 6^2 is a multiple of 4, then 6 is a multiple of 4" is false. Switching the hypothesis and conclusion is of sufficient importance to deserve its own name, the converse.

DEFINITION. The *converse* of $P \Rightarrow Q$ is $Q \Rightarrow P$.

EXERCISE. Write the converse of "If you study hard, then you will succeed in mathematics."

The middle rows of the truth table for the converse, $Q \Rightarrow P$, given next, reflect the statement's difference with $P \Rightarrow Q$. Notice that we do not switch all of the truth values in the converse; the first and last rows still have a T in them.

P	Q	$Q \Rightarrow P$
T	T	T
T	F	T
F	T	F
F	F	T

There are a number of propositional forms logically equivalent to $P \Rightarrow Q$. The most important is the *contrapositive*, defined next, which switches and negates both the hypothesis and conclusion. That is, "If the sidewalk is not wet, then it is not raining" is the contrapositive of our initial sentence "If it is raining, then the sidewalk is wet." Reflecting on such an example provides a valuable supplement to the truth table that follows, which shows the logical equivalence of an implication and its contrapositive.

DEFINITION. The *contrapositive* of $P \Rightarrow Q$ is $\neg Q \Rightarrow \neg P$.

Truth table of the contrapositive

P	Q	$\neg Q$	$\neg P$	$\neg Q \Rightarrow \neg P$
T	T	F	F	T
T	F	T	F	F
F	T	F	T	T
F	F	T	T	T

Working Negation of Implication

EXAMPLE 2

Verify that $\neg(P \wedge \neg Q)$ and $\neg P \vee Q$ are logically equivalent to $P \Rightarrow Q$.

Solution   The following truth tables make this clear:

P	Q	$P \Rightarrow Q$
T	T	T
T	F	F
F	T	T
F	F	T

P	Q	$\neg Q$	$P \wedge \neg Q$	$\neg(P \wedge \neg Q)$
T	T	F	F	T
T	F	T	T	F
F	T	F	F	T
F	F	T	F	T

P	Q	$\neg P$	$\neg P \vee Q$
T	T	F	T
T	F	F	F
F	T	T	T
F	F	T	T

The equivalence of $\neg(P \wedge \neg Q)$ and $P \Rightarrow Q$ in the previous example allows us to give a working negation of $P \Rightarrow Q$.

WORKING NEGATION. The working negation of $P \Rightarrow Q$ is $P \wedge \neg Q$.

This working negation fits perfectly with the idea of a counter-example, described in Example 1: An implication is false exactly when there is at least one instance making the hypothesis true and the conclusion false. It is worth emphasizing that the negation of an implication does not have an implication in it.

EXAMPLE 3

Find the working negation of $X \vee (X \Rightarrow (Y \wedge \neg X))$.

Solution   We start with $\neg(X \vee (X \Rightarrow (Y \wedge \neg X)))$ and move the "not" inside, step by step. In turn we get $\neg X \wedge \neg(X \Rightarrow (Y \wedge \neg X))$ and then $\neg X \wedge (X \wedge \neg(Y \wedge \neg X))$, which becomes $\neg X \wedge (X \wedge (\neg Y \vee X))$. We could write out a truth table for the original proposition as well as this negated one. However, with this working negation we have $\neg X$ and X, which can't both be true. Hence, we know that the working negation is always false and so the original proposition is a tautology, without bothering with a truth table. $\Diamond$

While the working negation of an implication is mathematically useful, it doesn't feel natural, partly because the mathematical meaning of "implication" has a certain artificialness: P and Q do not need to be related in any way for $P \Rightarrow Q$ to be true. For example, "If the earth has two moons, then $\pi = 3$" is true because the hypothesis is false. Nevertheless, we find this implication somewhat puzzling, even though we can accept the truth of the logically equivalent proposition "The earth doesn't have two moons or $\pi = 3$." Because we will focus on proofs in mathematics, these issues won't affect us. In a proof we don't need to worry about the situation when the hypothesis is false. And if there is no mathematical connection between an hypothesis and a conclusion, we won't be able to find a proof. On the positive side, as the examples of this section illustrate, our interpretation of $P \Rightarrow Q$ is the only one fitting our mathematical needs.

Alternative Wordings of Implications

Because implications occur so frequently in ordinary speech, English has many related constructions. The following list gives forms mathematicians consider logically equivalent, although the last three are easily misunderstood:

If P, then Q.

P implies Q.

Q is implied by P.

Q if P.

Q whenever P.

P is sufficient for *Q*.

Q is necessary for *P*.

P only if *Q*.

We will now discuss the last three constructions in the list. The sentence "*P* is sufficient for *Q*" says that if we have *P*, then we have enough evidence to force *Q*, the intended meaning of $P \Rightarrow Q$. However, we might be able to obtain *Q* without *P*, so *P* might **not** be necessary for *Q*. We say *Q* is necessary for *P* in the sense that we can't have *P* without already having (the conditions for) *Q*. This explanation may appear backwards, because the phrase "*Q* is necessary for *P*" sounds like *Q* happens before *P*, but implication is a logical connection, not a causal connection. The phrase "only if" matches the logical meaning of "is necessary for." However, in everyday English we would distinguish between the reasonable sentence "If you'll join me, I'll go" and the rather strange "You'll join me only if I go." Because the words "sufficient," "necessary," and "only if" often confuse students, we will use them sparingly in this book, with the exception of the situation discussed in the next subsection. Nevertheless, students should become comfortable with these ways of describing implication.

if and only if   iff

We often emphasize when both an implication and its converse hold by combining them into one proposition. The sentence "A quadrilateral is a square if and only if its diagonals are perpendicular bisectors of each other" makes two claims. It asserts both "If a quadrilateral is a square, then its diagonals are perpendicular bisectors of each other" and the converse "If the diagonals of a quadrilateral are perpendicular bisectors of each other, then the quadrilateral is a square." The frequency of the phrase "if and only if" leads mathematicians to abbreviate it as "iff." A common paraphrase of "*P* iff *Q*" is "*P* is necessary and sufficient for *Q*." Since both words "necessary" and "sufficient" appear, this phrase rarely confuses readers. The symbol for iff is $\Leftrightarrow$, whose truth table follows. The choice of $\Leftrightarrow$ follows naturally from combining $P \Rightarrow Q$ and its converse $Q \Rightarrow P$, rewritten as $P \Leftarrow Q$.

Truth table for "iff"

P	*Q*	$P \Leftrightarrow Q$
T	T	T
T	F	F
F	T	F
F	F	T

EXERCISE. Verify that $P \Leftrightarrow Q$ is logically equivalent to $(P \Rightarrow Q) \wedge (Q \Rightarrow P)$.

Reading Mathematics

In more advanced mathematics courses, instructors and authors expect students to study the text much more carefully than in previous courses. In problem solving courses like calculus, students often concentrate only on the worked-out examples and the specially displayed parts. However, courses emphasizing proof, abstraction, and theory depend heavily on the rest of the text as well. We consider several aspects of reading and writing mathematics here and in later sections.

First of all, reading mathematics is not a spectator sport. Always have a piece of paper and a pencil handy to work out the various steps and jot down questions you have. Even the best author can't convey mathematics without the reader actively doing the mathematics.

Examples are the easiest place to read actively. Instead of just following along, cover up all but the first part of the example. If the example solves or proves something, as Example 2 did, try sketching your answer before looking at the book's answer. Then look at the book's answer. Where your answer skips something or gets off track, ask yourself what previous material would have helped you. If the example motivates a new idea, as Example 1 did, follow along line by line. Verify each line. You can understand important or subtle examples better by summarizing the key parts of the example.

While summarizing important points in any part of the text is valuable, understanding mathematics requires concentrating on the details. The most crucial places are the statements of theorems, proofs, and definitions. We will consider definitions in Section 1.3 and proofs in Chapter 2. Let's consider one aspect of theorems here. The hypothesis of a theorem is crucial for knowing when the conclusion holds. For example, many high school students memorize the Pythagorean theorem as simply "$a^2 + b^2 = c^2$." Unfortunately, that equation by itself is meaningless without a context and even false if we don't have a Euclidean right triangle with c for the length of its hypotenuse. When you read an implication, you should certainly look for examples where the hypothesis is true and test whether the conclusion is also true. Theorems deserve even more scrutiny. Mathematicians rarely include unneeded hypotheses in theorems they prove, because they are interested in the most general setting. So look for examples where some of the hypotheses fail. Do the conclusions fail?

Mathematical notation presents another difficulty in reading mathematics. Notation allows us to communicate complicated ideas compactly and unambiguously. However, the reader needs to note the form of the notation carefully, sometimes symbol by symbol. For example, in calculus you probably saw parentheses used in an exponent, as in $f^{(2)}$, to indicate repeated derivatives of a function, not the power of the function. As another example, subscripts are a valuable notational device to keep track of a family of related objects. For instance, the subscript in the sentence "Let $a_n = \frac{1}{n}$" alerts us that we should consider the family of fractions $\frac{1}{1}, \frac{1}{2}, \frac{1}{3}$, and so on, not just a generic fraction. A third common notational device is the use of different alphabets and scripts to indicate different kinds of mathematical objects. In geometry, points are often denoted by capital Latin letters, lines are designated by small Latin letters, and transformations are indicated by small Greek letters. Although we will not use Greek letters in this text, a list of the Greek alphabet appears before the Index.

To help you develop reading skills, there are true–false questions based on the text at the start of each problem section in Chapters 1 and 2 and selected other sections.

PROBLEMS

*1. For each statement that follows, decide whether it is true or (at least sometimes) false. If false, explain why. Cite the part of the text supporting your conclusion.

a) The mathematical use of implications differs from everyday use of them.

b) The converse of a proposition is logically equivalent to the original proposition.

c) The contrapositive of a proposition is logically equivalent to the original proposition.

d) The working negation of an implication is an implication.

e) The truth of an implication depends solely on the truth value of the conclusion.

f) The mathematical abbreviation "iff" combines an "if, then" proposition and its converse.

g) The mathematical abbreviation "iff" combines an "if, then" proposition and its contrapositive.

2. a) Write the converse of each given proposition.

b) Write the contrapositive of each proposition.

*i) If you don't study, your grade will suffer.

ii) If I am worried, I don't sleep well.

iii) If $x^2 - 4x + 4 = 0$, then $x = 2$.

iv) If the diagonals of a parallelogram are congruent, the parallelogram is a rectangle.

v) If $-1 < x < 1$, then $x^4 < x^2$.

3. Determine whether each of the following propositions is logically equivalent to "If you study hard, then you will succeed in mathematics," or its converse, or neither:

*a) If you don't study hard, you won't succeed in mathematics.

b) If you succeed in mathematics, you studied hard.

*c) If you don't succeed in mathematics, you didn't study hard.

d) If you don't study hard, you will succeed in mathematics.

e) You'll succeed in mathematics only if you study hard.

f) You must have studied hard whenever you succeeded in mathematics.

g) Your succeeding in mathematics is implied by your studying hard.

h) You'll not succeed in mathematics only if you don't study hard.

i) Your studying hard is sufficient for you to succeed in mathematics.

j) Your succeeding in mathematics implies you studied hard.

k) You'll succeed in mathematics iff you study hard.

l) Your studying hard is necessary for you to succeed in mathematics.

4. a) Name the famous quote and the famous mathematician who said it and that are spoofed in the cartoon at the end of the section.

b) How is the insect's thought related to the quote in part a?

5. Explain why two propositional forms X and Y are logically equivalent iff $X \Leftrightarrow Y$ is a tautology.

6. *a) You are told that the four cards in this experiment have a letter on one side and a number on the other. You are to test the accuracy of the following rule for these four cards by turning over the **fewest** number of cards: Rule: If the letter is a vowel, then the number is odd. You see four cards showing **E, J, 7, 4**, respectively. Which cards should you turn over to determine if the rule is correct?

b) In a bar you are to enforce the law, which says if someone is drinking alcohol, that person must be at least 21; but the management doesn't want you to disturb patrons unnecessarily. In which of the following four situations do you need to inquire further?

–The person has a beer. –The person has water.

–The person is 25. –The person is 19.

*This problem is similar to experiments described in Reed, 331–333. While the two parts are logically equivalent, fewer than 10% of college students answered correctly in the abstract version a, whereas more than 70% correctly answered in experiments with familiar contexts b.

7. Write the truth tables for these propositional forms.

   a) $F \vee (G \Rightarrow H)$.

*b) $G \wedge (H \Rightarrow \neg F)$.

   c) $(H \Rightarrow G) \Rightarrow (F \wedge G)$.

   d) $(F \vee \neg H) \Rightarrow (G \Leftrightarrow \neg F)$.

   e) $(G \wedge (F \Rightarrow H)) \Rightarrow ((H \Rightarrow G) \wedge F)$.

8. For each pair of propositional forms given, determine whether the propositional forms are logically equivalent. Justify your answers.

*a) $(A \Rightarrow B) \Rightarrow C$ and $A \Rightarrow (B \Rightarrow C)$.

   b) $A \Rightarrow (B \Rightarrow C)$ and $B \Rightarrow (A \Rightarrow C)$.

*c) $C \Rightarrow (A \vee B)$ and $(C \Rightarrow A) \vee (C \Rightarrow B)$.

   d) $C \Rightarrow (A \wedge B)$ and $(C \Rightarrow A) \wedge (C \Rightarrow B)$.

   e) $(A \vee B) \Rightarrow C$ and $(A \Rightarrow C) \wedge (B \Rightarrow C)$ (referred to in Section 2.3).

   f) $(A \wedge B) \Rightarrow C$ and $(A \Rightarrow C) \wedge (B \Rightarrow C)$.

9. Determine whether the given propositional forms are tautologies, contradictions, or neither. Justify your answers.

   a) $U \vee (U \Rightarrow W)$.

   b) $(W \Rightarrow \neg U) \wedge (U \Rightarrow W)$.

   c) $(W \Rightarrow \neg W) \wedge W$.

*d) $(U \wedge \neg U) \Rightarrow W$.

   e) $(U \Leftrightarrow W) \wedge \neg (U \Rightarrow W)$.

   f) $(U \Leftrightarrow W) \Rightarrow (W \vee \neg U)$.

   g) $W \vee (U \Rightarrow W)$.

   h) $(U \Rightarrow W) \vee (U \Rightarrow \neg W)$.

10. Verify that the given propositional forms are tautologies. (Frege considered these forms as fundamental for logic. See van Heijenoort, 29–47.)

   a) $S \Rightarrow \neg \neg S$.

   b) $(\neg \neg R) \Rightarrow R$.

   c) $R \Rightarrow (S \Rightarrow R)$.

   d) $(R \Rightarrow S) \Rightarrow (\neg S \Rightarrow \neg R)$.

e) $(R \Rightarrow (S \Rightarrow T)) \Rightarrow (S \Rightarrow (R \Rightarrow T))$.

f) $(T \Rightarrow (S \Rightarrow R)) \Rightarrow ((T \Rightarrow S) \Rightarrow (T \Rightarrow R))$.

11. Write working negations of the following propositional forms:

*a) $X \Rightarrow (Y \vee Z)$.

b) $(Y \vee Z) \Rightarrow X$.

c) $X \Rightarrow (Y \wedge \neg Z)$.

d) $X \Rightarrow (Y \Rightarrow Z)$.

*e) $(X \Rightarrow Y) \Rightarrow Z$.

f) $(X \Rightarrow Y) \Rightarrow (X \vee Z)$

g) $\neg (X \Rightarrow Z) \Rightarrow (Y \Rightarrow (X \wedge Z))$

12. This problem explores a simplified version of the popular Sudoku puzzles. For each grid, fill in the numbers from 1 to 4 so that each row, each column, and each 2×2 corner box has each of these four numbers in it just once. (A standard Sudoku puzzle is a 9×9 grid, using the numbers from 1 to 9.) For each part, indicate which entries you made could be deduced directly from the given numbers.

*a)

1		2	
	3		1
4			

b)

	4		
		1	
3			
			2

c)

2	3		
			3
4			
			4

13. The popular computer game "Minesweeper" requires the player to determine, on the basis of information from uncovered cells, which hidden cells are safe and which contain a mine. The number in each uncovered cell indicates how many mines are in adjacent covered or uncovered cells. In each grid shown here, determine whether the question marks cover a safe square or a mine. The blanks represent unknown areas. For each part, indicate which entries you made could be deduced directly from the given information.

a)

0	1	?			
0	2	?			
1	3	?	?	1	1
?	3	3	2	1	0
?	?	?	?	1	0
		?	1	1	0

b)

0	0	1	?		
0	0	2	?		
1	1	2	?	3	1
?	1	?	?	2	0
?	?	2	?	2	0
	?	?	?	1	0

14. Each of the sentences a through g comes from the instructions for the 1040 tax form for 2003. (For these and other examples of complicated English sentences, see http://www.irs.gov/pub/irs-pdf/i1040.pdf.) Use letters to replace each basic sentence; then write the entire sentence, using logical symbols; finally, write a working negation of that sentence. (Basic sentences should not contain the logical words "not," "if," "or," "and," and so on.) If you find some of these sentences ambiguous, blame the IRS. I give my interpretation of part a.

a) "If you check 'Yes,' your tax or refund will not change." (*page 19*)

Partial solution. Let Y stand for "You check 'Yes'," T stand for "Your tax will change" and R for "Your refund will change." The whole proposition is then $Y \Rightarrow \neg (T \vee R)$.

    b) "[I]f you file a separate return, you cannot take the student loan interest deduction, the tuition and fees deduction, the education credits, or the earned income credit." (page 20)

    c) "A husband and wife may file a joint return even if only one had income or if they did not live together all year." (page 20)

  *d) "If your spouse is a nonresident alien and you file a joint or separate return, your spouse must have either an SSN or an ITIN." (page 19)

    e) "Check the box on line 6b if you file either (a) a joint return or (b) a separate return and your spouse had no income and is not filing a return." (page 21)

    f) "[Y]our child's gross income can be $3050 or more if he or she was either (a) under age 19 at the end of 2003 or (b) under age 24 at the end of 2003 and was a student." (page 21)

    g) "You may check the box on line 4 only if as of December 31, 2003, you were unmarried or legally separated ... and either 1 or 2 next applies to you." (page 20) Conditions 1 and 2 are omitted for ease—this is just a math book!

15. **a)** Explain why every propositional form is logically equivalent to a form using only combinations of $\neg$ and $\Rightarrow$.

    b) Define the logical connective $\downarrow$ ("nor") by $P \downarrow Q$ iff $\neg P \vee \neg Q$. Find the truth table of $P \downarrow Q$. Find a propositional form using only $\downarrow$ logically equivalent to $\neg P$ and another logically equivalent to $P \vee Q$. Explain why every propositional form is logically equivalent to one using only $\downarrow$.

    c) Find other sets of logical connectives from which all propositional forms can be formed.

REFERENCES

REED, STEPHEN K. 2004. *Cognition: Theory and applications*, 6th ed., Belmont, CA: Wadsworth.

VAN HEIJENOORT, JEAN. 1967. *From Frege to Gödel: A source book in mathematical logic, 1879–1931.* Cambridge, MA: Harvard University Press.

1.3 QUANTIFIERS AND DEFINITIONS

Mathematical statements need more subtlety than the propositions of the two previous sections can convey. In particular, variables play a central role. However, we can't decide whether "$x + y = 0$" is a correct statement unless we know more about x and y. Mathematicians use the *existential quantifier* "there exists" and the *universal quantifier* "for all" together with sets to bind variables into working relationships. Thus, for instance, "There exist real numbers x and y such that $x + y = 0$" is true because $5 + (-5) = 0$. On the other hand "For all real numbers x and y, $x + y = 0$" is false because 2 and 3 are real numbers, but $2 + 3 = 5$, not 0.

Aristotle anticipated quantifiers when he developed syllogisms—the oldest existing rules of valid arguments. He analyzed sentences of the form "some X are Y" and "all X are Y," corresponding to our more modern quantifiers. You may have heard his most famous syllogism: "All people are mortal; Socrates is a person; therefore Socrates is mortal." While Aristotle made important contributions, general reasoning and especially mathematical reasoning transcend syllogisms. As Augustus De Morgan (1806–1871) noted, even when we are given "All horses are animals," no syllogism allows us to deduce "A horse's tail is an animal's tail." De Morgan and others in the nineteenth century developed our modern notions of quantifiers, which have the flexibility and subtlety needed for mathematics. Once we introduce quantifiers, we discuss definitions in mathematics.

Existential Quantifiers

The first mathematical quantifier, "there exists," asserts that there is some appropriate mathematical object satisfying whatever follows this expression.

EXAMPLE 1

Consider the sentence "There is a number whose square is negative." While many people might say that sentence is clearly false, mathematics students realize that imaginary numbers, such as i, have negative squares. This discrepancy results from the sentence not specifying what sort of things count as "numbers." We need to indicate what things can be considered. The proposition "There is an integer whose square is negative" is false, whereas "There is a complex number whose square is negative" is true.   ◇

EXAMPLE 2

The proposition "there is a real number equal to half of its cube" is true. We need at least one instance to make the proposition true, but we can have multiple instances. In this case, each of the real numbers $-\sqrt{2}$, 0, and $\sqrt{2}$ satisfies $x = \frac{1}{2}x^3$.   ◇

Following English usage, we often indicate an existential quantifier with the word "some." We need to be careful, though. Informally, the English sentence "Some mathematicians use language precisely" carries the connotation that some mathematicians don't use language precisely. Logically however, if, as they should, all mathematicians use language precisely, the sentence "Some mathematicians use language precisely" is still true. A disclaimer is in order about the extent of existence claimed. To avoid philosophical disputes while working, mathematicians assert only an abstract mathematical existence. Debating the "reality" of $i = \sqrt{-1}$, continuous functions, or even the number 2 would distract us from our passion of laying bare the structure and reasoning at the heart of mathematics.

We use a backwards "E" to represent the existential quantifier and abbreviate "such that" by *s.t.* To specify the range of values, we can use sets, such as the real numbers, written $\mathbb{R}$, and the relation $\in$ for "element of." (See the descriptions of some familiar sets of numbers in the next subsection.) Thus, we could write "$\exists x \in \mathbb{R} \; s.t. \; x = \frac{1}{2}x^3$" for the proposition of Example 2. (We read this string of symbols as "There is an element x of $\mathbb{R}$ such that $x = \frac{1}{2}x^3$.") As noted earlier, changing the set can change the truth of a proposition with an existential quantifier. For instance, "$\exists x \in \mathbb{N} \; s.t. \; x = \frac{1}{2}x^3$" is false, where $\mathbb{N}$ is the set of natural numbers.

Descriptions of Sets of Numbers

$\mathbb{N}$ represents the *natural numbers*—the positive whole numbers, which start 1, 2, 3, and continue forever.

$\mathbb{Z}$ stands for the *integers*—the natural numbers, their negatives, and 0. (The choice of $\mathbb{Z}$ comes from "zahl," the German word for number.)

$\mathbb{Q}$ symbolizes the *rational numbers*—all numbers that can be represented as a fraction $\frac{p}{q}$, where p is an integer and q is a natural number. (The choice of $\mathbb{Q}$ comes from the word "quotient," since "$\mathbb{R}$" is reserved for the reals.)

$\mathbb{R}$ denotes the *real* numbers—all rationals and irrationals, such as π. We can match the real numbers with points on a line. Real numbers have finite or infinite decimal representations.

$\mathbb{C}$ represents the *complex numbers*—numbers of the form $x + iy$, where x and y are real numbers and i satisfies $i^2 = -1$.

Universal Quantifiers

The second quantifier, the universal quantifier, enables mathematicians to focus on properties that hold for everything in a set of possibilities. As with existential quantifiers, the validity of a universally quantified proposition can depend on the set over which the variable varies.

EXAMPLE 3

The proposition "For all integers x, $x \leq x^2$" is true. The proposition "For all real numbers x, $x \leq x^2$" is false, since "$x \leq x^2$" fails when, for example, $x = 0.5$. A universal assertion is falsified by just one specific failure, called a *counter-example*. In everyday English we may expect a few "exceptions that prove the rule," but mathematical propositions tolerate no exceptions. $\Diamond$

Mathematicians use an upside down "A" to represent the universal quantifier. Thus, "$\forall x \in \mathbb{Z}, (x \leq x^2)$" and "$\forall x \in \mathbb{R}, (x \leq x^2)$" represent the propositions of

Example 3. In everyday speech the quantifier often appears at the end of the sentence. For instance, someone might restate the first proposition of Example 3 as "$x \leq x^2$ for any integer x." While this order is just fine for ordinary communication, it has definite drawbacks in complicated sentences and in proofs.

More on Quantifiers

Quantifiers raise subtle points beyond the preceding discussion, some of which we consider briefly. An unquantified variable, such as the x in "$x \leq x^2$", is called a *free variable*. A sentence with one or more free variables in it is not a proposition; it is called a *predicate*. Such a sentence becomes a proposition when all of its variables are *bound* with a quantifier. Usually, we need to specify the set over which the variable ranges. Thus, "$\forall x \in X$, $P(x)$," where X is a set and $P(x)$ is a predicate, represents a propositional form with a universal quantifier. When two variables are from the same set, we often abbreviate our symbols, as "$\forall x, y \in \mathbb{R}$," rather than writing the more formal "$\forall x \in \mathbb{R}$, $\forall y \in \mathbb{R}$."

In set theory we will consider properties of all sets. Naively, we would like to write "for all sets S in the 'set of all sets'," just as we write "for all x in $\mathbb{R}$." Russell's Paradox, presented in Section 1.4, makes this impossible. We will avoid this problem by writing "for all sets S" or "there is a set S," as appropriate.

Students often substitute implications for universal quantifiers. After all, "If $x \in \mathbb{Z}$, then $x \leq x^2$" has the same meaning as the first proposition of Example 3. However, universal quantifiers have two advantages. First, as we will see shortly, universal and existential quantifiers are closely linked through negations. Introducing each variable with one of these two quantifiers enables us to maintain that vital link. Also, with more complicated propositions, the differences between "for all" and "if..., then ..." become essential to understanding.

Some students find it helpful to think of "for all" as an infinite "and" ranging over each possible value. Similarly, "there exists" acts as an infinite "or." For instance, the proposition "$\forall x \in \mathbb{Z}$, $(x \leq x^2)$" asserts "... and $-1 \leq (-1)^2$ and $0 \leq 0^2$ and $1 \leq 1^2$ and $2 \leq 2^2$ and...." Thus, the proposition is true because every individual instance is true. In the same way, the proposition "$\exists x \in \mathbb{N}$ $s.t.$ $x + x^2 = 90$" asserts "$1 + 1^2 = 90$ or $2 + 2^2 = 90$ or $3 + 3^2 = 90$ or...." The existential proposition is true because at least one of the individual instances is true (here, $9 + 9^2 = 90$). Of course, we can't make an infinite truth table to specify the meaning of "for all" or "there exists." Instead, we rely on our prior understanding.

Negating Quantifiers

We need to negate quantified statements in useful ways for proofs. The analogy of "for all" as an infinite "and" and "there exists" as an infinite "or" suggests we look to De Morgan's Laws for help in negating quantifiers. Recall the logical equivalence of $\neg(P \wedge Q)$ and $(\neg P \vee \neg Q)$: The negation of "and" involves an "or." Similarly, the negation of "for all" involves "there exists." Indeed, this matches our earlier remark that the existence of only one counter-example falsifies a universal statement. Analogously, the other De Morgan's Law, equating $\neg(P \vee Q)$ and $(\neg P \wedge \neg Q)$, leads to the negation of an existential quantifier requiring a universal quantifier. That is, "$\neg \, (\forall x \in X, P(x))$" is logically equivalent to "$\exists x \in X \, s.t. \, \neg \, P(x)$" and "$\neg \, (\exists x \in X \, s.t. \, P(x))$" is logically equivalent to "$\forall x \in X, \neg \, P(x)$."

WORKING NEGATION. The working negation of "$\forall x \in X, \, P(x)$" is "$\exists x \in X$ $s.t. \, \neg \, P(x)$."

The working negation of "$\exists x \in X \, s.t. \, P(x)$" is "$\forall x \in X, \neg \, P(x)$."

Note: We do not negate the set relationship $x \in X$ when we negate the quantifier.

EXAMPLE 4

The negation of Aristotle's premise "All people are mortal" is "There is some person who is immortal." To be overly formal we could write these sentences symbolically by designating P to be the set of all people and $M(x)$ to be the predicate "x is mortal." Then Aristotle's sentence becomes "$\forall x \in P, \, M(x)$" and its negation becomes "$\exists x \in P \, s.t. \, \neg \, M(x)$." $\Diamond$

EXAMPLE 5

Both propositions "There exists a purple unicorn" and "There is a unicorn that is not purple" are false. Their negations are therefore both true: "All unicorns are not purple" and "All unicorns are purple." These sentences may seem contradictory until we remember that there are no unicorns, so the "all" doesn't range over anything. We say these propositions are *vacuously true*. $\Diamond$

EXAMPLE 6

Decide which of the following propositions is false and write its working negation:

$$\exists x \in \mathbb{R} \, s.t. \, \forall y \in \mathbb{R}, \, (x < y \text{ or } y < x).$$

$$\exists x \in \mathbb{R} \, s.t. \, \forall y \in \mathbb{Z}, \, (x < y \text{ or } y < x).$$

Solution It is important to read the quantifiers in order from left to right. The first proposition is false, since whatever x is chosen, we could pick $y = x$. The meaning of the working negation, given next, matches this idea. Start from the outside to find the working negation of this proposition. Thus, we first convert the $\exists$, then the $\forall$, and finally the "or."

$$\neg \, (\exists x \in \mathbb{R} \, s.t. \, \forall y \in \mathbb{R}, \, (x < y \text{ or } y < x)) \text{ becomes}$$

$$\forall x \in \mathbb{R}, \, \neg \, \forall y \in \mathbb{R}, \, (x < y \text{ or } y < x), \text{ which becomes}$$

$$\forall x \in \mathbb{R}, \, \exists y \in \mathbb{R} \, s.t. \, \neg \, (x < y \text{ or } y < x). \text{ De Morgan's Laws give}$$

$$\forall x \in \mathbb{R}, \, \exists y \in \mathbb{R} \, s.t. \, (x \geq y \text{ and } y \geq x). \text{ This last is true since, as said earlier,}$$
$$\text{for any } x \text{ we can pick } y = x.$$

The second proposition is true. For instance, pick $x = \pi$. Then every integer y is either bigger than π or smaller than π. $\Diamond$

EXAMPLE 7

Write the working negation of "$\forall \epsilon > 0, \, \exists \delta > 0 \, s.t. \, \forall x \in \mathbb{R}, \, (|x - a| < \delta \Rightarrow |f(x) - L| < \epsilon)$."
The notation "$\forall \epsilon > 0$" is a short hand for "ϵ is a positive real number."

Solution   Fortunately, we do not need to understand this complicated proposition in order to write its working negation. We just bring the negation one further step "inside" with each line.

$$\neg\,(\forall\epsilon > 0,\ \exists\delta > 0\ s.t.\ \forall x \in \mathbb{R},\ (|x - a| < \delta \Rightarrow |f(x) - L| < \epsilon))$$

$$\exists\epsilon > 0\ s.t.\ \neg\,(\exists\delta > 0\ s.t.\ \forall x \in \mathbb{R},\ (|x - a| < \delta \Rightarrow |f(x) - L| < \epsilon))$$

$$\exists\epsilon > 0\ s.t.\ \forall\delta > 0,\ \neg\,(\forall x \in \mathbb{R},\ (|x - a| < \delta \Rightarrow |f(x) - L| < \epsilon))$$

$$\exists\epsilon > 0\ s.t.\ \forall\delta > 0,\ \exists x \in \mathbb{R}\ s.t.\ \neg\,(|x - a| < \delta \Rightarrow |f(x) - L| < \epsilon)$$

$$\exists\epsilon > 0\ s.t.\ \forall\delta > 0,\ \exists x \in \mathbb{R}\ s.t.\ (|x - a| < \delta \wedge |f(x) - L| \geq \epsilon).$$

REMARK.   The original proposition is the formal definition of $\lim_{x \to a} f(x) = L$.   ◇

Uniqueness

Frequently, mathematicians strengthen existence propositions by saying exactly how many items satisfy the conditions. In a modification of Example 2, there is exactly one integer satisfying $x = \frac{1}{2}x^3$, namely, $x = 0$. We use the notation $\exists!$ for "there exists a unique," so we write "$\exists!x \in \mathbb{Z}\ s.t.\ x = \frac{1}{2}x^3$." While there are symbols for "there exists exactly $n \ldots$," we will have no need of such sophistication. The virtue of the symbols $\forall$ and $\exists$ is to make logical formats clear, not to provide a shorthand for every possible situation.

EXAMPLE 8

Real functions are of vital importance in mathematics. Graphically, functions satisfy the "vertical line test": Every vertical line intersects the graph of the function in just one point. Logically speaking, f satisfies the property "For every $x \in \mathbb{R}$ there is a unique $y \in \mathbb{R}$ such that $f(x) = y$."   ◇

Order of Quantifiers

When a proposition contains both a universal quantifier and an existential quantifier, their order matters. In general, we apply the quantifiers from left to right, where earlier variables can affect the choices for later ones, as in the following example:

EXAMPLE 9

Compare "For all real numbers r, there is a real number s such that $r + s = 3$" with "There is a real number s such that for all real numbers r, we have $r + s = 3$." Symbolically, these propositions are "$\forall r \in \mathbb{R},\ \exists s \in \mathbb{R}\ s.t.\ r + s = 3$" and "$\exists s \in \mathbb{R}\ s.t.\ \forall r \in \mathbb{R},\ r + s = 3$."

Solution   The first proposition is true, since we can pick s in terms of r: Choose $s = 3 - r$. However, the second proposition is false because it asserts that one value of s works simultaneously for all r.   ◇

Formal mathematics requires quantifiers of variables to occur before the variables are used and in the proper order. However, ordinary English often places the quantifiers out of order, after the variables, and even omits some quantifiers. These informal practices risk confusion when sentences are more complicated. Readers must be wary, and writers should be helpful.

EXAMPLE 10

Rewrite "There is a rational number between any two rationals" symbolically, making the quantifiers explicit.

Solution   Of the three rational numbers involved, the first one described, with an existential quantifier, depends on the other two, which have understood universal quantifiers. Thus, the quantifier for the latter two needs to precede the other quantifier. We could write "$\forall q, r \in \mathbb{Q}, \exists s \in \mathbb{Q} \, s.t. \, (q \leq s \leq r) \vee (r \leq s \leq q)$." However, the "or" ($\vee$) isn't very satisfying, and even worse, we could pick $s = q$, which defeats the spirit of the mathematical idea. A more satisfying mathematical statement is "$\forall q, r \in \mathbb{Q}, (q < r \Rightarrow \exists s \in \mathbb{Q} \, s.t. \, q < s < r)$," which makes sure s is strictly between q and r. Because q and r are already general numbers, we do not need to include the option $r < q$ as well. A clear restatement in ordinary English might read "For any two distinct rational numbers there is another rational strictly between them." $\Diamond$

EXAMPLE 11

[Optional] Limits such as $\lim_{x \to 0} \frac{1}{x^2} = \infty$ are often described with phrases similar to the following: "By choosing x close enough to 0, we can make $\frac{1}{x^2}$ as large as we want." This sentence has three quantifiers hidden in it: one for the x; one for a variable δ, the traditional choice for a variable regulating the closeness of x to 0; and one for a variable, say, B, regulating how large $\frac{1}{x^2}$ is. Furthermore, this last quantifier logically must precede the others because how close x needs to be to 0 depends on how big we want $\frac{1}{x^2}$ to be. The correct sentence appears in Problem 9c, although you are not expected to be able to build such a sentence at this point. $\Diamond$

Common Errors with Quantifiers

Example 9 illustrates one common error: switching the order of $\forall$ and $\exists$. We list this warning again, give several others, and provide examples for these others on the basis of odd and even numbers. (The notation $P(x, y)$ refers to a generic predicate with two variables, such as "x is the student of y.")

"$\forall x \in X \, \exists y \in Y \, s.t. \, P(x, y)$" does **not** imply "$\exists y \in Y \, s.t. \, \forall x \in X \, P(x, y)$."

(See Example 9 for an example.)

"$(\exists x \in X \, s.t. \, P(x)) \wedge (\exists x \in X \, s.t. \, Q(x))$" does **not** imply "$\exists x \in X \, s.t. \, (P(x) \wedge Q(x))$."

The first proposition allows different values of x to satisfy the different predicates, whereas the second requires the same x to satisfy both. For instance, "There is an integer that is odd and there is an integer that is even" is true. However, "There is an integer that is odd and even" is false.

"$\forall x \in X \, (P(x) \vee Q(x))$" does **not** imply "$(\forall x \in X \, P(x)) \vee (\forall x \in X \, Q(x))$."

The first proposition claims that every x satisfies one of two options, so some x can satisfy P and others can satisfy Q. In the second one, every x must satisfy P or every x must satisfy Q. For instance, "For all integers x, (x is odd or x is even)" is true. However, "All integers are odd or all integers are even" is false.

"$\forall x \in X\ P(x) \Rightarrow \forall x \in X\ Q(x)$" does **not** imply "$\forall x \in X\ (P(x) \Rightarrow Q(x))$."

This last error is more subtle. Again, we use "x is odd" for $P(x)$ and "x is even" for $Q(x)$. The first proposition becomes "If all integers x are odd, then all integers are even." This proposition is true because the hypothesis is false. The second proposition becomes "For all integers x, if x is odd, then x is even," which is false. For instance, when $x = 3$, we have the false proposition "If 3 is odd, then 3 is even," providing a counter-example.

Definitions

Mathematical definitions differ in important ways from everyday definitions. In particular, a mathematical definition must give us criteria we can use in proofs, rather than an intuitive description of the object in question. Although we need intuition and examples to understand and use definitions, proofs must rely solely on the formal definitions. Consider the familiar concept of an angle and the less familiar formal definition: $\angle ABC$ is made up of the two rays $\overrightarrow{BA}$ and $\overrightarrow{BC}$. Instead of including infinitely long rays, your intuition of an angle probably includes just the parts of them "close to B." As in Figure 1, for D between A and B and E between C and B, $\angle DBE$ should be the same angle as $\angle ABC$. Conveniently and not coincidentally, as rays $\overrightarrow{BA} = \overrightarrow{BD}$ and $\overrightarrow{BC} = \overrightarrow{BE}$, so indeed the formal definition ensures that $\angle ABC = \angle DBE$. By encompassing the entire ray, this definition avoids needing to decide how much of the sides of the angle are needed.

In ordinary English a definition describes how a term is commonly used and can include any of the properties common to things satisfying that term. Mathematical definitions narrowly and even artificially stipulate how a term can be used. In particular, we explicitly choose the defining properties. For instance, consider the following definition: A quadrilateral is a *rectangle* iff it has four right angles. (We italicize the word we are defining to emphasize its role.) Even though all rectangles have many other properties, such as that their opposite sides are congruent and parallel, those properties are not part of this definition and would need to be proven. Further, as Aristotle pointed out, a definition does not guarantee the existence of things satisfying

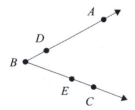

FIGURE 1

the definition. So existence needs to be proven. Indeed, in some non-Euclidean geometries there are no quadrilaterals satisfying our definition of a rectangle.

Mathematicians often use everyday words to describe technical terms. Typically, one of the meanings of the word suggests the new technical meaning. Unfortunately, English words often have multiple meanings, so students need to make sure that they connect the correct everyday intuition to the mathematical term. We have seen this already with logical words such as "or" and "if."

Mathematical definitions are generally "iff" propositions, as in the previous definition of a rectangle. However, many texts write definitions using just "if," giving in the previous case "A quadrilateral is a rectangle if it has four right angles." Such a definition unfairly leaves students to wonder whether there might be examples of rectangles with fewer than four right angles. There aren't. Logically, we need definitions to have an "iff," because we need to be able to go in both directions—from the term defined to the defining property and vice versa.

Many formal definitions require quantifiers, even though our intuition may hide the quantifier. For instance, children understand the notions of even and odd numbers (defined next) perfectly well without any understanding of quantifiers. However, proofs involving these definitions will require making these quantifiers explicit.

DEFINITION. An integer b is *even* iff there is an integer c such that $b = 2c$. An integer d is *odd* iff there is an integer e such that $d = 2e + 1$.

When you read a new definition, make sure you try out some examples. From the definition of "odd," we can verify that 17 is odd because $17 = 2 \cdot 8 + 1$. Similarly, $-2806 = 2 \cdot (-1403)$ confirms that -2806 is even. Also spend some time thinking about whether your intuition of the terms defined matches the formal definition. For instance, students generally come up with a description of "even" similar to our definition. However, they are inclined to define "odd" as "not even." Our definition doesn't tell us that any integer must be even or odd, but not both. That is a theorem needing proof, even if everyone already "knows" that fact. If we had defined "odd" as "not even," then we would need to prove that odd integers could be written in our defined form $2e + 1$. The previous definition will be more useful in proofs because it gives us an equation to work with, rather than just a negation.

The distinction between definitions and theorems deserves emphasis. Definitions are not theorems, and theorems are not definitions. We do not need to justify the definitions of "even" and "odd" given here. They are starting points. Theorems do, however, need to be proven, and those proofs generally make use of the definitions of the terms involved. Once we have proven a property, we are free to use it in another argument just as we use definitions. Similarly, once we have a definition we may use it in defining other terms.

Historical Remarks

In spite of the authority granted to Aristotle (384–322 B.C.), mathematicians from at least Euclid (300 B.C.) on have never really used syllogisms in their reasoning. Some Stoic philosophers in ancient Greece did describe various proof formats, including direct reasoning and proofs by contradiction. Unfortunately, their work was largely overlooked. Mathematicians in the nineteenth century reformulated logic on their own. In the process, they developed quantifiers and based mathematical reasoning on them. Gottlob Frege (1848–1925) developed the modern approach of using variables, sets, and quantifiers. He used what we call the working negation of the universal quantifier to define the existential quantifier. He also explicitly gave the mathematical understanding of implication presented in Section 1.2.

Euclid's geometry text contains many definitions, some meeting our modern expectations and some failing. For example, his definition of perpendicular lines as lines with adjacent angles congruent is readily used in proofs. However, his definition of a point as "that which has no part" is of no use in a proof. Few mathematicians before the nineteenth century seem to have realized the futility of trying to define everything. The starting terms, such as "point" and "line" in geometry or "set" in set theory, are better left undefined. In formal mathematics, we employ axioms to restrict how undefined terms are used. Then other terms are defined according to these and are used in proofs. We will be more informal with undefined terms from set theory throughout this text. For instance, we assume that the reader has a sufficiently clear understanding of the basic sets of numbers described earlier ($\mathbb{N}$, $\mathbb{Z}$, $\mathbb{Q}$, $\mathbb{R}$, and $\mathbb{C}$).

PROBLEMS

*1. For each statement that follows, decide whether it is true or (at least sometimes) false. If false, explain why. Cite the part of the text supporting your conclusion.

a) Every equation must be true or false.

b) The working negation of a universal quantifier has an existential quantifier in it.

c) The working negation of "There exists $x \in X$ such that $P(x)$" is "For all x not in X, $P(x)$."

d) The order of quantifiers can affect the meaning of a proposition.

e) If "For all x, $P(x)$" is true, then "There exists x such that $P(x)$" must be true.

f) If "$\exists! x \in A$ s.t. $P(x)$" is true, then "$\exists x \in A$ s.t. $P(x)$" is true.

g) Mathematical definitions do not need to give all of the properties of the object defined.

h) A mathematical definition must satisfy our intuitive understanding of the term, even if it is hard to use the definition in a proof.

2. Decide which of the propositions that follow are true and which are false. If a proposition is false, provide a counterexample to it.

*a) $\forall x \in \mathbb{N}$, $x^2 + 3x + 2 \geq 0$.

b) $\forall x \in \mathbb{Z}$, $x^2 + 3x + 2 \geq 0$.

c) $\forall x \in \mathbb{Q}$, $x^2 + 3x + 2 \geq 0$.

*d) $\forall x \in \mathbb{R}$, $x^2 + 3x + 2 \geq 0$.

3. Decide which of the following propositions are true and which are false:

a) $\exists x \in \mathbb{Q}$ s.t. $x^3 + 3 = 0$.

*b) $\exists x \in \mathbb{R}$ s.t. $x^3 + 3 = 0$.

    c) $\exists x \in \mathbb{C} \ s.t. \ x^3 + 3 = 0.$

  *d) $\exists! x \in \mathbb{Q} \ s.t. \ x^3 + 3 = 0.$

    e) $\exists! x \in \mathbb{R} \ s.t. \ x^3 + 3 = 0.$

    f) $\exists! x \in \mathbb{C} \ s.t. \ x^3 + 3 = 0.$

4. Decide which of the following propositions are true and which are false:

  *a) $\forall x \in \mathbb{N}, \ \exists y \in \mathbb{N} \ s.t. \ x \leq y.$

    b) $\forall x \in \mathbb{Z}, \ \exists y \in \mathbb{Z} \ s.t. \ x \leq y.$

    c) $\forall x \in \mathbb{Q}, \ \exists y \in \mathbb{Q} \ s.t. \ x \leq y.$

    d) $\exists x \in \mathbb{N} \ s.t. \ \forall y \in \mathbb{N}, \ x \leq y.$

    e) $\exists x \in \mathbb{Z} \ s.t. \ \forall y \in \mathbb{Z}, \ x \leq y.$

  *f) $\exists x \in \mathbb{Q} \ s.t. \ \forall y \in \mathbb{Q}, \ x \leq y.$

5. Use quantifiers, sentence symbols, and variables to write each given sentence. (You can designate sets such as P for all people.) Identify what each symbol and variable represents. Then write the negation of the sentence symbolically and in English. Lewis Carroll, the author of *Alice in Wonderland,* created these sentences for his book *Symbolic Logic* (Carroll, 1242–1245).

  *a) "Babies are illogical."

    b) "No ducks waltz."

    c) "Nobody is despised who can manage a crocodile."

6. The wording of the familiar phrase "All that glitters is not gold" doesn't match its intended meaning. Let T be the set of physical things, $GL(x)$ mean that "x glitters" and $G(x)$ mean "x is made of gold." Translate the following mathematical propositions into English. Then decide which option(s) best fit(s) the usual meaning of the phrase "All that glitters is not gold," which option(s) best fit(s) the exact wording of that phrase, and which option doesn't fit either. (For the record, Shakespeare's original quote in *The Merchant of Venice* was "All that glisters is not gold.")

  a) $\forall x \in T$, if $GL(x)$, then $\neg G(x)$.   **c)** $\forall x \in T, \ \neg$ if $GL(x)$, then $G(x)$.

  b) $\neg \forall x \in T$, if $GL(x)$, then $G(x)$.   **d)** $\exists x \in T \ s.t. \ GL(x)$ and $\neg G(x)$.

7. Rewrite the following propositions symbolically with the quantifiers explicit and in the correct place:

  *a) There is a unique line parallel to a given line through a point not on the given line.

    b) There is a unique point the same distance from any three given points of the plane W. (Denote the distance between P and Q as $d(P, Q)$.)

    c) Every natural number can be written as the sum of the squares of four integers.

    d) Two real numbers are equal iff the absolute value of their difference is less than any positive number.

8. If you hear someone say "Nothing is better than chocolate" and later "Brussels sprouts are better than nothing," does that person actually prefer Brussels sprouts to chocolate? Explain. Use a quantifier to rewrite the first sentence to make its meaning explicit.

9. Write working negations of each proposition given next. If the proposition is in words, write the negation in words; if it is in symbols, write the negation in symbols.

  *a) Every integer is even or odd.

   b) All triangles have inscribed and circumscribed circles.

   *c) (From Example 11) $\forall B \in \mathbb{R}, \exists \delta > 0 \, s.t. \, \forall x \in \mathbb{R}, (|x| < \delta \Rightarrow \frac{1}{x^2} > B)$.

   d) (From the definition of a Cauchy sequence) $\forall \epsilon > 0, \exists K \in \mathbb{N} \, s.t. \, \forall n, m \in \mathbb{N},$
     $((n > K \wedge m > K) \Rightarrow |x_n - x_m| < \epsilon)$.

   e) Your answers to Problem 7.

10. a) Find all values of c so that $2x^3 - 3x^2 - 12x + c = 0$ has exactly two roots.

   b) Find all values of c so that $2x^3 - 3x^2 - 12x + c = 0$ has exactly three roots.

   c) Find all values of c so that $2x^3 - 3x^2 - 12x + c = 0$ has exactly one root.

   d) Find all values of c so that $2x^3 - 3x^2 - 12x + c = 0$ has at least two roots.

11. For each of the given sentences, decide which of the pairs of quantifiers (or both or neither) "$\forall s \in \mathbb{R}, \exists t \in \mathbb{R} \, s.t.$" and "$\exists t \in \mathbb{R} \, s.t. \, \forall s \in \mathbb{R}$" makes the sentence into a true proposition. If a proposition fails, explain why.

   *a) $s + t = s^2$ **d)** $t^2 = s$

   b) $s^2 = t$ **e)** $\sqrt[3]{t} = s$

   c) $s \times t = 0$ **f)** $\dfrac{|st|}{s^2 + t^2 + 1} < 1$

12. The geometric mean of two positive numbers a and b is $\sqrt{ab}$. After experimenting with several choices for a and b, make a conjecture about the relationship of the geometric mean and the arithmetic mean (average) of two positive numbers. Make the quantifiers explicit and in the correct place in your conjecture.

13. Compare the reasoning in the cartoon at the end of this section with the syllogism at the start of this section. Describe the error in the dog's reasoning.

14. For each of the inadequate definitions of a quadrilateral that follow, draw an example of a non-quadrilateral satisfying that attempt, but not succeeding attempts. Then give a definition of a quadrilateral satisfied by exactly the shapes you consider quadrilaterals. Assume that the definition of a line segment is already known.

   a) A shape is a *quadrilateral* iff it has four sides.

   *b) A shape is a *quadrilateral* iff it is made of four straight line segments.

   c) A shape is a *quadrilateral* iff it is made of four straight line segments connected at their ends.

   d) A shape is a *quadrilateral* iff it has exactly four distinct vertices $A, B, C,$ and D and four line segments $\overline{AB}, \overline{BC}, \overline{CD},$ and $\overline{DA}$, any two of which intersect in at most one of the four vertices.

In Problems 15 to 18, some parts have more than one correct answer.

15. Write definitions for the following italicized geometrical terms:

   a) A *parallelogram*. (For parts a, b, and c, assume that quadrilaterals are defined. See Problem 14.)

   *b) A *trapezoid*. (See Figure 2.)

   c) A *kite*. (See Figure 3.)

   d) A *polygon* with n sides.

   e) A *regular* polygon.

   f) A *polyhedron*.

FIGURE 2

FIGURE 3

16. For each of the given sentences about integers, write definitions for the italicized terms, making any quantifiers explicit. In parts f and g, do not assume the existence of the number in your definition.

  a) An integer is a *square*.

  b) An integer is a *multiple* of an (other) integer.

  *c) An integer is a *divisor* of an (other) integer.

  d) An integer is a *common multiple* of two integers.

  e) An integer is a *common divisor* of two integers.

  f) An integer is (the) *least common multiple* of two positive integers.

  *g) An integer is (the) *greatest common divisor* of two integers.

17. For each of the given sentences about real numbers, write definitions for the italicized terms, making any quantifiers explicit. In parts e and f do not assume the existence of the number in your definition.

  a) The absolute value of a number.

  *b) A real number is an nth *root* of a given real number.

  *c) A real number is an *upper bound* of a set of real numbers.

  d) A real number is a *lower bound* of a set of real numbers.

  e) A real number is (the) *least upper bound* of a set of real numbers.

  f) A real number is (the) *greatest lower bound* of a set of real numbers.

18. (Linear algebra) Write definitions for the following italicized terms from linear algebra, making any quantifiers explicit:

  a) A vector $\vec{w}$ is a *linear combination* of the vectors $\vec{v_1}, \vec{v_2}, \ldots, \vec{v_n}$.

  b) A square matrix is *invertible*.

  c) A square matrix is *singular*. (Don't use part b.)

  d) A vector w is *spanned* by the vectors $\vec{v_1}, \vec{v_2}, \ldots, \vec{v_n}$.

  e) The vectors $\vec{v_1}, \vec{v_2}, \ldots, \vec{v_n}$ are *linearly independent*.

19. In English, we use other quantifiers, such as "most" or "a few." Let "$\widehat{M} x \in X$" symbolically represent "for most $x \in X$."

  a) Investigate the relationship of $\forall$, $\exists$, $\widehat{M}$, and the negation of $\widehat{M}$ in various situations.

  b) If we have both $\widehat{M} x \in X$, $(P(x) \Rightarrow Q(x))$ and $\widehat{M} x \in X$, $(Q(x) \Rightarrow R(x))$, what can we conclude about $(P(x) \Rightarrow R(x))$ on X?

  c) The quantifier "most" works well with finite sets, but is vague with infinite sets. Use $\widehat{V}$ as a symbol for the quantifier "all but finitely many" and redo parts a and b with $\widehat{V}$ in place of $\widehat{M}$. Consider situations similar to Problem 2.

20. Find the intended deductions from these syllogisms invented by Lewis Carroll (Carroll, 1242–1243).

  *a) "Babies are illogical; Nobody is despised who can manage a crocodile; Illogical persons are despised."

  b) "No ducks waltz; No officers ever decline to waltz; All my poultry are ducks."

REFERENCE

CARROLL, LEWIS. 1976. *Complete works*. New York: Vintage Books.

1.4 INTRODUCTION TO SETS

"No one shall be able to drive us from the paradise that
Cantor created for us."

—David Hilbert (1862–1943)

For the past century, the language of sets has provided a common and precise mathematical vernacular. Initially, you can naively imagine a set as something like a bag and its elements as the objects in the bag. While precise definitions characterize mathematics, since the nineteenth century people have widely realized they can't define everything or the defining will never stop. In a formal axiomatic theory mathematicians start with a few undefined terms, letting the axioms determine how these terms behave. Our informal study of sets also avoids defining a set. Instead of giving formal axioms, we start with the familiar sets $\mathbb{N}$, $\mathbb{Z}$, $\mathbb{Q}$, $\mathbb{R}$, and $\mathbb{C}$, described in Section 1.3. We trust these sets are unambiguous. For these sets, we assume the usual properties for $+$, $-$, $\times$, and $\div$ and, except for $\mathbb{C}$, the usual properties of $\leq$ and $<$. We rely on precise definitions for further terms on the basis of the fundamental sets and concepts of this section.

We can specify a set by explicitly listing its elements within braces. Thus, the set $A = \{1, \#, \$\}$ has three elements. We write $1 \in A$ to indicate that the object 1 is an *element* of the set A. (Read $\in$ as "is an element of.") We negate $\in$ as $\notin$; thus $3 \notin A$ and $A \notin A$. We extend this listing approach when we write $B = \{1, 2, \ldots, 100\}$ to mean that B has all of the natural numbers between 1 and 100. Since the ellipsis $\ldots$ risks introducing ambiguity, we try to avoid relying on such notation in proofs. Instead, to specify which elements x are in a set X, we use the notation $X = \{x \in Y : P(x)\}$, where Y is a known set and P is some property. (Read $\{x \in Y : P(x)\}$ as "the set of all x in Y such that $P(x)$.") For instance, we can describe the previous set B by $B = \{x \in \mathbb{N} : x \leq 100\}$. If the possibilities for the variable(s) before the colon (x in the previous examples) are clear from the condition after it, the known set can be omitted. For instance, $\mathbb{Z} = \{z : z \in \mathbb{N}, \text{ or } -z \in \mathbb{N}, \text{ or } z = 0\}$. The key qualification for a set is that a reader can determine exactly which elements belong to it and which do not.

Many familiar sets, such as $\mathbb{N}$ and $\mathbb{R}$, are infinite. The unending sequence of natural numbers led Aristotle to distinguish two senses of infinity: potentially infinite and actually infinite. In counting things, each particular number is finite, but there is always another one handy for as far as needed—the idea of a potential infinity. Thus, there are potentially infinitely many English sentences because we can keep making more. However, people have not written or spoken an actual infinity of English sentences. It is worth noting that there is no number "infinity." The symbol ∞ refers to an unending process, whether in counting with elements of $\mathbb{N}$ or in moving continuously using elements of $\mathbb{R}$. However, the sets $\mathbb{N}$ and $\mathbb{R}$ as sets are actually infinite. That is, unlike the situation with English sentences, there are not just potentially, but actually, infinitely many numbers.

DEFINITION.   Given sets X and Y, we say X is a *subset* of Y, written $X \subseteq Y$, iff for all $x \in X$, we have $x \in Y$.

EXAMPLE 1

The familiar sets of numbers satisfy many subset relations, such as $\mathbb{N} \subseteq \mathbb{Z}$, $\mathbb{Z} \subseteq \mathbb{Q}$, $\mathbb{Q} \subseteq \mathbb{R}$, and $\mathbb{R} \subseteq \mathbb{C}$. $\diamond$

One particular subset of any set deserves special mention, the *empty set*, written $\emptyset$, the set with no elements. For instance, $\{n \in \mathbb{N} : n < 0\} = \emptyset$ and $\{x \in \mathbb{Q} : x^2 = 2\} = \emptyset$. It is not obvious if these two empty sets are equal—even though they are equally empty, they are empty for different reasons. After all, our naïve intuition of a set as a bag suggests we could easily have many different empty sets, a bunch of different empty bags. The *axiom of extension*, given next, resolves this matter and gives the primary way to prove two sets equal. In effect, this axiom defines what "equals" means for sets. It also implies there is just one empty set.

AXIOM.   If A and B are sets, then $A = B$ iff $A \subseteq B$ and $B \subseteq A$; that is, two sets are equal when they contain exactly the same elements.

As a consequence of this axiom, the elements of a set do not have an order and can be repeated. Thus, $\{1, \#, \$\} = \{\$, 1, \#, 1, \#, \$, \$\}$.

EXAMPLE 2

The three sets $\{1, \#, \$\}$, $\{\{1\}, \#, \$\}$, and $\{\{1, \#, \$\}\}$ are all different. The third set has just one element, which is the first set. The first two sets each have three elements, but not the same elements, since $1 \neq \{1\}$. In a similarly playful vein, we can distinguish the sets $\emptyset$, $\{\emptyset\}$, $\{\{\emptyset\}\}$, $\{\emptyset, \{\emptyset\}\}$, and $\{\emptyset, \{\{\emptyset\}\}\}$—surely, "much ado about nothing." ◊

Bounded and unbounded intervals are familiar and important subsets of the real numbers $\mathbb{R}$. Recall that the symbols ∞ and $-\infty$, read "infinity" and "negative infinity," do not denote numbers. With intervals, they represent the unbounded process of heading indefinitely in the positive or negative directions, respectively. In particular, the convention of using a bracket to indicate that the endpoint belongs to the interval never applies to the part of an interval involving $\pm\infty$; instead, we always use a parenthesis with these symbols.

DEFINITION. For a, $b \in \mathbb{R}$, if $a < b$, then the *closed interval* $[a, b]$ is $\{x \in \mathbb{R} : a \leq x \leq b\}$ and the *open interval* (a, b) is $\{x \in \mathbb{R} : a < x < b\}$. For $a \in \mathbb{R}$, we denote $(-\infty, a) = \{x \in \mathbb{R} : x < a\}$ and $[a, \infty) = \{x : a \leq x\}$; and similarly for $(-\infty, a]$ and (a, ∞); we can write $(-\infty, \infty)$ for $\mathbb{R}$.

As students have already learned, the notation (a, b) unfortunately also denotes an ordered pair, discussed in Section 1.6. To minimize this ambiguity, a writer, whether a student or a textbook author, needs to indicate which meaning is intended. Another pitfall needs to be mentioned. The notation "$3 \leq x < 7$" for the interval $[3, 7)$ is actually an abbreviation of "$3 \leq x$ and $x < 7$." Thus, to describe the real numbers not in the interval $[3, 7)$, we cannot use "$3 > x \geq 7$," which no x can ever satisfy. Instead, we write "$3 > x$ or $x \geq 7$." In general, we build new sets from sets we have by using the logical words "and," "or," and "not."

New Sets from Old

Given two sets A and B, we can form several new sets from them, called their *intersection* $A \cap B$, their *union* $A \cup B$, and their *difference* $A - B$. The pictures in Figure 1, called *Venn diagrams*, capture the essential properties of $A \cap B$, $A \cup B$, and $A - B$, which are defined next.

DEFINITION. Given two sets A and B, define $A \cap B = \{x : x \in A \text{ and } x \in B\}$ and

$$A \cup B = \{x : x \in A \text{ or } x \in B\} \text{ and}$$

$$A - B = \{x : x \in A \text{ and } x \notin B\}.$$

| A | B | $A \cap B$ | $A \cup B$ | $A - B$ |

FIGURE 1

As the Venn diagram indicates, the definition of $A \cap B$ gives the set of objects common to both of the sets A and B. Similarly, $A \cup B$ includes everything in A, everything in B, as well as everything in common to both of them. The difference of the sets, $A - B$, contains the part of A not in B and so is always a subset of A. Note that set difference is quite different from the subtraction of numbers. In particular, we can't take away elements of B that aren't in A.

EXAMPLE 3

Find $C \cap D$ and $C \cup D$ if C is the closed interval $[1, 4]$ and D is the open interval $(2, 7)$. Repeat for $C - D$ and $D - C$.

Solution The intersection and union are *half open intervals*: $C \cap D = \{x : 2 < x \leq 4\} = (2, 4]$ and $C \cup D = \{x : 1 \leq x < 7\} = [1, 7)$. The differences are closed and open intervals, respectively: $C - D = [1, 2]$ and $D - C = (4, 7)$. $\diamondsuit$

The definitions of "intersection" and "union" make clear their connection with the logical words "and" and "or." Indeed, our symbols for these words, $\cap$ and $\cup$, are happily reminiscent of $\wedge$ and $\vee$. However, the difference of sets does not match the word "not." The concept of the *complement* of a set A, written $\overline{A}$, fills this role, as illustrated in the Venn diagram of Figure 2, but it requires a remark and a disclaimer. The remark is that no notation for complement is standard—some books use $\overline{A}$, and others use A^c, A', or $\tilde{A}$. Inherent in the picture of a Venn diagram is its boundary, restricting the complement to some relevant context, a *universal set*, as it is usually called. For instance, we would naturally expect the complement of the open interval $(2, 7)$ to be $\{x \in \mathbb{R} : x \notin (2, 7)\} = (-\infty, 2] \cup [7, \infty)$. The universal set here ($\mathbb{R}$) is given just before the colon. However, if we tried to define $\overline{(2, 7)}$ as $\{x : x \notin (2, 7)\}$, in addition to the desired numbers, this set would arguably hold everything that wasn't a number, such as the sun, $\emptyset$, $\mathbb{R}$, purple unicorns, and even more unimaginable objects. Actually, the logical need to restrict each discussion to an appropriate universal set stems from a more interesting issue: Russell's Paradox.

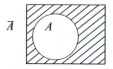

FIGURE 2

Russell's Paradox

We rephrase Bertrand Russell's argument, given in 1901, which shows that there cannot be any truly universal set. His idea shifts the self-contradictory nature of the sentence "This sentence is false" to the context of set theory. Suppose there were a set of all sets, called $\mathcal{U}$. "Define" the set $B = \{x \in \mathcal{U} : x \notin x\}$. That is, the things in B are sets that are not elements of themselves. (The false self-reference of "This sentence is false" becomes the negative self-reference $x \notin x$.) Almost every set you can imagine would be in B. For instance, the set of all people is not a person. Similarly, $\mathbb{R}$ would be in B, since $\mathbb{R}$ is a set, not a real number; and so $\mathbb{R} \notin \mathbb{R}$. Since B is supposed to be a set in $\mathcal{U}$, either $\mathcal{B} \in \mathcal{B}$ or $\mathcal{B} \notin \mathcal{B}$. However, Russell's clever definition of $\mathcal{B}$ guarantees a contradiction: Suppose, first, that $\mathcal{B} \in \mathcal{B}$. Then, by definition, $\mathcal{B}$ doesn't satisfy the definition of being in $\mathcal{B}$, so $\mathcal{B} \notin \mathcal{B}$. For the other possibility, suppose $\mathcal{B} \notin \mathcal{B}$. Then $\mathcal{B}$ satisfies the requirement of being in $\mathcal{B}$. So $\mathcal{B} \in \mathcal{B}$. Either way we get a contradiction. Thus, $\mathcal{B}$ can't be a set in $\mathcal{U}$. But it should be a perfectly fine subset of $\mathcal{U}$. The fault lies in supposing there is such a universal set $\mathcal{U}$. Without a truly universal set, the concept of a complement of a set always presumes a previously declared and appropriately limited universal set.

DEFINITION. Given a (universal) set U and a subset A of U, the *complement* of A (relative to U) is $\overline{A} = \{x \in U : x \notin A\}$.

EXAMPLE 4

Use Venn diagrams to determine whether $\overline{A - B}$ and $\overline{A} \cup B$ are equal.

$\overline{A - B}$

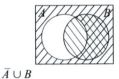

$\overline{A} \cup B$

FIGURE 3

Solution The first Venn diagram of Figure 3 uses dark shading for the region $A - B$ and shades its complement $\overline{A - B}$ lightly. In the second diagram, the regions $\overline{A}$ and B have different shading. Their union, $\overline{A} \cup B$, is the region with any shading. Since the lightly shaded region in the first diagram matches the entire shaded region in the second, we conclude $\overline{A - B} = \overline{A} \cup B$. It is worth noting that these Venn diagrams do not actually prove that these sets are equal. $\Diamond$

EXAMPLE 5

For sets J, K, and L, determine whether the sets $K \cup (J - L)$ and $(K \cup J) - (K \cup L)$ are equal or one set is a subset of the other or neither is a subset of the other.

$K \cup (J - L)$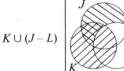

$(K \cup J) - (K \cup L)$

FIGURE 4

Solution The first Venn diagram in Figure 4 indicates the regions making up $K \cup (J - L)$, while the second diagram illustrates $(K \cup J) - (K \cup L)$. Clearly, the two regions differ. Indeed, it appears that $(K \cup J) - (K \cup L) \subseteq K \cup (J - L)$. We can reinforce this with an example. Let $J = \{1, 2, 3, 4, 5\}$, $K = \{1, 3, 5, 7, 9\}$, and $L = \{1, 4, 9, 16\}$. Then $(K \cup J) - (K \cup L) = \{1, 2, 3, 4, 5, 7, 9\} - \{1, 3, 4, 5, 7, 9, 16\} = \{2\}$, which is a subset of $K \cup (J - L) =$

$\{1, 3, 5, 7, 9\} \cup \{2, 3, 5\} = \{1, 2, 3, 5, 7, 9\}$. This example does not prove the general inclusion, just as Venn diagrams don't prove statements. However, this example does prove that the sets $K \cup (J - L)$ and $(K \cup J) - (K \cup L)$ are not always equal. One counter-example disproves a general proposition. ◇

Communicating Mathematics

As we shift from elementary logic to mathematical content and then to proofs, your work will involve more explaining. Your readers (including your instructor) deserve as much respect as you do as a reader of any text. Here are some starting suggestions; in Chapter 2, we'll give writing suggestions specifically for proofs.

As with any writing, be sure to use complete sentences, including a verb. While all mathematical propositions, such as "$2 < 4$," have verbs, you will often be talking about the expression, so the sentence needs its own English verb.

The amount of detail you need to include in an explanation or proof depends on your audience. The less experience they have with mathematics, the more carefully and completely you need to explain yourself. For a homework assignment, assume that your audience is a student who has just been introduced to the concepts of the current section, even though your instructor is probably the one actually reading your work. Thus, homework should refer explicitly to the definitions of the current section, but can assume a knowledge of basic concepts from previous sections.

Aim in your writing to be correct, clear, and complete. When you have mastered these attributes, you can move on to the other "c" of mathematical writing, conciseness, and even aim for the elusive goal of elegance.

Symbols, which figure prominently in mathematical writing, present special challenges to readers. For your personal notes, you may use them any way you find helpful, but formal writing needs careful use of symbols. Start sentences with words, and separate mathematical expressions with words whenever possible, unless the symbols form a list or together make a proposition. The words should talk about the symbols or explain the reasoning. In particular, formal mathematical writing does not mix mathematical symbols with words as a shorthand. For instance, "primes > 2 are odd" is unnecessarily confusing; instead, write "primes greater than 2 are odd." (Although formal English writes out numbers, mathematicians are free to use "2" instead of "two.") In formal writing, logical connectives such as "and" go between propositions. Thus, "The solutions to $\cos(x) = 0$ include $x = \frac{\pi}{2}$ and $x = -\frac{\pi}{2}$" is correct, whereas "The solutions ... include $x = \frac{\pi}{2}$ and $-\frac{\pi}{2}$" is sloppy.

Be especially careful in your use of the symbol $=$. The symbols directly preceding and following it must be literally equal as elements, whether as numbers, functions, or sets. In particular, it is very confusing to equate propositions. For instance, from $(x - y)^2 = 4$, we can deduce $x - y = \pm 2$, but writing $(x - y)^2 = 4 = x - y = \pm 2$ will confuse almost anyone except the author and is false.

Logical symbols, such as $\Rightarrow$, $\neg$, $\wedge$, $\vee$, $\forall$, $\exists$, and $\exists!$ should not be mixed with words in formal writing. Logicians use them to make the logical form explicit, rather than just to shorten sentences. (Instructors, including this author, often use these symbols and many others as abbreviations on the blackboard. The constraints of time in class often override the care mathematical communication deserves in formal writing.)

Historical Remarks

Mathematicians have informally used ideas of sets for centuries. As with many of the topics of this text, in the nineteenth century sets needed more careful consideration. Georg Cantor (1845–1918) deserves credit for founding set theory. His profound study of infinite sets moved Hilbert to make the statement at the start of this section. We'll consider Cantor's contributions more thoroughly in Chapter 5.

Others were more instrumental in developing the parts of elementary set theory discussed in this section. Venn popularized the diagrams named after him, although he was not the first to use such visual aids. By 1879, Gottlob Frege (1848–1925) drew the fundamental distinction between being an element of a set ($\in$) and being a subset ($\subseteq$). Even so, some others (but not Cantor) continued to confuse these two concepts. Contributing to the confusion at the time was a debate on when two sets were equal. Consider {the evening star} and {the second planet from the sun}. These sets contain the same object, namely, the planet Venus. However, talking about "the evening star" in a poem creates different images than one would get from substituting "the second planet from the sun." So are the sets equal or not? Ernst Zermelo (1871–1953) settled this question in favor of equality, with his publication in 1908 of an axiomatic foundation of set theory. (Our axiom giving set equality is his.) Perhaps surprisingly, he also needed to assert the existence of the empty set as an axiom, as well as the existence of an infinite set (built from the empty set).

George Boole's introduction of the symbol 0 in his logical calculus in 1847 was an important predecessor of the empty set. He gave this symbol various interpretations, including "false" as we used it in Problem 9 of Section 1.1. At other times he used it as "nothing" and the class of nothing—in other words, the empty set. Zermelo developed his axioms for set theory partially in response to set theory paradoxes, especially Russell's paradox. Prior to 1900, mathematicians thought sets were so obvious that no careful description of their properties was needed. The elementary nature of Russell's paradox convinced them that the flaws in their reasoning needed immediate attention. Zermelo was successful in giving a satisfactory foundation, although others continued to improve his approach, which is now standard.

PROBLEMS

*1. For each statement that follows, decide whether it is true or (at least sometimes) false. Cite the part of the text supporting your conclusion.

  a) The notation (a, b) always means an open interval.

  b) $a \in B$ iff $\{a\} \subseteq B$.

  c) $\frac{1}{\infty} = 0$.

  d) For any two sets J and K, one of the following is true: $J \subseteq K$ or $K \subseteq J$.

  e) $\{x \in \mathbb{R} : x^2 < 0\} = \{z \in \mathbb{Z} : 3z = 7\}$.

  f) $A \cap B$ means the same thing as $A \wedge B$.

  g) The complement of a set A is $\{x : x \notin A\}$.

 2. Consider $U = \mathbb{N} \cap [1, 10] = \{1, 2, \ldots, 10\}$ as the universal set and let $C = \{1, 4, 9\}$, $D = \{2, 3, 5, 7\}$, $E = \{2, 4, 6, 8, 10\}$, and $F = \{1, 2, 3, 4, 5\}$. List the elements of the following sets:

*a) $E \cup F$ d) $E - D$ g) $\overline{E}$

b) $E \cap F$ e) $D - E$ *h) $\overline{D \cup F}$

c) $C \cap D$ f) $E - (F - D)$ i) $(C \cap F) \cup (D - E)$

3. Consider $\mathbb{R}$ as the universal set and write the following sets as intervals or unions of intervals, where $G = [-2, 5]$, $H = (0, 2)$, $I = [3, 9)$ and $J = (1, 7]$:

   a) $G \cup I$ d) $G - H$ g) $(G \cup J) - (H \cup I)$

   b) $I \cap J$ *e) $J - I$ *h) $\overline{G \cap J}$

   c) $H \cap I$ f) $\overline{J}$ i) $\overline{H \cup I}$

4. Shade in the region of Figure 5 represented by each of the following expressions:

   a) $S \cap \overline{T} \cap W$ e) $(T - S) \cup (W - S)$

   b) $S \cup (\overline{T} \cap W)$ f) $(S \cap W) - (T \cap W)$

   c) $T \cup (S - W)$ g) $\overline{((S \cup T) \cap \overline{W})}$

   d) $W \cup (\overline{S \cup T})$ h) $\overline{(W \cap \overline{S}) \cup T}$

FIGURE 5

5. On the basis of Venn diagrams, determine which of the given equalities appear to be true. For each false one, determine whether one of the sets is a subset of the other.

   *a) $J \cup (K \cup L) = (J \cup K) \cap L$.

   b) $J \cap (K \cup L) = (J \cap K) \cup L$.

   *c) $K \cap (J - L) = (K \cap J) - (K \cap L)$.

   d) $(L \cap K) - J = (L - J) \cap (K - J)$.

   e) $(L \cup K) - J = (L - J) \cup (K - J)$.

   f) $(J - L) - K = J - (L - K)$.

   g) $(L \cup K) - (L \cap K) = (L - K) \cup (K - L)$.

6. Translate each pair of propositional forms from Problem 4 of Section 1.1 into an equation involving intersections and unions of sets. That is, replace $\wedge$ with $\cap$ and $\vee$ with $\cup$. Use Venn diagrams to verify that the resulting equalities seem to hold in general (referred to in Problem 10 of Section 1.5).

7. a) Translate De Morgan's Laws into equations about sets. On the basis of Venn diagrams, determine which of these laws seem(s) to hold for sets:

   b) Draw Venn diagrams that you think translate $P \Rightarrow Q$ and $P \Leftrightarrow Q$. Explain your choice.

   c) Find a propositional form corresponding to the set $P - Q$. Explain your choice.

   d) The previous parts relate logically equivalent propositions and equal sets. We used truth tables to determine whether two propositions were logically equivalent and Venn diagrams to determine whether sets were equal. Explain how truth tables and Venn diagrams express similar information. Hint: Relate the rows of a truth table and the regions of a Venn diagram.

8. *a) Suppose X is a set with 10 elements and Y is a set with 7 elements. What is the largest number of elements $X \cup Y$ can have? What is the smallest number? Explain (referred to in Sections 2.1 and 5.3).

   b) Repeat part a for $X \cap Y$.

   c) Repeat part a with X having x elements and Y having y elements.

   d) Repeat part c for $X \cap Y$.

   e) Let $|W|$ be the number of elements in the (finite) set W. Find a formula for $|A \cup B|$ involving $|A|$, $|B|$, and $|A \cap B|$. *Hint:* Look at a Venn diagram.

f) Generalize part e to determine a formula for $|A \cup B \cup C|$ involving $|A|$, $|B|$, $|C|$, and the sizes of their various intersections. (Hint: Look at a Venn diagram and some specific examples.) The formula involves quite a few sets built from A, B, and C.

9. ***a)*** List all of the subsets of $\{1, 2\}$.

   b) List all of the subsets of $\{1, 2, 3\}$.

   c) How many subsets are there of $\{1, 2, 3, 4\}$? Explain.

   d) Generalize part c.

10. **a)** The Venn diagrams in Figure 1 represent the interactions of generic sets. Draw Venn diagrams representing the special situations $B \subseteq C$ and $D \cap E = \emptyset$.

    b) Explain why Figure 1 has as many regions as could possibly occur when two sets interact. Repeat for Figure 5 and three sets.

    c) How many regions are needed for a Venn diagram with four generic sets? Explain. Design a Venn diagram for four sets S, T, U, and W so that each possible region is connected and appears exactly once. (Hint: You may need to use an unusually shaped region for at least one set.)

    d) Generalize your explanation in part b to larger numbers of sets.

11. ***a)*** How many elements are in $\{1, 2, \{3, 4, 5\}\}$?

    b) How many elements are in $\{1, \emptyset, \{2, \{3, 4\}, 5\}\}$?

12. For each pair of sets, determine whether the first is an element, a subset, or both of the second; the second is an element, subset, or both of the first; or none of these options.

    a) $\emptyset$ and $\{\emptyset\}$

    b) $[1, 4]$ and $\{\pi\}$

    c) $\{[1, 4]\}$ and $\{\pi\}$

    d) $\{\emptyset\}$ and $\{\emptyset, \{\emptyset\}\}$

    e) $\{\emptyset, \{\emptyset\}\}$ and $\{\{\emptyset\}\}$

    f) $\mathbb{R}$ and $\{\{\mathbb{R}\}\}$

    g) $\{\mathbb{R}, \{\mathbb{R}\}\}$ and $\{\mathbb{R}, \{\mathbb{R}\}, \{\mathbb{R}, \{\mathbb{R}\}\}\}$

13. Suppose A, B, and C are nonempty sets of positive real numbers, with $A \subseteq B \subseteq C$.

    a) Suppose for all $x \in B$, we have $x > \sqrt{x}$; that is, "$\forall x \in B\ (x > \sqrt{x})$" is true. Decide which of the following must be true, which might be true, and which must be false.

    i) $\forall x \in A\ (x > \sqrt{x})$ **iii)** $\exists x \in B\ s.t.\ (x > \sqrt{x})$ **v)** $\exists x \in A\ s.t.\ (x \le \sqrt{x})$
    ii) $\forall x \in C\ (x > \sqrt{x})$ **iv)** $\exists x \in B\ s.t.\ (x \le \sqrt{x})$ **vi)** $\exists x \in C\ s.t.\ (x \le \sqrt{x})$

    b) If the proposition "$\exists x \in B\ s.t.\ (x > \sqrt{x})$" is true, decide which of the following must be true, which might be true, and which must be false.

    i) $\exists x \in A\ s.t.\ (x > \sqrt{x})$ **iii)** $\forall x \in B\ (x > \sqrt{x})$ **v)** $\forall x \in A\ (x \le \sqrt{x})$
    ii) $\exists x \in C\ s.t.\ (x > \sqrt{x})$ ***iv)*** $\forall x \in B\ (x \le \sqrt{x})$ **vi)** $\forall x \in C\ (x \le \sqrt{x})$

    c) Which, if any, of your answers in part a) would change if we allowed the sets to be empty?

14. Algebraists find *symmetric difference*, written $A \triangle B$, an interesting operation for sets. We define $A \triangle B = (A - B) \cup (B - A)$.

    a) Use Venn diagrams to verify that $A \triangle B$ equals $(A \cup B) - (A \cap B)$.

    b) Use Venn diagrams to verify that $\triangle$ is commutative and associative; that is, $A \triangle B = B \triangle A$ and $(A \triangle B) \triangle C = A \triangle (B \triangle C)$.

    c) What is $A \triangle \emptyset$? $\emptyset \triangle A$? $A \triangle A$? We call $\emptyset$ the *identity* for $\triangle$ because it acts like the additive identity 0 does for addition and the multiplicative identity 1 does for multiplication. The result of $A \triangle A$ illustrates that each element acts as its own *inverse* for $\triangle$.

    d) Investigate whether $\cap$ distributes over $\triangle$; that is, does $X \cap (Y \triangle Z) = (X \cap Y) \triangle (X \cap Z)$?

e) Investigate whether $\cup$ distributes over Δ; that is, does $X \cup (Y \Delta Z) = (X \cup Y) \Delta (X \cup Z)$?

1.5 INTRODUCTION TO NUMBER THEORY

"Mathematics is the queen of the sciences and number theory is the queen of mathematics."

—Carl Friedrich Gauss (1777–1855)

We define and discuss several basic properties of natural numbers and integers that appear frequently throughout mathematics. While the sum, difference, and product of two integers are always integers, division is trickier and thus mathematically more interesting. The divisors of a number and particularly primes have intrigued mathematicians for centuries and are now vital in many applications. After investigating the relation "divides," we turn to the concepts of remainders and modular arithmetic. Finally, we explore recursive definitions, a common way to describe patterns closely linked with proofs by induction, discussed in Section 2.4.

Divides

Ancient Greek mathematicians generalized the distinction between odd and even (divisible by two), long before people considered fractions as numbers. Our term to this day (*k divides n*) still reflects this legacy, even though we more readily think of the operation "division" than the relation "divides." While we can certainly divide 5 by 3, we say that 3 doesn't divide 5. The relation "divides" fits well with the notion of factoring numbers. Relations, such as "divides" or "less than," are either true or false; whereas operations, such as $\div$ or $\cup$, give a new thing from two things. Because of these potentially confusing differences, I advise students to avoid using the operation "division" when working with the relation "divides." Also, pay attention to the existential quantifier in the next definition.

DEFINITION. An integer k *divides* an integer n iff there is an integer d such that $d \cdot k = n$. We say k is a *divisor* of n, or a *factor* of n, and abbreviate "k divides n" by $k|n$.

EXAMPLE 1

The integer 6 divides 24 because $4 \in \mathbb{Z}$ and $4 \cdot 6 = 24$. Similarly, $3 \cdot -5 = -15$ implies that both 3 divides -15 and -5 divides -15. However, 4 doesn't divide 10, because $d \cdot 4 = 10$ requires $d = 2.5$, which is not an integer. $\Diamond$

Note that 1 divides any integer n, since $n \cdot 1 = n$. From the same equation we see that every integer divides itself. The positive divisors of a positive integer n lie between 1 and n, although not every number between 1 and n needs to divide n. Anyone who has studied these positive divisors quickly notices the importance of what we call prime numbers. Primes act as the building blocks of $\mathbb{N}$ under multiplication.

DEFINITION. A natural number p is a *prime number* iff $p > 1$ and the only positive divisors of p are 1 and p.

EXAMPLE 2

Factor $10,800$ into primes.

Solution   We readily see that $10,800 = 108 \cdot 10 \cdot 10$. We can factor a 2 out of any even number, so we get $10,800 = (2 \cdot 54) \cdot (2 \cdot 5) \cdot (2 \cdot 5) = 2 \cdot 2 \cdot 27 \cdot 2 \cdot 5 \cdot 2 \cdot 5$. Finally, $27 = 3 \cdot 3 \cdot 3$. So, $10,800 = 2^4 3^3 5^2$. If we had factored in a different order, say, dividing by 3 first, would the answer differ? No, the Fundamental Theorem of Arithmetic guarantees that factoring always yields exactly the same primes, repeated the same number of times. We need to prohibit 1 from being a prime for this to be true. Otherwise, we could factor $10,800$ as $1^7 2^4 3^3 5^2$ or $1^5 2^4 3^3 5^2$ or infinitely many other factorizations. $\Diamond$

THEOREM: FUNDAMENTAL THEOREM OF ARITHMETIC. (Euclid).

Every integer greater than 1 is a prime or can be factored into primes in a unique way, up to the order of the primes.

(The proof appears in the Supplement to Section 2.4.)

EXERCISE.   What integers divide 0? What integers does 0 divide?

Mathematicians have extended the relation "divides" to common divisors and transformed that idea into the greatest common divisor (also called the greatest common factor). The word "greatest" implicitly needs existence and uniqueness to make sense. Unfortunately, 0 and 0 have no greatest common divisor, because every integer divides 0. With the exception of that unimportant case, the greatest common divisor of two integers exists and is unique.

DEFINITION.   An integer d is a *common divisor* of the integers a and b iff d divides a and d divides b. An integer g is a (the) *greatest common divisor* of the nonzero integers a and b iff g is a common divisor of a and b and for every common divisor d of a and b, we have $d \leq g$. In this case we write $\gcd(a, b) = g$.

EXAMPLE 3

Find the greatest common divisor of $10,800$ and $11,025$.

Solution   While we could search among the sixty divisors of $10,800$ and twenty-seven divisors of $11,025$ for the greatest common divisor, the Fundamental Theorem of Arithmetic suggests an easier way. Consider the common prime factors of these numbers. Since $10,800 = 2^4 3^3 5^2$ and $11,025 = 3^2 5^2 7^2$, a common divisor can have prime factors of only 3 and 5. Furthermore, we can't have more than two factors of each. That is, all common divisors are divisors of $3^2 5^2 = 225$, and so $\gcd(10,800, \ 11,025) = 225$. $\Diamond$

Example 3 suggests an important fact about the gcd of two numbers. Not only is it the greatest of the common divisors, but also, all of the common divisors divide it. However, that is not part of the definition; rather it is a theorem, and so will need to be proven.

Remainders

So far, we have considered when a number divides another. Now, 7 doesn't divide 33; indeed, dividing 33 by 7 gives a quotient of 4 with a remainder of 5, since $33 = 4 \cdot 7 + 5$. The generalization of this idea of remainders is still called the Division Algorithm, even though its statement isn't an algorithm in the computer science meaning. In computer science an algorithm is a step-by-step process, as Example 5 illustrates.

THEOREM: THE DIVISION ALGORITHM.   For any integer z and any natural number n there are unique integers q and r such that $z = qn + r$ and $0 \leq r < n$.

(The proof appears in Section 2.4.)

EXAMPLE 4

For the remainder of $z = -53$ divided by $n = 7$, it may seem natural to start with $53 = 7 \cdot 7 + 4$ and so write $-53 = (-7) \cdot 7 - 4$. However, the remainder can't be negative. Instead, we write $-53 = (-8) \cdot 7 + 3$. ◇

EXAMPLE 5

Use the Division Algorithm to find the greatest common divisor of 10,800 and 1764. Euclid described the method of this example, now called the Euclidean Algorithm, an algorithm in the modern sense.

Solution   The first time, let $z = z_1$ be the bigger of the two numbers (10,800) and $n = n_1$ the smaller (1764). From the Division Algorithm, $10{,}800 = 6 \cdot 1764 + 216$. To continue, we use the divisor $n_1 = 1764$ as z_2 and the remainder $r_1 = 216$ as n_2. Then $1764 = 8 \cdot 216 + 36$, and the third round gives $216 = 6 \cdot 36 + 0$. Thus, the last nonzero remainder, 36, divides 216 and in turn it divides 1764 and 10,800, making 36 a common divisor. In fact, $36 = \gcd(10{,}800, 1764)$. Euclid showed that this approach always produces the greatest common divisor. This repeated division seems an unenlightening way to find the gcd of two numbers, compared with the prime factorizations in Example 3. However, factoring very large numbers—even with computers—is extremely hard, whereas computers can readily implement the Euclidean Algorithm. ◇

EXAMPLE 6

Use the Division Algorithm to explain why every integer is either even or odd, but not both.

Solution   For any integer z, when we divide z by $n = 2$, the Division Algorithm tells us there are exactly two possibilities: $z = 2q + 0 = 2q$ or $z = 2q + 1$. These options match our definitions of even and odd, respectively. So either z is even or z is odd. Since $0 \neq 1$, no integer is both even and odd. ◇

Modular Arithmetic

In the 1700s, several mathematicians, including Leonhard Euler (1707–1783), generalized the notion of splitting the integers into evens and odds through the concept of congruence. Gauss (1777–1855) deserves the credit for making congruence (mod n) a powerful tool now used in many areas of mathematics. Two congruent numbers can be seen as similar in some ways, just as congruent triangles share properties. Our clocks group time together into 12-hour time slots, with an additional designation of A.M. and P.M. Noon one day is similar to noon another day, more similar in important ways than noon is to 8 P.M. on the same day, for instance.

EXAMPLE 7

Given an integer z and $n = 7$, the possible remainders are the integers from 0 to 6. Each row of Table 1 gives several numbers with the same remainder, which is listed on the left of the row. The integers with remainder 0 are clearly the multiples of 7, which are similar to one another in that sense. The ones with remainder, say, 3 are all shifted by 3 from the multiples of 7. In general, two numbers with the same remainder differ by a multiple of 7 or, equivalently, 7 divides their difference. We use this last property to define congruence (mod n).

TABLE 1

r	...		z			...
0	...	−7	0	7	14	...
1	...	−6	1	8	15	...
2	...	−5	2	9	16	...
3	...	−4	3	10	17	...
4	...	−3	4	11	18	...
5	...	−2	5	12	19	...
6	...	−1	6	13	20	...

$\Diamond$

DEFINITION. Given the natural number n, two integers a and b are *congruent* (mod n) iff n divides $a - b$. In this case, we write $a \equiv b (\bmod\, n)$.

EXAMPLE 8

Which of the congruences $37 \equiv 73\,(\bmod\,12)$, $25 \equiv -34\,(\bmod\,13)$, and $-235 \equiv -478\,(\bmod\,9)$ are true?

Solution We need to see whether n divides $a - b$ in each case. For the first one, $37 - 73 = -36$, and 12 divides -36, so $37 \equiv 73\,(\bmod\,12)$ is true. However, $25 - (-34) = 59 = 4 \cdot 13 + 7$ is not divisible by 13, so the second one is false. Finally, $-235 - (-478) = 243 = 9 \cdot 27$, so the third one is true.

EXAMPLE 9

Is there an integer x such that $x \equiv 4 (\bmod\,5)$ and $x \equiv 5 (\bmod\,6)$? Is there an integer x such that $x \equiv 4 (\bmod\,9)$ and $x \equiv 5 (\bmod\,6)$?

Solution The possible values of x satisfying $x \equiv 4(\mod 5)$ include 4, 9, 14, 19, 24, 29, and so on. Those satisfying $x \equiv 5(\mod 6)$ include 5, 11, 17, 23, 29, and so on. Since 29 is in both lists, it is a common solution. Among the infinitely many other solutions is $59 = 29 + 30$.

However, the other pair of congruences have no common solution. The key is that both 6 and 9 are multiples of 3. A solution to $x \equiv 4(\mod 9)$ satisfies $x - 4 = 9q$, for some $q \in \mathbb{Z}$. That is, $x = 9q + 4 = 3(3q) + 3 + 1 = 3(3q + 1) + 1$. Similarly, a solution to $y \equiv 5(\mod 6)$ satisfies $y = 6s + 5 = 3(2s + 1) + 2$. Since x and y have different remainders when divided by 3, the Division Algorithm implies that no integer can possibly satisfy both congruences. $\Diamond$

EXAMPLE 10

How does congruence $(\mod n)$ interact with addition?

Solution Consider $8 + 16 = 24$. The corresponding remainders $(\mod 7)$ are $8 \equiv 1(\mod 7)$, $16 \equiv 2(\mod 7)$, and $24 \equiv 3(\mod 7)$. How nice! The remainders of the first two add to give the remainder of their sum: $1 + 2 = 3$. In a sense, the equations $8 + 16 = 24$ and $1 + 2 = 3$ express the same information "mod 7." The general situation is only a bit more complicated. Consider 13 and 11 and their corresponding remainders 6 and 4 when divided by 7. Now $13 + 11 = 24$, which has a remainder of 3, whereas $6 + 4 = 10$, instead of 3. However, $10 \equiv 3(\mod 7)$. In general, as we will show in Section 4.2, if $a \equiv b(\mod n)$ and $c \equiv d(\mod n)$, then $a + c \equiv b + d(\mod n)$.

The interaction of congruence $(\mod n)$ with arithmetic operations provides valuable insights in algebra and combinatorics. Problem 13 investigates some of these interactions. $\Diamond$

Recursive Definitions

Some concepts about integers are awkward to define in general, even though the pattern is easy to describe. Recursive definitions define a concept for a general term based on previous ones, starting from one or more initial cases. Consider first the familiar notation for exponentials, defined next.

DEFINITION. For $x \in \mathbb{R}$, define $x^1 = x$ and for $n \in \mathbb{N}$, $x^{n+1} = x(x^n)$.

The definition starts with an initial case, here for the exponent 1. Then succeeding cases (higher powers here) are defined in terms of already specified ones. Newton's Method from calculus provides a more significant example of a recursive definition. It gives successive and, usually, increasingly accurate approximations of a root of an equation $g(x) = 0$, starting from an initial guess x_1 and the recursive formula $x_{n+1} = x_n - \frac{g(x_n)}{g'(x_n)}$.

EXAMPLE 11

The well-known Fibonacci numbers, defined next, require a more involved recursion based on two initial values. Leonardo of Pisa (1170–1250), better known as Fibonacci, posed an odd problem about rabbit reproduction. He started with one pair of rabbits and assumed that, after an initial wait of a month, each pair of rabbits produced another pair of rabbits each month and no rabbits died. He found the numbers of pairs of rabbits in succeeding months to be 1 1 2 3 5 and so on. One can find the next Fibonacci number by adding the two previous ones, which the recursive definition formalizes. Fibonacci found this summation property as follows: The number of rabbits at time $n + 2$ comes from the rabbits at time $n + 1$, who keep on living, plus the babies born to those old enough to have babies, which is the number at time n. While the original problem is only a historical curiosity, the Fibonacci numbers have received extensive investigation. These numbers actually appear naturally in the number of spirals in plants such as pine cones, sunflowers, and daisies. ◊

DEFINITION. The first two *Fibonacci numbers* are $f_1 = 1$ and $f_2 = 1$. The $(n + 2)^{nd}$ *Fibonacci number* is $f_{n+2} = f_n + f_{n+1}$.

EXAMPLE 12

From f_1, $f_1 + f_3$, $f_1 + f_3 + f_5$ and other sums of the first Fibonacci numbers with odd indices, make a general conjecture.

Solution We have $f_1 = 1$, $f_1 + f_3 = 1 + 2 = 3$, $f_1 + f_3 + f_5 = 1 + 2 + 5 = 8$. Each of these sums is itself a Fibonacci number: $3 = f_4$, $8 = f_6$, and 1 is both f_1 and f_2. We can guess that the sum of the first n Fibonacci numbers with odd indices is the next Fibonacci number with an even index. In symbols, $\sum_{i=1}^{n} f_{2i-1} = f_{2n}$. (Problem 18b asks you to define summations like this recursively.) ◊

PROBLEMS

*1. For each statement that follows, decide whether it is true or (at least sometimes) false. If false, explain why. Cite the part of the text supporting your conclusion.

a) $k \mid n$ iff $k \div n$.

b) Two integers don't always have a greatest common divisor.

c) Even when we divide a negative integer by a positive integer, the remainder can't be negative.

d) The smallest prime is 1.

e) For all $n \in \mathbb{N}$ and all $a, b \in \mathbb{Z}$, if $a = b$, then $a \equiv b \pmod{n}$.

f) For all $n \in \mathbb{N}$ and all $a, b \in \mathbb{Z}$, if $a \equiv b \pmod{n}$, then $a = b$.

g) A recursive definition defines a property for a potential infinity of cases.

2. a) Factor each of the following numbers into primes:

i) 45	**iii)** 98	**v)** 30
ii) 100	***iv)** 441	**vi)** 105

b) For each number in part a), determine the number of its positive divisors.

c) Make a conjecture involving primes about the number of positive divisors a natural number has.

3. List all positive common divisors of 225 and 375. Find their greatest common divisor.

4. a) Find the greatest common divisors of the following numbers:

*i) 48 and 72	**iii)** 245 and 35
ii) 75 and 56	**iv)** 324 and 432

b) Modify the definition of greatest common divisor to accommodate more than two numbers.

c) Find gcd(80, 96, 120) and gcd(378, 783, 837).

5. a) Define a common multiple of two positive integers a and b. Define lcm(a, b) as the least positive integer that is a common multiple of a and b.

b) Find these least common multiples:

  *i) lcm$(48, 72)$ **iii)** lcm$(245, 35)$

  ii) lcm$(75, 56)$ **iv)** lcm$(324, 432)$

c) Find lcm$(80, 96, 120)$ and lcm$(378, 783, 837)$.

d) Explain why we need to specify in part a that lcm(a, b) is positive. Why don't we need to specify that gcd(a, b) is positive?

6. Determine which of the propositions that follow are true. For those failing for integers, give a counterexample and determine whether they hold when we replace "integers" by "natural numbers." For those still failing for natural numbers, provide a counterexample.

  a) For all integers x, $x|x$.

  *b) For all integers y and z, if $y|z$, then $y \le z$.

  c) For all integers a, b, and c, if $a|b$ and $b|c$, then $a|c$.

  d) For all integers j and k, if $j|k$ and $k|j$, then $j = k$.

  e) For all integers q and r, if $q|r$, then $r|q$.

7. a) Suppose d is a common divisor of both 360 and 450. Does d divide their sum? their difference? their product? Explain.

b) Suppose we are given two numbers x and y and told d is a common divisor of their sum and their difference. Must d be a common divisor of x and y? Hint: Try some small values for x and y.

8. Generalize Example 6 for $n > 2$.

9. Find the remainders in the next examples and describe a general pattern. Then test your pattern with some other examples.

  *a) The remainder of 273 when divided by 11, and the remainder of -273 divided by 11.

  b) The remainder of 273 when divided by 12 and the remainder of 273 divided by 4. Note: 4 is a divisor of 12.

  c) The remainder of $273 = 21 \cdot 13$ when divided by 11, the remainder of 21 divided by 11, and the remainder of 13 divided by 11.

10. a) In Example 3 we used common prime factors to find the gcd of two numbers. Describe how to use prime factors to find the lcm of two natural numbers. Check your description with the numbers in Problem 4.

b) Find an arithmetic relationship among the positive integers $a, b,$ gcd(a, b), and lcm(a, b).

c) The greatest common divisor of two natural numbers seems similar to the intersection of two sets. The least common multiple and the union seem similarly related. Redo Problem 6 from Section 1.4, using gcd and lcm instead of intersection and union. Which of the equalities still hold?

11. Determine which of the following congruences are true:

  a) $29 \equiv 14 \pmod 5$ **c)** $-29 \equiv 147 \pmod 7$

  *b) $29 \equiv 147 \pmod 6$ **d)** $-297 \equiv -1473 \pmod 8$

12. Test out the given conjectures with examples. Which appear to be true in general? Explain.

  *a) If $a \equiv b \pmod n$, then $b \equiv a \pmod n$.

b) If $a \equiv b(\text{mod } n)$, then $-a \equiv -b(\text{mod } n)$.

c) If $a \equiv b(\text{mod } n \cdot k)$, then $a \equiv b(\text{mod } n)$ and $a \equiv b(\text{mod } k)$.

d) The converse of part c.

e) If $a \equiv b(\text{mod } n)$ and $b \equiv c(\text{mod } n)$, then $a \equiv c(\text{mod } n)$.

f) If $a \cdot c \equiv b \cdot c(\text{mod } n)$, then $a \equiv b(\text{mod } n)$.

g) The converse of part f.

13. *a) Investigate, as in Example 10, how congruence (mod n) interacts with subtraction. Explain.

b) Repeat part a with multiplication.

c) Repeat part a with division.

d) Repeat part a with exponentiation.

14. *a) Find the smallest positive x, if any, satisfying both conditions in each part.

    i) $x \equiv 3(\text{mod } 8)$ and $x \equiv 6(\text{mod } 9)$.   **iv)** $x \equiv 9(\text{mod } 11)$ and $x \equiv 6(\text{mod } 9)$.

    ii) $x \equiv 5(\text{mod } 6)$ and $x \equiv 6(\text{mod } 14)$.   **v)** $x \equiv 11(\text{mod } 12)$ and $x \equiv 6(\text{mod } 9)$.

    iii) $x \equiv 7(\text{mod } 9)$ and $x \equiv 6(\text{mod } 14)$.

b) Make a conjecture about when you can be sure two congruences, such as in part a, have a common solution. Give an upper bound to the smallest positive solution, provided there is a solution. Explain.

15. For some values of a, b, and n, we can make sense of division (mod n). We define ordinary division by $x \div y = w$ iff $y \cdot w = x$. Similarly, for integers a and b, we can say $a \div b = c(\text{mod } n)$ iff there is a unique integer c such that $0 \le c < n$ and $a = b \cdot c(\text{mod } n)$. For instance, $5 \div 2 = 7(\text{mod } 9)$ because $2 \times 7 = 14 \equiv 5(\text{mod } 9)$. However, no value c satisfies $5 \div 3 = c(\text{mod } 9)$. Further, $3 \div 6 = c(\text{mod } 9)$ has three possible values of c between 0 and 8, namely, $c = 2$, $c = 5$, and $c = 8$. Investigate for which values of n between 3 and 7 and $a, b \in \{1, \ldots, n - 1\}$ we have just one appropriate value c for $a \div b = c(\text{mod } n)$, for which values we have no values of c, and for which values we have more than one value of c. Make some conjectures about division (mod n).

16. *a) Find the first 12 Fibonacci numbers.

b) Make a conjecture about which Fibonacci numbers are even.

c) Make a conjecture about which Fibonacci numbers are divisible by 3.

d) Make a conjecture about the sum of the first n Fibonacci numbers in terms of a Fibonacci number.

e) Make a conjecture about the gcd of two successive Fibonacci numbers.

17. We can define sequences of numbers similar to the Fibonacci numbers.

a) Define the g-numbers recursively by $g_1 = 1$, $g_2 = 3$, and $g_{n+2} = g_1 + g_2$. Find the first six g-numbers. Make a conjecture about which g-numbers are even.

b) Which choices of the initial numbers for g_1 and g_2 in part a will result in the same pattern of even and odd g-numbers?

c) Define the h-numbers recursively by $h_1 = 1$, $h_2 = 1$, $h_3 = 1$, and $h_{n+3} = h_n + h_{n+1} + h_{n+2}$. Which h-numbers are even?

18. Give recursive definitions for the following concepts:

a) Factorials, written $n!$.

b) The sum of the numbers $a_1, \ldots, a_n$, written $\sum_{i=1}^{n} a_i$.

c) The product of the numbers $a_1, \ldots, a_n$, written $\prod_{i=1}^{n} a_i$.

d) The sequence $1\ 3\ 2\ 2.5\ 2.25\ \ldots$, where a term is the average of the two preceding terms after the initial terms.

e) A "tower of powers" $T_n(x)$ of n repeated exponentiations of x. For instance, $T_3(5) = 5^{(5^5)}$.

f) The sequence that starts out $\sqrt{2}$, $\sqrt{2\sqrt{2}}$, $\sqrt{2\sqrt{2\sqrt{2}}}$, and so on.

1.6 ADDITIONAL SET THEORY

Mathematicians use more extensive set theory than what was introduced in Section 1.4. In this section, we study several other frequently used kinds of sets. We start by presenting the familiar ideas of graphs, ordered pairs, and functions in the language of sets. Then we consider sets of sets, a construction unneeded in high school mathematics and calculus, but vital in higher mathematics.

Ordered Pairs and Cartesian Products

The familiar graphs of high school mathematics feature ordered pairs (x, y) as the names of points, with x indicating horizontal position and y vertical position. Logically, an ordered pair doesn't depend on horizontal and vertical axes, but depends just on the ability to distinguish which item is first and which is second. We leave a formal definition of an ordered pair to Problem 14, since working mathematicians do not actually think in terms of such a definition. Your intuition of ordered pairs suffices. We define only the situation when two ordered pairs are equal, on the basis of distinguishing the first and second elements of pairs. (You may recall in Section 1.4 we similarly defined when two sets were equal, but not what a set is.) More important, we define the Cartesian product of two sets by using ordered pairs. The familiar coordinate plane with its axes gives a picture of the Cartesian product $\mathbb{R} \times \mathbb{R}$, the setting where all the graphs of calculus functions "live." The adjective "Cartesian" honors René Descartes (1596–1650), one of the founders of analytic geometry.

DEFINITION. Given elements a, b, c, and d, the ordered pairs (a, b) and (c, d) are *equal* iff $a = c$ and $b = d$.

DEFINITION. Given two sets A and B, their *Cartesian product* is $A \times B = \{(a, b) : a \in A \text{ and } b \in B\}$.

EXAMPLE 1

Figure 1 illustrates the set $S \times T$ in $\mathbb{R} \times \mathbb{R}$, where $S = [-1, 2]$ and $T = [1, 2]$.

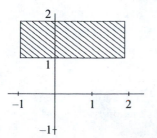

FIGURE 1. $S \times T$ in $\mathbb{R} \times \mathbb{R}$.   ◇

EXAMPLE 2

For $V = \{\clubsuit, \diamondsuit, \heartsuit, \spadesuit\}$ and $W = \{A, K, Q\}$, their Cartesian product has 12 elements: $V \times W = \{(\clubsuit, A), (\clubsuit, K), (\clubsuit, Q), (\diamondsuit, A), (\diamondsuit, K), (\diamondsuit, Q), (\heartsuit, A), (\heartsuit, K), (\heartsuit, Q), (\spadesuit, A), (\spadesuit, K), (\spadesuit, Q)\}$. Figure 2 gives an easier way to visualize this set of ordered pairs. The set $T = \{\clubsuit, \diamondsuit\}$, is a subset of V, and $S = \{A, Q\}$ is a subset of W. Note the Cartesian product $T \times S = \{(\clubsuit, A), (\clubsuit, Q), (\diamondsuit, A), (\diamondsuit, Q)\}$ is a subset of $V \times W$, but neither T nor S is a subset of $V \times W$.

Q	$(\clubsuit, Q)$	$(\diamondsuit, Q)$	$(\heartsuit, Q)$	$(\spadesuit, Q)$
K	$(\clubsuit, K)$	$(\diamondsuit, K)$	$(\heartsuit, K)$	$(\spadesuit, K)$
A	$(\clubsuit, A)$	$(\diamondsuit, A)$	$(\heartsuit, A)$	$(\spadesuit, A)$
	$\clubsuit$	$\diamondsuit$	$\heartsuit$	$\spadesuit$

FIGURE 2. $V \times W$.

Graphs from calculus focus, not on the entire Cartesian product $\mathbb{R} \times \mathbb{R}$, but rather on subsets. We will consider general subsets of arbitrary Cartesian products in Chapter 4, where we call such subsets *relations*. In Chapter 3 we consider *functions*, familiar from calculus, as special subsets of Cartesian products.

Sets of Sets

The easiest collection of sets to describe consists of all subsets of a given set X, called the *power set* of X.

DEFINITION. Given a set X, its *power set* is $\mathcal{P}(X) = \{A : A \subseteq X\}$, the set of all subsets of X.

EXAMPLE 3

Let $A = \{p, q\}$. Then $\mathcal{P}(A) = \{\emptyset, \{p\}, \{q\}, \{p, q\}\}$. Note that $\emptyset$ is a subset of any set. Similarly, if $B = \{p, q, r\}$, then $\mathcal{P}(B) = \{\emptyset, \{p\}, \{q\}, \{r\}, \{p, q\}, \{p, r\}, \{q, r\}, \{p, q, r\}\}$. The numbers of subsets of different sizes form an interesting pattern, explored in Problem 7. For $\mathcal{P}(B)$, the number of subsets of sizes 0, 1, 2, and 3 are 1, 3, 3, and 1, respectively. $\diamondsuit$

Just as a set is made up of individual elements, a power set is made up of sets. Intuitively, we know that the things in a set are simpler than the set. For instance, the number 7 is simpler than the set of all real numbers, $\mathbb{R}$. Similarly, a particular subset of $\mathbb{R}$, say, $[1, 7]$, is easier to grasp than the set of all such subsets, $\mathcal{P}(\mathbb{R})$. It might help to think of different levels for elements and a set containing those elements. The relation $\in$ ("is an element of") connects things at one level to things at the next "higher" level: $7 \in \mathbb{R}$ or $[1, 7] \in \mathcal{P}(\mathbb{R})$. The subset relation $\subseteq$ connects things at the same level: $[1, 7] \subseteq \mathbb{R}$.

Indexed Families of Sets

We often need to work with a whole family of related sets. It is useful to have names for these sets that resemble one another. One common situation using families of sets involves the union or intersection of many sets. The definitions of union and intersection of two sets from Section 1.4 work well when there are only a few sets. However, our imaginations and mathematical needs can easily expand to infinite collections of sets.

EXAMPLE 4

All the shaded intervals in Figure 3 together represent the union of the intervals $[1, 1.5], [2, 2.5],$ $[3, 3.5],$ and so on. To avoid depending on phrases like "and so on," we need to describe the infinitely many related sets at once and take their union or intersection. The intervals in Figure 3 all have the form $[n, n + 0.5]$ for $n \in \mathbb{N}$, suggesting we could talk about them collectively as $\{[n, n + 0.5] : n \in \mathbb{N}\}$. This set of sets is called an *indexed family* of sets, defined more generally next. Here the variable n is the *index*, with different values of n giving us different sets. We call $\mathbb{N}$ the *index set*. The notation for their union, $\bigcup_{n\in\mathbb{N}} [n, n + 0.5]$, is reminiscent of the notation for a summation, such as $\sum_{n=1}^{\infty} \frac{1}{3^n} = \frac{1}{2}$. We avoid the notation $_{n=1}^{\infty}$ for two reasons. First of all, we could use an index set other than $\mathbb{N}$ for which the symbol ∞ may be inappropriate. Second, the definitions of union and intersection that follow use quantifiers, which treat the index set as a whole set. In comparison, infinite sums are defined in terms of limits, which involve an unending process indicated with the symbol ∞.

FIGURE 3   ◇

DEFINITION. An *indexed family* of sets $\{S_i : i \in I\}$ is a set of sets S_i, one for each *index* i in the *index set* I.

It is important to remember the difference between an individual set S_i, the index set I, and the family of sets $\{S_i : i \in I\}$. Index sets are most often sets of numbers, although the definition places no such restriction. Usually, for different indices, say, $i \neq j$, the sets S_i and S_j are different; but this also is not required.

The definition of the union of two sets in Section 1.4 used the logical word "or." We now generalize unions to include infinitely many sets. Section 1.3 discussed the idea of the existential quantifier as an "infinite or." Thus, it is reasonable to use $\exists$ (there exists) in the definition of union that follows. Similarly, the "and" in the earlier definition of the intersection of two sets becomes $\forall$ (for all) in the definition of the intersection of arbitrarily many sets.

DEFINITION. Given a nonempty indexed family of sets $\{S_i : i \in I\}$, their *union* is $\bigcup_{i\in I} S_i = \{x : \exists i \in I \text{ s.t. } x \in S_i\}$ and their *intersection* is $\bigcap_{i\in I} S_i = \{x : \forall i \in I \ x \in S_i\}$.

EXAMPLE 5

The vertical cross sections of a region in $\mathbb{R} \times \mathbb{R}$ form an indexed family of sets of interest in calculus. In Figure 4, the index set is $[0, \pi]$. The highlighted vertical segment is one of the sets $S_x = \{(x, y) : \sin(x) \leq y \leq 2\sin(x)\}$, and the family is $\{S_x : x \in [0, \pi]\}$. The entire region

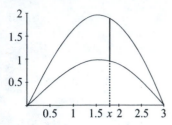

FIGURE 4

between the two curves is the union $\bigcup_{x \in [0,\pi]} S_x$. The intersection of all of these cross sections is empty. Even more, the intersection of any two different cross sections is empty. ◇

EXAMPLE 6

For $k \in \mathbb{N}$, let T_k be the interval $[1 + \frac{1}{k}, 4 + \frac{1}{k}]$. Find and illustrate T_1, T_2, T_{10}, $\bigcap_{k \in \mathbb{N}} T_k$ and $\bigcup_{k \in \mathbb{N}} T_k$.

Solution For $k = 1, 2$ and 10, we have $T_1 = [2, 5]$, $T_2 = [1.5, 4.5]$, and $T_{10} = [1.1, 4.1]$. Figure 5 illustrates these sets above a number line. The movement of these sets suggests that they are shifting bit by bit to the left. Since $\bigcap_{k \in \mathbb{N}} T_k$ contains those numbers in all of the T_k, this intersection starts at 2, the left endpoint of T_1. The right endpoint is the largest number in all of the T_k, which is 4. Hence, $\bigcap_{k \in \mathbb{N}} T_k = [2, 4]$. The union $\bigcup_{k \in \mathbb{N}} T_k$ includes all of $T_1 = [2, 5]$ and, as well, some numbers less than 2. Indeed, every number between 1 and 2 is in some T_k; so $\bigcup_{k \in \mathbb{N}} T_k = (1, 5]$. These sets appear below the number line in Figure 5.

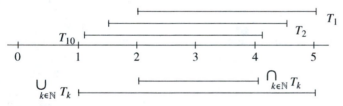

FIGURE 5

This set illustrates the dual nature of many mathematical concepts as a process and an object. It is easier to see the effects of taking the infinite union or intersection of sets as a process. However, the end result and the formal definition treat these as objects. It takes time and practice to be able to shift between these two aspects. ◇

Partitions

A partition, defined next, splits a set into separate pieces, as in Figure 6. For another instance, the family of all parallel lines with the same slope, illustrated in Figure 7 and explored in Example 9, partitions the plane into separate pieces.

DEFINITION. A *partition* of a set B is a family of sets $\{S_i : i \in I\}$ such that

    i) each S_i is nonempty and
    ii) for each $b \in B$, there is a unique S_i such that $b \in S_i$.

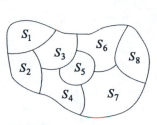

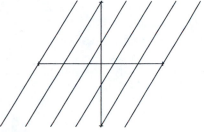

FIGURE 6   **FIGURE 7**

EXAMPLE 7

Let $\mathcal{E} = \{2x : x \in \mathbb{Z}\}$ be the set of even numbers and $\mathcal{O} = \{2x + 1 : x \in \mathbb{Z}\}$ be the set of odd numbers. Then $\{\mathcal{E}, \mathcal{O}\}$ is a partition of all of the integers $\mathbb{Z}$. Condition i simply says that there is at least one even integer and one odd integer. The existence part of the second condition says each integer is even or odd. The uniqueness part of that condition says that no integer is both even and odd. $\diamond$

EXAMPLE 8

In Example 5 the cross sections $S_x = \{(x, y) : \sin(x) \le y \le 2\sin(x)\}$ form a *partition* of the region shown in Figure 4. $\diamond$

The form of the previous definition of a partition is convenient for proofs, but a more common definition, given next, often matches students' visual intuition. This alternative definition uses the idea of disjoint sets, defined first.

DEFINITION.   Two sets are *disjoint* iff their intersection is empty. A family of sets $\{S_i : i \in I\}$ is *pairwise disjoint* iff for all $i, j \in I$, if $S_i \ne S_j$, we have $S_i \cap S_j = \emptyset$.

EXAMPLE 9

Let $A_1 = [1, 5]$, $A_2 = [2, 7]$ and $A_3 = [6, 9]$. Even though $\bigcap_{i \in \{1,2,3\}} A_i = \emptyset$, these sets are not pairwise disjoint. In particular, $A_1 \cap A_2 = [2, 5] \ne \emptyset$ and $A_2 \cap A_3 = [6, 7] \ne \emptyset$. The sets $A_1 = [1, 5]$, $A_3 = [6, 9]$, and $A_5 = [11, 15]$ are pairwise disjoint, since no two overlap at all. $\diamond$

ALTERNATIVE DEFINITION OF A PARTITION.   A *partition* of a set B is a family of sets $\{S_i : i \in I\}$ such that

   i) each S_i is nonempty,

   ii) $B = \bigcup_{i \in I} S_i$, and

   iii) the family $\{S_i : i \in I\}$ is pairwise disjoint.

EXAMPLE 10

Explore the two definitions of a partition, using the family of all lines parallel to a given line, which forms a partition of the Euclidean plane.

Solution   The set of points in the plane is the set B of the definition. The family of sets is the set of lines parallel to a given line, say, those parallel to the line L. The axiom of parallels in Euclidean geometry states, "For every line L and every point P there is a unique line K such that $K\,||L$ and P is on K." The wording "there is a unique line" in this axiom matches the phrase "there is a unique S_i" in part ii of our first definition of a partition. As Figure 7 indicates, this axiom promises first that all these parallel lines together cover the entire plane, which matches part ii of the alternative definition of a partition. Also, it promises that no two different parallels intersect, which matches part iii of the alternative definition of a partition. In both definitions part i simply says that all lines have points on them.

Let's rewrite these geometric ideas, using set theory language and a specific line. The whole plane is the set $\mathbb{R} \times \mathbb{R}$. Consider the specific line with equation $y = 2x + 3$. As a subset of $\mathbb{R} \times \mathbb{R}$, this is the set $\{(x, y) : y = 2x + 3\}$. The lines parallel to this line all have the same slope of 2. We can name them, using their y-intercepts: $L_b = \{(x, y) : y = 2x + b\}$. Our original line is thus L_3, and the family of sets is $\mathcal{L} = \{L_b : b \in \mathbb{R}\}$. Then part i of either definition is fulfilled because each L_b is nonempty; for instance, $(0, b) \in L_b$. Part ii of the first definition says, given a point (s, t), there is exactly one choice of b. The correct value of b is $t - 2s$ in order that (s, t) is on L_b. (That is, $2x + b = 2(s) + t - 2s = t$.) Part ii of the alternative definition asserts, $\mathbb{R} \times \mathbb{R} = \bigcup_{b \in \mathbb{R}} L_b$; that is, every point is on some line. Similarly, part iii becomes "For all $b, c \in \mathbb{R}$, if $L_b \neq L_c$, then $L_b \cap L_c = \emptyset$," which simply says that two different parallel lines never intersect.

The choice of the equation $y = 2x + 3$ was not important; any nonvertical equation $y = mx + b$ works the same way. The set of all vertical lines, with equations $x = b$, also forms a partition. $\Diamond$

The two definitions are just different ways of saying the same idea, emphasizing different ways of looking at the idea of a partition.

We have now seen a variety of types of sets: sets of numbers, sets of ordered pairs, and sets of sets. When you work with a set, you can avoid many mistakes by first asking what form its elements have.

Historical Remarks

We correctly credit René Descartes (1596–1650) and Pierre de Fermat (1601–1665) with the introduction of analytic geometry, which links algebraic formulas and geometric shapes. However, neither one of them used the modern idea of ordered pairs to name geometric points. Like many concepts in set theory and logic, the ordered pair arose in the latter part of the nineteenth century. Gottlob Frege (1848–1925) initiated the use of ordered pairs in 1893. In 1897, Giuseppe Peano (1858–1932) focused the requirements of an ordered pair to just the definition we gave for the equality of two ordered pairs. Kazimierz Kuratowski in 1921 invented the modern abstract definition given in Problem 14c. Ernst Zermelo (1871–1953) included axioms for the power set of a set, and the union of an arbitrary (indexed) family of sets appears as axioms in his groundbreaking 1908 paper giving axioms for set theory. Other sets we have discussed, such as the Cartesian product of any two sets, can be built from Zermelo's axioms.

PROBLEMS

*1. For each statement that follows, decide whether it is true or (at least sometimes) false. If false, explain why. Cite the part(s) of the text supporting your conclusion.

a) $\mathbb{Z} \times \mathbb{Z} \subseteq \mathbb{R} \times \mathbb{R}$.

b) For any sets A and B we have $A \times B = B \times A$.

c) If $x \in X$, then $\{x\} \subseteq \mathcal{P}(X)$.

d) If $x \in X$, then $\{x\} \in \mathcal{P}(X)$.

e) For any set A we have $A \in \mathcal{P}(A)$.

f) For the family of sets $\{S_i : i \in I\}$ we have $I \subseteq \bigcup_{i \in I} S_i$.

g) For the family of sets $\{S_i : i \in I\}$ we have $i \in \bigcup_{i \in I} S_i$.

h) For the family of sets $\{S_i : i \in I\}$ we have $S_1 \subseteq \bigcup_{i \in I} S_i$.

2. *a) For $T = \{3, 5, 7\}$, $V = \{\#, \$\}$ and $W = \{a, b, c, d\}$ list the elements in $T \times V$, $V \times T$, $V \times W$ and $V \times V$.

b) If X has x elements and Y has y elements, how many elements are in $X \times Y$ (referred to in Section 5.3)?

c) For W as in part a, what is $W \times \emptyset$? Does the number of elements of $W \times \emptyset$ fit with your answer in part b?

d) List the elements in $(T \times V) \cap (V \times T)$.

3. For $X = \{1, 2, 3\}$ and $Y = \{1, 2, 3, 4, 5, 6\}$ list the elements of these subsets of $X \times Y$.

*a) $F = \{(x, y) : y = \frac{4}{3-x}\}$.

b) $G = \{(x, y) : y = x!\}$.

c) $H = \{(x, y) : x = |4 - y|\}$.

d) $J = \{(x, y) : (x - 2)^2 = \sin(\frac{\pi y}{2})\}$.

4. Let $D = [1, 3]$, $E = [2, 4]$, $F = [5, 7]$ and $G = [6, 8]$. Investigate the relationship between the following pairs of sets.

*a) $(D \cup E) \times (F \cup G)$ and $(D \times F) \cup (E \times G)$

b) $(D \cap E) \times (F \cap G)$ and $(D \times F) \cap (E \times G)$

c) $(D - E) \times (F - G)$ and $(D \times F) - (E \times G)$

d) Write $(D \times E) \cap (E \times D)$ as the Cartesian product of two sets. Explain your choice and draw a picture.

e) How are $\overline{D} \times \overline{F}$ and $\overline{D \times F}$ related? (For the universal sets use $\mathbb{R}$ and $\mathbb{R} \times \mathbb{R}$, as appropriate.)

5. *a) Find $\mathcal{P}(C)$ if $C = \{x, y\}$.

b) Find $\mathcal{P}(\mathcal{P}(C))$.

c) If X has n elements, how many elements are in $\mathcal{P}(X)$?

d) If X has n elements, how many elements are in $\mathcal{P}(\mathcal{P}(X))$? Generalize.

e) If X has n elements and Y has k elements, how many elements are in $\mathcal{P}(X \times Y)$? In $\mathcal{P}(X) \times \mathcal{P}(Y)$?

6. Let $A = \{1, 2\}$ and $B = \{2, 3, 4\}$. For each pair of sets, determine whether they are equal or not. If they are unequal determine if one is a subset of the other.

*a) $\mathcal{P}(A) \cup \mathcal{P}(B)$ and $\mathcal{P}(A \cup B)$

b) $\mathcal{P}(A) \cap \mathcal{P}(B)$ and $\mathcal{P}(A \cap B)$

c) $\mathcal{P}(B) - \mathcal{P}(A)$ and $\mathcal{P}(B - A)$

d) $\mathcal{P}(A) \times \mathcal{P}(B)$ and $\mathcal{P}(A \times B)$

7. Make a table with the number of subsets of various sizes in sets of increasing sizes. For instance, by Example 3, the row for a set of three elements will have entries 1, 3, 3 and 1 for the number of subsets of sizes 0, 1, 2 and 3, respectively. Describe any patterns you find in this table.

8. Use examples to explore these conjectures. Explain your conclusions.

   a) Conjecture: If $S \subseteq T$, then $\mathcal{P}(S) \subseteq \mathcal{P}(T)$.

   b) Conjecture: The converse of part a.

9. Find the union and intersection of each indexed family of intervals. Hint: Graph several of the intervals.

   *a) Let $V_i = [1 + \frac{1}{i}, 7 - \frac{1}{i}]$, for $i \in \mathbb{N}$.

   b) Let $W_j = (2 + \frac{2}{j}, 6 + \frac{6}{j})$ for $j \in \mathbb{N}$.

   c) Let $X_k = [3 - \frac{3}{k}, 9 - \frac{3}{k})$ for $k \in \mathbb{N}$.

   d) Let $Y_m = (4 - \frac{4}{m}, 8 + \frac{8}{m}]$ for $m \in \mathbb{N}$.

   e) Let $Z_n = [3 - \frac{5}{n}, 5 - \frac{3}{n})$ for $n \in \mathbb{N}$.

10. For $i \in [1, 2]$ let R_i be the rectangle $[-i, i] \times [-3 + i, 3 - i]$.

    a) On a graph draw R_1, $R_{1.5}$, and R_2.

    b) On a graph draw $\bigcup_{i \in [1,2]} R_i$.

    c) On a graph draw $\bigcap_{i \in [1,2]} R_i$.

    d) Compare $\bigcup_{i \in [1,1.5]} R_i$ with your answer in part b.

    e) Compare $\bigcap_{i \in [1,1.5]} R_i$ with your answer in part c.

11. We generalize the partition in Example 7 using (mod n).

    a) Let $n = 3$. Let $T_0 = \{z \in \mathbb{Z} : z \equiv 0 \,(\text{mod } 3)\}$, $T_1 = \{z \in \mathbb{Z} : z \equiv 1 \,(\text{mod } 3)\}$ and $T_2 = \{z \in \mathbb{Z} : z \equiv 2 \,(\text{mod } 3)\}$. Explain why the set $\{T_0, T_1, T_2\}$ forms a partition of $\mathbb{Z}$.

    b) Define the indexed family of sets for the corresponding partition when $n = 4$.

    c) Define the indexed family of sets for the corresponding partition for a general n.

12. Which of the given families of sets form a partition of $\mathbb{R}$? For those that fail, explain what part(s) of the definition fail(s) and adjust the family so it becomes a partition.

    a) $\{[i, i + 1) : i \in \mathbb{Z}\}$.

    *b) $\{B_j : j \in \mathbb{N}\}$, where $B_j = [-j, -j + 1) \cup (j - 1, j]$.

    c) $\{[2k - 1, 2k + 1] : k \in \mathbb{Z}\}$.

    d) $\{S_m : m \in [-1, 1]\}$, where $S_m = \{x : \sin(x) = m\}$.

    e) $\{P_r : r \in \mathbb{R}\}$, where $P_r = \{x \in \mathbb{R} : x^2 = r\}$.

13. For each family of sets given, draw a graph showing several of the sets. Then decide which of these families of sets form a partition of $\mathbb{R} \times \mathbb{R}$. For those that fail, explain what part(s) of the definition fail(s). The shape of a typical element in each family is listed, although some special sets in some families differ.

    *a) (Circles) $\{C_r : r \in [0, \infty)\}$, where $C_r = \{(x, y) : x^2 + y^2 = r^2\}$.

*b) (Lines) $\{L_m : m \in \mathbb{R}\}$, where $L_m = \{(x, y) : y = mx\}$.

c) (Parabolas) $\{P_i : i \in \mathbb{R}\}$, where $P_i = \{(x, y) : y = i(x^2 + 1)\}$.

d) (Hyperbolas) $\{H_j : j \in \mathbb{R}\}$, where $H_j = \{(x, y) : xy = j\}$.

e) (Hyperbolas) $\{V_k : k \in \mathbb{R}\}$, where $V_k = \{(x, y) : y = \frac{k}{x}\}$.

f) (Hyperbolas) $\{W_q : q \in \mathbb{R}\}$, where $W_q = \{(x, y) : y = \frac{1}{x+q}\}$.

14. This problem investigates formal definitions of ordered pairs and triples.

a) Explain why the definition of $(a, b) = (c, d)$ would fail if we defined $(a, b) = \{a, b\}$.

b) Repeat part a using $(a, b) = \{a, \{b\}\}$. Hint: Use both x and $\{x\}$ as elements.

c) Explain why the standard definition $(a, b) = \{\{a\}, \{a, b\}\}$ fulfills our definition of $(a, b) = (c, d)$.

d) Define what $(a, b, c) = (d, e, f)$ should mean. Use the definition in part c for (a, b) to write $(x, (y, z))$ as a set. If we defined $(x, y, z) = (x, (y, z))$ would your definition of $(a, b, c) = (d, e, f)$ be satisfied?

e) Find alternative definitions of the ordered triple (x, y, z).

15. Is $\bigcup_{n \in \mathbb{N}} \mathcal{P}(\{1, 2, 3, \ldots n\})$ equal to $\mathcal{P}(\mathbb{N})$? Explain.

DEFINITIONS FROM CHAPTER 1

Section 1.3

▪ An integer b is *even* iff there is an integer c such that $b = 2c$. An integer d is *odd* iff there is an integer e such that $d = 2e + 1$.

Section 1.4

▪ Given sets X and Y, we say X is a *subset* of Y, written $X \subseteq Y$, iff for all $x \in X$, we have $x \in Y$.

▪ For a, $b \in \mathbb{R}$, if $a < b$, then the *closed interval* $[a, b]$ is $\{x \in \mathbb{R} : a \leq x \leq b\}$ and the *open interval* (a, b) is $\{x \in \mathbb{R} : a < x < b\}$. For $a \in \mathbb{R}$, we denote $(-\infty, a) = \{x \in \mathbb{R} : x < a\}$ and $[a, \infty) = \{x : a \leq x\}$. Similarly, for $(-\infty, a]$ and (a, ∞). We can write $(-\infty, \infty)$ for $\mathbb{R}$.

▪ Given two sets A and B, define $A \cap B = \{x : x \in A \text{ and } x \in B\}$ and

$A \cup B = \{x : x \in A \text{ or } x \in B\}$ and

$A - B = \{x : x \in A \text{ and } x \notin B\}$.

▪ Given a (universal) set U and a subset A of U, the *complement* of A (relative to U) is $\overline{A} = \{x \in U : x \notin A\}$.

Section 1.5

▪ An integer k *divides* an integer n iff there is an integer d such that $d \cdot k = n$. We say k is a *divisor* of n or a *factor* of n and abbreviate "k divides n" by $k|n$.

▪ A natural number p is a *prime number* iff $p > 1$ and the only positive divisors of p are 1 and p.

- An integer d is a *common divisor* of the integers a and b iff d divides a and d divides b. An integer g is a (the) *greatest common divisor* of the nonzero integers a and b iff g is a common divisor of a and b and for every common divisor d of a and b, we have $d \le g$. In this case, we write $\gcd(a, b) = g$.

- Given the natural number n, two integers a and b are *congruent* (mod n) iff n divides $a - b$. In this case, we write $a \equiv b(\text{mod } n)$.

- For $x \in \mathbb{R}$, define $x^1 = x$, and for $n \in \mathbb{N}$, $x^{n+1} = x(x^n)$.

- The first two *Fibonacci numbers* are $f_1 = 1$ and $f_2 = 1$. The $(n + 2)^{\text{nd}}$ *Fibonacci number* is $f_{n+2} = f_n + f_{n+1}$.

Section 1.6

- Given elements a, b, c, and d, the ordered pairs (a, b) and (c, d) are *equal* iff $a = c$ and $b = d$.

- Given two sets A and B, their *Cartesian product* is $A \times B = \{(a, b) : a \in A \text{ and } b \in B\}$.

- Given a set X, its *power set* is $\mathcal{P}(X) = \{A : A \subseteq X\}$, the set of all subsets of X.

- An *indexed family* of sets $\{S_i : i \in I\}$ is a set of sets S_i, one for each *index* i in the *index set* I.

- Given a nonempty indexed family of sets $\{S_i : i \in I\}$, their *union* is $\bigcup_{i \in I} S_i = \{x : \exists i \in I \text{ s.t. } x \in S_i\}$ and their *intersection* is $\bigcap_{i \in I} S_i = \{x : \forall i \in I \ x \in S_i\}$.

- Two sets are *disjoint* iff their intersection is empty. A family of sets $\{S_i : i \in I\}$ is *pairwise disjoint* iff, for all $i, j \in I$ with $i \ne j$, we have $S_i \cap S_j = \emptyset$.

- A *partition* of a set B is a family of sets $\{S_i : i \in I\}$ such that
  i) each S_i is nonempty and
  ii) for each $b \in B$, there is a unique S_i such that $b \in S_i$.

Alternatively,

- A *partition* of a set B is a family of sets $\{S_i : i \in I\}$ such that
  i) each S_i is nonempty,
  ii) $B = \bigcup_{i \in I} S_i$, and
  iii) the family $\{S_i : i \in I\}$ is pairwise disjoint. ∎

ALGEBRAIC AND ORDER PROPERTIES OF NUMBER SYSTEMS

Closure For all $a, b \in \mathbb{N}$, their sum $a + b$ and product $a \cdot b$ are in $\mathbb{N}$.
For all $a, b \in \mathbb{Z}$, their sum $a + b$, product $a \cdot b$, and difference $a - b$ are in $\mathbb{Z}$.

For all $a, b \in \mathbb{Q}$, their sum $a + b$, product $a \cdot b$, and difference $a - b$ are in $\mathbb{Q}$ and, if $b \neq 0$, their quotient $a \div b$ is in $\mathbb{Q}$.

The closure properties for $\mathbb{Q}$ also hold in $\mathbb{R}$ and $\mathbb{C}$.

Common Algebraic Properties   For all $a, b, c \in \mathbb{C}$ (or any of its subsets $\mathbb{N}$, $\mathbb{Z}$, $\mathbb{Q}$, and $\mathbb{R}$):

$a + b = b + a$	$a \cdot b = b \cdot a$	*commutativity*
$a + (b + c) = (a + b) + c$	$a \cdot (b \cdot c) = (a \cdot b) \cdot c$	*associativity*
$a \cdot (b + c) = a \cdot b + a \cdot c$	$(b + c) \cdot a = b \cdot a + c \cdot a$	*distributivity*

Identity Properties   In $\mathbb{N}$, $\mathbb{Z}$, $\mathbb{Q}$, $\mathbb{R}$, and $\mathbb{C}$, 1 is the *multiplicative identity*: If x is any number of the set, then $x \cdot 1 = x = 1 \cdot x$. In $\mathbb{Z}$, $\mathbb{Q}$, $\mathbb{R}$, and $\mathbb{C}$, 0 is the *additive identity*: If x is any number of the set, then $x + 0 = x = 0 + x$.

Inverse Properties   In $\mathbb{Z}$, $\mathbb{Q}$, $\mathbb{R}$, and $\mathbb{C}$, every number x has an *additive inverse* $-x$ satisfying $x + (-x) = 0 = -x + x$. In $\mathbb{Q}$, $\mathbb{R}$, and $\mathbb{C}$, every nonzero number x has a *multiplicative inverse* x^{-1} satisfying $x \cdot (x^{-1}) = 1 = x^{-1} \cdot x$.

Order Properties   For all $a, b, c \in \mathbb{R}$ (or any of its subsets $\mathbb{N}$, $\mathbb{Z}$ and $\mathbb{Q}$)

$a \leq a$	*reflexive*
If $a \leq b$ and $b \leq c$, then $a \leq c$.	*transitive*
If $a \leq b$ and $b \leq a$, then $a = b$.	*antisymmetry*
$a < b$ or $b < a$ or $a = b$.	*trichotomy*
If $a \leq b$, then $a + c \leq b + c$.	*additive order*
If $a \leq b$ and $0 \leq c$, then $a \cdot c \leq b \cdot c$.	*multiplicative order*

CHAPTER 2

PROOFS

"Proofs really aren't there to convince you that something is true—
they're there to show you why it is true."
—Andrew Gleason (1921–   )

MATHEMATICS DEMANDS a higher level of certainty than any other
human endeavor, making proofs a hallmark of mathematics. In a civil court case, the
jury decides on "the preponderance of the evidence," while in a criminal case, the
standard of evidence is "beyond all reasonable doubt." Mathematical proofs aim for
the ultimate standard: "logically necessary." Of course, in actual proofs the audience
determines the level of detail needed to be convinced that the result is logically
necessary. Nevertheless, whether the reader is a student or a professional
mathematician, the author must make all assumptions and reasoning explicit. This
chapter examines proofs, explains and illustrates a variety of proof formats, and
discusses the general qualities of good proof writing.

Before we plunge into the details of proofs, let's consider the value of proofs.
Why would mathematicians demand the standard of logical necessity? One reason
is an internal one. Mathematical conjectures generally cover infinitely many cases,
so no finite number of examples could ever provide the "preponderance of the
evidence." Only a logical proof can eliminate the possibility of an unknown
counterexample beyond what anyone has considered. (See Guy, 697–712, for an
account of plausible numerical patterns that fail.)

The importance of applications of mathematics provides a second reason for
proofs. Nonmathematicians rely on the certainty of mathematics in applications. A
mathematical model of a real situation enables understanding and prediction.
Mathematics alone doesn't make the predictions certain in the real world: No model
can include every aspect of reality. But the certainty of the mathematics within the
model redirects our attention to how well the assumptions of the mathematics fit the
situation we are modeling. For instance, the revolutionary shift in physics from
Newtonian mechanics to relativity and quantum mechanics in the twentieth century
never called into question the mathematics underlying any of these theories. Rather,
the issue was how the mathematics fit the new physical evidence.

Consider a third factor on the role of proofs: Only mathematics is abstract enough to enable us to prove results. Even physics, the most mathematical of the natural sciences, depends on experiments and so can't achieve such certainty. In a sense comforting to mathematicians, we can be more certain of the infinitude of prime numbers (see Theorem 2.5.1) than of the exact time of the next solar eclipse. Indeed, the computations involved in determining a real eclipse rely directly on the certainty of abstract mathematical objects. We'll leave others to ponder how abstract numbers can predict the motions of the sun and moon while we immerse ourselves in the intricacies and wonders of proofs.

2.1 PROOF FORMAT I: DIRECT PROOFS

"... the two operations of our understanding, intuition
and deduction, on which alone ... we must rely in the acquisition
of knowledge."

—René Descartes (1596–1650)

Mathematical propositions frequently take the form "for all __ , if __ , then __." In this section, we consider direct proofs, the easiest way to show the truth of propositions of this straightforward form, along with some variations. Later sections explore other proof formats. Our first example, even though it shows something grade school children know, illustrates two vital principles of proofs: following logical formats and using definitions. We first remind you of the definitions of even and odd numbers here to make the role of these definitions easier to understand. Other definitions from Chapter 1, as well as fundamental algebraic and order properties of numbers we will assume in this chapter, can be found at the end of Chapter 1.

DEFINITION. An integer b is *even* iff there is an integer c such that $b = 2c$. An integer d is *odd* iff there is an integer e such that $d = 2e + 1$.

EXAMPLE 1

The sum of two odds is even. We turn this informal fact into an explicit proposition, give a proof, and discuss the proof.

CLAIM. For all $x, y \in \mathbb{Z}$, if x and y are odd, then $x + y$ is even.

Proof. Let $x, y \in \mathbb{Z}$. Then $x + y \in \mathbb{Z}$. Assume that x and y are odd. From the definition of odd integers for x and y, there exist $j, k \in \mathbb{Z}$ such that $x = 2j + 1$ and $y = 2k + 1$. These equalities force $x + y = (2j + 1) + (2k + 1) = 2(j + k + 1)$. (This last equality follows from the algebraic properties of $\mathbb{Z}$, listed at the end of Chapter 1.) Since $j + k + 1 \in \mathbb{Z}$, we see that $x + y$ fulfills the definition of an even number. ∎

DISCUSSION. A proof is a one-sided conversation between the author and an idealized, skeptical reader. First of all, note how the logical format of the proof follows the form of the proposition. In particular, both begin by introducing the variables. The proposition starts "for all $x, y \in \mathbb{Z}$." In response, the proof starts "Let $x, y \in \mathbb{Z}$." The verb "Let" indicates that the author's argument will hold regardless of any choice the reader might consider for the variables x and y, matching the required generality of "for all." The proof quickly assumes the hypothesis: x and y are odd. Then the reasoning leads to the conclusion: $x + y$ is even.

Why does this format show the "logical necessity" of the entire implication? Briefly, it is a proof because we showed that whenever the given conditions hold, the conclusion must hold. In more depth, the truth table of the implication $P \Rightarrow Q$ in Table 1 shows that only one situation can put the truth of the whole proposition at risk. The only row of the truth table of $P \Rightarrow Q$ ending with an F is where P is true and Q false. Thus, our job in the proof is to ensure that this case logically cannot occur. (If it could, we'd have a counterexample, discussed in Sections 1.2 and 1.3 and subsequently here.) The verb "assume" in the proof preceding the hypothesis restricts us to the situation when the hypothesis (P) is true, the case we need to consider. The reasoning then must force the conclusion (Q) also to be true, eliminating the only troublesome row of the truth table.

TABLE 1.
Implication

P	Q	P $\Rightarrow$ Q
T	T	T
T	F	F
F	T	T
F	F	T

The reasoning from hypothesis to conclusion can rely only on what comes before—here, the hypothesis and the definitions of terms. Indeed, the definitions of "odd" and "even" provide the keys. The definition of "odd" supplies us with the variables j and k in the equations $x = 2j + 1$ and $y = 2k + 1$. (It bears special mention that we need two different variables j and k since we must not assume that x and y are related.) We then manipulate these equations algebraically to fit the definition of an even number: $x + y = 2(j + k + 1)$. Also, we made sure to say $x + y \in \mathbb{Z}$ and $j + k + 1 \in \mathbb{Z}$ in the proof to fulfill the later use of the definition of "even." (We can justify these statements from the closure property of $\mathbb{Z}$ described at the end of Chapter 1.)

The same basic proof can appear in different wordings, such as this succinct version of our first proof.

Proof. Let x and y be odd integers, say, $x = 2j + 1$ and $y = 2k + 1$. Then $x + y = 2j + 1 + 2k + 1 = 2(j + k + 1)$. Since all the numbers are integers, $x + y$ is even. ■

DISCUSSION. While this version is correct, it would not be a good model for someone beginning proofs, since it assumes the reader understands the role of definitions at each step. The audience for a proof matters. After all, someone comfortable with using mathematical definitions would likely find both of these proofs "obvious" (an overused term).

Finally, the end of the proof is noted with the symbol ■. In earlier centuries, mathematicians would write "Q. E. D." after a proof, abbreviating "quod erat demonstrandum," Latin for "which was to be demonstrated." ◇

EXERCISE. Fill in the given proof of the following proposition from Section 1.2: For all integers n, if n is a multiple of 4, then n^2 is a multiple of 4.

Proof. Let $n \in$___. Suppose ___. Then there is some integer x such that $n = 4x$. Hence, ____. Since ___$\in \mathbb{Z}$, we have shown our conclusion, namely, ___. ∎

The next example gives two different proofs for a property you may have noted about the sizes of unions and intersections of finite sets. (If needed, you should remind yourself of the definitions of union, intersection, and difference of sets in Section 1.4 before reading these proofs. Problem 8 in that section explored this property.) Suppose, for instance, G has 12 elements and H has 8 elements. Then their union has at least the 12 elements of G and at most 20 elements if G and H don't overlap. Similarly, their intersection has between 0 and 8 elements. However, the size of their union is closely related to the size of their intersection. The intersection measures the overlap of the two sets, so the bigger the intersection, the smaller the union will be (see Figure 1).

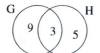

FIGURE 1

Suppose the intersection $G \cap H$ has 3 elements. Then G has those 3 elements plus 9 others to total 12. Similarly, H has those 3 common elements plus 5 others. So, altogether there must be $3 + 9 + 5 = 17$ elements. The idea of this example is the heart of the first proof. However, neither an example nor a Venn diagram qualify as a proof of a general proposition. In the discussion before the first proof we lay out how one might turn an idea into a proof. (In terms of the Descartes quote starting this section, we are linking our intuition with our deductions.) The second proof finds a different, but still general, way to count the elements of the union.

EXAMPLE 2

CLAIM. For all finite sets G and H, if G has k elements, H has n elements, and $G \cap H$ has j elements, then $G \cup H$ has $k + n - j$ elements.

DISCUSSION. We know the first and last sentences of the proof. We start with the hypothesis: "Let G and H be finite sets with k and n elements, respectively. Suppose $G \cap H$ has j elements." And we end with the conclusion: "Thus, the number of elements of $G \cup H$ must be $k + n - j$, as desired." Those sentences are often written first, with a big gap left between them. Our idea is to split G into two pieces, $G \cap H$ and what is left, which we defined to be $G - H$. We similarly split H into $G \cap H$ and $H - G$. Now we count and hope the mathematics all works.

First Proof. Let G and H be finite sets with k and n elements, respectively. Suppose $G \cap H$ has j elements. The set G splits into two separate parts, $G \cap H = \{x \in G : x \in H\}$

and $G - H = \{x \in G : x \notin H\}$. Since $G \cap H$ has j elements and G has a total of k elements, $G - H$ has $k - j$ elements. Similarly, we split H into $G \cap H$ with j elements and $H - G$ with $n - j$ elements. Now $G \cup H = \{x : x \in G \text{ or } x \in H\}$ splits into three separate parts: $G - H$, $G \cap H$, and $H - G$. Thus, the number of elements of $G \cup H$ must be $(k - j) + j + (n - j) = k + n - j$, as desired.   ∎

Second Proof.   Let G and H be finite sets with k and n elements, respectively. Suppose $G \cap H$ has j elements. For ease let $G = \{g_1, g_2, \ldots, g_j, g_{j+1}, \ldots, g_k\}$. Since there is no order to the elements of G or H, we can suppose that the elements these sets have in common are $G \cap H = \{g_1, g_2, \ldots, g_j\}$ and that $H = \{g_1, g_2, \ldots, g_j, h_{j+1}, \ldots, h_n\}$. Note that H must have $n - j$ elements other than the j elements in common with G in order to have n elements. Then $G \cup H = \{g_1, g_2, \ldots, g_j, \ldots, g_k, h_{j+1}, \ldots, h_n\}$, which has $k + (n - j)$ elements, as claimed.   ∎

DISCUSSION.   Both proofs start with the givens from the universal quantifier and the hypotheses. Then both reason to arrive at the conclusion. Although their approaches look different, they are equivalent. The first proof simply generalizes the example. When you are looking for a proof, a good example can often suggest a fruitful approach.

The use of subscripts, as in the second proof, is often unnatural for students. Why would one think of this approach? The proposition counts something, so the proof must reason about numbers. However, it would be inappropriate to assume, for example, that the elements of G are the numbers from 1 to k. The subscripts allow us to use properties of numbers to count the elements without restricting what the elements of G and H are. In effect, we get all of the benefits of numbers while still talking about general elements. There is a risk of confusion in using subscripts. For example, if $G \cap H$ were empty, then $j = 0$. Then the part $g_1, g_2, \ldots, g_j$ of the set $G = \{g_1, g_2, \ldots, g_j, g_{j+1}, \ldots, g_k\}$ isn't there. Similarly, if $G = G \cap H$, then $j = k$ and $g_{j+1}, \ldots, g_k$ wouldn't be there. Note that the reasoning of the proof works in these cases, just as it does when j is strictly between 0 and k, even if the notation looks strange.

I personally like the first proof better because I think visually, and I see the union in Figure 1 as composed of the three pieces the proof uses. Others might find the early shift to numbers easier to follow.   ◊

Variations

Not every mathematical proposition and proof fits the form just discussed. The remainder of this section considers closely related variations, starting with disproving propositions.

Counterexamples   In Chapter 1, we saw how universal statements and implications were false if even one exception (counter-example) occurred. To disprove a proposition of the form $\forall x \in X \ P(x) \Rightarrow Q(x)$, recall its working negation—the existential proposition $\exists x \in X \ s.t. \ P(x) \wedge \neg Q(x)$.

EXAMPLE 3

Disprove the following:

CLAIM.   For all $x \in \mathbb{R}$, if $x < x^2$, then $x^2 < x^3$.

Proof. Using the working negation, we show there is $x \in \mathbb{R}$ such that $x < x^2$ and $x^2 \geq x^3$. Consider the real number $x = -2$. Then $-2 < 4 = (-2)^2$, but $(-2)^2 = 4$ is larger than $(-2)^3 = -8$. Any negative real number similarly provides a counterexample. ∎

DISCUSSION. There is no logical need to explain how you found the counterexample or how broadly the proposition fails. However, such information can give a deeper understanding, as the last sentence of the preceding proof does. ◇

Derivations Derivations of formulas in algebra and trigonometry, although fairly straightforward, often tempt students to manipulate equations instead of proving them in general.

EXAMPLE 4

DERIVE: $\sec^2(\theta) = \tan^2(\theta) + 1$.

INCORRECT DERIVATION. The definitions of secant and tangent convert $\sec^2(\theta) = \tan^2(\theta) + 1$ to $\frac{1}{\cos^2(\theta)} = \frac{\sin^2(\theta)}{\cos^2(\theta)} + 1$. Multiply both sides by $\cos^2(\theta)$ to get $1 = \sin^2(\theta) + \cos^2(\theta)$. Finally, we know $\sin^2(\theta) + \cos^2(\theta) = 1$, so $1 = 1$, proving the formula. ▲

DISCUSSION. What went wrong? The previous argument went backwards, assuming the conclusion and "proving" the rather silly fact that $1 = 1$. (We will denote the end of an incorrect or questionable argument with the symbol ▲.) One reasonable approach starts with one side of the equality and converts it to the other.

DERIVATION. Suppose θ is an angle for which $\tan(\theta)$ exists. By the definition of tangent, $\tan^2(\theta) + 1 = \frac{\sin^2(\theta)}{\cos^2(\theta)} + 1$. For the tangent to exist, $\cos^2(\theta) \neq 0$, so we can rewrite $\frac{\sin^2(\theta)}{\cos^2(\theta)} + 1$ as $\frac{\sin^2(\theta)}{\cos^2(\theta)} + \frac{\cos^2(\theta)}{\cos^2(\theta)} = \frac{\sin^2(\theta) + \cos^2(\theta)}{\cos^2(\theta)}$. The identity $\sin^2(\theta) + \cos^2(\theta) = 1$ converts $\frac{\sin^2(\theta) + \cos^2(\theta)}{\cos^2(\theta)}$ to $\frac{1}{\cos^2(\theta)}$, which is the definition of $\sec^2(\theta)$. ∎

DISCUSSION. Theoretically, we could salvage the incorrect derivation by reversing the order of its steps, although several steps would appear quite mysterious to the first-time reader. Such a proof would defeat the point Gleason makes in the opening quote of this chapter. ◇

Proofs of "if and only if" Propositions The logical connective "iff" combines an implication with its converse. That is, $P \Leftrightarrow Q$ is logically equivalent to both $P \Rightarrow Q$ and $Q \Rightarrow P$. Thus, the default proof of an "iff" proposition shows each direction separately, as in Example 5, which follows. Sometimes, one direction has the same reasoning as the other, but with the order switched. Example 6 on set equality illustrates a quicker proof in such a situation.

EXAMPLE 5

Let $f(x) = x^3 + 3$.

CLAIM. For all $v, w \in \mathbb{R}$, $v = w$ iff $f(v) = f(w)$.

Proof. Let $f(x) = x^3 + 3$ and let v and w be real numbers. ($\Rightarrow$) Assume $v = w$. By substitution, $f(v) = v^3 + 3 = w^3 + 3 = f(w)$. ($\Leftarrow$) For the other direction, suppose $f(v) = f(w)$. That is, $v^3 + 3 = w^3 + 3$, or more simply, $v^3 = w^3$. Each real number has only one real cube root, so $v = \sqrt[3]{v^3} = \sqrt[3]{w^3} = w$. ∎

DISCUSSION. For the first direction, substitution works for any function. The other direction works only for certain functions, called one-to-one and studied in Chapter 3. Consider instead the function $g(x) = x^2 + 3$. Then the second direction would be false: $g(-2) = 7 = g(2)$, but $-2 \neq 2$. In effect, g is two-to-one, whereas f is one-to-one. Separating the two directions of this proof helps to emphasize the difference between what holds generally and what depends on a particular property of this function.   ◇

Proofs of Set Equality   By the axiom of set equality in Section 1.4, $A = B$ iff $A \subseteq B$ and $B \subseteq A$. In turn, the definition of $A \subseteq B$ is the implication "If $x \in A$, then $x \in B$." Thus, $A = B$ is really an "iff" proposition in disguise.

EXAMPLE 6

CLAIM. For all sets $R, S,$ and T, we have $R \cup (S \cap T) = (R \cup S) \cap (R \cup T)$.

REMARK. This is the distributive property of $\cup$ over $\cap$. Problem 4f in Section 1.1 gives the analogous property for $\vee$ (or) and $\wedge$ (and), which we use in the proof.

Proof. Let $R, S,$ and T be any sets. The definitions of union and intersection allow us to convert these set operations into the logical connectives "or" and "and." Thus, $x \in R \cup (S \cap T)$ iff $x \in R$ or $x \in S \cap T$, which is equivalent to $x \in R$ or ($x \in S$ and $x \in T$). The distributive property of "or" ($\vee$) over "and" ($\wedge$) implies that this last proposition is equivalent to ($x \in R$ or $x \in S$) and ($x \in R$ or $x \in T$). By the definitions of union and intersection, this last proposition is equivalent to $x \in R \cup S$ and $x \in R \cup T$, which is the same as $x \in (R \cup S) \cap (R \cup T)$. Hence, $R \cup (S \cap T) = (R \cup S) \cap (R \cup T)$. ∎

DISCUSSION. To avoid proving two directions for this proposition, we carefully used the words "iff" and "equivalent." We could have doubled the length by using implications throughout to show $R \cup (S \cap T) \subseteq (R \cup S) \cap (R \cup T)$ and then reversing the steps to show $(R \cup S) \cap (R \cup T) \subseteq R \cup (S \cap T)$. Readers might well find such repetition a burden rather than a help. At the other extreme, we could also have reduced this proof to only symbols, as follows: $x \in R \cup (S \cap T) \Leftrightarrow (x \in R \vee x \in S \cap T) \Leftrightarrow (x \in R \vee (x \in S \wedge x \in T)) \Leftrightarrow ((x \in R \vee x \in S) \wedge (x \in R \vee x \in T)) \Leftrightarrow x \in (R \cup S) \cap (R \cup T)$. This approach emphasizes the formal, symbolic relationship of $\cup$ with $\vee$ and $\cap$ with $\wedge$. However, it risks being unintelligible.   ◇

Suggestions

Learning to prove propositions, just like learning to solve problems, requires lots of practice. It also requires you to reflect on the process. So talk about proofs with other students as well as your instructor. Here are some general suggestions for getting started:

1. When you read a proposition you are to prove, first make sure you know all of the relevant definitions. They are essential in proofs. Mathematical definitions are written so that they can be used in proofs.

2. Identify the logical format of the proposition so that you know what you are given and what you need to prove. If the proposition is worded informally, consider writing it formally to reveal the format. If it is written formally, consider putting it into more conversational wording so that you understand its meaning.

3. Try an example. Of course, examples are not proofs. But you get a feel for the meaning of the proposition from an example and how the parts fit together. In addition, you sometimes see why it is true, the key to formulating a proof. Although it is sensible to pick an easy example at first, you may need to consider less familiar examples to discover important aspects of the proposition. For example, negative numbers sometimes work differently than positive numbers, but we more naturally plug in positive numbers.

4. If you know part of the proof, write it down. For example, I strongly encourage my students to start by writing down the givens (Let ___. Suppose ___.) Then they skip down a bit and write the conclusion. Often, the hypothesis and conclusion lead them to cite definitions, which might well be the second and next-to-last step.

Reading and Critiquing Proofs   The better you are at reading and critiquing proofs, the better you will be at writing good proofs. The problem sets of this chapter include examples of proofs to critique. It is usually easier to find fault with an anonymous attempt in a book than a fellow student's latest effort or your own work. Even so, finding errors is difficult, since jumps and errors in reasoning don't announce themselves.

EXAMPLE 7

Critique the following questionable "proofs" of the given claim, pointing out reasoning errors and unclear presentation.

CLAIM.   The square of an odd number is an odd number.

"Proof 1": $1^2 = 1, 3^2 = 9, 5^2 = 25, 7^2 = 49$, and $9^2 = 81$. No matter what odd number you take, its square ends in a 1, a 9, or a 5, which are all odd.   ▲

DISCUSSION.   This argument is flawed. Examples are not enough to prove a general proposition. The second sentence explains the general idea behind these examples. If one showed that every odd number was of the form $10k + 1$ or $10k + 3$ or $10k + 5$ or $10k + 7$ or $10k + 9$, one could use these examples to generalize to the squares of all odd numbers. However, such a proof involves significant work.

"Proof 2": Suppose the square of an integer is odd. Since an even integer squared is even, the original number must be odd.   ▲

DISCUSSION.   This argument is flawed. Most important, it tries to prove the converse of the claim instead of the claim. The format of the argument may be hard for a beginning student to recognize. (It is a proof by contradiction, discussed in the next section, but the writer doesn't make this explicit.) The lack of variables makes the proof more conversational and intuitive. However, avoiding the formal use of variables risks overlooking a reasoning error.

The following proof avoids the errors of the preceding arguments.

Proof. Let x be an odd integer. Then we can write x as $2k + 1$, where k is an integer. So $x^2 = (2k + 1)^2 = 4k^2 + 4k + 1 = 2(2k^2 + 2k) + 1$. Since $2k^2 + 2k$ is an integer, it follows that x^2 is odd. ■ ◇

Proof Formats

To show	Format
For all $x \in S$, ___.	Let $x \in S$ and show ___.
If P, then Q (direct).	Suppose P, show Q.
P iff Q	Prove "if P, then Q" and "if Q, then P."
$A \subseteq B$	Let $x \in A$, prove $x \in B$.
$A = B$	Prove $A \subseteq B$ and $B \subseteq A$.

PROBLEMS

*1. For each statement that follows, decide whether it is true or (at least sometimes) false. If false, explain why. Cite the part of the text supporting your conclusion.

   a) There is only one correct proof of a particular proposition.

   b) You can use an example to prove a general proposition.

   c) You can use an example to disprove a general proposition.

   d) You can use an example as a model for a general proof.

   e) All mathematical propositions are of the form "For all ___, if ___ , then ___ ."

   f) Set equality must be proven by a series of "iff" propositions.

2. Fill in the outlines to make proofs.

   *a) Claim: For all integers x and y, if x and y are even, then $x \cdot y$ is even.
     Proof. Let x and y be integers. Then $x \cdot y$ ___ . Suppose that ___ . From the definition of even, we can find ___ so that $x =$ ___ and $y =$ ___ . Then $x \cdot y =$ ___ $=$ ___ . Since ___ , ___ is even. ■

   b) Claim: For all sets A and B, $A \subseteq A \cup B$.
     Proof. Let ___ . From the definition of subset, $A \subseteq A \cup B$ means if $x \in A$, then ___ . So, suppose $x \in A$. Then $x \in A$ or ___ . From the definition of ___ we see $x \in$ ___ . Thus, $A \subseteq A \cup B$. ■

   c) Claim: For all sets C and D, we have $C = (C \cup D) \cap C$.
     Proof. Let ___ . To show set equality, we show that $C \subseteq (C \cup D) \cap C$ and, separately, that ___ . Now, $C \subseteq (C \cup D) \cap C$ means if $x \in C$, then ___ . So, suppose $x \in C$. Then we certainly have "$x \in C$ or $x \in D$," so $x \in C \cup D$. From the definition of ___ , we have $x \in$ ___ . That is, $C \subseteq (C \cup D) \cap C$. For the other direction, ___ means ___ . So, suppose $x \in$ ___ . The definition of ___ gives us $x \in C$. Then ___ . Since we have shown both directions, we can conclude $C = (C \cup D) \cap C$. ■

3. Prove these propositions:

   *a) For all integers x and y, if x and y are even, then $x - y$ is even.

   b) The product of two odd integers is odd.

c) The sum of an even and an odd integer is odd.

d) The product of an even and an odd integer is even.

4. Prove these propositions. Recall the set theory definitions in Section 1.4.

   *a) For all sets S and T, $S \cap T \subseteq S$.

   b) For all sets S and T, $S - T \subseteq S$.

   c) For all sets S, T and W, $(S - T) - W \subseteq S - (T - W)$.

   d) For all sets S, T and W, $(T - W) \cap S = (T \cap S) - (W \cap S)$.

5. Prove these propositions. Recall the definition of "divides" $(p|q)$ in Section 1.5.

   *a) For all integers p, q, r, if p divides q and q divides r, then p divides r (referred to in Section 4.1).

   b) For all $p, q, r \in \mathbb{Z}$, if $p|q$ and $p|r$, then p divides $q + r$.

   c) For all $p, q, r, a, b \in \mathbb{Z}$, if $p|q$ and $p|r$, then p divides $aq + br$ (referred to in the Supplement to Section 2.4).

   d) For all $p, q, r \in \mathbb{Z}$, if $p|q$, then $pr|qr$.

   e) Is the converse of part d true? If so, prove it; if not, give a counterexample.

6. Recall the definition of $(\bmod \, n)$ in Section 1.5.

   *a) Prove for all $a, b \in \mathbb{Z}$, if $a \equiv b(\bmod \, 12)$, then $a \equiv b(\bmod \, 6)$.

   b) Is the converse of part a true? If so, prove it; if not, give a counterexample.

   c) Generalize the statement in part a and prove it.

   d) Prove for all $a, b \in \mathbb{Z}$ and all $k, n \in \mathbb{N}$, if $a \equiv b(\bmod \, \mathrm{lcm}(n, k))$, then $a \equiv b(\bmod \, n)$ and $a \equiv b(\bmod \, k)$.

7. Prove the given propositions. Recall the definition of $A \times B$ in Section 1.6.

   a) For all sets D, E, F, and G, $(D \cap E) \times (F \cap G) = (D \times F) \cap (E \times G)$.

   b) For all sets D, E, F, and G, $(D - E) \times (F - G) \subseteq (D \times F) - (E \times G)$.

   c) In part b is the reverse inclusion $(D \times F) - (E \times G) \subseteq (D - E) \times (F - G)$ true? If so, prove it; if not, give a counterexample.

8. Derive these equalities and inequalities. For parts a, b, and c, use the angle addition formulas $\cos(\alpha + \beta) = \cos(\alpha)\cos(\beta) - \sin(\alpha)\sin(\beta)$ and $\sin(\alpha + \beta) = \sin(\alpha)\cos(\beta) + \cos(\alpha)\sin(\beta)$.

   *a) $\cos(3\theta) = \cos^3(\theta) - 3\cos(\theta)\sin^2(\theta)$.

   b) $\sin(3\theta) = 3\sin(\theta)\cos^2(\theta) - \sin^3(\theta)$.

   c) $\tan(2\theta) = \frac{2\tan(\theta)}{1 - \tan^2(\theta)}$, where these are defined.

   d) For all $t, u \in \mathbb{R}$, $t^2 + u^2 \geq 2tu$.

   e) (The Geometric–Arithmetic Mean Inequality) For all $x, y \in \mathbb{R}$, if $x \geq 0$ and $y \geq 0$, then $\sqrt{xy} \leq \frac{x+y}{2}$.

9. **a)** For all positive real numbers a, b, x, y, prove that if $a < x$ and $b < y$, then the rectangle with corners at $(x, y), (x, 0), (0, y)$, and $(0, 0)$ has area greater than ab.

   b) For all real numbers x and y, prove that if $x < -5$ and $y > 3$, then the Euclidean distance between the points (x, y) and $(3, -4)$ is greater than 10.

10. **a)** Disprove the proposition "For all $x \in \mathbb{N}$, $x^2 + x + 41$ is prime."

    b) For each $n \in \mathbb{N}$ with $n > 1$, show that the proposition "For all $x \in \mathbb{N}$ $x^2 + x + n$ is prime" is false.

c) Show the case $n = 1$ for part b is also false. (You will likely need a different argument from part b.)

11. For each "iff" proposition, give a proof for any direction that is true and a counterexample for any direction that fails.

a) For all $a, b, c, d \in \mathbb{N}$, $\frac{a}{b} < \frac{c}{d}$ iff $\frac{d}{c} < \frac{b}{a}$.

b) For all $a, b, c, d \in \mathbb{Z}$, if all are nonzero, then $\frac{a}{b} < \frac{c}{d}$ iff $\frac{d}{c} < \frac{b}{a}$.

c) For all $p, q \in \mathbb{Z}$, $\gcd(p, q) = p$ iff p divides q.

d) For all integers $x, y,$ and z, (x divides yz iff both $x|y$ and $x|z$).

12. For all $s, t \in \mathbb{R}$, assume that $st = 0$ iff ($s = 0$ or $t = 0$). Prove the Quadratic Formula: for all real numbers $a, b, c, x \in \mathbb{R}$, if $a \neq 0$, then $ax^2 + bx + c = 0$ iff $x = \frac{-b \pm \sqrt{b^2 - 4ac}}{2a}$. (Hint: First, divide $ax^2 + bx + c$ by a. Then "complete the square" of $x^2 + \frac{b}{a}x$. Next, recall factoring: $A^2 - B^2 = (A - B)(A + B)$).

13. Some theorems equate a list of propositions, such as "A, B, C, D, and E are equivalent." (For instance, linear algebra texts typically have a key theorem listing many properties equivalent to a square matrix being invertible.)

a) Write out all of the equivalences, such as $A \Leftrightarrow B$, that are part of the statement "A, B, C, D, and E are equivalent." Note that each such equivalence needs two proofs, one for each direction. How many implications are required with this approach?

b) Instead of proving all the implications from part a, mathematicians simply prove the cycle of implications $A \Rightarrow B$, $B \Rightarrow C$, $C \Rightarrow D$, $D \Rightarrow E$, and $E \Rightarrow A$. Explain why these five implications suffice to show the many more implications of part a.

c) Prove that the following are equivalent for any sets F and G (referred to in Section 7.4):

   i) $F \subseteq G$ ii) $F = F \cap G$ iii) $G = F \cup G$

14. Critique the following incorrect "proofs." Point out reasoning errors and unclear presentation. (If the claim is false, there must be an error in the argument. However, it is not enough to say that the claim is false. You need to find an error in the reasoning.)

*a) Claim: An even number divided by 2 is an odd number. "Proof": An even number is of the form $2n$. When you divide $2n$ by 2, you get just n, which is odd. ▲

b) Claim: Every integer is odd or even. "Proof": Adding 1 to an even number $(2j)$ gives us an odd number $(2j + 1)$, and adding 1 to an odd number $(2j + 1)$ gives us an even number $(2j + 1 + 1 = 2(j + 1))$. So the even and odd numbers switch back and forth, accounting for all of the numbers. ▲

c) Claim: For any sets A, B, C, and D, if $A \subseteq B$ and $C \subseteq D$, then $A \cap C \subseteq B \cap D$. "Proof": Suppose $A \subseteq B$ and $C \subseteq D$. Suppose that $x \in A$. Then $x \in B$. Similarly, suppose $y \in C$ so that $y \in D$. For both of these to happen, we have to have $A \cap C \subseteq B \cap D$. ▲

d) Claim: For any sets P and Q, if $P = Q$, then $P - Q = Q - P$. "Proof": From $P - Q = Q - P$, we can add Q and P to each side to get $2P = 2Q$. Then $P = Q$. ▲

e) Claim: If a set has n elements, then it has n^2 subsets of size two. "Proof": Suppose a set A has n elements. Then each subset of size two has n choices for its first element and n choices for its second element, giving n^2 subsets of size two. ▲

f) Claim: If a set has n elements, then it has $n(n - 1)(n - 2)$ subsets of size three. "Proof": Suppose a set A has n elements. Then each subset has n choices for its first element. The next element needs to differ from the first, so there are $n - 1$ choices for it. Similarly, the third differs from the first two, so there are only $n - 2$ choices for it. Thus, there are $n(n - 1)(n - 2)$ subsets of size three. ▲

g) Claim: For all integers i, j, and k, if i divides j and i divides k, then i divides $j - k$. "Proof": Suppose i divides both j and k. So we can say $j = ai$ and $k = bi$ for some a and b. In $j - k = ci$, substitute these values for j and k to get $ai - bi = ci$. Thus, $c = a - b$, showing i divides $j - k$. ▲

h) Claim: For all integers n, if n^2 is odd, then n is odd. "Proof": Let $n = 2k + 1$. Then $n^2 = 4k^2 + 4k + 1 = 2(2k^2 + 2k) + 1$, which is odd as well. So, if n^2 is odd, then n is odd. ▲

15. Critique the given questionable "proofs." Point out reasoning errors and unclear presentation. If the argument is a proof, say so. (If the claim is false, there must be an error in the argument. However, it is not enough to say that the claim is false. You need to find an error in the reasoning. Even if the claim is true, the argument need not be correct.)

*a) Claim: For any sets A and B, if $A \cup B = B$, then $A \cap B = A$. "Proof": For $A = \{1, 3, 5\}$ and $B = \{1, 2, 3, 4, 5\}$, $A \cup B = B$ because $A \subseteq B$. But then, $A \cap B = B$. ▲

b) Claim: No integer is both even and odd. "Proof": By definition, odd numbers have a $+1$ in them and even numbers don't. Therefore, no odd number is even. ▲

c) Claim: For all integers z, if z is even, then z^2 is even. "Proof": Suppose z is even. That is, $z = 2k$, for some integer k. Then $z^2 = (2k)^2 = 2(2k^2)$, which is even. ▲

d) Claim: For all integers s, t, and u, if st divides su, then t divides u. "Proof": Suppose st divides su. Now $\frac{su}{st}$ equals $\frac{u}{t}$ by canceling the s in the numerator and denominator. Hence, t divides u. ▲

e) Claim: For any sets P and Q, $(P - Q) \cap (Q - P) = \emptyset$. "Proof": Let P and Q be any sets and assume that $x \in (P - Q) \cap (Q - P)$. Then $x \in (P - Q)$ and $x \in (Q - P)$. Then $x \in P$ and $x \notin Q$ and $x \in Q$ and $x \notin P$, which is a contradiction. So, $x \in \emptyset$. ▲

f) Claim: For any sets J, K, and L, $(J \cap K) - L \subseteq J - (K \cap L)$. "Proof": Suppose $x \in (J \cap K) - L$. From the definition of set difference, $x \in J \cap K$ and $x \notin L$. In turn, that means $x \in J$ and $x \in K$ and $x \notin L$. If $x \notin L$, then $x \notin K \cap L$, which is a subset of L. Then we have $x \in J$ and $x \notin (K \cap L)$. Those are just what we need to conclude $x \in J - (K \cap L)$, showing that $(J \cap K) - L \subseteq J - (K \cap L)$. ▲

REFERENCE

Guy, R. 1988. The strong law of small numbers. *American Mathematical Monthly*, 95:697–712.

2.2 PROOF FORMAT II: CONTRAPOSITIVE AND CONTRADICTION

"[A proof by contradiction] is one of a mathematician's finest weapons. It is a far finer gambit than any chess play: a chess player may offer the sacrifice of a pawn or even a piece, but a mathematician offers the game."

—G. H. Hardy (1877–1947)

Some properties seem to defy a direct proof. Our study of logic in Chapter 1 suggests two alternative approaches, which we investigate in this section. These approaches are often helpful when we are trying to prove a conclusion with a negation in it.

Contrapositive

The contrapositive of the propositional form $P \Rightarrow Q$ is $\neg Q \Rightarrow \neg P$, and, as we saw in Section 1.2, these forms are logically equivalent. Thus, instead of assuming P and trying to prove Q, we could assume $\neg Q$ and try to prove $\neg P$.

EXAMPLE 1

CLAIM.   If a natural number written in base ten ends in a 7, it is not a perfect square.

DISCUSSION.   Your familiarity with numbers and a quick check will probably convince you that all squares end in 0, 1, 4, 5, 6, or 9. But showing that a number is not a square seems awkward. The contrapositive states, "If a natural number written in base ten is a perfect square, then it doesn't end with a 7." While "not ending in 7" also has a negation, it is easy to determine the possible last digits of squares in base ten. The following proof accomplishes this using the Division Algorithm, discussed in Section 1.5.

Proof by Contrapositive.   Let n be a natural number written in base ten. Thus, the last digit of n is one of the digits from 0 to 9. More formally, by the Division Algorithm, we can write $n = 10q + r$ for some integers q and r with r between 0 and 9. The remainder r is the last digit of n. Now $n^2 = (10q + r)^2 = 100q^2 + 20qr + r^2$. Since the terms $100q^2$ and $20qr$ affect just the ten's place and higher, only the value of r^2 can affect the last digit, the one's column. By inspection, the squares $0^2, 1^2 \ldots, 9^2$ end in 0, 1, 4, 9, 6, 5, 6, 9, 4, and 1, respectively. Thus, n^2 can't end with 7 (or 2, 3, or 8), showing the contrapositive of the original claim.   ∎

DISCUSSION.   Unless told otherwise, a reader will expect a direct proof; so alert your reader to the approach you will take. We will prove the Division Algorithm in Section 2.4.   ◇

Sometimes, one direction of an "iff" proposition can be done directly, while the other direction is more readily done by the contrapositive. Example 2 illustrates this situation.

EXAMPLE 2

CLAIM.   For x, any integer, x is odd iff x^2 is odd.

DISCUSSION.   The first direction ($\Rightarrow$) gives us as an initial assumption that a number x is odd, which we can fairly directly manipulate to tell us about its square. The other direction ($\Leftarrow$) gives us as a starting point that the square is odd. It is harder to see what to do with that fact to get back directly to the original. We'll try the contrapositive in the proof of this direction.

Proof.   Let $x \in \mathbb{Z}$. For the first direction, suppose that x is odd. So we can find an integer j such that $x = 2j + 1$. Then $x^2 = (2j + 1)^2 = 4j^2 + 4j + 1 = 2(2j^2 + 2j) + 1$. Since $2j^2 + 2j \in \mathbb{Z}$, we see x^2 is an odd integer.

For the other direction, we show the contrapositive. That is, we replace "If x^2 is odd, then x is odd" by "If x is not odd, then x^2 is not odd." So suppose that x is not odd. Example 6 of Section 1.5, a consequence of the Division Algorithm, simplifies this condition to x is even—say, $x = 2k$, where $k \in \mathbb{Z}$. Then $x^2 = (2k)^2 = 2(2k^2)$, which is an even integer because

$2k^2 \in \mathbb{Z}$. Again, from Example 6 of Section 1.5, x^2 is not odd. This shows the contrapositive and finishes the second direction of the proof.   ■

DISCUSSION. We are so familiar with the ideas of even and odd that it is tempting to think that the contrapositive of "If x^2 is odd, then x is odd" is "If x is even, then x^2 is even." While that isn't strictly correct, Example 6 of Section 1.5 assures us it is equivalent. Henceforth, for integers we will freely interchange "even" and "not odd" and, similarly, "odd" and "not even".   ◇

If the conclusion contains a negation, you should at least consider proving the contrapositive, since negations are often hard to prove. But even this technique may not seem powerful enough to link the hypothesis and conclusion. The next subsection discusses a closely related alternative, proof by contradiction.

Proof by Contradiction

In a proof of Q by contradiction, we assume the negation $\neg Q$ and look for any resulting contradiction (R and not R). Since contradictions can't happen in logic, there must be some error. If our reasoning after assuming $\neg Q$ is correct, the fault must be with the assumption of $\neg Q$. The only alternative is thus Q, our desired conclusion. This reasoning, in effect, depends on the fact that the proposition $(\neg Q \Rightarrow (R \wedge \neg R)) \Rightarrow Q$ is a tautology. We can use a proof by contradiction for a proposition of the form $P \Rightarrow Q$ as well. Recall the working negation of $P \Rightarrow Q$ is $P \wedge \neg Q$. Thus, we assume P and not Q and look for a contradiction. Again, the contradiction tells us our original assumption is wrong, forcing $P \Rightarrow Q$.

EXERCISE. Use a truth table to verify that $(\neg Q \Rightarrow (R \wedge \neg R)) \Rightarrow Q$ is a tautology.

EXAMPLE 3

CLAIM. The sum of a rational and an irrational is irrational.

DISCUSSION. It is hard to prove a number is irrational, which suggests it would be easier to assume the sum is rational. Of course, to talk about the sum being rational, we need first to introduce the two numbers.

Proof. Let q be rational and r irrational and suppose, for a contradiction, that $q + r$ were rational. Then $(q + r) - q = r$ is the difference of two rationals and so is a rational. But r is irrational, a contradiction. Thus, $q + r$ must be irrational.   ■

DISCUSSION. Warn your reader early in a proof when you use a proof by contradiction, since they are expecting you eventually to affirm the conclusion. We can readily convert the preceding argument into a proof by the contrapositive, given next. A proof by contrapositive requires an "if, then" format, so we need to restate the claim more explicitly: "For all $q \in \mathbb{Q}$ and all $r \in \mathbb{R}$, if $r \notin \mathbb{Q}$, then $q + r \notin \mathbb{Q}$." The contrapositive is "For all $q \in \mathbb{Q}$ and all $r \in \mathbb{R}$, if $q + r \in \mathbb{Q}$, then $r \in \mathbb{Q}$." Notice that the contrapositive has no negations in it, making it likely to be easier to prove than the original.

Proof.   Let $q \in \mathbb{Q}$ and $r \in \mathbb{R}$. To show the contrapositive, assume that $q + r$ is rational. Then $(q + r) - q = r$ is the difference of two rationals and so is a rational. Since the contrapositive is true, the original proposition is true. ∎ ◊

EXAMPLE 4

Given a checkerboard without its opposite corner squares, can we cover the remaining squares with thirty-one 2×1 rectangles?

    A checkerboard has eight squares on a side for a total of sixty-four squares (see Figure 1). Thus thirty-one 2×1 rectangles could, in principle, cover the $62 = 31 \times 2$ squares remaining after the opposite corners are removed. However, no matter how you attempt to arrange the rectangles, you will fail. But repeated failures hardly prove the impossibility. A proof by contradiction supposes a solution and looks for a contradiction. The following well-known proof cleverly uses the coloring on a checkerboard.

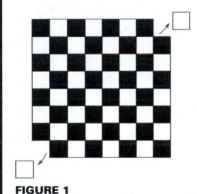

FIGURE 1

Proof.   For a contradiction, suppose that there were a covering of the trimmed checkerboard. Each rectangle would cover two adjacent squares, one white square and one black square. However, opposite corners of a checkerboard are the same color. So, the trimmed board has thirty-two squares of one color and only thirty of the other. We have a contradiction, showing the impossibility of such a covering. ∎ ◊

    It is often difficult to realize that a proof by contradiction is a good way to proceed or to determine where to look for the contradiction—after all, any contradiction will do.

EXAMPLE 5

CLAIM.   If r is a positive real number, then $(r + 1)/(r + 2) < (r + 3)/(r + 4)$.

DISCUSSION.   If we pick some positive values for r, we easily confirm that the inequality holds, but these examples don't seem to suggest any way to prove the proposition in general. Indeed it is not clear how to deduce an inequality. One advantage of a proof by contradiction (or by contrapositive) in this case is that we will get to start with an algebraic relation, something to work with.

Proof. Suppose that $r > 0$ and, for a contradiction, that $(r + 1)/(r + 2) \geq (r + 3)/(r + 4)$. Let's eliminate the denominators. Since $r > 0$, both denominators are also positive, so the inequality won't change. Thus, $(r + 1)(r + 4) \geq (r + 3)(r + 2)$. Multiplying out these terms, we get $r^2 + 5r + 4 \geq r^2 + 5r + 6$, which simplifies to $4 \geq 6$, which clearly contradicts $4 < 6$. Hence, we can conclude $(r + 1)/(r + 2) < (r + 3)/(r + 4)$. ■ $\Diamond$

Proofs are more than a string of logical deductions. They need to have an overall cohesiveness, almost a story line. Direct proofs have the simplest "plots"—they start at the beginning and go to the end. Proofs using contrapositive or contradiction tell a more complicated "tale," since the negation of the ending plays a major role. A good proof, like a good story, gives the reader clues about how the proof fits together and what aspects are important. We will encounter other proof formats in the later sections. Pay attention to the fit between the format and the proposition you are to prove.

Historical Remarks

While people in many cultures and over many millennia have computed and used mathematics in many ways, few have felt the need for proofs. No really convincing reasons explain why Greek mathematicians originated and valued proofs while other ancient cultures (and some not so ancient ones) didn't. Even today many nonmathematicians in Western culture are content to get answers without worrying about proofs of correctness.

The oldest written records of proofs or discussions about proofs come from ancient Greece. Greek sources credit the Pythagoreans between 500 B.C. and 400 B.C. with some deductive arguments, although the Pythagoreans didn't leave any writings. The following century saw a tremendous development of mathematics, especially including proofs. Few of those writings survive because they were surpassed by the greatest mathematics text of all time, Euclid's *Elements*, written in approximately 300 B.C. In a time when books had to be copied by hand, people chose to copy Euclid's masterwork rather than lesser, earlier works. Euclid already showed a mastery of the techniques of proof discussed in the first three sections of this chapter, although he doesn't give many uniqueness proofs. Euclid used more sophisticated proof formats as well, some anticipating modern proofs involving limits.

Proofs by contradiction depend on a characteristic rule of two-valued logic: Either a proposition is true or its negation is true, written $P \vee \neg P$. Essentially, no one questioned this rule once it was stated by Aristotle (384–322 B.C.) until the last century and a half. Since then, a few mathematicians and philosophers of mathematics, called intuitionists and constructivists, have doubted this rule, and so they doubted two-valued logic and proofs by contradiction. Mathematical logic developed as a discipline during this same time and has given careful consideration to this proof technique and all proof techniques. It has shown that two-valued logic is internally consistent and so proofs by contradiction do not put mathematics at risk. However, such internal consistency can't resolve the philosophical concerns of the intuitionists and constructivists. Even so, nearly all mathematicians and philosophers of mathematics confidently accept traditional two-valued logic and so proofs by contradiction. See Chapter 9 for a discussion on this and related issues.

Suggestions

Here are some suggestions beyond those in Section 2.1 for getting started on a proof:

1. Try rewording the proposition. Sometimes the contrapositive is clearer than the original.

2. Explore why all of the hypotheses are needed.

3. If you are stuck, try to think how the proposition could be false. That approach often helps you look at a proof in a different way. (For conjectures that might be true or false, it is particularly important to look at them both from the "true side" and the "false side.")

4. Introduce notation when it helps you concentrate on some aspect of the proposition (see Example 1).

Proof Formats

To show	Format
If P, then Q. (Contrapositive)	Suppose not Q, show not P.
If P, then Q. (Contradiction)	Suppose P and not Q, find any contradiction.

PROBLEMS

*1. For each statement that follows, decide whether it is always true or at least sometimes false. If false, explain why. Explain your answer or cite the part of the text supporting it.

 a) To use a proof by contrapositive, the proposition must be of the form "if P, then Q."

 b) To use a proof by contradiction, the proposition must be of the form "if P, then Q."

 c) A proof by contradiction might be able to be rewritten as a proof by contrapositive.

 d) A negation in the hypothesis is a good reason to try a proof by contradiction.

 e) A negation in the conclusion is a good reason to try a proof by contradiction or contrapositive.

 f) A contradiction occurs only when there is a mistake, so proofs by contradiction are wrong.

2. Fill in the outline proofs below.

 *a) Claim: For all real numbers r, if r is irrational, then $r - 14$ is irrational.
 Proof. Let ___ . For the contrapositive, we will show "If ___ , then ___ ." So, suppose ___ . Now 14 is also a rational and $r = (r - 14) + 14$. So ___ . ∎

 b) Claim: For all integers x and y, we have $6x + 9y \not\equiv 5 \pmod 3$. (The symbol $\not\equiv$ is the negation of $\equiv$.)
 Proof. Let ___ . For a contradiction, assume ___ . From the definition of $\equiv \pmod 3$, we have ___ $(6x + 9y) - 5$. That is, there is $k \in$ ___ such that $3k = 6x + 9y - 5$. Using algebra, we see that ___ , so 3 must divide 5, which is false. This gives a contradiction, and so ___ . ∎

 c) Claim: For all positive real numbers s and t, $\sqrt{st} \le \frac{s+t}{2}$. (The geometric mean is less than or equal to the arithmetic mean.)

Proof. Let ___ and suppose for a contradiction that ___ . Since everything is positive, when we square both sides, the direction of the inequality does not change. So ___ . Eliminate the denominator to get ___ . Put all terms on the same side and factor. Then $0 > (s - t)^2$, which gives a contradiction. Hence, ___ . ∎

3. Prove these propositions, using the contrapositive:

   *a) For any integer z, if 4 does not divide z^2, then z is not even.

   b) For all integers a and b, if $\gcd(a, b) = 1$, then at most one of the numbers is even.

   c) For all integers w and x, if wx is even, then w is even or x is even.

   d) For all $y \in \mathbb{Z}$, if y^6 is odd, then y is odd.

   e) For all $z \in \mathbb{Z}$, if z^5 is even, then z is even.

   f) For all $x \in \mathbb{Z}$, if $x^3 + x^4$ is odd, then x is odd.

4. Prove these propositions using contrapositive or contradiction. (A real number is *irrational* iff it is not rational.)

   *a) There are no positive real roots of $x^6 + 2x^3 + 4x + 5 = 0$.

   b) There is no negative rational number closest to 0.

   c) For all real numbers t, if t^2 is irrational, then t is irrational.

   d) The product of a nonzero rational and an irrational is irrational. What can you say about the product of two irrational numbers?

   e) If t and v are irrational numbers, then $t + v$ is irrational or $t - v$ is irrational.

   f) The equations $y = -x^2 + 2x + 2$ and $y = -x + 5$ have no real solutions in common.

   g) For all $r, s \in \mathbb{R}$, if both r and s are positive, then $\sqrt{r} + \sqrt{s} \neq \sqrt{r + s}$.

   h) For all $y, z \in \mathbb{R}$, if y and z are positive and $y \neq z$, then $4yz/(y + z) < y + z$.

5. a) In Example 4, replace the 8×8 checkerboard with an $n \times k$ rectangle. For which values of n and k does the argument there carry over? What happens with other sizes of rectangles?

   b) Consider a 6×10 rectangle with three of its corner squares removed. Show that this trimmed rectangle cannot be covered with nineteen 3×1 rectangles. (Hint: Try three colors.)

   c) For what values of n will your argument remain valid with $6 \times n$ trimmed rectangles? Explain.

   d) Generalize.

6. This problem assumes the Division Algorithm, stated in Section 1.5.

   a) Prove: The only set of three successive odd natural numbers that are all prime is $\{3, 5, 7\}$. (Hints: If the smallest of such a triple is x, what are the others? What are their possible remainders when divided by 3?)

   b) Prove: The only set of five primes so that successive primes differ by six is $\{5, 11, 17, 23, 29\}$.

   c) State a generalization of the previous part. Explain why you think your generalization should work. Remark: A recently submitted paper claims that there are infinitely many sequences of primes with a constant difference of any desired length. (The longer the desired length of primes, the bigger is the difference.) See Cipra, 1095.

7. The proof sketched in part a uses the Fundamental Theorem of Arithmetic to show that $\sqrt{2}$ is an irrational number. (See Section 1.5 for the statement of this theorem and Section 2.4 for its proof. The irrationality of $\sqrt{2}$ is used in Sections 4.4, 5.3, 8.1, and 8.3.) Fill out the proof of part a and modify it to prove the irrationality of other numbers in the later parts.

a) Proof. Suppose that $\frac{p}{q}$ is rational and, for a contradiction, $\frac{p}{q} = \sqrt{2}$. Why must $p^2 = 2q^2$? Why can we assert that both p and q are integers greater than 1? By the Fundamental Theorem of Arithmetic, we can factor p and q into primes, say, $p = p_1 \cdot p_2 \cdot \cdots \cdot p_n$ and $q = q_1 \cdot q_2 \cdot \cdots \cdot q_k$. Then $p^2 = p_1^2 \cdot p_2^2 \cdot \cdots \cdot p_n^2$. Thus, every prime in the factorization of p^2 appears an even number of times. Describe the factorization of $2q^2$ and prove it can't equal the factorization of p^2 from the uniqueness of the factorization in the Fundamental Theorem of Arithmetic. This contradicts $p^2 = 2q^2$, showing $\sqrt{2}$ must be irrational. ∎

*b) Prove that $\sqrt{3}$ is irrational.

c) Prove that $\sqrt{6}$ is irrational.

d) Explain why a modification of the proof in part a will not show that $\sqrt{4}$ is irrational.

e) Prove that if r is a prime, then $\sqrt{r}$ is irrational.

f) Prove that $\sqrt[3]{2}$ is irrational.

g) Generalize part e to include different roots, such as $\sqrt[3]{}$.

8. Critique the given incorrect "proofs." Point out reasoning errors and unclear presentation. (If the claim is false, there must be an error in the argument. However, it is not enough to say that the claim is false. You need to find an error in the reasoning.)

*a) Claim: For all $r \in \mathbb{R}$, if $r^2 < r^4$, then $-1 \leq r$. "Proof": Let $r \in \mathbb{R}$ and suppose for the contrapositive that $-1 \leq r$ is false. That is, $r < -1$. Because r and -1 are negative, when we square both sides, we switch the inequality. Thus, $1 < r^2$. Multiplying both sides by r^2, a positive number, gives $r^2 < r^4$, showing the contrapositive. ▲

b) Claim: For any sets S, T, and W, if $S \cap T \neq \emptyset$ and $T \cap W \neq \emptyset$, then $S \cap W \neq \emptyset$. "Proof": Suppose that $S \cap T \neq \emptyset$, $T \cap W \neq \emptyset$, and for a contradiction $S \cap W = \emptyset$. From the first two, we have some $t \in S \cap T$, and similarly $t \in T \cap W$. But then $t \in S$, $t \in T$, and $t \in W$. So $t \in S \cap W$, giving a contradiction. ▲

c) Claim: For all integers a, b, n, and k, if $a \not\equiv b \pmod{nk}$, then $a \not\equiv b \pmod{n}$ or $a \not\equiv b \pmod{k}$. (The symbol $\not\equiv$ is the negation of $\equiv$.) "Proof": We show the contrapositive. Suppose $a \equiv b \pmod{n}$ and $a \equiv b \pmod{k}$. Then both n and k divide $a - b$, which means that their product nk divides $a - b$, showing $a \equiv b \pmod{nk}$. ▲

d) Claim: For any sets P and Q, if $P \not\subseteq Q$, then $P - Q \neq \emptyset$. (The symbol $\not\subseteq$ is the negation of $\subseteq$.) "Proof": Suppose $P \not\subseteq Q$ and $P - Q = \emptyset$. Since $P - Q$ is empty, $x \notin P - Q$. Then, by the definition of the difference of sets, $x \notin P$ and $x \in Q$, showing that Q is bigger than P. That is, $P \subseteq Q$, a contradiction. ▲

e) Claim: For any sets C, D, and E, if $C \subseteq E$ and $D \subseteq E$, then $C \cup D \subseteq E$. "Proof": For the contrapositive, we start with $C \cup D \not\subseteq E$ or, in other words, $E \subseteq C \cup D$. Then, for $e \in E$, we have $e \in C \cup D$, giving $e \in C$ or $e \in D$. Thus, $E \subseteq C$ or $E \subseteq D$, which, by De Morgan's Laws, is the negation of $C \subseteq E$ and $D \subseteq E$. ▲

f) Claim: For any indexed family of sets $\{L_i : i \in I\}$, $\bigcap_{i \in I} L_i \subseteq \bigcup_{i \in I} L_i$. "Proof": For a contradiction, assume that $l \notin \bigcap_{i \in I} L_i$. Then, no matter which i we consider, $l \notin L_i$, which means that $l \notin \bigcup_{i \in I} L_i$, giving us a contradiction. ▲

9. Critique the given questionable "proofs." Point out reasoning errors and unclear presentation. If the argument is a proof, say so. (If the claim is false, there must be an error in the argument. However, it is not enough to say that the claim is false. You need to find an error in the reasoning.)

a) Claim: For all integers p and q, $\frac{p}{q} \in \mathbb{Q}$, if $p < q$, then $\frac{p}{q} < 1$. "Proof": For the contrapositive, suppose $\frac{5}{3} \geq 1$. Then $5 \geq 3$, the negation of $5 < 3$. This shows the contrapositive. ▲

*b) Claim: For all real numbers v and w, if v and w are irrational, then $v + w$ is irrational. "Proof": For a contradiction, suppose v and w are rational. Then $v + w$ would be rational, giving a contradiction. ▲

c) Claim: For all integers y and z, if $y + z$ is odd, then one of the integers is odd and one is even. "Proof": We prove the contrapositive. So, suppose the negation of the conclusion. That is, suppose either both y and z are even or both are odd. In the first situation we can write $y = 2j$ and $z = 2k$ for some integers j and k. But then $y + z = 2(j + k)$, an even number. Similarly, with the odd numbers we get $y = 2j + 1$, $z = 2k + 1$, and $y + z = 2(j + k + 1)$, an even number. This proves the contrapositive and so finishes the proof. ▲

d) Claim: For all integers s and t, if st is odd, then both s and t are odd. "Proof": For the contrapositive, suppose both s and t are even, say, $s = 2j$ and $t = 2k$. Then $st = 4jk$ is even, a contradiction. ▲

e) Claim: For any sets S and T, if $S \nsubseteq T$, then $S - T \neq \emptyset$. (The symbol $\nsubseteq$ is the negation of $\subseteq$.) "Proof": Let S and T be sets, and for the contrapositive suppose that $S - T = \emptyset$. Recall that $x \in S - T$ means that $x \in S$ and $x \notin T$. Now $S - T$ is empty, so for all x we can't have $x \in S$ and $x \notin T$. The negation of the proposition $P \wedge \neg Q$ is $P \Rightarrow Q$. So we have for all x, if $x \in S$, then $x \in T$. But that is exactly the definition of $S \subseteq T$. We have shown the contrapositive, finishing the proof. ▲

10. For the claim "For all $x \in \mathbb{Z}$, if $x^3 + x^4$ is odd, then x is odd," critique the following disproof: "Disproof": Let $x \in \mathbb{Z}$. Suppose $x^3 + x^4$ is odd. Whether x is even or odd, $x^3 + x^4$ is even, a contradiction. So the whole proposition must be false. ▲

REFERENCE

Cipra, B. 2004. Proof promises progress in prime progressions. *Science* 304(May 21):1095.

2.3 PROOF FORMAT III: EXISTENCE, UNIQUENESS, OR

"... mathematical proofs, like diamonds, are hard as well as clear, and will be touched with nothing but strict reasoning."

—John Locke (1632–1704)

This section broadens the logical format of propositions we can prove. While existence statements have a simple proof format, the interaction of existential and universal quantifiers deserves more careful investigation. We also consider the special formats for proofs of uniqueness and propositions involving the logical word "or."

Existence

Proving existence can be as easy as giving a specific instance, as in Example 1, to follow. Example 2 gives a more interesting situation, where we invoke a powerful theorem to show existence without determining an actual value. We call such a proof *nonconstructive* to distinguish it from *constructive* proofs that give an explicit candidate or a way to find one exactly. The extensive use of computers has increased the importance of constructive existence proofs.

EXAMPLE 1

CLAIM. There is $x \in \mathbb{R}$ such that $x^2 < x$.

Proof. Pick $x = 0.5$, which is a real number. Then $x^2 = 0.25 < 0.5$. ∎

DISCUSSION. The active verbs "pick" or "choose" seem to fit the existential quantifier, just as the more passive verb "let" matches the universal quantifier. ◇

EXAMPLE 2

(Calculus based)

CLAIM. There is $x \in \mathbb{R}$ such that $2x^5 - 5x^4 + 5 = 0$.

Proof. The function $f(x) = 2x^5 - 5x^4 + 5$, like all polynomials, is continuous. Further, $f(-1) = -2$ and $f(0) = 5$. By the Intermediate Value Theorem, for all y between -2 and 5, there is an x between -1 and 0 such that $f(x) = y$. We simply set $y = 0$ and invoke the theorem to guarantee that the desired root exists. ∎

DISCUSSION. The Intermediate Value Theorem gives little clue what the root is. While my calculator, with the aid of Newton's method, approximates x extremely well as -0.924358014926, this value isn't exact and doesn't prove that there is a mathematically exact value. You probably know the quadratic formula to find exact roots of second degree equations. Surprisingly, in 1824, the 19-year-old mathematician Niels Abel proved that we cannot find a corresponding algebraic formula to solve all fifth degree equations exactly and explicitly. The best we can do is use a constructive method to find increasingly accurate approximations and a nonconstructive proof of existence. ◇

Many definitions contain existential quantifiers. For instance, to prove that an integer x is odd, we need to prove there is an integer k such that $x = 2k + 1$. Fortunately, many theorems involving terms using existential quantifiers involve them in the hypothesis as well as the conclusion. Thus, we can use a value forced to exist from the hypothesis to find the value for the conclusion. Example 1 of Section 2.1 illustrates this situation with odd and even numbers.

Disproving Existential Propositions

In Section 2.1, we disproved a universal proposition by finding a counterexample. In terms of working negations, we proved $\neg \forall x \in X \ P(x)$ by proving $\exists x \in X \ s.t. \ \neg P(x)$. Similarly, we disprove the existential proposition $\exists x \in X \ s.t. \ P(x)$ by proving its working negation, the corresponding universal proposition $\forall x \in X \ \neg P(x)$.

EXAMPLE 3

Use calculus to disprove the

CLAIM. There is a real root of $x^4 + 4x + 5 = 0$.

DISCUSSION. The working negation of $\exists x \in \mathbb{R} \; s.t. \; x^4 + 4x + 5 = 0$ is $\forall x \in \mathbb{R}$ $x^4 + 4x + 5 \neq 0$. We will prove that every real x gives $x^4 + 4x + 5 > 0$.

Proof. Let $x \in \mathbb{R}$. The derivative of the function $f(x) = x^4 + 4x + 5$ is $4x^3 + 4$, which equals 0 iff $4x^3 = -4$ or $x = -1$. Further, when $x < -1$, the derivative $4x^3 + 4$ is negative, showing f is decreasing there. Similarly, f is increasing for $x > -1$. Hence, f has a minimum at $x = -1$, where $f(-1) = 2 > 0$. Since the minimum value is positive, all values of $x^4 + 4x + 5$ are positive. Thus, $x^4 + 4x + 5 = 0$ has no real root. ■   ◇

Combinations of Quantifiers

As discussed in Section 1.3, the order of quantifiers strongly affects the meaning of propositions. Proof formats follow the order of the quantifiers, as illustrated in Examples 4 and 5.

EXAMPLE 4

Decide whether the following propositions are true, and prove your answers:

a) For all $s, t \in \mathbb{R}$, there is $u \in \mathbb{R}$ such that the average of s and u is t.
b) There is $u \in \mathbb{R}$ such that for all $s, t \in \mathbb{R}$, the average of s and u is t.

   Substituting in some numbers for s and t suggests that the first proposition is true, while the second one is false.

Proof of a. Let $s, t \in \mathbb{R}$. Pick $u = 2t - s$. Since s and t are real numbers, u is as well. Further, $\frac{s+u}{2} = \frac{s+2t-s}{2} = t$. ■

DISCUSSION. The proof need not reveal how the value picked was found. Of course, I found the correct choice using algebra: Start with $\frac{s+u}{2} = t$ and solve for u. However, the algebra is not a proof of existence; it is the "scratch work" leading to the proof. In particular, when we start with the equation, we are assuming u already exists. Years of solving equations algebraically hide this important logical distinction.

   We prove the working negation of b: For all $u \in \mathbb{R}$, there are $s, t \in \mathbb{R}$ such that the average of s and u is not t.

Disproof of b. Let $u \in \mathbb{R}$. Pick $s = u$ and $t = u + 1$. Then $\frac{s+u}{2} = \frac{u+u}{2} = u \neq u + 1 = t$. ■

DISCUSSION. Any choice of s and almost any choice of t in terms of u and s would work, as a bit of scratch work reveals. The proposition in b fails because u is not allowed to depend on s, unlike the proposition in a.   ◇

   With good reason, students find propositions with both universal and existential quantifiers difficult to understand and prove. First of all, ordinary English doesn't pay strict attention to quantifiers or their order. Secondly, each alternation of quantifiers increases the sophistication of the proposition. Such sophisticated propositions occur especially frequently in analysis. Chapter 8 seeks to provide help in understanding such propositions. Here we will make a start on such proofs.

EXAMPLE 5

The formal definition of $\lim_{x \to 2} 4x - 5 = 3$ is "$\forall \epsilon > 0 \, \exists \delta > 0 \, s.t. \, \forall x \in \mathbb{R} \, (0 < |x - 2| < \delta \Rightarrow |(4x - 5) - 3| < \epsilon)$," quite a mouthful we won't fully explore here. Basically, we try to force $4x - 5$ to be "close enough" to 3 by choosing x to be "close enough" to 2. Logically, the choice of δ can depend on ϵ, since δ appears after ϵ in the definition. However, it is not clear how to choose δ in terms of ϵ. We'll consider several numerical instances in the table shown next. Following the order in the definition, we consider a value for ϵ to determine how close $4x - 5$ needs to be to 3. Then we pick a value for δ, determining how close x needs to be to 2. Finally, we consider the value of x. Then we plug the numbers into the inequalities to see which are true.

ϵ	δ	x	$|x - 2| < \delta$	$|(4x - 5) - 3| < \epsilon$	Implication	Remarks
0.2	0.1	2.09	true: $0.09 < 0.1$	false: $0.36 > 0.2$	False	δ too big
0.2	0.01	2.009	true: $0.009 < 0.01$	true: $0.036 < 0.2$	True	safe margin
0.01	0.00001	2.003	false: $0.003 > 0.00001$	false: $0.012 > 0.01$	True	poor choice for x
0.01	0.0025	1.99751	true: $0.00249 < 0.0025$	true: $0.00996 < 0.01$	True	δ "just right"

The remark "just right" in the last line comes from the following scratch work: We start with the desired inequality $|(4x - 5) - 3| < 0.01$ and simplify it to get $|4x - 8| < 0.01$. Factoring out a 4 gives $4|x - 2| < 4(0.0025)$. Thus, by requiring $|x - 2| < 0.0025$, we can be sure the inequality holds. Armed with this insight, let's turn to a proof.

CLAIM. $\lim_{x \to 2} 4x - 5 = 3$; that is, $\forall \epsilon > 0 \; \exists \delta > 0 \, s.t. \, \forall x \in \mathbb{R} \, (0 < |x - 2| < \delta \Rightarrow |(4x - 5) - 3| < \epsilon)$.

FORMAT. Even though it is complicated, the string of logic symbols in the definition of a limit suggest the structure of the proof. Here is a skeleton of the proof:

Let $\epsilon > 0$. Pick $\delta > 0$ somehow. Let $x \in \mathbb{R}$ and suppose $0 < |x - 2| < \delta$. Now somehow prove $|(4x - 5) - 3| < \epsilon$.

Proof. Let $\epsilon > 0$ and pick $\delta = \frac{\epsilon}{4}$, which is greater than 0. Let $x \in \mathbb{R}$ and suppose $0 < |x - 2| < \delta$. Then $|(4x - 5) - 3| = |4x - 8| = 4|x - 2| < 4\delta = 4\frac{\epsilon}{4} = \epsilon$, which proves the limit. ∎

DISCUSSION. The only imaginative part of this proof involves picking δ in terms of ϵ, but our scratch work based on examples suggested the successful choice $\delta = \frac{\epsilon}{4}$. Picking a smaller δ, such as $\delta = \frac{\epsilon}{13}$, would also work. Students in a first course in analysis often spend considerable time improving their ability to pick δ. The rest of the proof follows the format of the logic words in the formal definition of a limit. The condition $0 < |x - 2|$ does not really affect our proof, since the condition $|x - 2| < \delta$ was sufficient to prove the required inequality. See Chapter 8 for an explanation of the importance of including $0 < |x - a|$ in the definition of a limit. ◇

EXERCISE. For $\epsilon = 0.000002$, what value does the proof give for δ?

EXERCISE. Verify that the proof still holds using $\delta = \frac{\epsilon}{13}$.

Uniqueness

We have but one choice for u in Example 4 part a—only $u = 2t - s$ will work. That is, there is a unique u such that the average of s and u is t. Proving uniqueness of something involves an entirely separate process from proving existence. In effect, existence says, "There is at least one object," and uniqueness says, "There is at most one." Equivalently, we can say, "There are not two different objects," suggesting the format of Example 6 for proving uniqueness.

EXAMPLE 6

CLAIM. For all $s, t \in \mathbb{R}$, there is a unique $u \in \mathbb{R}$ such that the average of s and u is t.

Proof. We have already shown existence in Example 4. For uniqueness, suppose that both w and x satisfy the property. That is, the average of s and w is t, and the average of s and x is also t. Then $\frac{s+w}{2} = t = \frac{s+x}{2}$. Simple algebra gives us $s + w = s + x$ and so $w = x$; or in other words, there is at most one solution. ∎

DISCUSSION. The uniqueness proof doesn't depend on the actual solution $u = 2t - s$ or even on there being a solution. ◊

When uniqueness is false, we need to know how to disprove it.

EXAMPLE 7

Discuss the following false

CLAIM. For any point P and circle C in the Euclidean plane, there is a unique line tangent to C passing through P.

DISCUSSION. Figure 1 illustrates that the proposition can be wrong for two different reasons. The point S in the interior of the circle has no tangent passing through it. The point T exterior to the circle has two tangents passing through it. Either one shows the claim is false. The working negation of "there exists a unique" has to reflect these two ways it can be false: Either nothing "works" or at least two things "work." The working negation of the claim starts out "There is a point P and a circle C in the plane such that ...," so giving specific examples for P and C, as suggested in Figure 1, suffice to disprove the claim. The working negation of "There is a unique line" is more complicated.

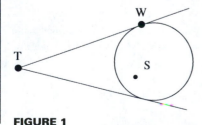

FIGURE 1

We can write the working negation of "For any point P and circle C in the Euclidean plane there is a unique line tangent to C passing through P" as follows: "There is a point P and a circle C in the plane such that (for every line k, k is not a tangent to C through P) or (there are two tangents m and n to C that pass through P and $m \neq n$)." As stated in the preceding discussion, the point S in Figure 1 fulfills the first option in the negation of uniqueness and the point T fulfills the second option in the negation of uniqueness.

Of course, for a point on the circle, such as W, there is a unique tangent, although the proof of uniqueness would require more geometry than this text assumes. (See Kay, 198.) $\diamond$

We state the general working negation for uniqueness next.

WORKING NEGATION.   The working negation of $\exists! x \in X \text{ s.t. } P(x)$ is $(\forall x \in X, \neg P(x)) \vee (\exists u, w \in X \text{ s.t. } P(u) \wedge P(w) \wedge u \neq w)$.

Or   An "or" occurring in the hypothesis or conclusion of a proposition alters the form of the proof. We start with an "or" in the hypothesis. By Problem 8e from Section 1.2, the propositional form $(A \vee B) \Rightarrow C$ is logically equivalent to $(A \Rightarrow C) \wedge (B \Rightarrow C)$. This reformulation allows us to prove the proposition with an "or" by using separate cases, even if the cases are not apparent initially, as in Example 8. Note: The $\wedge$ in $(A \Rightarrow C) \wedge (B \Rightarrow C)$ means we must prove each case separately. Difficult theorems can sometimes be proven by analyzing all possible cases. As an extreme case, the proof of the Four Color Map Theorem in 1976 required massive computer time to check more than a thousand cases that mathematicians had shown covered every possibility.

EXAMPLE 8

CLAIM.   For all $x \in \mathbb{R}$, if $x^2 - 9 > 0$, then $x^2 - x - 2 > 0$.

Proof.   Let $x \in \mathbb{R}$ and suppose that $x^2 - 9 > 0$. Now, $x^2 - 9 = (x - 3)(x + 3)$ and a product is positive iff both factors are positive or both are negative. These are the two cases we will consider separately. We also need the similar factorization $x^2 - x - 2 = (x - 2)(x + 1)$. *Case 1.* Suppose that $(x - 3) > 0$ and $(x + 3) > 0$. Then $x > 3$ and $x > -3$, which reduce to $x > 3$. Since $x > 3$, both $x - 2 > 0$ and $x + 1 > 0$ are true, showing $x^2 - x - 2 > 0$. *Case 2.* Similarly, from $x - 3 < 0$ and $x + 3 < 0$, we have $x < -3$. In turn, both $x - 2$ and $x + 1$ are negative, making $x^2 - x - 2 > 0$ in this case as well.   ∎

DISCUSSION.   Your format should tell your reader what the cases are and when each case is finished. Make sure you consider all of the possible cases. Here we needed to analyze the possibilities for when a product is positive. $\diamond$

We depend on a different division into cases when an "or" occurs in the conclusion.

EXERCISE.   Use truth tables to verify that $B \vee C$ is logically equivalent to $(\neg B \Rightarrow C)$ and $A \Rightarrow (B \vee C)$ is logically equivalent to $(A \wedge \neg B) \Rightarrow C$.

The previous exercise leads us, after assuming A, to prove the conclusion "B or C" by also assuming not B and then proving C. The step of assuming the falsity of one of the possible conclusions is sufficiently counterintuitive that the author of the proof should give the reader ample warning.

EXAMPLE 9

CLAIM.   For all $a, b \in \mathbb{R}$, if $ab = 0$, then $a = 0$ or $b = 0$.

Proof.   Let $a, b \in \mathbb{R}$, and suppose $ab = 0$. If $a = 0$, we're done. So suppose $a \neq 0$. Then $1/a \in \mathbb{R}$, and we can multiply both sides of $ab = 0$ by $(1/a)$ to get $(1/a)(ab) = (1/a)0$. This reduces to $b = 0$, as desired.   ∎

DISCUSSION.   The assumption $a \neq 0$ comes from the format, but it is exactly what we need to cancel the a in the given equation $ab = 0$. We could have just as easily worked with b instead of a. We depend on this proposition whenever we factor a polynomial to find its roots.   ◇

EXAMPLE 10

CLAIM.   For any system of two linear equations in two variables with constant terms equal to zero, there exist one or infinitely many solutions.

Proof.   Let $\begin{cases} ax + by = 0 \\ px + qy = 0 \end{cases}$ be any such system with variables x and y. The system always has the solution $x = 0$ and $y = 0$, so there certainly exists a solution. If it has only one solution, we're done. So suppose it has another, nonzero solution $x = x_1$ and $y = y_1$. Then I claim, for any real number s, that $x = sx_1$ and $y = sy_1$ also give a solution. Indeed, $a(sx_1) + b(sy_1) = s(ax_1 + by_1) = s0 = 0$, and similarly for the second equation. Since there are infinitely many choices for s, we have infinitely many candidates for solutions in this situation. Further, at least one of the values x_1 and y_1 is nonzero, so its multiples by s are all different. Thus, in general, the system has one solution or infinitely many solutions.   ∎

DISCUSSION.   The third and fourth sentences of this proof make the format of the proof clear by casting it as a proof by cases. However, the cases are logically "one solution" and "more than one solution," rather than the desired conclusion "infinitely many solutions." So the remainder of the proof takes us from "more than one" to "infinitely many."

Logically, we could have switched the roles of "one solution" and "infinitely many solutions." That is, we could have said in the third and fourth sentences of the proof, "If there are infinitely many solutions, we are done. So suppose that there are only finitely many solutions." I have no idea how one would argue from finitely many solutions to just one without sneaking in the preceding proof. Successfully deciding which approach leads to an easier proof requires experience, intuition, and, sometimes, luck.   ◇

What's in a Name?

Up to this point, none of the propositions we have proven were particularly memorable, so they were simply called claims to be proven. However, mathematics is full of theorems, corollaries, lemmas, and statements with other names. Here is a brief description of some of the commonly used different names for propositions:

> **Theorem.** A proven proposition of definite importance.
>
> **Lemma.** A proven proposition of lesser importance, usually a step towards proving a theorem.
>
> **Corollary.** A proven proposition that is a relatively easy consequence of a theorem.
>
> **Axiom.** An assumed proposition. From ancient Greece until at least 1800, an axiom was considered a "self-evident truth" and so needed no proof. In modern mathematics, axioms do not need to be self-evident, but they are accepted as starting points. In general, axioms should not be provable from other axioms or previously proven propositions. Synonyms for "axiom" include "postulate," "principle," and "common notion."
>
> **Conjecture.** A currently unproven proposition, usually based on some evidence.
>
> **Hypothesis.** There are two uses: "the 'if' part of an implication" and "a conjecture."
>
> **Paradox.** A proven proposition with counterintuitive meaning or consequences.

Suggestions

1. Before starting a proof involving cases, list all the cases to yourself. There can be more than two.

2. Feel free to do lots of scratch work to decide what value to pick for a variable with an existential quantifier. But remember that algebra involves assuming that a solution exists. Thus, the scratch work can't be part of the proof of existence. However, the proof might well use the scratch work backwards, as Examples 4 and 5 did.

3. When you use a variable, make sure you "properly introduce" it to your reader. Don't use a variable first in some expression and then explain what the symbol represents. In particular, make it immediately clear whether the variable comes from a universal quantifier ("let x ... ") or from an existential quantifier ("pick y ... ").

4. When the proposition is complicated, as in Example 5, lay out the entire logical format before filling in the details. You will often see where to concentrate your creative energy.

5. Uniqueness requires a separate proof from existence, with a different format. To show uniqueness, assume that two things "work" and show that they are equal.

Proof Formats

To show	Format
If P or Q, then R.	Case 1. Suppose P, prove R. Case 2. Suppose Q, prove R.
If P, then Q or R.	Suppose P. If Q, we are done. So suppose not Q and prove R.
There is $x \in S$ s.t. ___. (Constructive)	Pick $x \in S$ and prove ___.
There is a unique $x \in S$ s.t. ___.	Show existence. Uniqueness: Suppose y and z in S, satisfy ___, and prove $y = z$.

PROBLEMS

*1. For each statement that follows, decide whether it is true or (at least sometimes) false. If false, explain why. Explain your answer or cite the part of the text supporting it.

   a) An example never qualifies as a proof.

   b) To prove existence, you must construct a specific example.

   c) You can prove uniqueness without proving existence.

   d) The position of an "or" in a proposition can affect the format of the proof.

   e) A proof by cases always involves exactly two cases.

2. Fill in these outline proofs:

   a) Claim: Every rectangle has a square with the same area.
   Proof: Suppose the sides of the given rectangle are x and y, which are positive real numbers. Pick $s =$ ___ for the side of the square. Then the area of the square is ___ , which equals the area of the rectangle. ∎

   *b) Claim: For all $s \in \mathbb{R}$, there is a unique $t \in \mathbb{R}$ such that $\ln(t) = s$.
   Proof: Let ___ . For existence, pick $t =$ ___ . Then $\ln(t) =$ ___ $= s$. For uniqueness, suppose v and w both satisfy ___ . Then $e^{\ln(v)} = e^{\ln(w)}$. But $e^{\ln(v)} =$ ___ , and similarly, ___ . So ___ , showing uniqueness. ∎

   c) Claim: For all integers z, $z^2 \not\equiv 2 \pmod 4$.
   Proof: Let ___ . By the Division Algorithm, we can write $z = 4q + r$, where $0 \le r < 4$. Then $z^2 =$ ___ . Consider each of the possible values of r. *Case 1.* If $r = 0$, $z^2 =$ ___ $\equiv$ ___ $\pmod 4$. (Fill in all the other cases.) Since in each case $z^2 \not\equiv 2 \pmod 4$, the proposition is proven. ∎

3. Prove these propositions:

   *a) There is a real number r so that $r^2 < 3r - 2$.

   b) There is an integer n such that for all $k \in \mathbb{Z}$, k divides n.

   c) For every rational q, if $q \ne 1$, there is a unique $t \in \mathbb{Q}$ such that $qt = q + t$.

   d) There is a unique rational s such that for all $t \in \mathbb{Q}$, $st - t = t$.

   e) For all $b, s \in \mathbb{R}$, if $b > 0$ and $s > 0$, then there is a unique $h \in \mathbb{R}$ such that the area of a triangle with base b and height h equals the area of a square with side s.

   f) (Calculus needed) There is a unique real number b such that $f(x) = 2x^2 - 3x + 4$ has a derivative of 0 at $x = b$.

g) (Calculus needed) For any cubic function $g(x) = ax^3 + bx^2 + cx + d$ with $a \neq 0$, there is a unique real number k such that $g''(k) = 0$.

h) Let S be any set. Then there exist at least two elements of $\mathcal{P}(\mathcal{P}(S))$.

4. Prove the following propositions:

   *a) For all integers j and k, if $j + k$ is even, then both j and k are even or both are odd.

   b) If $x \in A \cup B$, but $x \notin A \cap B$, then $x \in A - B$ or $x \in B - A$.

   c) For any real numbers x and y, if $x^4 - y^4 = 0$, then $x = y$ or $x = -y$.

   d) For any complex numbers x and y, if $x^4 - y^4 = 0$, then $x = y$ or $x = -y$ or $x = yi$ or $x = -yi$.

   e) For all $n \in \mathbb{Z}$, $(n^2 + n)/2$ is an integer.

   f) For all $n \in \mathbb{Z}$, $(n^3 - n)/6$ is an integer.

   g) If 3 divides the integer z, then z is odd or 6 divides z.

5. Write and prove the working negation of each proposition:

   *a) For all $s \in \mathbb{R}$, there is $t \in \mathbb{R}$ such that $st = \pi$.

   b) For all $u \in \mathbb{N}$, $u^{u+2} \neq (u+2)^u$.

   c) For all $v \in \mathbb{N}$, $(v-1)! < v!$.

   d) For all $w \in \mathbb{R}$, if $w^3 < w^5$, then $w^2 < w^4$.

   e) For all $x > 0$ and $y > 0$, $\sqrt{xy} < \frac{x+y}{2}$.

   f) For every set A, if A is nonempty, then A is not a subset of its power set $\mathcal{P}(A)$.

   g) For every $b \in \mathbb{R}$, there is a unique root of $x^2 - bx = 0$.

6. Decide whether each of the following propositions is true and prove your answer:

   *a) There is $a \in \mathbb{Z}$ such that for all $b \in \mathbb{N}$, we have $b^a = 1$.

   b) For all $b \in \mathbb{N}$, there is $a \in \mathbb{Z}$ such that $b^a = 1$.

   c) There is $a \in \mathbb{Z}$ such that for all $b \in \mathbb{Z}$, we have $b^a = 1$.

7. Prove these propositions. Recall that $x \mid y$ means "x divides y" (referred to in Section 4.4).

   *a) For any integer p, we have $p \mid p$.

   b) For all integers p, q, and r, if $p \mid q$, then $pr \mid qr$.

   c) For all integers x, y, and z, if $x \mid y$ and $y \mid z$, then $x \mid z$.

   d) For all $i, j \in \mathbb{N}$, if $i \mid j$ and $j \mid i$, then $i = j$.

8. One proposition given next is true and the other is false. Prove the true one and prove the working negation of the false one.

   a) For all integers a, b, and c, if a divides bc and $\gcd(b, c) = 1$, then a divides b or a divides c.

   b) For all $r, a, c \in \mathbb{R}$, if $c \neq 0$ and $r \neq \frac{a}{c}$, then there is a unique $x \in \mathbb{R}$ such that $\frac{ax+1}{cx} = r$.

9. a) Complete the following definition for absolute value: $|x| = \begin{cases} - & \text{if } 0 \leq x \\ - & \text{if } x < 0 \end{cases}$.
   Prove the following propositions, using cases:

   b) For all $x \in \mathbb{R}$, $|x| = \sqrt{x^2}$.

   c) For all $x, y \in \mathbb{R}$, $|x \cdot y| = |x| \cdot |y|$.

   d) For all $s, t \in \mathbb{R}$, $|s + t| \leq |s| + |t|$.

   e) For all $x, y \in \mathbb{R}$, $|x| - |y| \leq |x - y|$.

   f) For all $i, j \in \mathbb{Z}$, if $i \mid j$ and $j \mid i$, then $|i| = |j|$.

10. Prove these propositions:

   *a) For any sets S and T, $S \subseteq S \cup T$.

   b) For any sets T and W, $T = (T \cap W) \cup T$.

   c) For any sets T and W, $T = (T \cup W) \cap T$.

   d) For any sets S, T, and W, $(W - S) \cup T = (W \cup T) - (S - T)$.

   e) For any sets S, T, and W, $(S \cup T) \times W = (S \times W) \cup (T \times W)$.

11. This problem assumes the Fundamental Theorem of Arithmetic, stated in Section 1.5.

   a) Prove: The difference of the squares of two integers is never 2. (Hints: Use letters for the integers, write the difference algebraically, and factor. Consider cases.)

   b) Repeat part a with 6 in place of 2.

   c) What other integers are impossible to obtain as the difference of the squares of two integers? Prove your answer. (Hint: Consider cases involving even and odd.)

12. For each pair of sets, decide whether they are always equal, one is always a subset of the other, or neither need be a subset of the other. Prove and give counterexamples as appropriate.

   *a) $(D \cup E) \times (F \cup G)$ and $(D \times F) \cup (E \times G)$.

   b) $\mathcal{P}(X) \cap \mathcal{P}(Y)$ and $\mathcal{P}(X \cap Y)$.

   c) $\mathcal{P}(X) \cup \mathcal{P}(Y)$ and $\mathcal{P}(X \cup Y)$ (referred to in Section 2.5).

   d) $\mathcal{P}(X) - \mathcal{P}(Y)$ and $\mathcal{P}(X - Y)$.

   e) $\mathcal{P}(X) \times \mathcal{P}(Y)$ and $\mathcal{P}(X \times Y)$.

13. For any indexed family of sets $\{G_i : i \in I\}$ with $I \neq \emptyset$ and any set H, prove that $\bigcup_{i \in I} (H \cap G_i) = H \cap (\bigcup_{i \in I} G_i)$ and $\bigcup_{i \in I} (H \cup G_i) = H \cup (\bigcap_{i \in I} G_i)$.

14. Prove the given limits, using the formal definition that $\lim_{x \to a} f(x) = L$ iff $\forall \epsilon > 0 \ \exists \delta > 0 \ s.t. \ \forall x \in \mathbb{R} \ (0 < |x - a| < \delta \Rightarrow |f(x) - L| < \epsilon)$. This problem is referred to in Problem 9 of Section 2.5.

   *a) $\lim_{x \to 3}(2x + 4) = 10$.

   b) $\lim_{x \to -1}(\frac{x}{4} + 3) = 2.75$.

   c) $\lim_{x \to 5}(-3x + 7) = -8$.

   d) $\lim_{x \to 3}(-12x + 41) = 5$.

   e) For all m, b, $a \in \mathbb{R}$, if $m \neq 0$, then $\lim_{x \to a}(mx + b) = ma + b$. (Hint: Does your proof work for $m < 0$?)

   f) For all b, $a \in \mathbb{R}$, $\lim_{x \to a}(0x + b) = b$.

   g) Write the working negation for the definition of $\lim_{x \to a} f(x) = L$.

   h) Use the negation in part g to prove that $\lim_{x \to 2}(x - 3) \neq 4$.

15. Critique the incorrect "proofs" that follow. Point out reasoning errors and unclear or incomplete presentations. (If the claim is false, there must be an error in the argument. However, it is not enough to say that the claim is false. You need to find an error in the reasoning.)

   a) Claim: For all $t \in \mathbb{R}$, there is a $u \in \mathbb{R}$ such that $t \div u = -1$. "Proof": Pick $u = -t$. Then $t \div (-t) = -1$. ▲

   b) Claim: There is a real number v such that for all $w \in \mathbb{R}$, $(w^2 + 1)v = 1$. "Proof": For any $w \in \mathbb{R}$, we know $w^2 + 1 \neq 0$, so we can choose $v = \frac{1}{w^2+1}$. Then $(w^2 + 1)v = \frac{w^2+1}{w^2+1} = 1$. ▲

   c) Claim: For all $x \in \mathbb{R}$, there is $y \in \mathbb{R}$ such that $x + y = \pi$. "Proof": Let $x \in \mathbb{R}$. For existence, pick $x = \pi$ and $y = 0$. Then $x + y = \pi + 0 = \pi$. ▲

*d) Claim: If $x^3 = x$, then $x = -1$ or $x = 0$ or $x = 1$. "Proof": If $x = -1$, then $x^3 = -1$ as well, so we'd be done. Otherwise, $x = 0$ or $x = 1$. But if $x = 0$, then $x^3 = 0$; and if $x = 1$, then $x^3 = 1$. This shows the result. ▲

e) Claim: If p and q are primes, then there is a unique integer k dividing both. "Proof": Suppose p and q are primes. By definition, the only divisors of p are 1 and p. Similarly, the only divisors of q are 1 and q. Obviously, the only number dividing both p and q is 1. ▲

f) Claim: There is a unique three-digit number with its digits adding to 8 and its digits multiplying to 10. "Proof": Suppose the digits are a, b, and c. So $a + b + c = 8$ and $a \cdot b \cdot c = 10$. For the product to be 10, the only choices for a, b, and c are positive divisors of 10, which are 1, 2, 5, and 10. But 10 is too big for the sum to be 8. So the digits are 1, 2, and 5, giving the number 125. ▲

16. Critique the questionable "proofs" that follow. Point out reasoning errors and unclear or incomplete presentations. If the argument is a proof, say so. (If the claim is false, there must be an error in the argument. However, it is not enough to say that the claim is false. You need to find an error in the reasoning.)

a) Claim: There is exactly one real root of $x^2 + 9 = 6x$. "Proof": Shift the $6x$ to the other side to get $x^2 - 6x + 9 = 0$, which factors to $(x - 3)(x - 3) = 0$, showing that $x - 3 = 0$, or $x = 3$, is the unique solution. ▲

b) Claim: There is exactly one root of $x^2 - 4x + 4 = 0$. "Proof": Clearly, $x = 2$ shows existence. For uniqueness, we know either $x = 2$ or $x \neq 2$. These give $x - 2 = 0$ and $x - 2 \neq 0$. Multiplying these together, we get $(x - 2)(x - 2) \neq 0$, or $x^2 - 4x + 4 \neq 0$. But this contradicts our equation, showing uniqueness. ▲

*c) Claim: For all sets X and Y, $\mathcal{P}(X \cup Y) \subseteq \mathcal{P}(X) \cup \mathcal{P}(Y)$. "Proof": Let $A \in \mathcal{P}(X \cup Y)$. Then $A \subseteq X \cup Y$ and so $A \subseteq X$ or $A \subseteq Y$. In the first case, $A \in \mathcal{P}(X)$. In the second case, $A \in \mathcal{P}(Y)$. Either way, $A \in \mathcal{P}(X) \cup \mathcal{P}(Y)$. ▲

d) Claim: For all sets K and L, $(K - L) \cup (L - K) \subseteq K \cup L$. "Proof": Let K and L be any sets, and $x \in (K - L) \cup (L - K)$. So, $x \in K - L$ or $x \in L - K$. That is, $(x \in K$ and $x \notin L)$ or $(x \in L$ and $x \notin K)$. We need to show that $x \in K \cup L$. If $x \in K$, we're done, so suppose $x \notin K$. But that means that we are in the case $(x \in L$ and $x \notin K)$. Then $x \in L$, and it follows that $x \in L \cup K$. ▲

e) Claim: For all x, y, $z \in \mathbb{Z}$, if $x|yz$, then $x|y$ or $x|z$. "Proof": Suppose for integers x, y, and z that x divides yz. Then there is some $k \in \mathbb{Z}$ such that $xk = yz$. If there is $j \in \mathbb{Z}$ such that $xj = y$, we're done. So, suppose $xj \neq y$. We must have $j < k$, since y is less than yz. So xj divides $xk = yz$. Since x doesn't divide y, j must divide y. So, divide both xj and yz by j to see that x divides z. ▲

f) Claim: For all $x \in \mathbb{Z}$, there is $y \in \mathbb{Z}$ such that x divides y. "Proof": Let $x \in \mathbb{Z}$ and $y = 0$. Then $x \cdot 0 = 0 = y$. So x divides y. ▲

g) Claim: For all sets A, B, C, and D, we have $(A \cup B) \times (C \cup D) \subseteq (A \times C) \cup (B \times D)$. "Proof": Let $(x, y) \in (A \cup B) \times (C \cup D)$. Then $x \in (A \cup B)$ and $y \in (C \cup D)$. Thus, $(x \in A$ or $x \in B)$ and $(y \in C$ or $y \in D)$. By the distributivity of "and" over "or," we have $(x \in A$ and $y \in C)$ or $(x \in B$ and $y \in D)$. That is, $(x, y) \in (A \times C) \cup (B \times D)$. ▲

REFERENCE

KAY, D. 1994. *College geometry: A discovery approach.* New York: Harper Collins.

2.4 PROOF FORMAT IV: MATHEMATICAL INDUCTION

"Mathematics is the science which draws necessary conclusions."

—Benjamin Peirce (1809–1890)

The unique structure of the natural numbers supports a special proof technique called mathematical induction. Although it shares a name with scientific induction, mathematical induction differs radically. Induction in science refers to the process of describing a general pattern on the basis of a number of individual observations, similar to making a conjecture in mathematics. Mathematical induction, henceforth shortened to induction, is a legitimate, rigorous way of proving that a statement holds for all natural numbers. The order on $\mathbb{N}$ provides the key to induction:

*$\mathbb{N}$ has a first element 1, each natural number k has a unique successor $k + 1$, and **every** element of $\mathbb{N}$ is in the sequence of 1 and its successors.

Equivalently, suppose S is a subset of $\mathbb{N}$ satisfying (i) $1 \in S$ and (ii) If $k \in S$, then $k + 1 \in S$. Then $S = \mathbb{N}$.

We repackage the previous idea as the Principle of Mathematical Induction and use it as an axiom. The word "every" in the previous starred sentence is written in bold type to emphasize the key assumption: Starting with 1 and adding 1 over and over will eventually take us to every natural number. After long deliberations, mathematicians have decided that this is a fundamental intuition about the natural numbers, not something open to proof.

THE PRINCIPLE OF MATHEMATICAL INDUCTION (PMI). Suppose for a property $P(x)$,

i) $P(1)$ is true and

ii) if $P(k)$ is true, then $P(k + 1)$ is true.

Then for all $n \in \mathbb{N}$, $P(n)$ is true.

EXAMPLE 1

Figure 1 suggests that the sum of the first n numbers is $\frac{n(n+1)}{2}$. For instance, $1 + 2 + 3 + 4 = 10 = \frac{4(4+1)}{2}$. In general, for all $n \in \mathbb{N}$, we have $1 + 2 + \ldots + n = \sum_{i=1}^{n} i = \frac{n(n+1)}{2}$. That is, $P(n)$ is the proposition $\sum_{i=1}^{n} i = \frac{n(n+1)}{2}$. Of course, our one example doesn't prove a general proposition.

Proof. For the initial step of induction, let $n = 1$. We easily verify that $\sum_{i=1}^{1} i = 1$ and $\frac{1(1+1)}{2} = 1$, as required.

For the induction step, suppose for some $n = k$ in $\mathbb{N}$ that $\sum_{i=1}^{k} i = \frac{k(k+1)}{2}$. Consider $n = k + 1$. We need to convert $\sum_{i=1}^{k+1} i$ to the desired form $\frac{(k+1)((k+1)+1)}{2}$, somehow using what we assumed for $n = k$. Now $\sum_{i=1}^{k+1} i = 1 + 2 + \ldots + k + (k + 1) = \left(\sum_{i=1}^{k} i \right) + k + 1 = \left(\frac{k(k+1)}{2} \right) + k + 1 = \frac{k(k+1)+2k+2}{2} = \frac{k^2+3k+2}{2} = \frac{(k+1)(k+2)}{2}$, as desired. By the Principle of Mathematical Induction, the equation $\sum_{i=1}^{n} i = \frac{n(n+1)}{2}$ holds for all $n \in \mathbb{N}$. ■

DISCUSSION. The preceding proof is completely rigorous, but many people find the "picture proof" of Figure 1 more insightful. If you use the notation $P(n)$ in an induction proof, be sure to identify the proposition this notation represents. Here, $P(n)$ would stand for $\sum_{i=1}^{n} i = \frac{n(n+1)}{2}$.

$n + \cdots + 3 + 2 + 1$

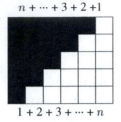

$1 + 2 + 3 + \cdots + n$

FIGURE 1. We count the squares in two ways. The rectangle has $n(n+1)$ squares. There are $1 + 2 + \cdots + n$ stacked white squares and $n + \cdots + 2 + 1$ black ones. So $n(n+1) = 2\sum_{i=1}^{n} i$. Now divide by 2 to get the equation. ◊

Intuitively, the two parts of an induction proof, as in Example 1, give a constructive way to verify the statement of the claim, $P(n)$, for any given n. First we verify that $P(1)$ is true. Then we use the truth of $P(1)$ and the induction step to establish the truth of $P(2)$. In turn, the truth of $P(2)$ and the induction step suffice to prove $P(3)$, and so on. However, saying "and so on" is not very precise. Instead, we invoke the PMI to prove all infinitely many cases at once. To invoke the PMI, we need to assume the generic case $n = k$, called the induction hypothesis, and show how to get to the next case ($n = k + 1$). We also need the initial step in an induction proof, as the following incorrect argument illustrates:

EXAMPLE 2

CLAIM. For all $n \in \mathbb{N}$, $\sum_{i=1}^{n} 2i = n^2 + n + 7$.

"Proof." Suppose for $n = k$ that $\sum_{i=1}^{k} 2i = k^2 + k + 7$. For $n = k + 1$, we have $\sum_{i=1}^{k+1} 2i = (\sum_{i=1}^{k} 2i) + 2(k+1) = k^2 + k + 7 + 2k + 2 = k^2 + 3k + 9$. When $n = k + 1$, the other side becomes $n^2 + n + 7 = (k+1)^2 + (k+1) + 7 = k^2 + 2k + 1 + k + 1 + 7 = k^2 + 3k + 9$. Since the two sides are equal, the case $n = k + 1$ also holds. By the PMI, the property holds for all $n \in \mathbb{N}$. ∎

DISCUSSION. Even though all of the preceding algebra is correct, without the initial step the previous argument completely collapses. The summation never equals $n^2 + n + 7$. It equals $n^2 + n$ instead. The algebra in the induction step couldn't catch that error. Only the initial step of a particular case guards against such uncritical manipulation of formulas. ◊

Of course, verifying many instances doesn't prove a property for every natural number, as the accompanying cartoon attests.

Some mathematicians prefer a different style for induction proofs, built on the substance of the third paragraph of this section. Let S be the subset of all natural numbers satisfying the desired property. We show that $1 \in S$, and if $n \in S$, then $n + 1 \in S$. Then the PMI guarantees $S = \mathbb{N}$. Example 3 illustrates this format.

EXAMPLE 3

CLAIM. For all $n \in \mathbb{N}$, a set with n elements has 2^n subsets.

DISCUSSION. In Example 2 of Section 1.6, we saw that $\mathcal{P}(\{p, q\}) = \{\emptyset, \{p\}, \{q\}, \{p, q\}\}$ had $4 = 2^2$ elements and $\mathcal{P}(\{p, q, r\}) = \{\emptyset, \{p\}, \{q\}, \{r\}, \{p, q\}, \{p, r\}, \{q, r\}, \{p, q, r\}\}$ had $8 = 2^3$ elements. An induction proof needs to show how to go from one level to the next. How do we get the eight subsets of $\{p, q, r\}$ from the four subsets of $\{p, q\}$? Note that each subset of $\{p, q\}$ is already a subset of $\{p, q, r\}$. The remaining four subsets all have r as an element. In fact, each of these four subsets is a subset of $\{p, q\}$ together with the element r. This insight suggests the following proof:

Proof. Let $S = \{n \in \mathbb{N} :$ if a set has n elements, it has 2^n subsets$\}$. We'll show $S = \mathbb{N}$. A set with one element, say, $\{a\}$, has two subsets, namely, $\{a\}$ and $\emptyset$. Thus, $1 \in S$.

For the induction hypothesis, suppose $k \in S$. Now consider any set with $k + 1$ elements, say, $A = \{a_1, a_2, \ldots, a_k, a_{k+1}\}$. Let $B = \{a_1, a_2, \ldots, a_k\}$, a set with k elements. So B has 2^k subsets. For each subset C of B, we find two different subsets of A, namely, C and $C \cup \{a_{k+1}\}$. This construction gives us $2 \cdot 2^k = 2^{k+1}$ candidates for subsets of A, the desired number.

We need to show that all of these candidates differ from one another and that there are no other subsets of A. Given two different subsets C and D of B, there must be an element in one that is not in the other. This difference remains between the subsets C, $C \cup \{a_{k+1}\}$, D, and $D \cup \{a_{k+1}\}$, so the 2^{k+1} subsets are all different. Now, let E be any subset of A. Either $a_{k+1} \notin E$ or $a_{k+1} \in E$. In the first case, $E \subseteq B$ and so already appears among our candidates. In the second case, $E - \{a_{k+1}\}$ is a subset of B and $E = (E - \{a_{k+1}\}) \cup \{a_{k+1}\}$ and so is one of our candidates. Thus, A has 2^{k+1} subsets and $k + 1 \in S$.

Since $1 \in S$ and (for all $k \in \mathbb{N}$, if $k \in S$, then $k + 1 \in S$), by the Principle of Mathematical Induction, $S = \mathbb{N}$, showing the claim. ∎

DISCUSSION. Saying $n \in S$ here is equivalent to saying $P(n)$ is true in the Principle of Mathematical Induction. The example suggested one key to this proof: turning a subset of the smaller set B into two subsets of the bigger set A. But we still needed to show that our procedure accounted for all of the subsets, without duplication. ◇

Initially, some students find the format of induction circular, since $P(k)$, which we assume true, sounds like $P(n)$, which we are supposed to show is true. However, k and n have different roles in induction. The induction step assumes that some particular, but generic, case $P(k)$ is true so as to show the next case is also true. The conclusion applies to all $n \in \mathbb{N}$ at once, a much stronger statement. The distinction between potentially infinite and actually infinite gives another way to see the different roles of k and n. The induction step addresses the potentially infinite: Wherever we are (step k), we can go one further (step $k + 1$). The conclusion holds for all of $\mathbb{N}$, an actual infinity.

While induction works easily on simple equations, as in Example 1, it applies more generally and with more involved proofs as well, as the first paragraph of the next proof illustrates. We stated and used the Division Algorithm in Section 1.5, and we repeat it here.

THEOREM 2.4.1. (The Division Algorithm). For all $z \in \mathbb{Z}$ and all $n \in \mathbb{N}$, there are unique integers q and r such that $z = nq + r$ and $0 \le r < n$.

DISCUSSION. We have two variables, z and n. Since induction is made for the natural numbers, it might appear that we should hold z fixed and do induction on n. However, there is little connection between dividing z by n and dividing z by $n + 1$, so the induction step would be hard. Instead, from $z = nq + r$ we can have $z + 1 = nq + (r + 1)$, suggesting that induction might work on z, at least for positive values of z. Then we'll address the negative values and, finally, the uniqueness. In effect, the three parts of this proof are quite different.

Proof. Let $n \in \mathbb{N}$ and $z \in \mathbb{Z}$. For existence, let's first restrict z to $\mathbb{N}$ in order to use induction on z. For $z = 1$, the initial step, either $n = 1$ and $z = n1 + 0$ or $n > 1$ and $z = n0 + 1$. Either way, existence for the initial step holds.

For the induction step, assume for $z = k$ that integers q and r satisfy $k = nq + r$, with $0 \le r < n$. For $z = k + 1$, we have $k + 1 = nq + r + 1$. If $r + 1 < n$, our induction step is satisfied. Since $r < n$, the only other possibility is $r + 1 = n$. But then $k + 1 = nq + n = n(q + 1) + 0$, which also satisfies our induction step. By the PMI, the existence of q and r holds for all $z \in \mathbb{N}$.

Turn now to existence for $z \le 0$. For $z = 0$, we can write $0 = n0 + 0$, so $q = 0$ and $r = 0$. For $z < 0$, we have $-z > 0$, and from the previous paragraph, we can find integers s and t with $-z = ns + t$, where $0 \le t < n$. Then $z = n(-s) - t$. Unfortunately, $-t$ might not be between 0 and n. So we need a different value for q than $-s$, as well as a different value for r. We shift to the next more negative value for q, which is $s - 1$. To compensate, we shift r up by n to $-t + n$. That is, $z = n(-s) - t = n(-s) - n + n - t = n(-s - 1) + n - t$. Now $0 \le n - t < n$ follows from $0 \le t < n$, showing existence in this case as well.

Finally, for uniqueness, let q, r, s, and t be integers satisfying $z = nq + r$ and $z = ns + t$, with $0 \le r < n$ and $0 \le t < n$. Then $nq + r = ns + t$ or $n(q - s) = t - r$. Further, r and t can't be farther apart than $n - 1$, since they are both between 0 and $n - 1$. That is, $-(n - 1) \le t - r \le n - 1$. But $t - r = n(q - s)$ is a multiple of n, and the only multiple of n between $-(n - 1)$ and $n - 1$ is 0. Thus $t - r = 0$, or $t = r$. Similarly, $n \neq 0$ and $n(q - s) = 0$ force $q = s$, showing uniqueness. ∎

DISCUSSION. When you use induction, tell your reader early in the proof. Otherwise, the verification in the initial step may look like an example instead of the start of a proof. Also, if there are several variables, such as n and z in this theorem, tell your reader which one is the induction variable.

Variations on Induction

The initial case in an induction proof need not be 1. The modified principle that follows allows us to start at any integer, even a negative integer or zero. For instance, the proof in Example 1 could have started with $n = 0$, although the picture proof wouldn't make much sense with $n = 0$.

The Modified Principle of Mathematical Induction   Suppose, for a property $P(x)$, that

  i) $P(a)$ is true for an integer a and
  ii) if $k \geq a$ and $P(k)$ is true, then $P(k + 1)$ is true.

Then for all $n \in \mathbb{Z}$, if $n \geq a$, then $P(n)$ is true.

EXAMPLE 4

Find a formula for the number of diagonals in a convex n-sided polygon (n-gon), where $n \geq 3$, and prove it. (A convex polygon has no "dents," so the diagonals are inside.)

Solution   Let $D(n)$ be the number of diagonals of a convex n-gon. Table 1 gives various values of $D(n)$ as well as the differences $d_n = D(n) - D(n - 1)$. The differences increase by 1 each time, just like the terms we added in Example 1. So we should expect that the sum of the differences, $D(n)$, should have a formula similar to the formula for the sum in Example 1, which was $0.5n^2 + 0.5n$. Let's look for a second degree equation $an^2 + bn + c$. (Problem 21 provides a different approach.) Our chart gives us $D(3) = 0 = 9a + 3b + c$, $D(4) = 2 = 16a + 4b + c$, and $D(5) = 5 = 25a + 5b + c$. We solve these equations to find $D(n) = 0.5n^2 - 1.5n$.

TABLE 1

n	3	4	5	6	7
$D(n)$	0	2	5	9	14
$d(n)$	–	2	3	4	5

Proof by Induction.   Our initial case $n = 3$ is a triangle, which has no diagonals, and $0.5(3^2) - 1.5(3) = 0$. For the induction step, suppose that every convex n-gon has $D(n)$ diagonals and that we have a convex $(n + 1)$-gon (see Figure 2). Label the vertices of the polygon $V_1, V_2 \ldots, V_n$, and V_{n+1}. Because the $(n + 1)$-gon is convex, the segment $\overline{V_1 V_n}$ is inside

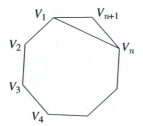

FIGURE 2

it, so this segment is a diagonal, and we have an n-gon with the vertices $V_1, V_2, \ldots, V_n$. From our induction assumption, this n-gon has $0.5n^2 - 1.5n$ diagonals, all of which are diagonals for the bigger polygon. In addition, we have the diagonal $\overline{V_1 V_n}$ and all of the diagonals from the last vertex V_{n+1}. Since V_{n+1} is already connected to V_1 and V_n by edges of the polygon, there are $n - 2$ diagonals from V_{n+1}. Thus, $D(n + 1) = 0.5n^2 - 1.5n + 1 + n - 2 = 0.5n^2 - 0.5n - 1$. Also, $0.5(n + 1)^2 - 1.5(n + 1) = 0.5n^2 - 0.5n - 1$ as well, proving the induction step. By the Modified Principle of Mathematical Induction, our formula holds for all convex polygons. ∎

DISCUSSION. Students working on problems like Example 4 usually notice how the numbers increase, here by $n - 1$. Sometimes, though, their induction step amounts to verifying that $0.5n^2 - 1.5n + (n - 1) = 0.5(n + 1)^2 - 1.5(n + 1)$. However, this problem concerns diagonals, so the algebra isn't enough. It is essential first to relate geometrically this increase with the number of diagonals in an n-gon and an $(n + 1)$-gon.

Note also how the induction part is worded. We assume the case $n = k$ and then start over with a polygon having $k + 1$ sides. A common mistake students make early on is to try to extend an n-gon to an $(n + 1)$-gon. Recall that we must show that our induction argument holds for all $n + 1$-gons, so we need to start with a general one, not one that conveniently fits our approach. The following incorrect wording illustrates this mistake:

Attempted Start of a Proof by Induction. Our initial case $n = 3$ is a triangle, which has no diagonals, and $0.5(3^2) - 1.5(3) = 0$. For the induction step, suppose that any convex n-gon with vertices $V_1, V_2 \ldots, V_n$ has $D(n)$ diagonals. Add the vertex V_{n+1} to make a convex polygon with $n + 1$ sides (etc.). ▲

DISCUSSION. At this point in the argument, the reader has a right to wonder whether every convex $n + 1$-gon can be built this way. While they can all be built this way, the proof would need to show that fact. The earlier proof avoids such an unnecessary step by starting with a general convex $n + 1$-gon. ◊

Strong Induction

Occasionally, the proof for the case $n = k + 1$ needs to rely on more than just the previous case $n = k$. The Principle of Strong Induction, stated next, justifies this practice. Intuitively, we know that strong induction is similar to ordinary induction, discussed previously. As before, we verify the first step and in turn use the first step to verify the second. Since we now have the first and second steps, we can use both of them in the induction step to verify the third step. Similarly, we can use as many of the first k steps as we need to verify the $k + 1^{st}$ step. This informal reasoning can be formalized into a proof that ordinary induction and strong induction are logically equivalent, although such a proof is probably unhelpful at this level. We will also adopt the Principle of Strong Induction as an axiom. (See Smith et al., 104–105, for a proof of the equivalence of these two types of induction.)

PRINCIPLE OF STRONG INDUCTION (PSI). Suppose for a property $P(x)$,

 i) $P(a)$ is true for some integer a and

 ii) if, for all $i \in \mathbb{Z}$, when $a \leq i \leq k$, $P(i)$ is true, then $P(k + 1)$ is true;

then for all $n \in \mathbb{Z}$, if $n \geq a$, then $P(n)$ is true.

Strong induction is particularly useful with terms defined recursively, since a given term may depend on several preceding terms, as in Example 5. (Section 1.5 discusses recursive definitions.)

EXAMPLE 5

Let $t_1 = 2$, $t_2 = 4$, and $t_3 = 8$, and for $n \geq 1$, define $t_{n+3} = t_{n+2} + t_{n+1} + 2t_n$. Find a pattern for t_n and prove your answer.

The recursion formula gives $t_4 = 8 + 4 + 2 \cdot 2 = 16$, another power of 2. So we guess $t_n = 2^n$.

Proof. To use strong induction, we need to establish the first three cases individually. We see immediately that $t_1 = 2 = 2^1$, $t_2 = 4 = 2^2$, and $t_3 = 8 = 2^3$. Suppose, for i between 1 and k, with $k \geq 3$, that $t_i = 2^i$. Then $t_{k+1} = t_k + t_{k-1} + 2 \cdot t_{k-2} = 2^k + 2^{k-1} + 2 \cdot 2^{k-2} = 2 \cdot 2^{k-1} + 2^{k-1} + 2^{k-1} = 4 \cdot 2^{k-1} = 2^{k+1}$, showing the induction step. ∎ By PSI, $t_n = 2^n$ for all $n \in \mathbb{N}$.

DISCUSSION. In the induction hypothesis we need the assumption $k \geq 3$ in order to use the three prior steps for the recursive definition. In turn, in the initial step we needed to verify individually all the instances before that point. ◊

EXAMPLE 6

Show that you can find a combination of dimes and quarters to equal any multiple of 5 cents larger than 15 cents.

Proof. For the initial steps, we see that 2 dimes are worth 20 cents and a quarter is worth 25 cents. Suppose we can make combinations of dimes and quarters equal to all multiples of 5 cents from 20 cents up to $5k$ cents, where $k \geq 5$. For $5(k + 1)$ cents, write $5(k + 1) = 5(k - 1) + 10$. Since $k \geq 5$, $k - 1 \geq 4$. So $5(k - 1) \geq 20$. By assumption, we have a combination of dimes and quarters worth $5(k - 1)$ cents. Now just add a dime to get $5(k + 1)$ cents. By strong induction, we can achieve all multiples of 5 cents larger than 15 cents. ∎

DISCUSSION. In the induction step, the case $k + 1$ depended only on the case $k - 1$. Because this dependence alternates between two strings of numbers (even and odd values of k), we needed two cases (20 and 25) to get us started and we needed to specify $k \geq 5$ so that we could back up one and still have a previously solved case. ◊

EXAMPLE 7

Can we write every integer greater than 2 as the sum of distinct Fibonacci numbers?

DISCUSSION. Recall the first six Fibonacci numbers are $1, 1, 2, 3, 5$, and 8. Let's first look at several integers. $3 = 2 + 1$, $4 = 3 + 1$, $5 = 3 + 2$, and $6 = 5 + 1$, as well as $6 = 3 + 2 + 1$. So far, so good. To get a feel about how to proceed in general, let's look at a bigger value, say, 31. Perhaps the easiest way to split 31 is to start with the biggest Fibonacci number less than 31, which is 21. Now we need to write what remains, 10, in terms of Fibonacci numbers. Now $10 = 8 + 2$, so $31 = 21 + 8 + 2$. The general situation simply uses the induction step of the induction proof to reduce a given case to an earlier case. We will need strong induction because we might need to go back quite far.

CLAIM. Every integer greater than 2 can be written as the sum of at least two distinct Fibonacci numbers.

Proof. For the initial step of the strong induction proof, note that $3 = 2 + 1$ and 2 and 1 are Fibonacci numbers.

For the induction step, assume that every integer from 3 to k can be written as the sum of distinct Fibonacci numbers. Consider $k + 1$. Let j be the largest Fibonacci number strictly less than $k + 1$. Since $k + 1$ is at least 4, j is at least 3. Consider the difference $d = (k + 1) - j \geq 1$. If $d = 1$ or $d = 2$, then $k + 1 = j + d$, which is a sum of distinct Fibonacci numbers. Otherwise, $3 \leq d \leq k$. By our strong induction hypothesis, we can write d as the sum of distinct Fibonacci numbers. As long as none of the terms in the sum for d is j, we are done with the induction step.

I claim that j is more than half of $k + 1$, so this won't happen. By the definition of Fibonacci numbers, if $j = F_{n+2}$, then $j = F_n + F_{n+1}$, and $j > F_{n+1}$. Suppose, for a contradiction, that $k + 1 \geq 2j$. But then, $k + 1 > j + F_{n+1}$, which would be a larger Fibonacci number than j—a contradiction, since j is the largest such number. So, the terms in the sum for d are all smaller than j, and we have written $k + 1$ as a sum of distinct Fibonacci numbers. By the Principle of Strong Induction, the claim holds for all integers greater than 2. ∎   ◇

The Well Ordering of $\mathbb{N}$

The ordering of the natural numbers has a special property, called well ordering, that sometimes makes an induction proof easier. If you consider any nonempty set of natural numbers, whether finite or infinite, you can always find the smallest one of them, but not necessarily a biggest one. For instance, there is a smallest prime (2), but not a biggest prime (see Theorem 2.5.1). The set $\mathbb{Z}$ of all integers, while ordered, is not well ordered in this sense. The set of odd integers illustrates this, since there is no smallest odd integer. The guarantee of a smallest natural number of any given type depends on how $\mathbb{N}$ is arranged. Every natural number is generated from the first one by adding one. Example 8 uses this well ordering to give a proof of a simplified version of the Fundamental Theorem of Arithmetic, whose proof appears after the problems of this section.

THEOREM 2.4.2. ($\mathbb{N}$ is Well Ordered). If S is any nonempty subset of $\mathbb{N}$, then S has a least element.

Proof. See Problem 22.   ∎

The intuitive idea behind Theorem 2.4.2 is readily understood. Given a subset S of natural numbers, we can seek its smallest element as follows: We first ask "Is $1 \in S$?" If the answer is yes, we're done. Otherwise, we ask "Is $2 \in S$?" In general, when we get another "no" we move to the next-smallest number—corresponding to the induction step. If $S \neq \emptyset$, eventually we get a "yes," and the first such time tells us the smallest element. If we never get a "yes," then S must have been empty. This intuitive idea is not a practical approach (especially for computers). After all, if we get "no" even one trillion times, we can't yet tell whether S is empty or its smallest element is still to come. Of course, the point of the theorem is to avoid this potentially unending process.

EXAMPLE 8

CLAIM. For all natural numbers n, if $n > 1$, then n has a prime factor.

Proof. Let $n \in \mathbb{N}$. Suppose $n > 1$. Let $S = \{x \in \mathbb{N} : x > 1$ and x is a factor of $n\}$. Now, S is nonempty because $n \in S$. By the well ordering of $\mathbb{N}$, S has a least element, which we'll call p. If p is a prime, we're done. For a contradiction, suppose that p is not prime and so has some positive factor f besides 1 and p. That is, f divides p and $1 < f < p$. Since f divides p and p divides n, we see f divides n. These conditions give us $f \in S$ and $f < p$, contradicting the assumption that p is the least element of S. So n has a prime factor. ∎

We can also use strong induction to prove this claim.

Proof. To start our strong induction proof, consider $n = 2$. Since 2 is a prime, we have a prime factor. Now assume that all integers from 2 to k have a prime factor and consider $n = k + 1$. If $k + 1$ is a prime, we are done. Otherwise, $k + 1$ can be factored into integers, say, $k + 1 = c \cdot d$, where $2 \le c \le k$ and $2 \le d \le k$. By our strong induction hypothesis, d has a prime factor e. Since e divides d and d divides $k + 1$, we know e divides $k + 1$, giving a prime factor for $k + 1$. By PSI, every natural number greater than one has a prime factor. ∎

DISCUSSION. Both proofs use the same key idea in different ways: If x divides y and y divides z, then x divides z. We will study this property, called *transitivity*, in Chapter 4. Repeating Example 8 will factor an integer greater than 1 into primes. The Fundamental Theorem of Arithmetic additionally guarantees the uniqueness of this factorization up to the order of the factors. Because this theorem needs several lemmas, we put its proof after the problem set. ◇

Sometimes, as in Example 9, we use the well-ordering property backwards to show that some auxiliary set S is empty. Then $\mathbb{N} - S = \mathbb{N}$.

EXAMPLE 9

CLAIM. For all $n \in \mathbb{N}$, 7 divides $10^n - 3^n$.
For instance, $10^1 - 3^1 = 7$, $10^2 - 3^2 = 100 - 9 = 91 = 7 \cdot 13$, and $10^3 - 3^3 = 1000 - 27 = 973 = 7 \cdot 139$.

Proof. Let $S = \{n \in \mathbb{N} : 7$ does not divide $10^n - 3^n\}$. We'll use well ordering and a proof by contradiction to show that S is empty, which is equivalent to the claim. Suppose, for a contradiction, that S is nonempty. Then it has a smallest element, say, k. Since k is the smallest natural number in S, we see that $k - 1$ is not in S. However, $k > 1$ because we saw $10^1 - 3^1 = 7$. So $k - 1 \in \mathbb{N}$ and 7 divides $10^{k-1} - 3^{k-1}$, say, $10^{k-1} - 3^{k-1} = 7j$. Then $10^{k-1} = 7j + 3^{k-1}$. In turn, $10^k = 10 \cdot 10^{k-1} = 10(7j + 3^{k-1})$. We need to write 10^k as $7h + 3^k$ for some integer h, so we play with the right side of the equation. We have $10(7j + 3^{k-1}) = 7 \cdot 10j + 10 \cdot 3^{k-1} = 7 \cdot 10j + (7 + 3)3^{k-1} = 7(10j + 3^{k-1}) + 3^k$. So, $10^k - 3^k = 7(10j + 3^{k-1})$. But then $k \notin S$. This contradiction forces S to be empty, showing the claim. ∎

DISCUSSION. The connection between the $k - 1$ and k terms in the preceding proof could be turned into the induction step of a regular induction proof. The proof using the well ordering property for Lemma 2.4.4 in the Appendix is much harder to turn into an induction proof. ◇

While induction and its variations provide convincing proofs of the truth of propositions, they sometimes fail to give any insight about why the propositions should be true. The proof in Example 1 seems to have this failing, compared with Figure 1. However, the proof in Example 8 helps clarify why the claim is true. Since mathematicians value both understanding and proof, they sometimes prefer other proof techniques over induction proofs. Nevertheless, induction is a valuable technique for proving properties about integers.

Historical Remarks

Mathematical induction developed more slowly than the proof formats discussed in the first three sections of this chapter. The Arabic mathematician al-Karaji, who died in 1019, gave one of the oldest arguments based on something very like induction. His argument for the formula of the sum of ten cubes ($\sum_{i=1}^{10} i^3$ in modern notation) depended on reducing each such sum to a smaller sum and showing the base case. It is worth pointing out that algebraic notation wasn't developed until more than five hundred years later. In 1321, Levi ben Gerson, a Jewish mathematician, gave several induction arguments in a format close to our modern one, although still without algebraic symbols. There is little evidence of later mathematicians building on these earlier efforts. With the work of Blaise Pascal (1623–1662) proofs by induction enter standard mathematical practice. Giuseppe Peano (1858–1932) laid out axioms for the natural numbers and formalized induction, which had become universally accepted much earlier.

Some historians of mathematics consider some proofs by contradiction forerunners of induction proof. In particular, they cite Euclid's proof of the infinitude of primes, given at the start of Section 2.5.

Induction Proof Formats

Type	Initial step	Induction step
$\forall n \in \mathbb{N}\ P(n)$. (Ordinary)	Show $P(1)$.	Assume $P(k)$, prove $P(k+1)$.
$\forall n \in \mathbb{Z}$ if $n \geq b$, then $P(n)$. (Modified)	Show $P(b)$.	Assume $P(k)$, prove $P(k+1)$.
$\forall n \in \mathbb{Z}$ if $n \geq b$, then $P(n)$. (Strong)	Show $P(b)$.	Assume $P(j)$, for $b \leq j \leq k$. prove $P(k+1)$.

PROBLEMS

*1. For each statement that follows, decide whether it is true or (at least sometimes) false. If false, explain why. Explain your answer or cite the part of the text supporting it.

a) Since scientific induction doesn't prove anything, neither does mathematical induction.

b) The initial step of an induction proof could start with a negative integer.

c) Strong induction is logically stronger than regular induction.

d) Every subset of $\mathbb{N}$ has a smallest element.

e) Induction proofs sometimes sacrifice insight for rigor.

f) The real numbers are well ordered using $\leq$.

2. Fill in the blanks to complete the given proofs. For part a, check out the first few cases to verify the formula. See Problem 21 for a way to find the formula.

a) Claim: For all $n \in \mathbb{N}$, $1 + 4 + 7 + 10 + \ldots + (3n + 1) = \sum_{i=0}^{n}(3i + 1) = \frac{3}{2}n^2 + \frac{5}{2}n + 1$.
Proof by induction. For $n = 1$, we have ___ . Suppose for $n = k$ that ___ . For $n = k + 1$, we have $\sum_{i=0}^{k+1}(3i + 1) = \sum_{i=0}^{k}(3i + 1) + (3(k + 1) + 1) = $ ___ . Using algebra, this becomes $\frac{3}{2}k^2 + \frac{11}{2}k + 5$. Now consider $\frac{3}{2}(k + 1)^2 + \frac{5}{2}(k + 1) + 1$, which equals ___ . By ___, we can conclude___ . ∎

*b) Claim: For n a nonnegative integer, $n < 2^n$ (referred to in Section 8.4).
Proof by induction. For the initial case, we use $n = $ ___, in which case ___ . Suppose for $n = k$, with $k \geq 0$, that ___ . Then $k + 1 \leq k + k = 2k < $ ___ . By ___, we can conclude ___ . ∎

c) Define $s_1 = 1$, $s_2 = 4$, and $s_3 = 9$ and for $n \geq 1$, $s_{n+3} = s_n - s_{n+1} + s_{n+2} + 2(2n + 3)$.
Claim: For all $n \in \mathbb{N}$, $s_n = n^2$.
Proof: For $n = 1, 2, 3$ and 4 ___ . Suppose $s_n = n^2$ for $1 \leq n \leq k + 3$ and $k \geq 1$. Then $s_{(k+1)+3} = $ ___ $= (k + 1)^2 - $ ___ $+ $ ___ $+ 2(2(k + 1) + 3) = k^2 + $ ___ $= ($ ___ $)^2$. Then by ___ . ∎

3. Prove each proposition by using induction. You may find it helpful to write out the first few terms, as is done in part a.

   *a) For all $n \in \mathbb{N}$, $2 + 5 + 8 + \ldots + (3n - 1) = \sum_{i=1}^{n}(3i - 1) = (3n + 1)n/2$.

   b) For all $k \in \mathbb{N}$, $\sum_{j=0}^{k}(4j + 3) = 2k^2 + 5k + 3$.

   c) For all $n \in \mathbb{N}$, $\sum_{k=0}^{n} 2^k = 2^{n+1} - 1$.

   d) For all $s \in \mathbb{N}$, $\sum_{i=0}^{s}(\frac{3}{2})^i = 2(\frac{3}{2})^{s+1} - 2$.

   e) For all $t \in \mathbb{N}$, $\sum_{i=1}^{t} i^2 = t(t + 1)(2t + 1)/6$.

   f) The sum of the first n odd squares is $\frac{4}{3}n^3 - \frac{1}{3}n$.

   g) For all $k \in \mathbb{N}$, $(1 + \frac{1}{2})^k \geq 1 + \frac{k}{2}$.

   h) For all $n \in \mathbb{N}$, $(1 + \frac{5}{3})^{2n} \geq 1 + \frac{10n}{3}$.

   i) (Geometric series) For all $n \in \mathbb{N}$ and all $a, r \in \mathbb{R}$, if $r \neq 0$ and $r \neq 1$, then $\sum_{i=0}^{n} ar^i = a\frac{(r^{n+1} - 1)}{r - 1}$ (referred to in Section 8.4).

   j) For all $j \in \mathbb{N}$, 3 divides $2j^3 + 4j + 9$.

   k) For all $k \in \mathbb{N}$, 4 divides $5^k - 1$.

   *l) For all $n \in \mathbb{N}$, 7 divides $11^n - 4^n$.

   m) For all $m \in \mathbb{N}$, if m is even, then $5^m \equiv 1 \pmod 6$.

   n) For all $r, s \in \mathbb{Z}$, and all $t \in \mathbb{N}$, if $r > s$, then $r^t \equiv s^t \pmod{r - s}$.

   o) (De Moivre's Theorem) For all $n \in \mathbb{N}$ and all $\alpha \in \mathbb{R}$, $(\cos(\alpha) + i \sin(\alpha))^n = \cos(n\alpha) + i \sin(n\alpha)$. Recall that i is the complex square root of -1. (Hint: See Problem 8 of Section 2.1.)

4. Prove each proposition, using the modified principle of induction.

   *a) For all $n \in \mathbb{N}$, if $n \geq 5$, then $2^n > n^2$.

   b) For all $k \in \mathbb{N}$, if $k \geq 4$, then $2^k < k!$.

   c) For all $z \in \mathbb{Z}$, if $z \geq 0$, then $\begin{bmatrix} 1 & 0 \\ 1 & 1 \end{bmatrix}^z = \begin{bmatrix} 1 & 0 \\ z & 1 \end{bmatrix}$.

   d) Part c holds for all $z \in \mathbb{Z}$. Extend your proof in part c to z negative by another induction on z.

   e) For all $n \in \mathbb{N}$, if $n > 5$, then $\sum_{i=1}^{n} \ln(i) > n$.

   f) Determine the smallest value of $c \in \mathbb{N}$ such that for all $n \in \mathbb{N}$, if $n > c$, then $n! > 3^n$. Prove your answer.

5. From the partial sums of the first few terms for each sequence, write the general sum, make a conjecture, and prove it by using induction.

   a) $1, 1 + 3, 1 + 3 + 5, 1 + 3 + 5 + 7, \ldots$

   b) $1, 1 + 8, 1 + 8 + 27, \ldots$

   c) $\frac{1}{1 \cdot 2}, \frac{1}{1 \cdot 2} + \frac{1}{2 \cdot 3}, \frac{1}{1 \cdot 2} + \frac{1}{2 \cdot 3} + \frac{1}{3 \cdot 4}, \ldots$

   d) $1 \cdot 1!, 1 \cdot 1! + 2 \cdot 2!, 1 \cdot 1! + 2 \cdot 2! + 3 \cdot 3!, \ldots$

6. Assume the following calculus properties:

   i) The derivative of $j(x) = x$ is 1.

   ii) Product rule for derivatives: If the derivatives of $g(x)$ and $h(x)$ are $g'(x)$ and $h'(x)$, respectively, then the derivative of $g(x) \cdot h(x)$ is $g'(x) \cdot h(x) + h'(x) \cdot g(x)$.

   Prove, for all $n \in \mathbb{N}$, that the derivative of x^n is nx^{n-1}.

7. *a) We can divide a large square into an $n \times n$ "checkerboard" of smaller squares. How many squares of any size do the lines of the $n \times n$ checkerboard form? Find and prove a formula for this number.

   b) Repeat part a for rectangles of any size formed by the lines of the $n \times n$ square checkerboard.

   c) We can similarly divide an equilateral triangle into many smaller equilateral triangles with n small triangles along each side of the original triangle. How many triangles of any size are formed by the lines so drawn? Find and prove a formula for this number.

   d) We can similarly divide a large cube into smaller cubes, n to a side. How many cubes of any size are thus formed from the lines dividing the big cube into smaller cubes? Find and prove a formula for this number.

8. We can build a (triangular) pyramid out of balls by forming a triangle of balls for a base with n balls on each side of the triangle. Then in the "hollows" of that layer we can place balls forming a smaller triangle with $n - 1$ balls on a side. We continue until we end up with just one ball in the top layer. Find and prove a formula in terms of n for the total number of balls in such a pyramid.

9. Prove the following properties for the Fibonacci numbers, f_n, defined in Section 1.5:

   *a) The sum of the first n Fibonacci numbers equals $f_{n+2} - 1$.

   b) The greatest common divisor of f_n and f_{n+1} is 1.

   c) The greatest common divisor of f_n and f_{n+2} is 1.

   d) The nth Fibonacci number f_n is even iff 3 divides n.

   e) $f_{n+6} = 4f_{n+3} + f_n$.

10. Define the *alternating Fibonacci numbers* a_n by $a_1 = 1, a_2 = -1$, and $a_{n+2} = a_n - a_{n+1}$.

    a) Find the value of a_n from $n = 3$ to $n = 7$.

    b) Prove for all $n \in \mathbb{N}$, $a_{2n-1} > 0$ and $a_{2n} < 0$.

*11. Define the numbers b_n recursively as follows: $b_1 = 1, b_2 = 2, b_3 = 3$; and for $n \geq 1$, $b_{n+3} = b_n + b_{n+1} + b_{n+2}$. Prove for all $n \in \mathbb{N}$ that $b_n \leq 2^{n-1}$.

12. Define the numbers c_n recursively as follows: $c_1 = 1, c_2 = 1$; and for $n \geq 1, c_{n+2} = 1/(c_n + c_{n+1})$. Prove for all $n \in \mathbb{N}$ that $\frac{1}{2} \leq c_n \leq 1$.

13. Define the numbers d_n recursively as follows: $d_1 = 7, d_2 = 6$; and for $n \geq 1, d_{n+2} = 2d_{n+1} - d_n$. Find a formula for d_n and prove your formula holds for all $n \in \mathbb{N}$.

14. Define the numbers e_n recursively as follows: $e_1 = 1$, $e_2 = 2$; and for $n \geq 1$, $e_{n+2} = (e_n + e_{n+1})/2$, the average of the two previous terms. Prove that the odd terms are increasing, the even terms are decreasing, and an odd term is less than the successor even term. That is, for all $n \in \mathbb{N}$, prove that $e_{2n-1} \leq e_{2n+1}$, $e_{2n} \geq e_{2n+2}$, and $e_{2n-1} < e_{2n}$. To what number do the terms e_n get increasingly close?

*15. An eccentric post office has only stamps worth 4 cents and stamps worth 7 cents. Find the largest integer stamp value that can't be made with these stamps, and prove that all larger integer values can be made.

16. Prove that the number of prime factors of an integer n greater than 1 is at most $\log_2 n$. (For instance, 12 has three factors, since $12 = 2 \cdot 2 \cdot 3$ and $3 < \log_2 12 \approx 3.585$.)

17. **a)** Use induction instead of the Division Algorithm to prove for all $n \in \mathbb{N}$, that n is even or n is odd, but not both.

   b) Extend your proof to all integers.

18. We know that every rational number is equal to infinitely many fractions. For instance, $\frac{2}{3} = \frac{10}{15} = \frac{8}{12}$. Use the well ordering of $\mathbb{N}$ to prove that for every rational q there is a smallest natural number n and some integer k such that $q = \frac{k}{n}$.

19. In a round robin tournament, every player plays every other. Assume that no match can end in a draw. Such a tournament doesn't have to have someone winning every match she or he plays. Define a person A to be an *almost winner* iff, for every B in the tournament, with $B \neq A$, then A beats B directly or there is some C such that A beats C and C beats B. Use the well ordering of $\mathbb{N}$ to show that every such tournament has an almost winner.

20. Some books define a set S of real numbers to be *inductive* iff for all $r \in \mathbb{R}$, if $r \in S$, then $r + 1 \in S$. Which of the given sets are inductive? For which ones could you use an induction proof to prove a property? Explain.

   *a) $\mathbb{R}$

   *b) $[1, 100]$

   c) $\emptyset$

   d) $\mathbb{Z}$

   e) $[10, \infty)$

   f) $\mathbb{N} \cup \{-3\}$.

   g) $[-123, \infty) \cap \mathbb{Z}$

   h) For any sets S and T, if S and T are inductive, is $S \cup T$? $S \cap T$? $S - T$? Explain.

21. In this problem we investigate the relationship between a function on integers and the "derived function" of its differences. For instance, let $f(n) = n^2$ and define the differences by $d(n) = f(n) - f(n-1)$. Then $d(n) = n^2 - (n-1)^2 = n^2 - (n^2 - 2n + 1) = 2n - 1$. While $2n - 1$ is not the derivative of n^2, it is related.

   *a) Let $f(n) = an^2 + bn + c$. Find the formula for $d(n) = f(n) - f(n-1)$.

   b) Suppose we know for a formula $f(n)$ that $f(0) = r$ and $d(n) = pn + q$. Use induction to show that $f(n) = \frac{p}{2}n^2 + (\frac{p}{2} + q)n + r$.

   c) Let $f(n) = an^3 + bn^2 + cn + d$. Find the formula for $d(n) = f(n) - f(n-1)$.

   d) Suppose for a formula $f(n)$ we know that $f(0) = s$ and $d(n) = pn^2 + qn + r$. Find a formula for $f(n)$ and prove your formula by induction. (A similar technique works for any degree function $d(n)$ of differences. Intelligence tests often have questions asking for the next number in a sequence. The pattern of differences is often easier to find and can reveal the original pattern.)

22. Complete the following outline to use the Principle of Strong Induction to prove the well ordering of $\mathbb{N}$ (Theorem 2.4.2):

Outline of Proof. Let S be a nonempty subset of $\mathbb{N}$ and, for a contradiction, assume that it has no least element. We consider the set $\mathbb{N} - S$. Why is $1 \in \mathbb{N} - S$? Suppose all natural numbers less than or equal to k are in $\mathbb{N} - S$. Prove $k + 1 \in \mathbb{N} - S$. What does the Principle of Strong Induction tell us about $\mathbb{N} - S$? Why does this give us a contradiction about S?

23. Critique the incorrect "proofs" that follow. Point out reasoning errors and unclear or incomplete presentations. (If the claim is false, there must be an error in the argument. However, it is not enough to say that the claim is false. You need to find an error in the reasoning.)

a) Claim: For n, any natural number, $\sum_{i=1}^{n} 2i = n(n + 1)$. "Proof": For the initial step, $\sum_{i=1}^{1} 2i = 2 = 1(1 + 1)$. Similarly, $\sum_{i=1}^{2} 2i = 2 + 4 = 6 = 2(3)$. For the induction step, $2 + 1 = 3$ and $\sum_{i=1}^{3} 2i = \sum_{i=1}^{2} 2i + 6 = 2(3) + 6 = 12 = 3(4)$. By the PMI, the formula holds for all n. ▲

b) Claim: A set with n elements has $\frac{n(n-1)}{2}$ distinct subsets with two elements. "Proof": For the initial step for induction, let $n = 1$. Then $\frac{1(1-1)}{2} = 0$. For the induction step, suppose a set with k elements has $\frac{k(k-1)}{2}$ distinct subsets with two elements. Then, for a set with $k + 1$ elements, we have $\frac{k(k-1)}{2} + k = \frac{k^2 - k + 2k}{2} = \frac{(k+1)k}{2}$ distinct subsets with two elements, finishing the induction step. By PMI, the formula holds for all n. ▲

*c) Claim: For all $n \in \mathbb{N}$, the sum of the first n powers of 2 is $2 \cdot 2^n$. "Proof": Suppose $\sum_{i=1}^{n} 2^i = 2 \cdot 2^n$. Then $\sum_{i=1}^{n+1} 2^i = \sum_{i=1}^{n} 2^i + 2^{n+1} = 2 \cdot 2^n + 2^{n+1} = 2 \cdot 2^{n+1}$. By PMI, the result holds for all $n \in \mathbb{N}$. ▲

d) Claim: For all $n \in \mathbb{N}$, $\begin{bmatrix} 1 & 1 \\ 0 & 1 \end{bmatrix}^n = \begin{bmatrix} 1 & n \\ 0 & 1 \end{bmatrix}$. "Proof": Let S be the set of all n in $\mathbb{N}$ for which $\begin{bmatrix} 1 & 1 \\ 0 & 1 \end{bmatrix}^n \neq \begin{bmatrix} 1 & n \\ 0 & 1 \end{bmatrix}$. If $S = \emptyset$, we're done. So by the well ordering of $\mathbb{N}$, suppose s is the least element of S. Then $\begin{bmatrix} 1 & 1 \\ 0 & 1 \end{bmatrix}^{s+1} \neq \begin{bmatrix} 1 & s \\ 0 & 1 \end{bmatrix}\begin{bmatrix} 1 & 1 \\ 0 & 1 \end{bmatrix} = \begin{bmatrix} 1 & s + 1 \\ 0 & 1 \end{bmatrix}$, which is a contradiction. Hence, the property holds for all $n \in \mathbb{N}$. ▲

e) Claim: All mathematicians own the same number of books. "Proof" by induction on the number of mathematicians: If there is just one mathematician, clearly all mathematicians have the same number of books. Now suppose that in every collection of k mathematicians they all have the same number of books and consider $k + 1$ mathematicians. Split off one mathematician, say, X, so that the others all have the same number of books. Now pick a different mathematician Y to split off. By the induction hypothesis again, all the mathematicians except Y have the same number of books, including X. However, Y already has the same number of books as all but perhaps X, so in fact all $k + 1$ mathematicians have the same number of books. By PMI, all mathematicians have the same number of books. ▲

24. Critique the questionable "proof" that follows. Point out reasoning errors and unclear or incomplete presentations. If the argument is a proof, say so. (If the claim is false, there must be an error in the argument. However, it is not enough to say that the claim is false. You need to find an error in the reasoning.)

a) Claim: For all $n \in \mathbb{N}$, $\sum_{k=1}^{n} k! < (n + 1)!$. "Proof": Clearly, $1! = 1 < 2 = 2!$, so the base case holds. Suppose now that $\sum_{k=1}^{j} k! < (j + 1)!$ and we consider the $j + 1$

case. Note that $(j+2)! = (j+1+1)(j+1)! = (j+1)(j+1)! + (j+1)!$. Then $\sum_{k=1}^{j+1} k! = \sum_{k=1}^{j} k! + (j+1)! < (j+1)! + (j+1)!$. Since $j \geq 1$, this last sum is less than $(j+2)!$, from the earlier equality. Hence, the PMI shows the property holds for all $n \in \mathbb{N}$. ▲

b) Claim: The sum of the first n odd squares is $\frac{4}{3}n^3 - \frac{1}{3}n$. "Proof" by induction: For $n = 1$, $1^2 = 1$ and $\frac{4}{3}(1) - \frac{1}{3}(1) = 1$. Suppose for $n = k$ that the sum of the first k odd squares is $\frac{4}{3}k^3 - \frac{1}{3}k$. Then the first $k+1$ odd squares gives $\frac{4}{3}k^3 - \frac{1}{3}k + (k+1)^2 = \frac{4}{3}(k+1)^3 - \frac{1}{3}(k+1)$, as desired. By the PMI, the property holds for all n. ▲

c) Claim: For all natural numbers n, if $n > 1$, then $\sum_{i=1}^{n}(4i - 1) = 2n^2 + n - 1$. "Proof": Let $n \in \mathbb{N}$. For the initial step for induction $n = 1$, the hypothesis is false; so even though the conclusion is false, the whole proposition is true. For the induction step, assume for $n = k$ that $\sum_{i=1}^{k}(4i - 1) = 2k^2 + k - 1$ and consider $n = k + 1$. Then $\sum_{i=1}^{k+1}(4i - 1) = \sum_{i=1}^{k}(4i - 1) + (4(k+1) - 1) = (2k^2 + k - 1) + 4k + 3 = 2k^2 + 5k + 2$. Also, $2(k+1)^2 + (k+1) - 1 = 2(k^2 + 2k + 1) + k = 2k^2 + 5k + 2$, showing the induction step holds. By PMI, the equality is proven. ▲

d) Define $b_1 = 1$, $b_2 = 2$ and $b_{n+2} = \frac{b_n + b_{n+1}}{2}$, the average of the two preceding terms. Claim: The odd terms are increasing. "Proof": To start, we note that $b_1 = 1 < 1.5 = \frac{1+2}{2} = b_3$, so the first pair of odd terms increases. Suppose for $n = k$ that the kth pair of odd terms increases: $b_{2k-1} < b_{2k+1}$. Consider the next odd pair: b_{2k+1} and b_{2k+3}. Now b_{2k+3} is the average of b_{2k+1} and b_{2k+2}. The terms are increasing by the induction hypothesis. So, $b_{2k+1} < b_{2k+2}$. Since the average is between the two numbers being averaged, we have $b_{2k+1} < b_{2k+3}$. By PMI, the inequality holds for all $n \in \mathbb{N}$. ▲

*e) Claim: For all $n \in \mathbb{N}$, $n! + 1$ is prime. "Proof": When $n = 1$, we see that $1! + 1 = 2$, which is prime. Suppose for the induction step that $k! + 1$ is a prime. Consider $(k+1)! + 1$. Since every number from 1 to $k + 1$ divides $(k+1)!$, they don't divide $(k+1)! + 1$. Thus, $(k+1)! + 1$ is not divisible by any number from 1 to $(k+1)!$. Hence, it is prime. By PMI, the property holds for all natural numbers n. ▲

f) Claim: $\sqrt{2}$ is irrational. "Proof": We use the well ordering of $\mathbb{N}$. Let $S = \{q \in \mathbb{N} : \exists p \in \mathbb{N} \, s.t. \, \frac{p}{q} = \sqrt{2}\}$. We need to show that S is empty. Suppose, for a contradiction, that it isn't. Then it has a smallest element q and a corresponding p so that $\frac{p}{q} = \sqrt{2}$. Squaring both sides and getting rid of the denominator gives us $p^2 = 2q^2$. Now p is either odd or even. Since p^2 is even (it is twice q^2), p is even. So we can write $p = 2r$ for some $r \in \mathbb{N}$. Thus, $2q^2 = p^2 = (2r)^2 = 4r^2$. Thus, $q^2 = 2r^2$. In turn, this means q is even, say, $q = 2s$. But then $\frac{p}{q} = \frac{2r}{2s} = \frac{r}{s}$, and so q isn't the smallest element of S, a contradiction. So $S = \emptyset$ and $\sqrt{2}$ is irrational. ▲

THE FUNDAMENTAL THEOREM OF ARITHMETIC

The proof of the Fundamental Theorem of Arithmetic requires a number of preparatory theorems. Rather than overload the text with these number theory results, I have chosen to place them in this section. Lemmas 2.4.4 and 2.4.6 are valuable in proving many other results.

LEMMA 2.4.3. For all $x, y \in \mathbb{N}$, if x divides y, then $x \leq y$.

Proof. Suppose that $x, y \in \mathbb{N}$ and x divides y. Then there is some $z \in \mathbb{Z}$ such that $xz = y$. Since both x and y are positive, z must be as well. Thus, $1 \leq z$. Multiplying both sides of $1 \leq z$ by the positive number x gives $x \leq xz = y$, as desired. ∎

LEMMA 2.4.4.   For all $a, b \in \mathbb{Z}$, with $a \neq 0$, there are $k, n \in \mathbb{Z}$ such that $\gcd(a, b) = ka + nb$.

Proof. Let $a, b \in \mathbb{Z}$ and suppose $a \neq 0$. Then $\gcd(a, b)$ exists. To use the well ordering of $\mathbb{N}$, consider $S = \{ia + jb : i, j \in \mathbb{Z} \text{ and } ia + jb > 0\}$, the positive integers "built" from a and b. (For instance, for $a = 6$ and $b = 10$, $2 \in S$ because $2 = 2 \cdot 6 - 1 \cdot 10$. Similarly, $3 \cdot 6 + 5 \cdot 10 = 68 \in S$.) We need to show that $\gcd(a, b) \in S$. First, to show that S is nonempty, consider $i = a$ and $j = b$, giving $a^2 + b^2$, which is positive since $a \neq 0$. By Theorem 2.4.2, S has a least element, which we'll call d. Thus, d can be "built" from a and b. That is, there are integers k and n such that $d = ka + nb$. We claim that $d = \gcd(a, b)$. By the Division Algorithm, we can write $a = dq + r$ with $0 \leq r < d$. Solving for r gives $r = a - dq = a - (ka + nb)q = (1 - k)a + (-nq)b$. If $r > 0$, this rewriting would force r to be in S as well as $r < d$. But d was assumed to be the smallest element. Hence, $r = 0$. In turn, this tells us that d divides a. Similarly, d must divide b, making d a common divisor, and so $d \leq \gcd(a, b)$. Since $\gcd(a, b)$ divides both a and b, by Problem 5c of Section 2.1, it divides every element of S, including d. Thus $\gcd(a, b) \leq d$, and so $\gcd(a, b) = d \in S$. ∎

LEMMA 2.4.5.   For all $a, b, c \in \mathbb{Z}$, if a divides bc and $\gcd(a, b) = 1$, then a divides c.

Proof. Let $a, b, c \in \mathbb{Z}$ and suppose $a|bc$ and $\gcd(a, b) = 1$. From Lemma 2.4.4, we can find $k, n \in \mathbb{Z}$ such that $1 = ka + nb$ or $c = c1 = c(ka + nb) = cka + cnb$. From $a|bc$ we can find an integer s such that $as = bc$. Substituting, we have $c = cka + nas = a(ck + ns)$, showing that a divides c. ∎

LEMMA 2.4.6.   For all p prime and all $b, c \in \mathbb{Z}$, if p divides bc, then p divides b or p divides c.

Proof. Let p be a prime and b and c integers with $p|bc$. If p divides b, we are done. So suppose p doesn't divide b. The only positive divisors of p are 1 and p, so these are the only candidates for positive common divisors of p and b. Since p is eliminated, $\gcd(p, b) = 1$. From Lemma 2.4.5, p divides c. ∎

THEOREM 2.4.7. (The Fundamental Theorem of Arithmetic). Every natural number n greater than 1 is prime or can be factored into primes in a unique way. That is, if $n = q_1 q_2 \ldots q_k$ and $n = r_1 r_2 \ldots r_j$, then $k = j$; and if we list the primes in increasing order, for each i we have $q_i = r_i$.

Proof. For the initial step in strong induction, the first integer greater than 1 is $n = 2$. Since 2 is prime, the only factors of 2 are itself and 1. So this factorization is unique. Now suppose for $2 \leq j \leq k$, we can factor j uniquely into primes. Consider $k + 1$. If $k + 1$ is prime, it is already factored into primes and its only factors are itself and 1, so the prime factorization is unique. So we may suppose $k + 1 = qr$, where $1 < q, r < k + 1$. The strong induction hypothesis allows us to factor q and r uniquely into primes, say, $q = q_1 q_2 \ldots q_k$ and $r = r_1 r_2 \ldots r_j$. Clearly, $k + 1 = q_1 q_2 \ldots q_k r_1 r_2 \ldots r_j$ is a factoring into primes. Suppose we have another prime factorization, $k + 1 = s_1 s_2 \ldots s_n$. Now s_1 is a prime and it divides qr. By Lemma 2.4.6, either s_1 divides q or it divides r, making s_1 one of the prime factors q_i or r_h. Without loss of generality, we can say $s_1 = q_1$. Now consider $(k + 1)/s_1 = s_2 \ldots s_n = q_2 \ldots q_k r_1 r_2 \ldots r_j$, an integer greater than 1. By the strong induction hypothesis, the lists of primes $s_2 \ldots s_n$ and $q_2 \ldots q_k r_1 r_2 \ldots r_j$ are identical except for their order. When we add s_1 back in, we see that the two factorizations of $k + 1$ also have the same factors except for their order. By PSI, our theorem holds for all integers n greater than 1. ∎

REFERENCE

Smith, D., M. Eggen, and R. St. Andre, 2001. *A transition to advanced mathematics*, 5th ed. Pacific Grove, CA.: Brooks/Cole.

2.5 FURTHER ADVICE AND PRACTICE IN PROVING

"A good proof is one that makes us wiser."

—Yu I. Manin (1937–)

Before we look at more aspects of theorem proving, let's pause to enjoy one of the rewards of mathematics—a beautiful proof. Euclid (circa 300 B.C.), in the most famous mathematics book of all time, *Elements*, uses some of our previous results to show the infinitude of primes in a most elegant way. (More accurately, Euclid showed that there are more primes than any given finite number since the Greeks thought in terms of potential infinities.)

THEOREM 2.5.1. (Euclid).   There are infinitely many primes.

Proof.   For a contradiction, suppose that there were only finitely many primes, say, $p_1, p_2 \ldots p_n$. Each of these primes divides their product $P = p_1 \cdot p_2 \cdot \ldots p_n$. Then, by the Division Algorithm (Theorem 2.4.1), dividing $P + 1$ by any of the primes p_i gives a remainder of 1, so no p_i divides $P + 1$. But by Example 8 of Section 2.4, $P + 1$ has a prime factor, which can't be in our list, giving us a new prime and a contradiction. Thus, there must be infinitely many primes. ∎

DISCUSSION.   For the first two primes $p_1 = 2$ and $p_2 = 3$, the preceding construction gives $P + 1 = 2 \cdot 3 + 1 = 7$, a prime. The first six primes $p_1 = 2$, $p_2 = 3$, $p_3 = 5$, $p_4 = 7$, $p_5 = 11$, and $p_6 = 13$ give $P + 1 = 2 \cdot 3 \cdot 5 \cdot 7 \cdot 11 \cdot 13 + 1 = 30031 = 59 \cdot 509$, and 59 and 509 are both primes. This proof is not naïvely constructive, since it doesn't give us an explicit list of infinitely many primes or a practical way to find more primes. Yet it is completely convincing because it confronts the concept of infinity head on through a proof by contradiction.

The irregular appearance of primes continues to fascinate some people, who now use thousands of linked computers to find ever bigger ones. The current biggest known prime (see Problem 11) has over nine million digits, but computers can't help us jump from individual primes, no matter how large, to the actual infinitude of primes. Euclid's beautiful old proof thus easily surpasses the work of the fastest, most modern computers, illustrating the Manin quote at the beginning of this section.

The Role of Figures in Proofs

Manin's quote expects proofs to do more than just convince us of the truth of results— they ought to explain why the results are true. In geometry and some other areas, such understanding sometimes comes more readily from studying a figure than by following abstract reasoning. (Some examples in previous sections used figures for just this purpose.) Yet, figures contain their own pitfalls. Most importantly, we easily overlook hidden assumptions implicit in the figure. This tension between understanding and rigor in proof stretches back to the ancient Greeks, who were the first to strive to prove theorems. Indeed, the very first proof in Euclid's great book doesn't follow from his postulates (axioms), as Example 1 explains.

EXAMPLE 1

Euclid based his first proof on these two postulates:

- "To draw a straight line [segment] from any point to any point."
- "To describe [draw] a circle with any center and distance [radius]."

PROPOSITION 1, BOOK I (Euclid). "On a given finite straight line, to construct an equilateral triangle." (Heath, 195, 199, 241)

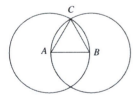

FIGURE 1

As in Figure 1, Euclid started with a line segment $\overline{AB}$ and used his circle postulate twice to construct circles with radius AB and centers A and B. He then chose one of the points of intersection of these two circles, say, C. Next, he used the first postulate to draw the line segments $\overline{AC}$ and $\overline{BC}$. Finally, he reasoned that all three line segments are equal in length, making an equilateral triangle. ▲

Euclid's postulates came from the tradition of compass and straight edge constructions. Unfortunately, none of his postulates allowed him to deduce the intersection of the two circles, which depended on a hidden assumption, but was obvious from the figure. This assumption apparently remained hidden for two thousand years after Euclid (see [Heath, 234–242]).   ◊

We shouldn't let modern standards of rigor diminish our appreciation of Euclid's great achievements, but we need to protect our own reasoning from inappropriate dependence on figures. How can we avoid such slips? At the extreme, mathematical logic formalizes proofs as strings of symbols with strict rules of deduction. Such formalism enables logicians to prove theorems about proofs, but it totally defeats the purposes of providing proofs for humans and of making us wiser. Instead, we must balance our needs for intuition and rigor. We can first grasp the heart of a proof by using the figure. Then we need to study the proof critically to ensure that its reasoning depends only on what is written, not the arrangement of the figure, as illustrated in Example 2.

EXAMPLE 2

Using calculus, prove that for all $n \in \mathbb{N}$, $\ln(n) \leq \sum_{i=1}^{n} \frac{1}{i} \leq 1 + \ln(n)$.

Checking a few early terms indicates these inequalities are correct without revealing why: $\ln(1) = 0 \leq 1 \leq 1 + \ln(1)$ and $\ln(4) \doteq 1.38 \leq 1 + \frac{1}{2} + \frac{1}{3} + \frac{1}{4} \doteq 2.08 \leq 1 + \ln(4) \doteq 2.38$. The two graphs in Figure 2 reveal the key idea from calculus: We can relate the sum to the area under a sequence of rectangles and the natural logarithms to the area under the curve $y = \frac{1}{x}$. Once we have the idea, the proof is not too hard to follow.

Proof. The curve $y = \frac{1}{x}$ is decreasing from $x = 1$ to $x = n$. Therefore, the left-hand approximations of $\int_1^n \frac{1}{x} dx$ will be too big and the right-hand approximations too small. That is, for all x in an interval $[i, i + 1]$, we have $\frac{1}{i+1} \leq \frac{1}{x} \leq \frac{1}{i}$ and $\sum_{i=1}^{n-1} \frac{1}{i+1} \leq \int_1^n \frac{1}{x} dx \leq \sum_{i=1}^{n-1} \frac{1}{i}$. Now $\int_1^n \frac{1}{x} dx = \ln(n)$, so $\sum_{i=1}^{n-1} \frac{1}{i+1} \leq \ln(n) \leq \sum_{i=1}^{n-1} \frac{1}{i}$. If we add 1 to the left and middle terms, we get $\sum_{i=1}^{n-1} \frac{1}{i+1} \leq 1 + \ln(n)$. Now we shift the variable i by one to get $\sum_{i=1}^{n} \frac{1}{i} \leq 1 + \ln(n)$, which is the second of the desired inequalities. For the remaining inequality, look at the other approximation $\ln(n) \leq \sum_{i=1}^{n-1} \frac{1}{i}$. By adding one more term to the right side, we get the first desired inequality $\ln(n) \leq \sum_{i=1}^{n} \frac{1}{i}$. ∎

DISCUSSION. The inequalities $\sum_{i=1}^{n-1} \frac{1}{i+1} \leq \int_1^n \frac{1}{x} dx \leq \sum_{i=1}^{n-1} \frac{1}{i}$ in the preceding proof come directly from Figure 2, or more accurately, the approximations the figure represents. The rest of the proof depends on comparing these inequalities with the desired ones in order to see what adjustments need to be made.

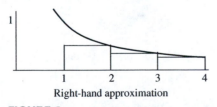

| Right-hand approximation | Left-hand approximation |

FIGURE 2 ◇

Arguments Involving Symmetry

Mathematicians use other ways than figures to convey the heart of proofs. One approach focuses on the key or representative case, rather than giving all possible cases when the proofs are essentially the same. The phrase "without loss of generality," sometimes abbreviated WLOG, often introduces such a focus. The reader needs to reflect on what the other options are and why this one adequately represents all possibilities. Other terms indicating the reduction of cases include "similarly" and "by symmetry." Of course, leaving out cases may seem like laziness on the author's part, but this sort of laziness can be a virtue if it promotes understanding. Sometimes the representation chosen greatly simplifies the proof, as in Example 3.

EXAMPLE 3

CLAIM.   Given n real numbers between 0 and 1, with $n \geq 2$, two of them differ by at most $\frac{1}{n-1}$.

Proof.   Without loss of generality, we may assume that the n numbers are listed in nondecreasing order: $0 \leq a_1 \leq a_2 \leq \ldots \leq a_n \leq 1$. The difference $a_n - a_1$ is at most $1 = 1 - 0$ and is the sum of differences $a_n - a_1 = (a_n - a_{n-1}) + (a_{n-1} - a_{n-2}) + \ldots + (a_3 - a_2) + (a_2 - a_1)$. Since there are $n - 1$ terms on the right adding up to at most 1, the average difference is at most $\frac{1}{n-1}$. Hence, the smallest of these differences is, as claimed, at most $\frac{1}{n-1}$.   ∎

DISCUSSION.   The phrase "without loss of generality" replaces a potentially distracting argument, which might go as follows: "Since there are only finitely many numbers, one of them is the smallest, and we can call it a_1. Once we have found and named the smallest k numbers, the remaining numbers have a smallest one, which we call a_{k+1}. Thus, we have the n numbers listed in increasing order: $0 \leq a_1 \leq a_2 \leq \ldots \leq a_n \leq 1$."

The trick of writing the difference $a_n - a_1$ as a sum of intermediate differences used to be called "telescoping" because of the analogy to old-style hand-held telescopes whose sections slid together, leaving just the ends and one section visible (see Figure 3). Mathematically, the long sum "telescopes" together because each intermediate term appears twice, once negatively and once positively, and so cancels out. While the ends do not justify the means in ethics, mathematics allows unmotivated steps if they lead to clear, convincing proofs. Unfortunately, unfamiliar maneuvers, such as telescoping, often seem like magic tricks to students.

FIGURE 3.   Telescoping.   ◇

EXAMPLE 4

Use analytic geometry to determine the possible distances between vertices of a cube in terms of the length of a side s.

Solution   Because cubes have wonderful symmetry, we can pick a convenient positioning of the coordinate axes. Place the origin at the center of the cube, with each axis coming out of the centers of two opposite faces (see Figure 4). Then the vertices of the cube have coordinates (x, y, z), where each coordinate is $\pm a$. Then each side has length $s = \sqrt{(a - (-a))^2 + 0^2 + 0^2} = 2a$, since adjacent vertices differ by a sign in just one coordinate. By symmetry, it doesn't matter which coordinate differs. Vertices along the diagonal of a face differ in signs in two coordinates. Thus, the distance between them is $\sqrt{(a - (-a))^2 + (a - (-a))^2 + 0^2} = 2a\sqrt{2} = s\sqrt{2}$.

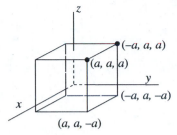

FIGURE 4

Similarly, directly opposite vertices differ in all three coordinates, so their distance is
$\sqrt{(a-(-a))^2 + (a-(-a))^2 + (a-(-a))^2} = 2a\sqrt{3} = s\sqrt{3}$.

The determinations of the coordinates of the vertices of a cube in general position would be quite tedious and would definitely obscure any understanding this problem might otherwise convey.   ◇

Implications of Implications

The heart of a proof can be hard to find when the proposition piles up the logical words in sophisticated ways. In Section 2.3, we considered the complication of multiple quantifiers. Here we consider multiple implications. Such forms often frustrate the natural approach of transforming the hypothesis directly into the conclusion. Implications in both the hypothesis and the conclusion, giving the form $(A \Rightarrow B) \Rightarrow (C \Rightarrow D)$, occur often enough to deserve our attention. Example 5 has this form because the definition of a subset turns "$X \subseteq Y$" into the implication "if $x \in X$, then $x \in Y$."

With such a complicated conclusion, it is generally more successful, after simply assuming the hypothesis, to focus immediately on the conclusion, hoping that the role of the hypothesis becomes clear along the way. Think of the hypothesis as a bridge over the logical "chasm" in Figure 5. We are given the bridge $(A \Rightarrow B)$ and our starting point C, and we need to travel to D. Since we can't move the bridge, we must travel from C to A, cross the bridge to B and then travel to D. The traveling corresponds to the logical transforming we do in the proof.

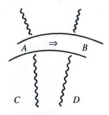

FIGURE 5

EXAMPLE 5

CLAIM. For any sets J, K, and L, if $J \subseteq K \cup L$ and $J \cap K = \emptyset$, then $J \subseteq L$.
(Hint: Draw a Venn diagram illustrating this claim.)

Proof. Let J, K, and L be any sets and suppose that $J \subseteq K \cup L$ and $J \cap K = \emptyset$. Now let $x \in J$. By our first hypothesis, $x \in K \cup L$, so $x \in K$ or $x \in L$. But $x \in J$ and our other hypothesis $J \cap K = \emptyset$ eliminate $x \in K$. So we have to have $x \in L$, showing $J \subseteq L$.   ∎

DISCUSSION.   Here the "bridge" is the compound hypothesis $J \subseteq K \cup L$ and $J \cap K = \emptyset$. Since we are showing $J \subseteq L$, the letter C in Figure 5 is the statement "Let $x \in J$," which is the second sentence of the proof. The letter D is "$x \in L$," which appears in the last sentence of the proof. The intervening sentences link these two by means of the hypothesis. It would be difficult to imagine how to transform that hypothesis $J \subseteq K \cup L$ and $J \cap K = \emptyset$ directly into $J \subseteq L$.   ◇

To Prove or Not to Prove

Logic may tell us that a proposition must be either true or false, but it doesn't tell us which it is. Professional mathematicians and students alike face conjectures without knowing whether to look for a proof or a counterexample. No algorithm resolves this problem; only experience and imagination seem to provide useful guides. (However, students learn to suspect that conjectures in homework problems tend to be weighted in favor of counterexamples.) If, after reading or making a conjecture, you have an intuition about its truth or falsity, follow your intuition. If you don't have a feel for the conjecture, you probably need to work more examples. You may well stumble on a counterexample or see why the conjecture must be true. If you are still in the dark, critique your examples. Did you pick generic values for the variables? Are there special, easily overlooked cases? In particular, zero and negative numbers can be easily missed exceptions. Similarly, the empty set has peculiar properties. Other times, one more example changes everything.

EXAMPLE 6

In 1650, Pierre de Fermat considered primes of the form $2^n + 1$. In 1801, Carl F. Gauss related these Fermat primes to regular polygons constructible by straight edge and compass.

FERMAT'S CONJECTURE.   If n is a power of 2, then $2^n + 1$ is prime.
We can factor reasonably small numbers into primes without using computers. Let's try a few. $2^1 + 1 = 3$ is prime, $2^2 + 1 = 5$ is prime, $2^4 + 1 = 17$ is prime, and $2^8 + 1 = 257$ is prime. By diligence, Fermat found that $2^{16} + 1 = 65537$ was prime as well. (Incidentally, if $n \geq 1$ and $2^n + 1$ is prime, we can show that n must be a power of 2. The only other known Fermat prime is $2^0 + 1 = 2$.) Just about halfway between Fermat's conjecture and Gauss's use of these primes, Euler disproved the conjecture by factoring (by hand!) $2^{32} + 1 = 4,294,967,297$ into 641 and $6,700,417$. Even with the advent of modern computers, people have searched totally unsuccessfully for any other Fermat primes.   ◇

Investigating when some property holds often depends on an examination of examples, as in Example 7.

EXAMPLE 7

Determine when $\mathcal{P}(X) \cup \mathcal{P}(Y) = \mathcal{P}(X \cup Y)$.

Solution In Problem 12c of Section 2.3, we saw that for all sets X and Y, $\mathcal{P}(X) \cup \mathcal{P}(Y) \subseteq \mathcal{P}(X \cup Y)$. Let's pick two small sets to get a better feel for this problem. For $X = \{1, 2\}$ and $Y = \{2, 3\}$, we have $\mathcal{P}(X) \cup \mathcal{P}(Y) = \{\emptyset, \{1\}, \{2\}, \{1, 2\}, \{3\}, \{2, 3\}\}$, but $\mathcal{P}(X \cup Y)$ also includes $\{1, 3\}$ and $\{1, 2, 3\}$. The existence of the numbers 1 and 3, one in each set, seems the key to the strict inequality. Said another way, if all members of X were in Y or the other way around, we should have equality. More formally, for all sets X and Y, $\mathcal{P}(X) \cup \mathcal{P}(Y) = \mathcal{P}(X \cup Y)$ iff $X \subseteq Y$ or $Y \subseteq X$. Problem 7j asks you to prove this conjecture. $\Diamond$

Style

By this time, you should have written a number of proofs and developed a feel for the three essential "c's" of proof writing: correctness, clarity, and completeness. These are essential because the first goal of a proof is to convince the reader of the logical necessity of the proposition. The fourth "c" of proof writing is "conciseness," which to some degree is in tension with "completeness." The amount of explanation needed depends on the audience. In this book, I have favored completeness over conciseness, whereas in a research paper, I switch priorities. In any case, a good proof makes explicit the aspects of particular interest in the proof, especially what is new relative to the audience, as well as the general format and approach of the proof. For instance, in a proof by contradiction, the negation of the conclusion is always new. For student proofs, recent definitions are new and so should be explicitly mentioned in proofs. The more involved the proof is, the more important the choice becomes of what to include. Both too much detail and too little detail make it hard to understand the heart of a proof. As you are no doubt aware by now, what some mathematicians call "obvious" may be just plain mysterious to a student. Unfortunately, the other direction happens as well.

Beyond the four "c's" of proof writing, proofs in different areas of mathematics have their own peculiar characteristics, some of which we will consider in later chapters. Also, individual mathematicians develop their own styles. All of us strive for the ultimate in style: elegance, a difficult quality to describe, but one through experience we come to recognize and value. Euclid's proof of Theorem 2.5.1 is my favorite example, with its imaginative choice and use of the number $P + 1$. An elegant proof ought to reveal why a proposition is true in a transparent way.

PROBLEMS

*1. For each statement that follows, decide whether it is true or (at least sometimes) false. If false, explain why. Explain your answer or cite the part of the text supporting it.

   a) A proof cannot depend logically on a figure.

   b) Figures should never be used with proofs.

   c) The phrase "without loss of generality" means that the reader has to fill in all of the missing cases.

   d) A conjecture is a wild guess.

   e) Symmetry arguments can organize and simplify proofs.

   f) The four "c's" of proof writing are equally important.

*2. Fill in the outline to complete the proof.

Claim: For all sets A, B, and C, if $B \subseteq C$, then $A \cup B \subseteq A \cup C$. Proof. Let ___ and suppose ___ . Now let $x \in$ ___ . Then, by the definition of union, $x \in$ ___ or $x \in$ ___ . In the first case, $x \in A \cup C$ immediately. In the second case, our hypothesis gives us ___ and so ___ . Thus, either way, if ___ , then ___ . ∎

The following problems use a mixture of approaches discussed throughout this chapter:

3. a) Figure 6 illustrates three lines dividing the plane into seven regions. From analogous figures with one to four lines, make and prove a conjecture about the maximum number of regions in the plane formed by n lines.

FIGURE 6. Seven regions from three lines.

b) Repeat part a for planes dividing three-dimensional space.

c) Repeat part a for great circles on a sphere. (Hint: Great circles intersect in 2 diametrically opposite points.)

d) Figure 7 illustrates that two equilateral triangles can have six points of intersection. Find the maximum number of points of intersection an equilateral triangle can make with a regular n-gon. Prove your answer.

FIGURE 7. Two equilateral triangles.

e) Find the maximum number of regions a regular k-gon and a regular n-gon can form, where $k < n$. Prove your answer.

4. Prove: For all $x = a + bi$, $y = c + di \in \mathbb{C}$, if both $x + y$ and $x \cdot y$ are real numbers, then both x and y are real or they are conjugates. (The conjugate of $a + bi$ is $a - bi$.)

5. a) For any sets P, Q, and R, show that if $P \cap Q \neq P \cup R$, then R is not a subset of P or P is not a subset of Q.

b) Write the converse for part a) and either prove it or disprove it.

6. If S and T are subsets of some universal set U, prove the following:

*a) $\overline{S \cup T} = \overline{S} \cap \overline{T}$.

b) $\overline{S \cap T} = \overline{S} \cup \overline{T}$.

c) $S \subseteq T$ iff $\overline{T} \subseteq \overline{S}$.

7. Prove the following propositions:

a) For any sets J, K, and L, if $J \subseteq K$ and $K \subseteq L$, then $J \subseteq L$ (referred to in Section 4.4).

*b) For any sets J, K, and L, if $L \subseteq K$ and $J \subseteq K$, then $J \cup L \subseteq K$.

c) For any sets E, F, G, and H, if $E \subseteq F$ and $G \subseteq H$, then $E \cup G \subseteq F \cup H$.

d) For all sets S, T, and W, if $S \subseteq T$, then $S - W \subseteq T - W$.

e) For all sets S, T, W, and X, if $S \subseteq T$ and $W \subseteq X$, then $S \cap W \subseteq T \cap X$.

f) For all sets S, W, and X, if $W \subseteq X$, then $S - X \subseteq S - W$.

g) For any sets M, P, and Q, if $M \cap P = M \cap Q$ and $P - M = Q - M$, then $P = Q$.

h) For any sets R, S, and T, $S \cup T = S \cap R$ iff $T \subseteq S$, and $S \subseteq R$.

i) For any sets V and W, $V \subseteq W$ iff $\mathcal{P}(V) \subseteq \mathcal{P}(W)$.

j) For any sets X and Y, $\mathcal{P}(X) \cup \mathcal{P}(Y) = \mathcal{P}(X \cup Y)$ iff $X \subseteq Y$ or $Y \subseteq X$.

8. Prove the following propositions:

a) For all k, $n \in \mathbb{Z}$, $(k - n)$ divides $k - 1$ iff $(k - n)$ divides $n - 1$.

b) For all x, $y \in \mathbb{Z}$, if x divides y, then $|x| \leq |y|$.

c) For all integers s, t, u, if u is odd and u divides both $s + t$ and $s - t$, then $u|s$ and $u|t$.

d) For all integers d, p, q, x, and y, if $px + qy = 1$ and $d > 1$, then d does not divide x or d does not divide y.

9. a) Let a, c, $L \in \mathbb{R}$, with $c > 0$, and f is a function defined on $\mathbb{R}$. Prove that if $\lim_{x \to a} f(x) = L$, then $\lim_{x \to a} c \cdot f(x) = c \cdot L$. (Hint: Note that the definition of a limit given in Problem 14 of Section 2.3 has an implication in it.)

b) Let a, L, $M \in \mathbb{R}$, and f and g be functions defined on $\mathbb{R}$. Prove that if $\lim_{x \to a} f(x) = L$ and $\lim_{x \to a} g(x) = M$, then $\lim_{x \to a} (f(x) + g(x)) = L + M$.

10. Prove these propositions about the nonempty indexed family of sets $\{S_i : i \in I\}$:

a) For any set B, $B - \bigcup_{i \in I} S_i = \bigcap_{i \in I} (B - S_i)$.

b) For any set B, $\left(\bigcup_{i \in I} S_i \right) - B = \bigcup_{i \in I} (S_i - B)$.

c) $\bigcap_{i \in I} S_i \subseteq \bigcup_{i \in I} S_i$.

*d) Suppose that $J \subseteq I$ and $J \neq \emptyset$. Then $\bigcup_{j \in J} S_j \subseteq \bigcup_{i \in I} S_i$.

e) Suppose that $J \subseteq I$ and $J \neq \emptyset$. Then $\bigcap_{i \in I} S_i \subseteq_{j \in J} S_j$.

f) For any indexed family of sets $\{S_i : i \in I\}$, $\bigcup_{i \in I} S_i - \bigcap_{i \in I} S_i = \bigcup_{i \in I, k \in I, i \neq k} S_i - S_k$. (Hint: Draw a Venn diagram with sets S_1, S_2, and S_3 to understand what this equation is saying.)

g) Which of your proofs use the assumption $I \neq \emptyset$? Why is it needed in these proofs?

11. a) Prove: For all $x \in \mathbb{N}$, if $2^x - 1$ is prime, then x is odd or $x = 2$. (Primes of the form $2^x - 1$ are called Mersenne primes. As of 2006, the largest known prime is $2^{32582657} - 1$, which has $9,808,358$ digits. Computers can check in a reasonable amount of time whether a given number $2^x - 1$ is prime, which partially explains why most of the really large primes are Mersenne primes.)

b) Show that the converse of part a is false by giving a counterexample.

c) A stronger version of part a is true: For all $x \in \mathbb{N}$, if $2^x - 1$ is prime, then x is prime. Find a counterexample to the converse of this proposition.

12. a) For all $n \in \mathbb{N}$, prove that n is prime or n is a perfect square or n divides $(n - 1)!$.

b) Using examples, verify that the condition "n is a perfect square" in part a is needed only for $n = 4$. For all $n \in \mathbb{N}$, prove that n is prime or $n = 4$ or n divides $(n - 1)!$.

13. a) For all $n \in \mathbb{N}$ and all z, q, $r \in \mathbb{Z}$, if $z = nq + r$, prove that $\gcd(z, n) = \gcd(n, r)$.

b) Use part a and the well ordering of $\mathbb{N}$ to show that the Euclidean Algorithm in Example 5 of Section 1.5 does indeed give the greatest common divisor of two natural numbers.

c) Generalize your proof in part b to allow n to be a nonzero integer.

14. By a number like 253 in base ten, we mean $2 \times 10^2 + 5 \times 10^1 + 3 \times 10^0$. In general,
    $a_n a_{n-1} \ldots a_0 = n \sum_{i=0}^{n} a_i \times 10^i$.

    *a) Show that 2 divides a number $a_n a_{n-1} \ldots a_0$ in base ten iff 2 divides a_0.

    b) Show that $2^2 = 4$ divides a number $a_n a_{n-1} \ldots a_0$ in base ten iff 2^2 divides $a_1 a_0$.

    c) Show that 2^k divides a number $a_n a_{n-1} \ldots a_0$ in base ten iff 2^k divides $a_{k-1} a_{k-2} \ldots a_1 a_0$.

    d) Find and prove corresponding results for divisibility in base fifteen and other bases.

15. For $n \in \mathbb{N}$, let $s(n)$ be the number of subsets of the first n natural numbers $\{1, 2, \ldots, n\}$ with no pairs of consecutive numbers.

    a) List the five subsets of $\{1, 2, 3\}$ with no pairs of consecutive numbers.

    b) List all the subsets of $\{1, 2\}$ with no pairs of consecutive numbers.

    c) List all the subsets of $\{1, 2, 3, 4\}$ with no pairs of consecutive numbers.

    d) Make a conjecture about $s(n)$ and prove it.

16. a) Let f_n be the nth Fibonacci number. Prove that the sum of the first n odd Fibonacci numbers, $f_1 + f_3 + \ldots + f_{2n-1} = \sum_{i=1}^{n} f_{2i-1}$, is f_{2n}.

    b) For all $n \in \mathbb{N}$, prove that $\sum_{i=1}^{n} (f_i)^2 = f_n f_{n+1}$.

    c) For all $n \in N$, if $n > 2$, prove that $(f_n)^2 - f_{n+1} f_{n-1} = (-1)^{n+1}$.

17. Define $A_0 = \emptyset$ and $A_{n+1} = A_n \cup \{A_n\}$.

    *a) Write out the sets A_1, A_2, and A_3. How many elements does each have?

    b) For all $n \in \mathbb{N}$, prove that A_n has n elements, $A_n \in A_{n+1}$ and $A_n \subseteq A_{n+1}$.

    c) For all $k, n \in \mathbb{N}$, prove that $k < n$ iff $A_k \in A_n$.

       (John Von Neumann [1905–1957] used A_n as his definition for the number n, as part of the effort to create the "universe" of sets from nothing—that is, from $\emptyset$.)

18. In map-coloring problems, adjacent regions are given different colors so that they can be distinguished. Regions touching only at a corner may be the same color (see Figure 8). A map has a *2-coloring*, provided that we can find a way to color its regions with just two colors so that adjacent regions have different colors.

FIGURE 8

Claim. If the regions of a map are drawn by the use of only lines crossing the entire map, then the map is 2-colorable.

Prove the claim, using the well ordering of $\mathbb{N}$. (Hint: On a copy of Figure 8 draw an additional line. Then find a systematic way to 2-color this new map on the basis of the coloring in Figure 8. Now recast this idea in the correct form to use the well ordering of $\mathbb{N}$). That is, start with a map made with the fewest number of lines that is not 2-colorable and omit a specific line to get a reduced map. Why can you 2-color the reduced map? Then argue how to color the original map.

19. Define a natural number to be *balanced* iff it is the average of two different prime numbers. For instance, 9 is balanced because it is the average of 7 and 11. We relate the following propositions:

> *P*: Every prime greater than 3 is balanced.
>
> *G* (Goldbach's conjecture): Every even integer greater than 2 can be written as the sum of two (not necessarily distinct) primes.
>
> *E*: Every even integer greater than 6 can be written as the sum of two distinct primes.
>
> *T*: Every integer greater than 5 can be written as the sum of at most three (not necessarily distinct) primes.
>
> *B*: Every integer greater than 3 is balanced.

> **a)** Prove: If *E*, then *P* and *G*.
>
> **b)** Prove the converse of part a.
>
> **c)** Prove: If *G*, then *T*.
>
> **d)** Prove: *E* iff *B*.

Goldbach made his conjecture in 1742. It remains unproven, as do the other propositions. It is known that every sufficiently large integer can be written as the sum of at most four primes (see Eves, 435–436).

20. Critique the incorrect "proofs" that follow. Point out reasoning errors and unclear presentation. (The truth of a claim is not relevant to your answer.)

> ***a)** Claim: Every even number greater than 2 is the sum of two primes (possibly the same prime twice). "Proof": From logic, either every even number greater than 2 is the sum of two primes or every even number greater than 2 is not the sum of two primes. But $4 = 2 + 2$, $6 = 3 + 3$, $8 = 3 + 5$, and so on. Clearly, the second alternative is false. Therefore, every even number greater than 2 is the sum of two primes. ▲
>
> **b)** Claim: $169 = 168$. "Proof": A square with sides of 13 has an area of $13^2 = 169$. Divide this square into the four pieces of Figure 9. Rearrange these four pieces as in Figure 10, giving a rectangle with sides 8 and 21. (The two trapezoids needed to be rotated.) Now $8 \cdot 21 = 168$. Hence, $169 = 168$. ▲

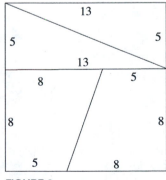

FIGURE 9

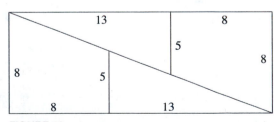

FIGURE 10

> **c)** Claim: For any sets *P* and *Q*, if $P - Q = Q - P$, then $P = Q$. "Proof": If *x* is in $P - Q$, then *x* is not in *Q*. From $P - Q = Q - P$, this means *x* is not in *P*. Similarly, if *x* is not in *P*, it is not in *Q*. By the contrapositive, $P = Q$. ▲

21. Critique the questionable "proofs" that follow. Point out reasoning errors and unclear presentation. If the argument is a proof, say so. (If the claim is false, there must be an error

in the argument. However, it is not enough to say that the claim is false. You need to find an error in the reasoning.)

*a) Claim: If a natural number n is prime, then $n + 13$ is not prime. "Proof": Any prime bigger than 2 is odd. But the sum of two odds is even, and so isn't prime. On the other hand, $2 + 13 = 15$ isn't prime. ▲

b) For any sets S, T, and V, $S \subseteq T \cap V$ iff $S \subseteq T$ and $S \subseteq V$. "Proof": By definition, $s \in T \cap V$ iff $s \in T$ and $s \in V$. Since $s \in S$, we get $S \subseteq T \cap V$ iff $S \subseteq T$ and $S \subseteq V$. ▲

c) Claim: There is an irrational number r such that $r^{\sqrt{2}}$ is rational. "Proof": Consider $r = \sqrt{2}^{\sqrt{2}}$ because then $r^{\sqrt{2}} = (\sqrt{2}^{\sqrt{2}})^{\sqrt{2}} = (\sqrt{2})^{\sqrt{2} \cdot \sqrt{2}} = \sqrt{2}^2 = 2$, which is rational. If r is irrational, we're done. Otherwise, it is rational and we can use, instead, $\sqrt{2}$ for r. ▲

d) Claim: For any sets P and Q, if $P - Q = Q - P$, then $P \subseteq Q$. "Proof": Suppose $P - Q = Q - P$ for sets P and Q. For $x \in P$, if $x \in Q$, we're done. So suppose $x \notin Q$. Then $x \in P - Q = Q - P$. But then $x \in Q$. Either way, $x \in Q$, showing $P \subseteq Q$. ▲

REFERENCES

EVES, H. 1976. *An introduction to the history of mathematics*, 4th ed. New York: Holt, Rinehart and Winston.

HEATH, T. 1956. *The thirteen books of Euclid's elements*, vol. 1. New York: Dover.

PROOF FORMATS

We summarize some of the proof formats we have encountered.

To show	Format
For all $x \in S$, ___.	Let $x \in S$ and show ___.
If P, then Q. (Direct)	Suppose P, show Q.
If P, then Q. (Contrapositive)	Suppose not Q, show not P.
If P, then Q. (Contradiction)	Suppose P and not Q, find a contradiction.
P iff Q.	Prove "if P, then Q" and "if Q, then P"
$A \subseteq B$.	Let $x \in A$ and prove $x \in B$.
$A = B$ (for sets).	Prove $A \subseteq B$ and $B \subseteq A$.
If P or Q, then R.	Case 1. Suppose P, prove R. Case 2. Suppose Q, prove R.
If P, then Q or R.	Suppose P. If Q, we are done. So suppose not Q and prove R.
There is $x \in S$___.	Pick x in S and prove ___.
There is a unique $x \in S$___.	Existence: Pick x in S and prove ___. Uniqueness: Suppose y and z in S satisfy ___ and prove $y = z$.
Induction For all $n \in \mathbb{N}$, $P(n)$.	Show $P(1)$ is true. Assume $P(k)$ is true. Use $P(k)$ to prove $P(k + 1)$ is true.

CHAPTER 3

FUNCTIONS

"By relieving the brain of all unnecessary work, a good notation sets it free to concentrate on more advanced problems...."
—Alfred Whitehead (1861–1947)

FUNCTIONS HAVE PLAYED a central role in mathematics for nearly 400 years. In particular, students and mathematicians since Newton have used calculus to investigate many individual functions and properties of general families of them. We build on the familiarity calculus provides to explore the general concept of a function. The first section of this chapter uses many examples to motivate turning familiar concepts into formal definitions. The second and third sections use these definitions to prove properties about functions. Our abstract, set theoretic conception and notation for functions relieve our brains of unnecessary work, to paraphrase Whitehead's quote. And indeed, mathematicians heavily use functions to develop more advanced concepts.

3.1 DEFINITIONS, NOTATION, AND EXAMPLES

In calculus, a function is often described as a process or rule converting input numbers into outputs, such as the sine function, which takes a number (or angle) x and gives us another number $\sin(x)$. We make this understanding explicit in Definition 1, which follows. In that definition, the word "rule" is undefined, and so perhaps this definition is not as easy to use in proofs as the second one. We'll show, however, that the two definitions are equivalent.

DEFINITION 1. A *function f* from a set X into a set Y is a rule that assigns to each $x \in X$ a unique $y = f(x) \in Y$.

Armed with calculators, students seem quite comfortable with almost any formula playing the role of such a rule, even a monstrosity like $\sqrt{e^x + \sin^2(1/(x^2 + 1))}$. Most eighteenth-century mathematicians fully expected that every function had a reasonable formula. Mathematicians' imaginations in the nineteenth century quickly

121

outstripped the supply of well determined formulas. To accommodate the much broader concept of a function, mathematicians focused on the graph of a function, using Georg Cantor's newly developed set theory. Recall that the graph of a function f consists of the points $(x, f(x))$. In the set theory language of Section 1.6, $(x, f(x))$ is an ordered pair. A function then becomes for us the set of ordered pairs in its graph.

DEFINITION 2.   A *function f* from a set X into a set Y is a set of ordered pairs $f = \{(x, y) : x \in X \text{ and } y \in Y\}$ such that for all $x \in X$, there is a unique $y \in Y$ such that $(x, y) \in f$. We almost always use the familiar $f(x) = y$ rather than the formal $(x, y) \in f$. We write $f : X \to Y$ to indicate that f is a function from X into Y. We call X the *domain* of f and Y its *codomain*. The *image* of x is $f(x)$. The *range* of f is $\{y \in Y : \exists x \in X \text{ s.t. } f(x) = y\}$. That is, the range is the set of all images of elements of X.

EXAMPLE 1

The sine function, with the formula $f(x) = \sin(x)$ and shown in Figure 1, is made up of the points $(x, \sin(x))$. Each x-value has exactly one y-value paired with it. High school texts sometimes describe this property as the "vertical line test," since every vertical line intersects the graph exactly once. Both definitions of a function formalize the idea of this vertical line test. Let's illustrate the additional terms of the second definition with the sine function. The domain of the sine function is $\mathbb{R}$, and its range is $[-1, 1]$. The image of, say, $x = \pi/2$ is $\sin(\pi/2) = 1$. Note that $\sin(x)$ is a number, the image of x. In particular, $\sin(x)$ is not a function; the function is called sine. The word "codomain" is probably less familiar than the other terms. The usual choice for the codomain of functions in calculus is all of $\mathbb{R}$, even if a particular function doesn't use all of it for its range. We could use any set containing at least $[-1, 1]$ for the codomain of this function. In fact, because a function is defined as the set of ordered pairs, we get the same function whether we write $f : \mathbb{R} \to \mathbb{R}$ or $f : \mathbb{R} \to [-1, 1]$, since the function is just the set of ordered pairs. Changing the domain, however, changes the function, since every element in the domain has a unique image.

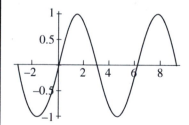

FIGURE 1. $\{(x, y) : y = \sin(x)\}$.   ◊

EXAMPLE 2

The unit circle in Figure 2 is a subset of $\mathbb{R} \times \mathbb{R}$, $\{(x, y) : x^2 + y^2 = 1\}$, but fails to be a function from $\mathbb{R}$ to $\mathbb{R}$ for two reasons. For x-values strictly between -1 and 1, there are two y-values satisfying the equation, instead of just one. For instance, both $(0.6, 0.8)$ and $(0.6, -0.8)$ lie on the circle. For x-values less than -1 and values greater than 1, such as

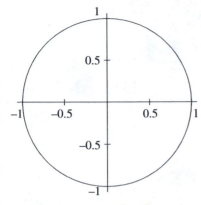

FIGURE 2. $\{(x, y) : x^2 + y^2 = 1\}$.

$x = 2$, there are no matching y-values, also violating the definition of a function. If we change the domain to $[-1, 1]$ we eliminate the second problem. By restricting the codomain, say, to $(-\infty, 0]$, we finally get a function we'll call g, the bottom half of the circle. We usually specify the function $g : [-1, 1] \to (-\infty, 0]$ by $g(x) = -\sqrt{1 - x^2}$. We could have restricted the codomain further, say, to $[-1, 0]$, the range of g. To get the function h representing the top half of the circle, we would need a different codomain, such as $[0, \infty)$, and a different formula, $h(x) = \sqrt{1 - x^2}$. $\Diamond$

REMARKS. The unit circle in Example 2 as well as all functions, including the sine function of Example 1, are *relations*, which are any subsets of a Cartesian product. Relations are the subject of Chapter 4. The equivalence of our two definitions, given next, is so straightforward that we'll call it a lemma, something used to prove more significant propositions. Thanks to this lemma, you may use either definition.

LEMMA 3.1.1. The two definitions of a function are equivalent.

Proof. We need to show that something satisfying either one of the definitions satisfies the other definition. Suppose in the first definition f is a rule assigning to each $x \in X$ a unique $y = f(x) \in Y$. Then the set of such ordered pairs $\{(x, y) : y = f(x)\}$ exactly fulfills the conditions to be a function by the second definition. For the other direction, suppose we have a function g as a set of ordered pairs (x, y). We use the following rule to determine y uniquely from x: Let $y = g(x)$ iff (x, y) is an ordered pair in g.   ∎

We want to study functions and their properties in general, so we won't use only $\mathbb{R}$ for our domain and codomain. In particular, we won't look at calculus properties such as continuity and differentiability until Chapter 8. What properties apply more broadly? Staying with the familiar a little longer, consider the functions $p : \mathbb{R} \to \mathbb{R}$, $q : \mathbb{R} \to \mathbb{R}$, and $r : \mathbb{R} \to R$ given by $p(x) = x^3$, $q(x) = x^3 - 3x^2$, and $r(x) = e^x$ (See Figure 3). Since all functions satisfy the vertical line test, it isn't too much of a stretch to ask if a "horizontal line test" tells us something. The function p fully

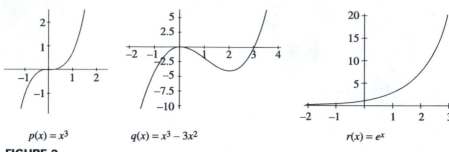

$p(x) = x^3$ $q(x) = x^3 - 3x^2$ $r(x) = e^x$

FIGURE 3

qualifies: All horizontal lines intersect the graph exactly once. The other two satisfy different parts of this "test." With q, every horizontal line intersects the graph, although sometimes more than once. The graph of r never intersects a horizontal line more than once, but it misses some horizontal lines. The differences among these functions with regard to the "horizontal line test" motivate the important definitions of onto and one-to-one functions.

DEFINITION. A function $f : X \rightarrow Y$ is *onto* iff for all $y \in Y$, there is $x \in X$ such that $f(x) = y$. That is, the range of f is all of Y. (Such a function is also called *surjective*.)

In terms of the "horizontal line test," an onto function intersects every horizontal line at least once. Thus the functions p and q of Figure 3 are onto. The function r has range $(0, \infty)$, not all of $\mathbb{R}$, so it is not onto.

DEFINITION. A function $f : X \rightarrow Y$ is *one-to-one* iff for all $s, t \in X$, if $f(s) = f(t)$, then $s = t$. (Such a function is also called *injective*.)

In terms of the "horizontal line test," a one-to-one function intersects each horizontal line at most once. Thus, p and r are one-to-one, but q is not one-to-one. The wording of the definition of one-to-one may make it hard at first to see how it matches with the description of intersecting each horizontal line at most once. The working negation of the definition of one-to-one may make this idea clearer: f is **not** one-to-one iff there are $s, t \in X$ such that $f(s) = f(t)$ and $s \neq t$. For instance, the function q has two distinct x-values, $s = 0$ and $t = 3$, so that $q(s) = 0 = q(t)$, but $s \neq t$, so q isn't one-to-one. For these values, q is "two-to-one." From the graph, sometimes q is three-to-one, sometimes it is two-to-one, and sometimes it is one-to-one. The functions p and r are always one-to-one.

The sine function, in Figure 1, is neither onto nor one-to-one. In fact, for y between -1 and 1, it is "infinite-to-one," since infinitely many x-values can map to the same y-value. For instance, all integer multiples of π map to 0.

Let's try out these definitions on some functions whose domains aren't sets of numbers.

EXAMPLE 3

Let $S = \{a, b, c, d\}$, $T = \{m, n, p\}$, and $U = \{w, x, y, z\}$. Figure 4 pictures three functions, $h : S \rightarrow T$, $j : T \rightarrow S$, and $k : S \rightarrow U$. More formally, we can define these functions as sets: $h = \{(a, m), (b, p), (c, n), (d, p)\}$, $j = \{(m, c), (n, d), (p, b)\}$, and $k = \{(a, x), (b, z), (c, w), (d, y)\}$. Then h is onto all of T, but is not one-to-one since $h(b) = p = h(d)$. In contrast, j is one-to-one, but not onto, since a is not the image of any element of T. Finally, k is both one-to-one and onto.

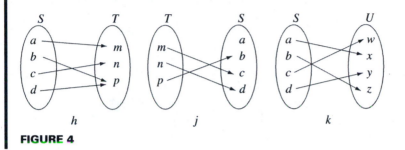

FIGURE 4

$\Diamond$

EXAMPLE 4

For any nonempty set S, let $g : S \rightarrow \mathcal{P}(S)$ be $g(x) = \{x\}$. Verify that g is a one-to-one function, but not onto.

Solution   For g to be a function, we need to start with a general element x of the domain S. Then g sends x to $\{x\}$, which is a subset of S and so is an element of the power set $\mathcal{P}(S)$. Thus, we have existence. The way we write formulas for functions in terms of the input guarantees uniqueness: Once we know the input x, we have no choice what the output is. In this case the output is $\{x\}$, the set with just one element, x.

Caution: Just writing a formula doesn't guarantee existence. The formula might not make sense for a particular value. In Example 2, for instance, the formula $g(x) = -\sqrt{1 - x^2}$ doesn't work when $x > 1$ or $x < -1$.

For one-to-one, we start with two general elements y and z and assume that their images are equal: $\{y\} = \{z\}$. But each of these sets has just one element, so $y = z$ and g is one-to-one.

Finally, to disprove onto, we need to find something in the codomain $\mathcal{P}(S)$ not in the range. It is tempting to think of something with two elements, like $\{a, b\}$, which certainly could not be any $\{x\}$, a set with just one element. Unfortunately, we don't know any particular element of S, let alone two. Further, the claim should work even if S had just one element. Consider for a moment $S = \{1\}$. Then $\mathcal{P}(S)$ has two elements: $\{1\}$ and $\emptyset$. Clearly, $\emptyset$ can't be an image under g because images have one element. Even more, $\emptyset$ is an element of any power set, since whatever S is, $\emptyset \subseteq S$. So, we have our counterexample: For any S, pick $\emptyset \in \mathcal{P}(S)$. Then for all $x \in S$, we know $g(x) = \{x\} \neq \emptyset$; so g is not onto. $\Diamond$

EXAMPLE 5

Define $a : \mathbb{Z} \times \mathbb{Z} \rightarrow \mathbb{Z}$ by $a((x, y)) = x + y$. This function is named a for "addition." A rule combining two elements from a set to get a third from that set is called an *operation*. Other familiar operations include subtraction, multiplication, and averaging of numbers. Division is not an operation on $\mathbb{Z}$ or $\mathbb{R}$, since division by 0 is not defined. In general, an operation is a function whose domain is the Cartesian product of a set with itself and whose codomain is that set. (We'll study operations more deeply in Chapter 7.) Addition of integers is an onto function, since, for any desired sum z, we can easily find an ordered pair, say, $(z, 0)$, so that $a((z, 0)) = z + 0 = z$. However, addition is not one-to-one: $(1, 3) \neq (2, 2)$, yet $a((1, 3)) = 4 = a((2, 2))$.　　$\Diamond$

EXAMPLE 6

The familiar concept of a derivative is an example of a function. Its inputs are functions, say, $g : \mathbb{R} \to \mathbb{R}$ given by $g(x) = x^3 - \sin(x)$, and its outputs are other functions, in this case, $g' : \mathbb{R} \to \mathbb{R}$ given by $g'(x) = 3x^2 + \cos(x)$. If we used D to represent the function of taking a derivative, we'd have $D(g) = g'$.

We can select the domain of D to be the set of differentiable real functions. We can use the set of all real functions as the codomain. As calculus students learn quickly, the function D is not one-to-one, since two different functions can have the same derivative. Indeed, the derivative of the functions given by $h(x) = x^2$ and $j(x) = x^2 + 7$ are the same. Is D onto? This is a noticeably harder question, but of definite mathematical interest. We can readily find an "antiderivative" for easy functions. For instance, for k given by $k(x) = x^3/3$ and h given by $h(x) = x^2$, we have $D(k) = h$. The Fundamental Theorem of Calculus assures us that if f is continuous on a closed interval, then there is some function F so that $D(F) = f$ on that interval. (There is no promise that we can find a formula for F. Even fancy techniques of integration fail with many formulas, such as $f(x) = \cos(\sqrt{1 + x^3})$.) Discontinuous functions, however, do not need to be derivatives of anything. So, in general, D is not onto.　　$\Diamond$

New Functions from Old Ones

Composition　Calculus frequently works with compound functions like $f : \mathbb{R} \to \mathbb{R}$ defined by $f(x) = e^{x^3} = r(p(x))$, built from the easier functions p and r introduced earlier. We call f the *composition* of the other two, writing $f = r \circ p$. (Composition is an operation. See Example 5 for a discussion of operations.)

DEFINITION.　Given functions $f : X \to Y$ and $g : Y \to Z$, their *composition* is $g \circ f$, defined by $g \circ f = \{(x, z) : \exists\, y \in Y \text{ s.t. } f(x) = y \wedge g(y) = z\}$. In more informal terms, we write $g \circ f(x) = g(f(x))$.

EXAMPLE 7

Figure 5 illustrates the composition $j \circ h : S \to S$, where j and h are the functions from Example 3. The first drawing shows how the two functions h and j are linked through the

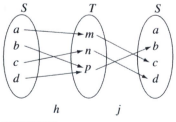

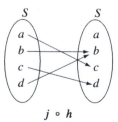

FIGURE 5

intermediary set T. For instance, $h(a) = m$ and $j(m) = c$. So, $j \circ h(a) = c$. The second drawing gives the function $j \circ h$ with only its domain and codomain showing. ◇

The notation $g \circ f(x) = g((f(x))$ suggests that the composition is a function. It is, and it takes elements of X to Z, but we will need to prove these facts in Section 3.2. We will investigate other properties of the composition of functions as well.

Inverses Once mathematicians understand something, they often turn it around to see what happens. For instance, the function $p : \mathbb{R} \to \mathbb{R}$ given by $p(x) = x^3$ suggests the cube root function $c : \mathbb{R} \to \mathbb{R}$, defined by $c(x) = \sqrt[3]{x}$, which reverses what p did. We call c the *inverse function* of p. Note that $y = p(x) = x^3$ iff $x = c(y) = \sqrt[3]{y}$. However, we need to be careful about inverses of functions. For instance, the squaring function $s : \mathbb{R} \to \mathbb{R}$ defined by $s(x) = x^2$ presents a problem. Because both $(-3)^2 = 9$ and $3^2 = 9$, we can't naïvely define a square root function taking 9 to both -3 and 3. According to the next definition, the squaring function s has an inverse, but it doesn't have an inverse function. Of course, the notion of a square root is so important that mathematicians have defined a related function by restricting the domain and codomain (see Figure 6).

DEFINITION. Given a function $f : X \to Y$, its *inverse* is $\{(y, x) : y \in Y, x \in X$ and $(x, y) \in f\}$. If this set of ordered pairs is a function from Y to X, we call it the *inverse function* of f and write f^{-1}.

Images and Pre-images of Sets Mathematicians often push a concept beyond its original setting to look for new insights. Rather than just putting elements into a function, we can try putting in sets of elements. This extension provides many insights, some discussed in Section 3.3. Just as $f(x)$ is the image of the element x, we'll call $f[A]$ the *image* of the set A. We place brackets instead of parentheses around the set A to emphasize that A is not an element in the domain of f. We will also consider the reverse process of mapping subsets of the codomain to their *pre-images* in the domain. Images and pre-images of sets play an important role in further mathematics, including linear algebra, abstract algebra, analysis, and topology.

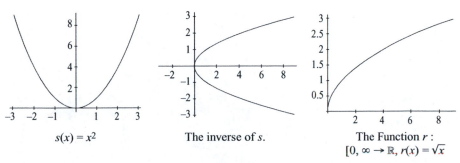

$s(x) = x^2$ The inverse of s.   The Function r : $[0, \infty \to \mathbb{R}, r(x) = \sqrt{x}$

FIGURE 6

DEFINITION. Let $f : X \to Y$ be a function. If $A \subseteq X$, then $f[A] = \{y \in Y : \exists\, a \in A\ s.t.\ y = f(a)\}$. We call $f[A]$ the *image* of A. If $B \subseteq Y$, then $f^{-1}[B] = \{x \in X : f(x) \in B\}$. We call $f^{-1}[B]$ the *pre-image* of B.

REMARK. Even if the function $f : X \to Y$ does not have an inverse function, its pre-images are still defined for subsets of Y. The bracket notation $f^{-1}[B]$ serves to separate the idea of a pre-image from the more restrictive situation of an inverse function.

EXAMPLE 8

For $s : \mathbb{R} \to \mathbb{R}$ given by $s(x) = x^2$, we have $s(3) = 9$ and $s(-3) = 9$, so $s[\{3, -3\}] = \{9\}$ and $s^{-1}[\{9\}] = \{3, -3\}$. Find the images and pre-images of the intervals $[1, 2]$, $[-5, -3)$, and $(-2, 3)$.

Solution First let's consider these intervals as subsets of the domain and find their images. From the first graph in Figure 7, we see that the images of the x-values from 1 to 2 start at $y = 1$ and increase to $y = 4$. Thus, the image of $[1, 2]$ is the interval $[1, 4]$. Following our notation, we write $s[[1, 2]] = [1, 4]$. Similarly, $s[[-5, -3)] = (9, 25]$, also shown in the first graph of Figure 7. From the second graph in Figure 7, images for the interval $(-2, 3)$ start at almost $y = 4$, drop to $y = 0$, and increase almost to $y = 9$. So $s[(-2, 3)] = [0, 9)$.

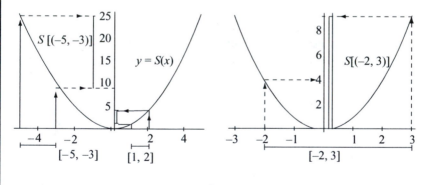

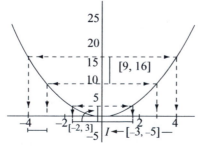

FIGURE 7

Next, consider these intervals as subsets of the codomain. With the squaring function s, we can write $s^{-1}[\{9\}] = \{-3, 3\}$, even though $s^{-1}(9)$ isn't defined. The x-values mapped to the interval $[9, 16]$ include the values between 3 and 4. They also include the x-values between -4 and -3. So the pre-image is $[-4, -3] \cup [3, 4]$, illustrated in the third graph of Figure 7. Using the notation in the definition, we write $s^{-1}[[9, 16]] = [-4, -3] \cup [3, 4]$. Since no real number has a negative square, $s^{-1}[[-5, -3]] = \emptyset$. The pre-image of $(-2, 3)$ is the same as the pre-image of $[0, 3)$ because negative values are not in the range. Thus, $s^{-1}[(-2, 3)] = (-\sqrt{3}, \sqrt{3})$, illustrated in the final graph of Figure 7. $\Diamond$

REMARKS. For $f : X \to Y$, we have $f^{-1}[Y] = X$ and $f[X]$ is the range of f. The existential quantifier in the definition of $f[A]$ is often needed in proofs. Some books use the easier and more elegant definition $f[A] = \{f(a) : a \in A\}$, which too easily leaves students unsure how to prove properties.

Historical Remarks

By 1600, mathematicians and scientists started describing one quantity in terms of another, the basic idea underlying functions. Notably, Galileo Galilei (1564–1642) used geometry to describe many interactions, such as the distance an object falls in terms of the time of the fall. However, functions were often mixed together with more general curves, equations, and other relationships. René Descartes (1596–1650) and Pierre de Fermat (1601–1665) developed analytic geometry in the 1630s, joining algebraic equations with geometric shapes. They were interested in equations of curves, whether or not they were modern functions. For example, they studied ellipses like $2x^2 + 5y^2 = 1$, as well as polynomial functions like $y = x^3 + 2x^2 + x$. In addition, from the advent of calculus with Newton (1642–1727), mathematicians studied infinite series such as $\sum_{n=0}^{\infty} \frac{x^n}{n!}$.

It was openly discussed in the eighteenth century whether calculus studied analytic expressions or curves or infinite series. Bit by bit, functions increased in importance and mathematicians' understanding of them grew more abstract. Joseph Fourier (1768–1830) showed how to use trigonometric series to give a huge variety of modern-day functions. (A trigonometric series is an infinite sum of sines and cosines, such as $\sum_{n=1}^{\infty} \frac{1}{n} \sin(n(x))$.) These series described functions beyond the usual idea of a curve, since these functions didn't need to be smooth or even continuous in the modern sense. Nor did there need to be any algebraic or analytic equation matching these series. In response, mathematicians through the nineteenth century increasingly used the word "function" as a rule assigning exactly one image to each value in the domain, regardless of what that rule might be.

However, this very general definition called into question mathematicians' understanding of calculus terms about functions, such as "continuous" and "differentiable." In the late nineteenth century, strange examples of functions discontinuous everywhere and others continuous everywhere, but nowhere differentiable, stretched mathematicians' intuition far beyond earlier ideas. The advent of set theory and the set theory definition of a function further expanded horizons by allowing the domain and codomain to be any sets, rather than sets of numbers. Indeed, Ernst Zermelo (1871–1953), in the first paper giving axioms for set theory (1908), gives a completely general abstract description of functions.

PROBLEMS

*1. For each statement that follows, decide whether it is true or (at least sometimes) false. If false, explain why. Explain your answer or cite the part of the text supporting it.

a) A function is a formula.

b) e^x is a function from $\mathbb{R}$ to $\mathbb{R}$.

c) For any function, its range is a subset of its codomain.

d) A one-to-one function is always an onto function.

e) If the domain and codomain of f are equal, then $f \circ f$ is defined.

f) If f is a function from X to Y and Y is a subset of Z, then f is a function from X to Z.

g) If $f : X \to Y$ is a function, then $f[X] = Y$.

h) If $f : X \to Y$ is a function, then $f^{-1}[Y] = X$.

2. Which of the given expressions are functions with domain $X = \{1, 2, 3\}$ and codomain $Y = \{1, 2, 3, 4, 5, 6\}$? For the functions, give their range; for the others, explain why they fail.

*a) f defined by $f(x) = \frac{4}{3-x}$

*b) g defined by $g(x) = x!$

c) h defined by $h(x) = \frac{6}{4-x}$

d) $j = \{(2, 5), (3, 4), (1, 5)\}$

e) $k = \{(1, 3), (3, 5), (5, 2)\}$

f) $m = \{(1, 4), (2, 6), (3, 3), (2, 5)\}$

3. For each formula, use a graph to find the largest subset of $\mathbb{R}$ that can be the domain of a function with that formula, assuming the codomain is $\mathbb{R}$. For that domain, determine the range.

*a) $g(x) = \sqrt{x^2 - 4}$

b) $h(x) = \frac{2x+1}{x-3}$

c) $j(x) = \cot(x) = \frac{\cos(x)}{\sin(x)}$

d) $k(x) = \frac{(x-1)^2}{x^2-2x}$

4. Decide which of the following descriptions define functions:

*a) $g : \mathbb{R} \to \mathbb{R}$, where $g(x) = \frac{x}{x^2+1}$.

*b) $h : \mathbb{R} \to \mathbb{R}$, where $h(x) = \frac{x^2}{x+1}$.

c) $j : \mathbb{R} \to \mathbb{Z}$, where $j(x) = z$ iff $x - 1 < z \leq x$. (Referred to in Problem 3 of Section 3.2.)

d) $k : \mathbb{Q} \to \mathbb{Z}$, where $k(\frac{p}{q}) = p + q$.

e) $m : \mathbb{Z} \times \mathbb{Z} \to \mathbb{Z}$, where $m(x, y) = \gcd(x, y)$.

f) $n : \mathbb{N} \times \mathbb{N} \to \mathbb{N}$, where $n(x, y) = \gcd(x, y)$.

g) $p : \mathbb{Z} \to \mathcal{P}(\mathbb{Z})$, where $p(z) = \{x \in \mathbb{Z} : x|z\}$.

5. The definition of a function $f : X \to Y$ can be written symbolically as $\forall x \in X \; \exists! \; y \in Y$ s.t. $f(x) = y$. For a function f, what do the following symbols represent? $\forall y \in Y \; \exists! \; x \in X$ s.t. $f(x) = y$.

6. For each function defined from $\mathbb{R}$ to $\mathbb{R}$, decide whether it is one-to-one, onto, both, or neither. Explain. (Use a graph or calculus as needed.)

a) $f(x) = 3x^5 - 5x^3$

*b) $g(x) = 3x^5 + 5x^3$

*c) $h(x) = 2x^4 - 4x^2$

d) $j(x) = \frac{x}{x^2+1}$

e) $k(x) = \frac{x^3}{x^2+1}$

f) $m(x) = \frac{x|x|}{x^2+1}$

g) $p(x) = x \sin(x)$

h) $q(x) = \frac{1}{2^x}$

7. Using the functions from Problem 6, decide whether each composition is one-to-one, onto, both, or neither.

a) $f \circ f$ **d)** $p \circ f$ **g)** $m \circ q$

*b) $g \circ g$ **e)** $q \circ f$ **h)** $m \circ g$

c) $h \circ h$ ***f)** $k \circ f$ **i)** $g \circ k$

8. For each part, give an example of functions $f : \mathbb{R} \to \mathbb{R}$ and $g : \mathbb{R} \to \mathbb{R}$ satisfying the conditions there.

*a) f is onto, but $g \circ f$ is not onto.

b) f is one-to-one, but $g \circ f$ is not one-to-one.

c) g is onto, but $g \circ f$ is not onto.

d) g is one-to-one, but $g \circ f$ is not one-to-one.

e) $g \circ f$ is onto, but f is not onto.

f) $g \circ f$ is one-to-one, but g is not one-to-one.

9. **a)** Use the definition of one-to-one and the examples from Problems 7 and 8 to conjecture what conditions ensure that the composition of two functions is one-to-one.

b) Use the definition of onto and the examples from Problems 7 and 8 to conjecture what conditions ensure that the composition of two functions is onto.

10. Which of the functions that follow have an inverse function? Give a formula for this inverse for those with inverse functions. Explain what fails for those without inverses functions.

a) $f : \mathbb{Z} \to \mathbb{Z}$, where $f(x) = 2x$.

*b) $g : \mathbb{N} \to \mathbb{N}$, where $g(x) = \sqrt{x^2}$.

c) $h : \mathbb{Z} \to \mathbb{Z}$, where $h(x) = \sqrt{x^2}$.

d) $j : \mathbb{Z} \to \mathbb{N}$, where $j(x) = \begin{cases} 2x & \text{if } x > 0 \\ 1 - 2x & \text{if } x \leq 0 \end{cases}$.

e) $k : \mathbb{Z} \to \mathbb{Z}$, where $k(x) = \begin{cases} 2x & \text{if } x \text{ is odd} \\ \frac{x}{2} & \text{if } x \text{ is even} \end{cases}$.

11. Use the definition of an inverse function and the examples in Problem 10 to conjecture what conditions a function needs to have in order for it to have an inverse function.

12. Let $g : \mathbb{R} \to \mathbb{R}$ and $h : \mathbb{R} \to \mathbb{R}$ be given by $g(x) = 4x - x^2$ and $h(x) = \cos(x)$. Find the following images and pre-images:

*a) $g[[-1, 1]]$ e) $g^{-1}[[-2, 5]]$ i) $h[\{2k\pi : k \in \mathbb{Z}\}]$

b) $g[[-2, 5]]$ f) $g^{-1}[\{2, 4, 6, 8\}]$ j) $h^{-1}[\mathbb{R}]$

c) $g[[-3, 0] \cup (3, 6)]$ g) $h[[0, \frac{\pi}{2}]]$ k) $h^{-1}[[2, 5]]$

*d) $g^{-1}[[-1, 1]]$ *h) $h[\mathbb{R}]$ l) $h^{-1}[\{-1\}]$

13. The *floor* function takes a real number x to the greatest integer less than or equal to x, which we write as $\lfloor x \rfloor$. Thus, $\lfloor \pi \rfloor = 3$, $\lfloor 2 \rfloor = 2$ and $\lfloor -1.4 \rfloor = -2$. Figure 8 gives the graph of $y = \lfloor x \rfloor$. (Some texts call this function the *greatest integer function* and write $[x]$. The term *floor function* is preferable, since mathematicians also use the *ceiling* function, where $\lceil x \rceil$ is the smallest integer greater than or equal to x; this problem is referred to in Sections 3.2, 4.2, and 8.5.)

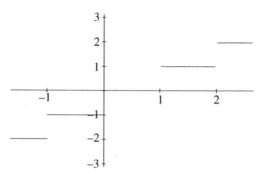

FIGURE 8. The graph of $\lfloor x \rfloor$.

a) Graph these functions: $f(x) = \lfloor x^2 \rfloor$ and $g(x) = \lfloor x \rfloor^2$
b) Graph $h(x) = x - \lfloor x \rfloor$.
c) Find $f[[-2.5, 2.5]]$, $g[[-2.5, 2.5]]$, and $h[[-2.5, 2.5]]$.
d) Find $f[\mathbb{R}]$, $g[\mathbb{R}]$, and $h[\mathbb{R}]$.
e) Find $f^{-1}[\{0\}]$, $g^{-1}[\{0\}]$, and $h^{-1}[\{0\}]$.
f) Find $f^{-1}[\{4\}]$, $g^{-1}[\{4\}]$, and $h^{-1}[\{4\}]$.

14. This problem investigates how unions, intersections, and other set operations interact with images of sets. Let $f : \mathbb{R} \to \mathbb{R}$ be the squaring function: $f(x) = x^2$ and let $A = [-4, 2]$ and $B = [-1, 3]$.

a) Find $f[A \cup B]$ and $f[A] \cup f[B]$.
*b) Find $f[A \cap B]$ and $f[A] \cap f[B]$.
c) Find $f[A - B]$ and $f[A] - f[B]$.
d) Find $f[\overline{A}]$ and $\overline{f[A]}$.
e) Use part a and other similar examples to make a conjecture about the relationship between $f[X \cup Y]$ and $f[X] \cup f[Y]$ for a general function f and general subsets X and Y.
f) Use part b and other similar examples to make a conjecture about the relationship between $f[X \cap Y]$ and $f[X] \cap f[Y]$ for a general function f and general subsets X and Y.
g) Use part c and other similar examples to make a conjecture about the relationship between $f[X - Y]$ and $f[X] - f[Y]$ for a general function f and general subsets X and Y.

h) Use part d and other similar examples to make a conjecture about the relationship between $f[\overline{X}]$ and $\overline{f[X]}$ for a general function f and general subset X.

i) Use examples to investigate how $f[X]$ and $f[Y]$ are related if $X \subseteq Y$.

15. Repeat Problem 14 for inverse images.

16. We can think of union and intersection as operations on the power set of a set. For instance, let $I : \mathcal{P}(\mathbb{R}) \times \mathcal{P}(\mathbb{R}) \to \mathcal{P}(\mathbb{R})$ be $I(A, B) = A \cap B$. Then $I([1, 5], (3, 9]) = (3, 5]$. Explain why I is onto. Is the corresponding function for union also onto? Are either one-to-one? Explain.

17. Let $A = \{a, b, c\}$ and $S = \{s, t\}$. We can describe one function f from A to S by $f(a) = t$, $f(b) = s$, and $f(c) = s$.

   *a) How many functions are there from A to S?

   b) How many functions are there from S to A?

   c) Suppose X has n elements and Y has k elements. How many functions are there from X into Y?

18. a) How many one-to-one functions are there from a set with two elements onto itself?

   b) How many one-to-one functions are there from a set with three elements onto itself?

   c) How many one-to-one functions are there from a set with n elements onto itself?

19. a) How many one-to-one functions are there from a set with three elements into a set with four elements?

   *b) How many one-to-one functions are there from a set with three elements into a set with five elements?

   c) How many one-to-one functions are there from a set with four elements into a set with seven elements?

   d) How many one-to-one functions are there from a set with k elements into a set with n elements, provided $k < n$?

3.2 COMPOSITION, ONE-TO-ONE, ONTO, AND INVERSES

Before we turn to proving general properties about functions in this section, it may be worthwhile to discuss some reasons mathematicians find the properties of this section so important.

In geometry, people have studied congruent figures for over two thousand years. In the last two hundred years, mathematicians have realized the power of using functions (geometric transformations) mapping a figure to a congruent one. Since we can go between congruent figures in either direction, these functions have to have inverse functions. As Theorem 3.2.3 shows, these transformations are one-to-one and onto. Further, they can be combined. Theorems 3.2.1 and 3.2.2 guarantee that compositions of such transformations are again one-to-one onto transformations. This approach has wonderfully transformed our understanding of geometry, many other areas of mathematics, and several areas of physics and chemistry.

Linear algebra is full of functions, called linear transformations and represented as matrices. Students quickly realize some matrices work more nicely than others. These matrices correspond to linear transformations that are one-to-one and onto.

Nearly every linear algebra text lists many properties equivalent to the one-to-one and onto characterization, including that the matrix is square and has an inverse. Matrix multiplication is designed to match composition of functions.

In algebra, topology, and other fields, mathematicians have found a fruitful way to show that two structures are similar. A one-to-one onto function with some additional relevant property provides the key.

Let's turn now to proving several properties of functions. As with most theorems in mathematics, the discovery of these properties grew out of studying examples, such as those in the previous problem set. Refer to Section 3.1 as needed for the definitions of the terms and to Chapter 2 for proof formats.

THEOREM 3.2.1. Suppose $f : X \to Y$ and $g : Y \to Z$ are functions. Then $g \circ f$ is a function from X into Z.

Proof. Suppose $f : X \to Y$ and $g : Y \to Z$ are functions. By definition of $g \circ f$, the domain and codomain are correct. We must show that $g \circ f$ satisfies the "vertical line test": For all $x \in X$, there is a unique $w \in Z$ such that $(x, w) \in g \circ f$.

Existence Let $x \in X$. Since f is a function, there is some $y \in Y$ such that $(x, y) \in f$ or, more familiarly, $f(x) = y$. Now g is a function, so for this y there is some $z \in Z$ such that $g(y) = z$. Thus, we can choose $w = z \in Z$ and obtain $g \circ f(x) = g(f(x)) = g(y) = z = w$.

Uniqueness Suppose for some $x \in X$, both s and t in Z satisfy $(x, s) \in g \circ f$ and $(x, t) \in g \circ f$. By definition of $g \circ f$, we have $u, v \in Y$ such that $f(x) = u$, $g(u) = s$, $f(x) = v$, and $g(v) = t$. Since f is a function, $u = f(x) = v$. In turn, since g is a function $s = g(u) = g(v) = t$. So, $s = t$, showing uniqueness. ∎

While $g \circ f$ is a function from X to Z in Theorem 3.2.1, $f \circ g$ need not be a function at all. For $f \circ g$ to be a function, the codomain of g would have to equal the domain of f; that is, $X = Z$. Let's investigate $f \circ g$ and $g \circ f$ when they are both functions. Are they equal? Rarely. For instance, consider the real functions f and g with $f(x) = x^2 + 1$ and $g(x) = x + 2$. Then $f \circ g(x) = (x + 2)^2 + 1 = x^2 + 4x + 5$, whereas $g \circ f(x) = (x^2 + 1) + 2 = x^2 + 3$. From these different formulas, you are probably convinced that the compositions differ. In particular, $f \circ g(0) = 5$, whereas $g \circ f(0) = 3$. This example raises the question: How do we prove two functions are the same or different? The following statement follows directly from the definition of a function as a set, so it hardly qualifies as a theorem:

Function Equality Two functions $f : A \to B$ and $g : C \to D$ are *equal* iff $A = C$ and for all x in A, $f(x) = g(x)$.

EXERCISE. Use the previous statement and working negations to determine when two functions are not equal.

EXAMPLE 1

Students often think the functions given by the equations $h(x) = \frac{x^2-2x}{x}$ and $j(x) = x - 2$ are the same since we can cancel an x in the numerator and denominator of $\frac{x^2-2x}{x}$ to get $x - 2$. However, this cancellation assumes that $x \neq 0$. If the domain of j is all of $\mathbb{R}$, then these functions are different because $j(0)$ exists, whereas $h(0)$ is undefined. If the common domain for h and j omits 0, then, indeed, the functions are equal because in $\mathbb{R} - \{0\}$ the two functions agree. $\Diamond$

REMARK. While codomains are not directly involved in determining function equality, codomains are essential in determining when functions are onto.

One-to-one onto functions are sufficiently important in mathematics that they have their own special terminology.

DEFINITION. A function $f : X \to Y$ is a *bijection* iff f is both one-to-one and onto.

If $X = Y$, then a bijection f is a *permutation* of X.

For instance, the function $p : \mathbb{R} \to \mathbb{R}$ with $p(x) = x^3$ is a bijection and a permutation of $\mathbb{R}$.

THEOREM 3.2.2.

a) Suppose $f : X \to Y$ and $g : Y \to Z$ are onto functions. Then $g \circ f : X \to Z$ is onto Z.

b) Suppose $f : X \to Y$ and $g : Y \to Z$ are one-to-one functions. Then $g \circ f : X \to Z$ is a one-to-one function from X into Z.

c) Suppose $f : X \to Y$ and $g : Y \to Z$ are bijections. Then $g \circ f : X \to Z$ is a bijection from X onto Z.

Proof.

a) Suppose $f : X \to Y$ and $g : Y \to Z$ are onto functions. By the previous result, we know $g \circ f$ is a function. To show $g \circ f$ is onto, let c be a general element of Z. We need to find a corresponding element d in X. Because g is onto Z, there is some $b \in Y$ such that $g(b) = c$. In turn, f is onto Y, so there is $a \in X$ so that $f(a) = b$. Pick $d = a$. Then $g \circ f(d) = g(f(a)) = g(b) = c$. Hence, $g \circ f$ is onto.

b) Suppose $f : X \to Y$ and $g : Y \to Z$ are one-to-one functions. To fulfill the definition of one-to-one for $g \circ f$, let $s, t \in X$ and suppose $g \circ f(s) = g \circ f(t)$. That is, $g(f(s)) = g(f(t))$. Since g is one-to-one, we can conclude that $f(s) = f(t)$. In turn, f is one-to-one, so $s = t$, showing that $g \circ f$ is one-to-one.

c) Since a bijection is a one-to-one onto function, we simply use parts a and b. ∎

EXAMPLE 2

The drawing on the left of Figure 1 shows a bijection g from a set S onto a set T. If we reverse the arrows, as in the second drawing, we get the inverse function $g^{-1} : T \to S$. Figure 2 shows a similar situation for the permutation h on a set U and its inverse. Theorem 3.2.3 shows that inverses are closely related to bijections.

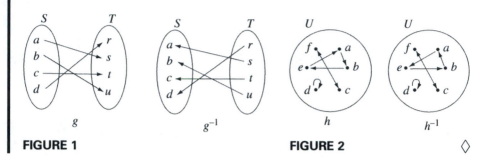

FIGURE 1   FIGURE 2 $\Diamond$

THEOREM 3.2.3. Let $f : X \to Y$ be a function. Then f^{-1} is a function from Y into X iff f is one-to-one and onto. Further, f^{-1} is one-to-one and onto.

Proof. Suppose $f : X \to Y$ is a function. ($\Rightarrow$) First, assume f^{-1} is a function from Y to X. To show f is onto, let $y \in Y$. Because f^{-1} is a function, there is a unique $x \in X$ such that $f^{-1}(y) = x$. But this existence gives $f(x) = y$ and f is onto. Further, the uniqueness of images for f^{-1} exactly fits one-to-one: Suppose $f(s) = f(t) = y$. Then $f^{-1}(y) = s$ and $f^{-1}(y) = t$, so $s = t$ and f is one-to-one.
　　($\Leftarrow$) See Problem 4 for this direction and the rest of the proof. ∎

In calculus, we talk about the inverse function even of functions that are not one-to-one and onto. To do so, we need to restrict the domain and codomain to fit the conditions of the previous theorem. Consider the exponential function defined by $f(x) = e^x$, whose range is the positive reals, $(0, \infty) = \mathbb{R}^+$, not all of $\mathbb{R}$. To get an inverse function, we simply restrict the codomain of f to its range: $f : \mathbb{R} \to \mathbb{R}^+$. Then the natural logarithm function is indeed the inverse function: $f^{-1}(x) = \ln(x)$. For functions that are not one-to-one, such as the sine function, we need to restrict both the domain and the codomain in order to get a one-to-one onto function. If we restrict the sine function to go from $[-\frac{\pi}{2}, \frac{\pi}{2}]$ to $[-1, 1]$, then its inverse function goes from $[-1, 1]$ to $[-\frac{\pi}{2}, \frac{\pi}{2}]$.

Counting Functions

How many functions are there from a set with n elements to a set with k elements? How about one-to-one functions? bijections? We answer these counting questions.

THEOREM 3.2.4. If X is a set with n elements and Y is a set with k elements, then there are k^n functions from X to Y.

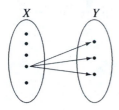

FIGURE 3. Each arrow represents one of the k choices each of the n elements of X has for its image.

Proof. A function is determined by where it sends each element of X. Now each of the n elements of X has k possible images (see Figure 3). Further, the image chosen for an element a in X has no effect on the image chosen for $b \in X$. That is, for each of the k choices for a, there are k choices for b, and so $k \cdot k = k^2$ choices for those two elements. Since there are n such elements in X, there are k^n ways to choose images for all elements of X, each of which gives a function. ∎

THEOREM 3.2.5. If X and Y are sets with n elements, then there are $n!$ bijections from X onto Y.

Proof. We will use induction on the number n of elements in X and Y. When $n = 1$, there is just one choice for the image of the one element of X, namely, the only element of Y. Thus, there is just one function, which is clearly one-to-one and onto, and so is a bijection. Of course, $1! = 1$, so the theorem is correct in this case.

For the induction step, suppose that for any two sets with $n = k$ elements there are $k!$ bijections from one to the other. Let X and Y each have $k + 1$ elements, say, $X = \{x_i : i \le n \text{ and } i \in \mathbb{N}\}$ and $Y = \{y_j : j \le n \text{ and } j \in \mathbb{N}\}$. From the proof of Theorem 3.2.4, we know x_1 has $k + 1$ possible choices for its image in Y. Let's call the image $f(x_1) = y_p$. Now consider $X' = X - \{x_1\}$ and $Y' = Y - \{y_p\}$. These are sets with k elements. For $f : X \to Y$ to be a bijection, it needs to map X' one-to-one onto Y', as well as take x_1 to y_p. By the induction hypothesis, we know there are $k!$ ways to map X' one-to-one onto Y' and still have $f(x_1) = y_p$. Hence, there are $(k + 1)k! = (k + 1)!$ possible bijections from X onto Y. By the Principle of Mathematical Induction, the theorem holds for all $n \in \mathbb{N}$. ∎

COROLLARY 3.2.6. If X is a set with n elements, there are $n!$ permutations on X.

Proof. Use Theorem 3.2.5 with $X = Y$. ∎

We leave the proof for the number of one-to-one functions to Problem 8.

PROBLEMS

*1. For each statement that follows, decide whether it is true or (at least sometimes) false. Explain your answer or cite the part of the text supporting it.

a) The composition of any two functions is a function.

b) If $f : A \to B$, $g : B \to C$, and $h : C \to D$ are one-to-one, then $f \circ (g \circ h)$ is one-to-one.

c) If $f : A \to B$, $g : B \to C$, and $h : C \to D$ are one-to-one, then $h \circ (g \circ f)$ is one-to-one.

d) Bijections always have inverse functions.

e) The number of functions from a set with r elements to itself is $r!$.

2. ***a)** Prove $a : \mathbb{Z} \to \mathbb{Z}$ defined by $a(x) = 23 - x$ is one-to-one and onto.

b) Prove $b : \mathbb{R} \to \mathbb{R}$ defined by $b(x) = 7x - 5$ is one-to-one and onto.

c) Prove $c : \mathbb{Z} \to \mathbb{Z}$ defined by $c(x) = 7x - 5$ is one-to-one, but not onto.

d) Prove $d : \mathbb{R} \to \mathbb{R}$ defined by $d(x) = x^4 - x^2$ is neither one-to-one nor onto.

e) Prove $e : \mathbb{N} \to \mathbb{N}$ defined by $e(x) = x^2$ is one-to-one, but not onto.

f) Prove $f : \mathbb{R} \to \mathbb{R}$ defined by $f(x) = x^5 + 7$ is one-to-one and onto.

g) Prove $g : \mathbb{R} \to \mathbb{Z}$ defined by $g(x) = \lfloor x \rfloor$ is onto, but not one-to-one. (The floor function $\lfloor x \rfloor$ is defined in Problem 13 of Section 3.1.)

h) Prove $h : \mathbb{Z} \to \mathbb{N}$, where $h(x) = \begin{cases} 2x & \text{if } x > 0 \\ 1 - 2x & \text{if } x \le 0 \end{cases}$ is one-to-one and onto.

i) Prove $i : [1, \infty) \to [-1, \infty)$ defined by $i(x) = x^2 - 2x$ is one-to-one and onto. Hints: Use the quadratic formula and note the actual domain and codomain.

3. **a)** Prove that the function j defined in Problem 4c of Section 3.1 equals the floor function $\lfloor \ \rfloor$ defined in Problem 13.

b) We define $\lceil x \rceil$ to be the least integer greater than or equal to x. We call $\lceil \ \rceil : \mathbb{R} \to \mathbb{Z}$ the *ceiling function*. Prove for all $x \in \mathbb{R}$ that $\lceil x \rceil = -\lfloor -x \rfloor$ and $\lfloor x \rfloor = -\lceil -x \rceil$.

4. Complete the proof of Theorem 3.2.3.

5. For a function $f : X \to Y$ and $W \subseteq X$, define the *restriction* of f to W to be $f|_W : W \to Y$ given by $f|_W(w) = f(w)$ for all $w \in W$ (referred to in Section 5.2).

***a)** For $f : \mathbb{R} \to \mathbb{R}$ given by $f(x) = \cos(x)$ and $W = [0, \pi]$, graph $f|_W$. Why might one be interested in this restriction in a calculus class?

b) If $f : X \to Y$ is a function and $W \subseteq X$, prove $f|_W$ is a function from W into Y.

c) Suppose $f : X \to Y$ is one-to-one and $W \subseteq X$. Prove $f|_W$ is one-to-one.

d) Give an example to show $f|_W$ need not be onto even if f is onto.

e) Suppose $f : X \to Y$ is a function, $W \subseteq X$, and $f|_W$ is onto Y. Prove f is onto.

6. For a function $f : \mathbb{R} \to \mathbb{R}$, define f to be *increasing* on $\mathbb{R}$ iff for all $s, t \in \mathbb{R}$, if $s < t$, then $f(s) < f(t)$. For instance, f given by $f(x) = 3x + 2$ is increasing.

***a)** Prove that if f is increasing on $\mathbb{R}$, then f is one-to-one.

b) Assume that f is increasing on $\mathbb{R}$ and it has an inverse function. Prove that its inverse function is increasing on $\mathbb{R}$.

c) Prove that the composition of increasing functions is increasing.

d) Define a *decreasing* function on $\mathbb{R}$.

e) Redo part a, changing increasing to decreasing.

f) Redo part b, changing increasing to decreasing.

g) What can you say about the composition of two decreasing functions? three decreasing functions? Prove your answers and generalize.

7. Modify the proof of Theorem 3.2.4 to give an alternative proof of Theorem 3.2.5. Hint: How many choices are there for the image of a particular element of X? Once that image is chosen, how many choices are there for a second element of X?

8. Modify the reasoning from Problem 7 to show that the number of one-to-one functions from a set X with k elements to a set Y with n elements, $k < n$, is the product $n(n-1)(n-2)\cdots(n-k+1)$, which is often rewritten as $\frac{n!}{(n-k)!}$.

9. Suppose $f : X \to Y$ and $g : Y \to Z$ are functions and $g \circ f$ is one-to-one.

   a) Determine which one of the functions f and g must be one-to-one, and prove your answer.

   b) Give an example showing that the other function need not be one-to-one.

10. Suppose $f : X \to Y$ and $g : Y \to Z$ are functions and $g \circ f$ is onto.

    a) Determine which of the functions f and g must be onto, and prove your answer (referred to in Problem 16 of Section 5.1).

    b) Give an example showing that the other function need not be onto.

11. For a set X, define the *identity* i_X on X by $i_X(w) = w$, where w is any element of X. Recall that a permutation on a set X is a one-to-one function from X onto X.

    a) For any set X, prove that i_X is a permutation on X.

    b) The composition of two permutations on X is a permutation on X (referred to in Sections 7.1 and 7.2).

    *c) Show, for any function $f : X \to X$, that $f \circ i_X = f = i_X \circ f$. (These equations justify the name identity. In arithmetic, 0 is the additive identity because $0 + w = w = w + 0$. Similarly, 1 is the multiplicative identity.)

    d) Show that if f is a permutation on X, then f^{-1} exists and is a permutation and $f \circ f^{-1} = i_X = f^{-1} \circ f$.

    e) Show that if f, g, and h are permutations on X, then $f \circ (g \circ h) = (f \circ g) \circ h$. Hint: What is the image of x for each of these compositions?

        (The previous parts show that the set of all permutations forms a "group," an important type of structure in abstract algebra. See Chapter 7.)

    f) Suppose that f and g are permutations on X. Prove $(f \circ g)^{-1} = g^{-1} \circ f^{-1}$. Hint: Let $p \in X$ and $f^{-1}(p) = q$. Now give a name to $g^{-1}(q)$ and see what $f \circ g$ does to that element.

12. Suppose $f : X \to Y$ and $g : Y \to X$ are functions and $g \circ f = i_X$, the identity function on X. What must be true about f? about g? Prove your answers.

13. Critique the incorrect "proofs" that follow. Point out reasoning errors and unclear or incomplete presentations. (If the claim is false, there must be an error in the argument. However, it is not enough to say that the claim is false. You need to find an error in the reasoning.)

    *a) Suppose $f : X \to Y$ and $g : Y \to Z$ are onto functions. Claim: Then $g \circ f$ is onto Z. "Proof": Suppose that f and g are functions and let $x \in X$. Because f is onto, we get $y \in Y$ with $f(x) = y$. Again, g is onto, so we have $z \in Z$ with $g(y) = z$. Thus, $g(f(x)) = g(y) = z$, showing onto. ▲

    b) Claim: The function $h : \mathbb{N} \times \mathbb{N} \to \mathbb{N}$ given by $h(k, n) = k - n$ is one-to-one. "Proof": Let $j, k, n \in \mathbb{N}$, and suppose $h(j, n) = h(k, n)$. Then $j - n = k - n$, which forces $j = k$. Similarly, let $k, n, p \in \mathbb{N}$, and suppose $h(k, n) = h(k, p)$. Then $k - n = k - p$, which forces $n = p$. Either way, we have one-to-one. ▲

c) Suppose $f : X \to Y$ and $g : Y \to Z$ are functions and $g \circ f$ is one-to-one. Claim: Both f and g are one-to-one. "Proof": Suppose $f : X \to Y$ and $g : Y \to Z$ are functions and $g \circ f$ is one-to-one. Let $s, t \in X$, and suppose $g \circ f(s) = g \circ f(t)$. Because $g \circ f$ is one-to-one, we know $s = t$. But then, $f(s) = f(t)$ in Y. Thus, we have both $s = t$ and $f(s) = f(t)$, showing f is one-to-one. Now let's show g is one-to-one: Let $f(s), f(t) \in Y$, and suppose $g(f(s)) = g(f(t))$. Since we already showed $f(s) = f(t)$, g is one-to-one. Thus, both f and g are one-to-one. ▲

d) Suppose X is a finite set with k elements, Y is a finite set with n elements, and $k < n$. Claim: Then the number of one-to-one functions from X to Y equals the number of onto functions from Y to X. "Proof": Suppose X is a finite set with k elements, Y is a finite set with n elements, and $k < n$. Suppose $f : X \to Y$ is a one-to-one function. Simply switch the order of all the ordered pairs (x, y) in f to get the ordered pairs (y, x) from Y to X. Because f is one-to-one, each $y \in Y$ is matched to a unique $x \in X$, ensuring we have a function (which is just f^{-1}). Further, the domain of f is all of X, so the switched function f^{-1} is onto all of X. ▲

14. Critique the questionable "proofs" that follow. Point out reasoning errors and unclear or incomplete presentations. If the argument is a proof, say so. (If the claim is false, there must be an error in the argument. However, it is not enough to say that the claim is false. You need to find an error in the reasoning.)

a) Claim: For $f : X \to X$ and $g : X \to X$, $f \circ g = g \circ f$. "Proof": Consider $f(x) = 5x$ and $g(x) = 3x$. Then for any $x \in \mathbb{R}$, $g \circ f(x) = 3(5x) = 15x = 5(3x) = f \circ g(x)$. ▲

*b) Suppose $f : X \to Y$ and $g : Y \to X$ are functions and $g \circ f$ is the identity on X. Claim: g is the inverse function of f. "Proof": Suppose $f : X \to Y$ and $g : Y \to X$ are functions and $g \circ f$ is the identity on X. That is, $g \circ f = i_X$. We compose both sides with f^{-1} to get $(g \circ f) \circ f^{-1} = i_X \circ f^{-1}$. From Problem 11c, the right side becomes $i_X \circ f^{-1} = f^{-1}$. Similarly, we use Problem 11d and e for the left side: $(g \circ f) \circ f^{-1} = g \circ (f \circ f^{-1}) = g \circ i_Y = g$. Then $(g \circ f) \circ f^{-1} = i_X \circ f^{-1}$ reduces to $g = f^{-1}$. ▲

c) Define $f : \mathbb{R} \to \mathbb{R}$ by $f(x) = \sqrt[3]{x} - 5$. Claim: Then f is onto all of $\mathbb{R}$. "Proof": For a contradiction, suppose f is not onto. Then there is some y that isn't the image of any x. But consider $x = y^3 + 5$. $f(x) = \sqrt[3]{(y^3 + 5) - 5} = \sqrt[3]{y^3} = y$—contradiction. So f is onto. ▲

d) Suppose $f : X \to Y$ and $g : Y \to Z$ are one-to-one functions. Claim: Then $g \circ f$ is a one-to-one function from X into Z. "Proof": Suppose $f : X \to Y$ and $g : Y \to Z$ are one-to-one functions. Let s and t be different elements of X. By the contrapositive of the definition of one-to-one, $f(s) \neq f(t)$. Similarly, $g(f(s)) \neq g(f(t))$. That is, if $s \neq t$, then $g \circ f(s) \neq g \circ f(t)$, showing $g \circ f$ is one-to-one. ▲

e) Define $f : (0, 1) \to (0, 1)$ by $f(x) = x^x$. Claim: f is one-to-one. "Proof": For f as given in the problem, the numbers we are raising to powers are between 0 and 1. Now for $0 < a < 1$ and $b < c$, we have $a^b > a^c$. (Consider $0.5^2 \doteq 0.25 > 0.125 = 0.5^3$.) Thus, if $b < c$, then $b^b > b^c > c^c$. Hence, f is decreasing, as defined in Problem 6. By part e of that problem, f is one-to-one. ▲

3.3 IMAGES AND PRE-IMAGES OF SETS

This section focuses on properties of images $f[A]$ and pre-images $f^{-1}[B]$. Recall that any function has pre-images, even if it doesn't have an inverse function. After restating the definitions of image and pre-image, we'll use examples to conjecture some properties involving images and pre-images with unions and intersections of sets. After the corresponding theorems and proofs, we will provide a taste of why mathematicians find images and pre-images of sets interesting and important.

DEFINITION. Let $f : X \to Y$ be a function. If $A \subseteq X$, then $f[A] = \{y \in Y : \exists a \in A \ s.t. \ y = f(a)\}$. We call $f[A]$ the *image* of A. If $B \subseteq Y$, then $f^{-1}[B] = \{x \in X : f(x) \in B\}$. We call $f^{-1}[B]$ the *pre-image* of B.

EXAMPLE 1

Let $f : [0, 8] \to [0, 18]$ be $f(x) = 8x - x^2$ and $A = [0, 3]$ and $B = [2, 7]$. Compare $f[A] \cup f[B]$ and $f[A \cup B]$. Compare $f[A] \cap f[B]$ and $f[A \cap B]$.

Solution From the graph in Figure 1, we see $f[A] = [0, 15]$ and $f[B] = [7, 16]$, which means $f[A] \cup f[B] = [0, 16]$. Similarly, $f[A \cup B] = f[[0, 7]] = [0, 16]$. We might guess this holds in general for any sets A and B and any function f. Theorem 3.3.1 confirms this conjecture.

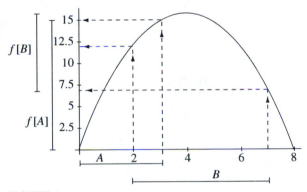

FIGURE 1

The situation is different with intersection. From $f[A] = [0, 15]$ and $f[B] = [7, 16]$, we know $f[A] \cap f[B] = [7, 15]$. In contrast, $A \cap B = [2, 3]$ and $f[A \cap B] = [12, 15]$, which doesn't equal $f[A] \cap f[B]$. The best we could hope for in general is $f[A \cap B] \subseteq f[A] \cap f[B]$. Theorem 3.3.1 turns this hope into a proof. $\diamondsuit$

EXAMPLE 2

Let $f : [0, 8] \to [0, 18]$ be $f(x) = 8x - x^2$, as in Example 1, and $C = [7, 18]$ and $D = [-2, 12]$. Compare $f^{-1}[C] \cup f^{-1}[D]$ and $f^{-1}[C \cup D]$. Compare $f^{-1}[C] \cap f^{-1}[D]$ and $f^{-1}[C \cap D]$.

Solution Figure 2 illustrates that $f^{-1}[C] = [1, 7]$ and $f^{-1}[D] = [0, 2] \cup [6, 8]$, implying $f^{-1}[C] \cup f^{-1}[D] = [0, 8]$. Since $C \cup D = [-2, 18]$ covers the entire codomain, $f^{-1}[C \cup D] = [0, 8]$, the entire domain. For the intersection, $C \cap D = [7, 12]$ and both $f[C \cap D]$ and $f^{-1}[C] \cap f^{-1}[D]$ equal $[1, 2] \cup [6, 7]$. Theorem 3.3.2 confirms that the equalities $f^{-1}[C] \cup f^{-1}[D] = f^{-1}[C \cup D]$ and $f^{-1}[C] \cap f^{-1}[D] = f^{-1}[C \cap D]$ hold in general.

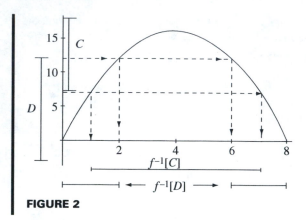

FIGURE 2 ◇

THEOREM 3.3.1. Let $f : X \to Y$ be a function and A and B be subsets of X. Then

   a) If $A \subseteq B$, then $f[A] \subseteq f[B]$.

   b) $f[A \cup B] = f[A] \cup f[B]$.

   c) $f[A \cap B] \subseteq f[A] \cap f[B]$.

Proof. Let $f : X \to Y$ be a function and A and B be subsets of X.

   a) Suppose $A \subseteq B$. To show the containment in the conclusion, let $y \in f[A]$. By definition of $f[A]$, we can write $y = f(a)$ for some $a \in A$. Now, $a \in B$, since $A \subseteq B$. So $y \in f[B]$, showing that $f[A] \subseteq f[B]$.

   b) We need to show containment in each direction. First for $f[A \cup B] \subseteq f[A] \cup f[B]$, let $y \in f[A \cup B]$. Then there is some $x \in A \cup B$ such that $f(x) = y$. Then $x \in A$ or $x \in B$, leading to two cases. Case 1. Suppose $x \in A$. Then $y = f(x) \in f[A]$, so $y \in f[A] \cup f[B]$. Case 2, $x \in B$, is similar.

      Next, we show that $f[A] \cup f[B] \subseteq f[A \cup B]$. Note that $A \subseteq A \cup B$, so, from part a, $f[A] \subseteq f[A \cup B]$. Similarly, from $B \subseteq A \cup B$, we have $f[B] \subseteq f[A \cup B]$. Since everything in $f[A] \cup f[B]$ is in $f[A]$ or in $f[B]$, all of $f[A] \cup f[B]$ is in $f[A \cup B]$. Since both containments hold, the two sets are equal.

   c) See Problem 5. ∎

THEOREM 3.3.2. Let $f : X \to Y$ be a function and C and D subsets of Y.

   a) If $C \subseteq D$, then $f^{-1}[C] \subseteq f^{-1}[D]$.

   b) $f^{-1}[C \cup D] = f^{-1}[C] \cup f^{-1}[D]$.

   c) $f^{-1}[C \cap D] = f^{-1}[C] \cap f^{-1}[D]$.

Proof. We prove one direction of part c. See Problem 7 for the rest.

To show $f^{-1}[C] \cap f^{-1}[D] \subseteq f^{-1}[C \cap D]$, let $x \in f^{-1}[C] \cap f^{-1}[D]$. Then $x \in f^{-1}[C]$ and $x \in f^{-1}[D]$. By definition of pre-images, $f(x) \in C$ and $f(x) \in D$. Then $f(x) \in C \cap D$. Again, the definition of pre-images assures us that $x \in f^{-1}[C \cap D]$, finishing this direction. ∎

Finally, we combine images and pre-images together in the following theorem:

THEOREM 3.3.3. Let $f : X \rightarrow Y$ be a function, $A \subseteq X$, and $C \subseteq Y$. Then

a) $A \subseteq f^{-1}[f[A]]$.

b) If f is one-to-one, then $A = f^{-1}[f[A]]$.

c) $f[f^{-1}[C]] \subseteq C$.

d) $f[f^{-1}[C]] = C$ iff C is a subset of the range of f.

Proof. See Problem 12. ∎

Images of sets give us a way to reinterpret two famous theorems of calculus: the Intermediate Value Theorem and the Extreme Value Theorem. (See Problems 12 and 14, respectively, of Section 8.3 for their statements.) Suppose $f : \mathbb{R} \rightarrow \mathbb{R}$ is a continuous function on a closed bounded interval $[a, b]$. The Intermediate Value Theorem guarantees that every y-value between $f(a)$ and $f(b)$ is in the range of f. That is, the range contains that interval. Since this holds for all choices a and b, intervals must map to intervals. The Extreme Value Theorem promises that such a function achieves the maximum and minimum of its range. That is, the interval of the range is a closed bounded interval. In short, $f[[a, b]] = [\min(f), \max(f)]$, where $\min(f)$ is the minimum of f on $[a, b]$ and $\max(f)$ is the maximum of f on $[a, b]$.

More advanced mathematics makes extensive use of images and pre-images of sets. Analysis, the theoretical foundation of calculus, also uses pre-images. Pre-images of intervals need not be intervals. For instance, for $f(x) = \sin(x)$, we have $f^{-1}[(0, 2)] = \bigcup_{n \in \mathbb{Z}} (2n\pi, 2n\pi + \pi)$, as the graph in Figure 3 suggests. Analysts characterize continuous functions as those for which the pre-image of any open set is an open set. As shown in analysis, open sets in $\mathbb{R}$ are unions of open intervals. The pre-image description enables mathematicians to study continuous functions in many contexts beyond $\mathbb{R}$, settings where the limit characterization of continuity used in calculus might not make sense.

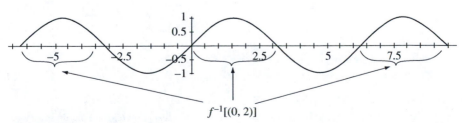

FIGURE 3. $f(x) = \sin(x)$.

In linear algebra, we use a function $f : V \rightarrow W$, called a linear transformation, to map a vector space V to another vector space W. One can show that if S is a subspace of V, then $f[S]$ is a subspace of W. Also, if T is a subspace of W, then $f^{-1}[T]$ is a subspace of V. Abstract algebra generalizes and explores this connection between special types of subsets in different systems.

PROBLEMS

*1. For each statement that follows, decide whether it is true or (at least sometimes) false. If false, explain why. Explain your answer or cite the part of the text supporting it.

   a) For a continuous function, the pre-image of an interval is an interval.

   b) For a continuous function, the image of an interval is an interval.

   c) For all functions $f : J \rightarrow K$ and all subsets S, T, and W of J, $f[S \cup T \cup W] = f[S] \cup f[T] \cup f[W]$.

   d) For all functions $f : J \rightarrow K$ and all subsets L of K, $L = f[f^{-1}[L]]$.

2. Let $h : \mathbb{R} \rightarrow \mathbb{R}$ be given by $h(x) = 2x^2 - 4$.

   *a) Find $h[[-1, 3]]$.

   b) Find $h[(-4, 2)]$.

   c) Find a set A so that $h[[1, 3]] = h[A]$ and $A \cap [1, 3] = \emptyset$.

   d) Find $h^{-1}[[-6, 0]]$.

   e) Find $h^{-1}[(-2, 2)]$.

   f) Find two nonempty sets A and B so that $A \cap B = \emptyset$ and $h^{-1}[A] = h^{-1}[B]$.

3. a) For $g : \mathbb{R} \rightarrow \mathbb{R}$ given by $g(x) = 2x - x^2$, determine $g[\emptyset]$, $g[\mathbb{R}]$, $g^{-1}[\emptyset]$, and $g^{-1}[\mathbb{R}]$.

   b) For $h : \mathbb{R} \rightarrow \mathbb{R}$ given by $h(x) = \cos(x)$, determine $h[\emptyset]$, $h[\mathbb{R}]$, $h^{-1}[\emptyset]$, and $h^{-1}[\mathbb{R}]$.

4. Let $f : X \rightarrow Y$ be any function.

   a) Determine $f[\emptyset]$, $f^{-1}[\emptyset]$, and $f^{-1}[Y]$. Prove your answers.

   b) What condition on f permits you to show that $f[X] = Y$? Prove your answer.

   c) Describe all subsets B of Y for which $f^{-1}[B] = \emptyset$. Prove your answer.

   d) Describe all subsets B of Y for which $f^{-1}[B] = X$. Prove your answer.

5. *a) Give an example of a function $f : X \rightarrow Y$ and subsets A and B of X, showing that Theorem 3.3.1 part c can't in general be strengthened to an equality.

   b) Prove Theorem 3.3.1 part c.

   c) Show that if f is one-to-one, then $f[A \cap B] = f[A] \cap f[B]$.

6. Let f be a function $f : X \rightarrow Y$ and A and B subsets of X.

   *a) Prove that $f[A - B] \subseteq f[A]$.

   b) Give an example of f, A, and B, where $f[A]$, $f[B]$ and $f[A - B]$ are all nonempty sets, but $f[A] - f[B] = \emptyset$.

   c) Show, for all f, A, and B, that $f[A] - f[B] \subseteq f[A - B]$.

   d) Show that if f is one-to-one, then $f[A - B] = f[A] - f[B]$.

7. Prove the rest of Theorem 3.3.2.

8. Either prove or find a counterexample for the following conjecture: For all functions $f : X \rightarrow Y$ and all subsets C and D of Y, $f^{-1}[C - D] = f^{-1}[C] - f^{-1}[D]$.

9. Let f be a function $f : X \to Y$ and $\{A_i : i \in I\}$ be a family of subsets of X.

   a) Generalize and prove part b of Theorem 3.3.1, using $\bigcup_{i \in I} A_i$ in place of $A \cup B$.

   b) Generalize and prove part c of Theorem 3.3.1, using $\bigcap_{i \in I} A_i$ in place of $A \cap B$.

   c) Generalize your example from Problem 5a by giving an infinite family of subsets $\{A_i : i \in I\}$ instead of just two subsets.

10. Let f be a function $f : X \to Y$ and $\{D_i : i \in I\}$ be a family of subsets of Y.

    a) Generalize and prove part b of Theorem 3.3.2, using $\bigcup_{i \in I} D_i$ in place of $C \cup D$.

    b) Generalize and prove part c of Theorem 3.3.2, using $\bigcap_{i \in I} D_i$ in place of $C \cap D$.

11. For $f : X \to Y$, the set function $f[A]$ can be thought of as the function $F : \mathcal{P}(X) \to \mathcal{P}(Y)$ defined, for any subset A of X, by $F(A) = f[A]$.

    *a) If f is one-to-one, is F as well? If so, prove it; if not, give a counterexample.

    b) If f is onto, is F as well? If so, prove it; if not, give a counterexample.

    c) If F is one-to-one, is f as well? If so, prove it; if not, give a counterexample.

    d) If F is onto, is f as well? If so, prove it; if not, give a counterexample.

12. Prove Theorem 3.3.3.

13. Suppose that $\{S_i : i \in I\}$ is a partition of a nonempty set Y and that $f : X \to Y$ is a function from a nonempty set X onto Y. Prove $\{f^{-1}[S_i] : i \in I\}$ is a partition of X.

14. Critique the incorrect "proofs" that follow. Point out reasoning errors and unclear or incomplete presentations. (If the claim is false, there must be an error in the argument. However, it is not enough to say that the claim is false. You need to find an error in the reasoning.)

    *a) Let $f : X \to Y$ and A and B be subsets of X. Claim: If $f[A] = f[B]$, then $A = B$. "Proof": Suppose $f[A] = f[B]$, where $f : X \to Y$ and $A \subseteq B \subseteq X$. From Theorem 3.3.1 and $A \subseteq B$, we have $f[A] \subseteq f[B]$. Now $f[A] = f[B]$ also means $f[B] \subseteq f[A]$, which means $B \subseteq A$. But from $A \subseteq B$ and $B \subseteq A$, we get $A = B$, finishing the proof. ▲

    b) Let $f : X \to Y$ and A be a subset of X. Claim: $f^{-1}[f[A]] \subseteq A$. "Proof": Let $a \in f^{-1}[f[A]]$, where $f : X \to Y$ and $A \subseteq X$. Then $f(a) \in f[A]$, which in turn means $a \in A$. Thus, $f^{-1}[f[A]] \subseteq A$. ▲

    c) Let $f : X \to Y$ and $\{A_i : i \in I\}$ be a family of subsets of X. Claim: $\bigcap_{i \in I} f[A_i] \subseteq f[\bigcap_{i \in I} A_i]$. "Proof": Let $f : X \to Y$ and $\{A_i : i \in I\}$ be a family of subsets of X. Suppose $y \in \bigcap_{i \in I} f[A_i]$. Then for all $i \in I$, $y \in f[A_i]$. By definition of $f[A_i]$, there is $x \in A_i$ such that $y = f(x)$. Since this holds for each $i \in I$, $x \in \bigcap_{i \in I} A_i$, which shows $y \in f[\bigcap_{i \in I} A_i]$. This completes the proof. ▲

15. Critique the questionable "proofs" that follow. Point out reasoning errors and unclear or incomplete presentations. If the argument is a proof, say so. (If the claim is false, there must be an error in the argument. However, it is not enough to say that the claim is false. You need to find an error in the reasoning.)

    a) Let $f : X \to Y$ and C and D be subsets of Y. Claim: $f^{-1}[C \cup D] = f^{-1}[C] \cup f^{-1}[D]$. "Proof": For $f : X \to Y$, and C and D subsets of Y, we have
    $$f^{-1}[C \cup D] = \{x : f(x) \in C \cup D\} = \{x : f(x) \in C \text{ or } f(x) \in D\}$$
    $$= \{x : f(x) \in C\} \cup \{x : f(x) \in D\} = f^{-1}[C] \cup f^{-1}[D]. ▲$$

    b) Let $f : X \to Y$ and A and B be subsets of X. Claim: $f[A \cup B] = f[A] \cup f[B]$. "Proof": Let $f : X \to Y$ and A and B be subsets of X. We apply the definitions of $f[S]$ and of union:

$$f[A \cup B] = \{f(x) : x \in A \cup B\} = \{f(x) : x \in A \text{ or } x \in B\}$$
$$= \{f(x) : x \in A\} \cup \{f(x) : x \in B\} = f[A] \cup f[B]. \ \blacktriangle$$

*c) Let $f : X \to Y$ and C and D be subsets of Y. Claim: If $C \subseteq D$, then $f^{-1}[C] \subseteq f^{-1}[D]$. "Proof": Suppose $f : X \to Y$ and $C \subseteq D \subseteq Y$. Then $y \in C \Rightarrow y \in D$. Suppose $f(x) = y$. Then $f(x) \in C \Rightarrow f(x) \in D$. By definition of f^{-1}, this gives us $x \in f^{-1}[C] \Rightarrow x \in f^{-1}[D]$. But that simply means $f^{-1}[C] \subseteq f^{-1}[D]$, as desired. $\blacktriangle$

DEFINITIONS FROM CHAPTER 3

Section 3.1

- A *function f* from a set X into a set Y is a rule that assigns to each $x \in X$ a unique $y = f(x) \in Y$.

  A *function f* from a set X into a set Y is a set of ordered pairs $f = \{(x, y) : x \in X \text{ and } y \in Y\}$ such that for all $x \in X$, there is a unique $y \in Y$ such that $(x, y) \in f$. We almost always use the familiar $f(x) = y$ rather than the formal $(x, y) \in f$. We write $f : X \to Y$ to indicate that f is a function from X into Y. We call X the *domain* of f and Y its *codomain*. The *image* of x is $f(x)$. The *range* of f is $\{y \in Y : \exists x \in X \text{ s.t. } f(x) = y\}$. That is, the range is the set of all images of elements of X.

- A function $f : X \to Y$ is *onto* iff for all $y \in Y$, there is $x \in X$ such that $f(x) = y$. That is, the range of f is all of Y. (Such a function is also called *surjective*.)

- A function $f : X \to Y$ is *one-to-one* iff for all $s, t \in X$, if $f(s) = f(t)$, then $s = t$. (Such a function is also called *injective*.)

- Given functions $f : X \to Y$ and $g : Y \to Z$, their *composition* is $g \circ f$ defined by $g \circ f = \{(x, z) : \exists y \in Y \text{ s.t. } f(x) = y \wedge g(y) = z\}$. In more informal terms, we write $g \circ f(x) = g(f(x))$.

- Given a function $f : X \to Y$, its *inverse* is $\{(y, x) : y \in Y, x \in X \text{ and } (x, y) \in f\}$. If this set of ordered pairs is a function from Y to X, we call it the *inverse function* of f and write f^{-1}.

- Let $f : X \to Y$ be a function. If $A \subseteq X$, then $f[A] = \{y \in Y : \exists a \in A \text{ s.t. } y = f(a)\}$. We call $f[A]$ the *image* of A. If $B \subseteq Y$, then $f^{-1}[B] = \{x \in X : f(x) \in B\}$. We call $f^{-1}[B]$ the *pre-image* of B.

Section 3.2

Function Equality

- Two functions $f : A \to B$ and $g : C \to D$ are equal iff $A = C$ and for all x in A, $f(x) = g(x)$.

- A function $f : X \to Y$ is a *bijection* iff f is both one-to-one and onto. If $X = Y$, then f is a *permutation* of X.

CHAPTER 4

RELATIONS

"Mathematics is concerned only with the enumeration and comparison of relations."

—Carl Friedrich Gauss (1777–1855)

IN **GEOMETRY,** two lines can be related in special ways, such as being parallel or perpendicular. These are examples of relations and, as specific examples, were investigated even in antiquity. While mathematicians have studied general functions for over 350 years, the corresponding idea of a mathematical relation is more recent. In the second half of the nineteenth century, with the advent of an abstract approach to mathematics, mathematicians saw the benefit of considering formal properties. Further, the development of set theory and logic at the same time provided a language to formulate relations in general.

4.1 RELATIONS

"We think in generalities, but we live in details."

—Alfred North Whitehead (1861–1947)

In previous mathematics courses, you have encountered many specific relations, including $<$ (less than), $\parallel$ (parallel), and $\subseteq$ (subset). However, you may not have studied relations or their properties in general, which we undertake in this chapter. Although relations seem very different from functions, our definition of a relation will confirm that all functions are relations. Let's start by describing how functions and relations appear different, before we connect them. Functions take inputs and give outputs: The squaring function takes 3 and gives 9. However, "less than" and other relations don't seem to "give" anything in response to inputs. Rather, "$2 < 5$" and "$\{2, 3\} \subseteq \{1, 3, 5\}$" are propositions, the first true and the second false. We can think of functions this way as well: "$3^2 = 9$" is a true proposition.

Set theory gives a useful mathematical language for both functions and relations. From the set theory definition of a relation, given next, the relation $<$ is simply the collection of instances where it is true. That idea might not seem helpful, but it matches our definition of a function as a set of ordered pairs (with other conditions). Thus, the squaring function consists of all the pairs making the equation true, such as $(3, 9)$ and $(-2, 4)$. After the formal definition of a relation, we give examples of some relations and their associated sets. Since all of the relations we will consider in this text are binary relations—ones with two inputs—we will simply omit the adjective "binary" after the definition. (Ternary relations have three inputs, such as betweenness: "4 is between 1 and 6." Unary relations have one input, such as "5 is prime.")

DEFINITION. A *(binary) relation R* from a set X to a set Y is a subset of $X \times Y$. A subset of $X \times X$ is called a *relation on X*. We generally write *aRb*, rather than the more formal $(a, b) \in R$, to indicate that a and b are related by R.

EXAMPLE 1

Numerical Relations. Less than ($<$), equals ($=$), divides ($|$), and congruence ($\equiv (\bmod n)$) can be defined as relations on $\mathbb{N}$ or $\mathbb{Z}$. (Of course, $<$ is a relation on any subset of $\mathbb{R}$, and $=$ is a relation on any set.)

Our usual way of writing relations, such as "$2 < 5$," with the relation between the elements, motivates the convention to write *aRb* in the preceding definition. We can modify $<$ in various ways to get other relations, such as $\le$, $>$, and $\ge$.

According to our definition of a relation, the relation $<$ on $\{1, 2, 3\}$ is the set $\{(1, 2), (1, 3), (2, 3)\}$. Such a formal definition completely disguises the meaning of "less than," although it is hard to fault it logically. This formal designation does allow us to say that the relation $\le$ is the union of the relations $<$ and $=$ on the set $\mathbb{R}$. $\Diamond$

EXAMPLE 2

Geometric Relations. Parallel ($\|$) and perpendicular ($\perp$) are relations on the set of all lines in the Euclidean plane. Congruence ($\cong$) and similarity ($\sim$) are relations on the set of all triangles in the Euclidean plane. (We can define these relations for other sets as well.) Incidence (a point "is on" a line) is a relation from the set of points to the set of lines. The condition of a line being tangent to a circle can be seen as a relation from the set of lines to the set of circles. $\Diamond$

EXAMPLE 3

Set Theoretic Relations. We can think of subset ($\subseteq$) as a relation on the power set $\mathcal{P}(A)$ of a set A. Belongs ($\in$) is a relation that can be from a set A to its power set $\mathcal{P}(A)$. Similarly, the relation of not belonging ($\notin$) can be seen as a relation from a set A to its power set $\mathcal{P}(A)$. (We generalize the idea of $\in$ and $\notin$ in Problem 12.) Every function $f : X \to Y$ is a relation from the set X to the set Y. $\Diamond$

EXAMPLE 4

As in Figures 1, 2, and 3, we can picture a relation on a set by using dots to represent the elements of the set and arrows to indicate which elements are related. Figure 1 illustrates the relation D for "divides" on the set $\{1, 2, 3, 4, 6, 12\}$. Since any number divides itself, each dot has an arrow looping to itself. Figure 2 illustrates the made-up relation N, for "nearby," where aNb iff $|a - b| = 1$ on the set $\{1, 2, 3, 4, 5\}$. Since $|a - b| = |b - a|$, when there is an arrow from one dot to another, there is an arrow the other way as well. To reduce the number

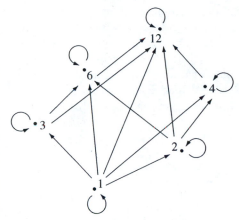

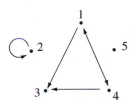

FIGURE 1. *D* on {1, 2, 3, 4, 6, 12}.

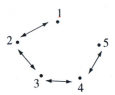

FIGURE 2. *N* on {1, 2, 3, 4, 5}.

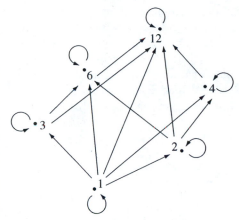

FIGURE 3. *S* on {1, 2, 3, 4, 5}.

of arrows, we simply use a double-headed arrow. Figure 3 has no particular significance; the relation *S* is just a subset of $F \times F$, where $F = \{1, 2, 3, 4, 5\}$. ◇

Properties of Relations

When mathematicians compared a variety of familiar relations, they noted some common features appearing repeatedly. We consider the four most important of these properties, which all apply to relations on sets. While these properties can be studied individually, their full meaning comes from combinations of them. We will first illustrate these properties through numerous examples and then consider how they fit together.

DEFINITION.

A relation R on a set X is *reflexive* iff for all $x \in X$, xRx.

A relation R on a set X is *symmetric* iff for all $x, y \in X$, if xRy, then yRx.

A relation R on a set X is *transitive* iff for all $x, y, z \in X$, if xRy and yRz, then xRz.

A relation R on a set X is *antisymmetric* iff for all $x, y \in X$, if xRy and yRx, then $x = y$.

The names of these properties fit the formal definitions. When you reflect, you reflect on yourself, and a relation is reflexive if each element is related to itself: xRx. We call something symmetric if its different parts can switch. In a symmetric relation, we can switch the roles of elements: xRy and yRx. The word "transit" means to move something across. In a transitive relation, we can move the relation "across" the common element y to go from xRy and yRz to xRz. Mathematicians coined the word "antisymmetric" to convey the sense of close to the opposite of symmetric. In the most important situations where symmetry fails, the reflexive property holds; so the antisymmetric property was designed to allow the reflexive property while symmetry is broken.

All of these definitions use the universal quantifier. Hence, even one exception means that a property fails.

EXAMPLE 5

Consider the relations illustrated in Figures 1, 2, and 3. The arrows indicate which elements are related. For instance, the arrow from 6 to 12 in Figure 1 indicates 6 divides 12, or $6D12$.

The relation D is reflexive, since each element has an arrow looping to itself. That is, for each number, xDx. The relations N and S are not reflexive.

The relation N is symmetric, because all arrows are double arrows. The other two relations are not symmetric.

Transitivity is a harder property to visualize from the arrows. If we can follow arrows from the element x to the element z, say, by way of y, then there is an arrow directly from x to z. The relation D is transitive, while the other two are not transitive. To illustrate, the arrows from 1 to 2 and 2 to 4 in Figure 1 correspond to the terms xRy and yRz in the definition of "transitive." The arrow from 1 to 4 matches the required xRz in the definition. Of course, we would need to check out every possible triple to show that D is transitive. Better, we could give a general proof, as in Problem 5a of Section 2.1. In Figure 3, transitivity works sometimes, as with $1S4$ and $4S3$ giving $1S3$. But other times, it fails, as with $4S1$ and $1S4$, but not $4S4$. So S is not transitive.

The relation D is antisymmetric, but the other two are not. Any double arrow between two different elements fulfills the hypothesis of xRy and yRx, but causes the conclusion $x = y$ to fail. $\Diamond$

Table 1 indicates which of these four properties hold for 11 familiar relations discussed in Examples 1, 2, and 3. (A "Y" indicates "yes," and an "N" indicates "no.") These relations are intentionally grouped to emphasize common properties.

Equivalence Relations

The first five of the relations in Table 1 ($=$, $\parallel$, $\cong$, $\sim$ and $\equiv$ (mod n)) are all reflexive, symmetric, and transitive. Relations satisfying all three of these properties collect together elements that are sufficiently alike that they can often be treated the same way. In that sense, they generalize equality, since equal items, like the fractions $\frac{1}{2}$ and $\frac{3}{6}$, can be used interchangeably. It is no coincidence, for instance, that mathematicians use the word "similar" for similar triangles; two such triangles are very much alike in geometric properties. This collection of properties occurs so often in

TABLE 1. Properties of Familiar Relations

Property Relation	Reflexive	Symmetric	Transitive	Antisymmetric
$=$ (1)	Y	Y	Y	Y
$\parallel$ (2)	Y	Y	Y	N
$\cong$ (3)	Y	Y	Y	N
$\sim$ (4)	Y	Y	Y	N
$\equiv$ (mod n) (5)	Y	Y	Y	N
$\leq$ (6)	Y	N	Y	Y
$\subseteq$ (7)	Y	N	Y	Y
$\mid$ on $\mathbb{N}$ (8)	Y	N	Y	Y
$<$ (9)	N	N	Y	Y
$\mid$ on $\mathbb{Z}$ (10)	Y	N	Y	N
$\perp$ (11)	N	Y	N	N

such important relations that we give the common name of "an equivalence relation" to such relations. We will study equivalence relations in detail in Section 4.2.

DEFINITION. A relation R on a set X is an *equivalence relation* iff R is reflexive, symmetric, and transitive.

EXAMPLE 6

Consider the relation S on the set $X = \{1, 2, 3, 45, 67, 890\}$, where aSb iff a and b have the same number of digits. We can use the arrows in Figure 4 to verify that S is an equivalence relation. Each element has an arrow that loops to itself, indicating reflexive: xSx. Each arrow between elements is double headed, indicating that if xSy, then ySx and S is symmetric. Transitivity holds because, if there is a path from one element x to another one z, say, through y, then there is a direct path: If xSy and ySz, then xSz.

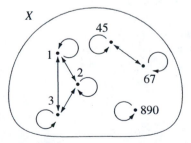

FIGURE 4   ◇

Partial Order

The sixth, seventh, and eighth relations in the table ($\leq$, $\subseteq$ and $\mid$ on $\mathbb{N}$) are reflexive, transitive, and antisymmetric. All of these relations impose an ordering on the set, designating some elements as "bigger" in some sense than others. Relations with these three properties occur frequently enough in important situations that mathematicians have given them a common name: partial order. Partial orders are the focus of Section 4.4. (The ninth and tenth relations, $<$ and $|$ on $\mathbb{Z}$, have some, but not all of these properties. They provide some sense of order as well.)

DEFINITION.   A relation R on a set X is a *partial order* (or *partial ordering*) iff R is reflexive, transitive, and antisymmetric.

EXAMPLE 7

Figure 5 illustrates the made-up relation $\sqsubseteq$ on the set $X = \{a, b, c, d, e, i\}$, where $x \sqsubseteq z$ iff x comes alphabetically before z and both are vowels or both are consonants. As in Example 6, the arrows indicate that $\sqsubseteq$ is reflexive and transitive. Further, because none of the arrows is double-headed, $\sqsubseteq$ is antisymmetric and so is a partial order. The adjective "partial" is particularly apt here, since many of the elements can't be compared to each other. For instance, because e is a vowel and d is a consonant, neither $e \sqsubseteq d$ nor $d \sqsubseteq e$ is true.

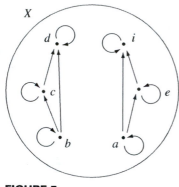

FIGURE 5   ◇

It is worth noting that the definition of equality of sets in Section 1.4 corresponds exactly to the antisymmetric property for the relation $\subseteq$. The relation $\subseteq$ gives some sense of bigger or smaller for sets, but not as completely as $\leq$ does for numbers. For instance, while $\{1, 2\} \subseteq \{1, 2, 3, 4\}$, we have no subset relation between $\{1, 2\}$ and $\{1, 3, 4, 5\}$. Thus, $\subseteq$ is a "partial" order. In contrast, given any two real numbers x and y, one of them is greater than or equal to the other. Therefore, $\leq$ is an example of a *linear* order, a special kind of partial order.

Historical Remarks

Symbols for particular relations and the general idea of a relation developed slowly over many centuries. Robert Recorde chose the now familiar equal sign $=$ in 1557 because he thought parallel lines were the most nearly alike things. The symbol $\in$ didn't appear until 1889 in the work of Giuseppe Peano (1858–1932), although set theory was developing throughout the second half of the nineteenth century. Indeed, throughout that time, a number of mathematicians confused the relations we now denote by $\in$ and $\subseteq$. General relations emerged in 1879 from Gottlob Frege's logical investigations. In 1897, Peano went the next step, describing relations as sets of ordered pairs. Richard Dedekind (1831–1916) studied equivalence relations and, by 1900, considered the idea of nonlinear orderings. However, it wasn't until 1914 that Felix Hausdorff (1868–1942) gave the general definition of a partial order.

PROBLEMS

*1. For each statement that follows, decide whether it is true or (at least sometimes) false. If it is false, explain why. Explain your answer or cite the part of the text supporting it.

a) Every function is a relation.

b) Every relation is a function.

c) According to the definition of transitive, the three elements must be different.

d) If a relation is antisymmetric, it can't be symmetric.

e) Both $\leq$ and $\geq$ are partial orders on $\mathbb{R}$.

f) If R is an equivalence relation on X and aRb, then a and b are the same in every respect.

2. For each of the following relations on the set of all human beings, decide whether the relation is reflexive, symmetric, transitive, and/or antisymmetric.

a) is the sister of

b) is a sibling of

*c) has the same mother as

d) is married to

e) is the parent of

f) is an ancestor of

g) has a common ancestor with

*3. For each relation determined by the drawings in Figure 6, determine which of the properties of reflexive, symmetric, transitive, and antisymmetric hold.

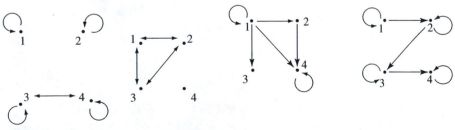

FIGURE 6. Four relations on {1, 2, 3, 4}.

4. The order for words in the dictionary is called the *lexicographic order*.

a) Describe lexicographic order for words.

b) How would a lexicographic order for the numbers from 1 to 100 differ from the usual order?

5. We generalize the definitions of the domain and range of a function from Section 3.1. Given a relation R from X to Y, define the domain of R and the range of R. Remark: The definition of a function from X to Y ensured that its domain is all of X. The domain of a relation need not be all of X.

6. a) Give three integers for x such that $x \equiv 0 \,(\text{mod } 5)$. What property do all such x satisfy?

   b) Give three integers for y such that $y \equiv 1 \,(\text{mod } 5)$. Describe the form of all such y.

   c) Give three integers for z such that $z \equiv 3 \,(\text{mod } 6)$. Describe the form of all such z.

7. *a) On $\mathbb{N}$ define $\equiv \,(\text{mod } 4 \text{ and } 10)$ by $a \equiv b \,(\text{mod } 4 \text{ and } 10)$ iff $a \equiv b \,(\text{mod } 4)$ and $a \equiv b \,(\text{mod } 10)$. Find two positive numbers congruent to 0 (mod 4 and 10).

   b) On $\mathbb{N}$ define $\equiv \,(\text{mod } 4 \text{ or } 10)$ by $a \equiv b \,(\text{mod } 4 \text{ or } 10)$ iff $a \equiv b \,(\text{mod } 4)$ or $a \equiv b \,(\text{mod } 10)$. Find two positive numbers congruent to 0 (mod 4 or 10).

   c) Which of $\equiv \,(\text{mod } 4 \text{ and } 10)$ and $\equiv \,(\text{mod } 4 \text{ or } 10)$ is/are reflexive?

   d) Repeat part c for symmetric.

   e) Repeat part c for transitive.

8. Given a relation R from a set X to a set Y, define the *inverse relation* R^{-1} from Y to X by $sR^{-1}t$ iff tRs. For instance, the inverse of the relation "x divides y" on $\mathbb{Z}$ is the relation "y is a multiple of x."

   a) For each relation in Examples 1, 2, and 3, describe its inverse, using a common name if it has one.

   b) What can you say about a relation R on a set X satisfying $R = R^{-1}$?

   c) Suppose a relation R is reflexive and antisymmetric on the set X. For what pairs x, $y \in X$ do we have xRy and $xR^{-1}y$?

9. Partial orders naturally give rise to a variety of terms. From the following examples, write general definitions for the italicized terms:

   a) For a partial order $\sqsubseteq$ of a set X, define when it is a *linear order*. For instance, $\leq$ on $\mathbb{R}$ is a linear order, but $\mid$ on $\mathbb{N}$ is not. (See comments after the definition of a partial order.)

   b) For a partial order $\sqsubseteq$ of a set X, define a *greatest element M* of X. For instance, for $\subseteq$ on $\mathcal{P}(\mathbb{R})$ the greatest element is $\mathbb{R}$. For $\leq$ on $\mathbb{R}$ there is no greatest element.

   c) For a partial order $\sqsubseteq$ of a set X, define an *upper bound u* of a subset S of X. For instance, 3 and $\sqrt{13}$ are upper bounds of the subset $[-1, 3)$ for the relation $\leq$ on $\mathbb{R}$.

   d) For a partial order $\sqsubseteq$ of a set X, define a *least upper bound b* of a subset S of X. For instance, 3 is the least upper bound of $[-1, 3)$ in part c. For $\leq$ on $\mathbb{R}$ the subset $\mathbb{N}$ has no least upper bound.

   e) For a partial order $\sqsubseteq$ of a set X, define a *least element m* of X. (Compare part b.)

   f) For a partial order $\sqsubseteq$ of a set X, define a *lower bound l* of a subset S of X. (Compare with part c.)

   g) For a partial order $\sqsubseteq$ of a set X, define a *greatest lower bound g* of a subset S of X. (Compare with part d.)

   h) For a partial order $\sqsubseteq$ of a set X, define the *successor s* of an element r of X. For instance, for $\leq$ on $\mathbb{Z}$ the only successor of 3 is 4. On $\mathbb{Q}$ with $\leq$ there is no successor of 3. For $\mid$ on $\mathbb{N}$ there are many successors of 6, including 12, 18, and 30, but not 24 or 36.

   i) Make the following concept precise, without using the undefined term "between": A partial order $\sqsubseteq$ on a set X is *dense*, provided that "between" two related elements lies a third, distinct element. For instance, $\leq$ is dense on $\mathbb{Q}$ and $\mathbb{R}$, but not on $\mathbb{N}$ or $\mathbb{Z}$.

10. The relation $<$ is transitive and antisymmetric, but not reflexive (nor symmetric). We can say something stronger than "not reflexive": No value x is ever less than itself. This property is called *irreflexive*.

   a) Give a formal definition of "irreflexive."

   *b) Find several, preferably familiar, examples of irreflexive relations.

   c) Find examples of irreflexive transitive relations. Are they symmetric? antisymmetric? Are they related to partial orders? Explain.

11. For each of the given relations on $\mathbb{N} \times \mathbb{N}$, decide which of the properties of a partial order hold. If a property holds for a relation, give an example; if not, give a counterexample.

   a) $(a, b) \sqsubseteq (c, d)$ iff $a \le b$ and $c \le d$.

   b) $(a, b) \sqsubseteq (c, d)$ iff $a \le c$ and $b \le d$.

   c) $(a, b) \preceq (c, d)$ iff $a \le c$ or $b \le d$.

   d) $(a, b) \trianglelefteq (c, d)$ iff $a < c$ or $(a = c$ and $b \le d)$.

   e) $(a, b) \Subset (c, d)$ iff $ad \le bc$.

12. Given a relation R on X, define the *complement* of R to be $\overline{R}$, where $a\overline{R}b$ iff $a, b \in X$ and not (aRb). Equivalently, $\overline{R} = \{(a, b) \in X \times X : (a, b) \notin R\}$. (The relation $\overline{R}$ is, as a set, the complement of the set R in the universal set $X \times X$.) For each part that follows, if your answer is "no," provide a relation on a set and elements to give a counterexample:

   a) If R is reflexive, is $\overline{R}$ reflexive?

   b) If R is not reflexive, must $\overline{R}$ be reflexive?

   c) Repeat part a for symmetric.

   d) Repeat part a for transitive.

   e) Repeat part a for antisymmetric.

13. Find a relation on the set $\{1, 2, 3, 4\}$ that is symmetric and antisymmetric, but is not the equality relation $=$.

14. Find eight different relations on $\mathbb{Z}$, one for each of the following eight options:

   a) Reflexive, symmetric, and transitive

   b) Reflexive, symmetric, but not transitive

   c) Reflexive and transitive, but not symmetric

   d) Symmetric and transitive, but not reflexive

   e) Reflexive, but neither symmetric nor transitive

   f) Symmetric, but neither reflexive nor transitive.

   g) Transitive, but neither reflexive nor symmetric.

   h) Neither reflexive nor symmetric nor transitive.

15. ***a)** How many relations are there on a set with 2 elements?

   b) How many reflexive relations are there on a set with 2 elements?

   c) How many symmetric relations are there on a set with 2 elements?

   d) How many antisymmetric relations are there on a set with 2 elements?

16. **a)** How many relations are there on a set with n elements?

   b) How many reflexive relations are there on a set with n elements?

   c) How many symmetric relations are there on a set with n elements?

   d) How many antisymmetric relations are there on a set with n elements?

4.2 EQUIVALENCE RELATIONS

Section 4.1 defined equivalence relations, based on examples, but we postponed proofs of those examples or of properties of equivalence relations. We start with congruence (mod n), arguably one of the most useful equivalence relations in mathematics. Our proof that it is an equivalence relation allows us to model how to prove the three key properties—reflexive, symmetric, and transitive. We then explore congruence more fully. After that, we'll turn to other equivalence relations and equivalence classes.

Congruence

THEOREM 4.2.1. For every $n \in \mathbb{N}$, congruence (mod n) is an equivalence relation on $\mathbb{Z}$.

Proof. Let $n \in \mathbb{N}$. Recall, for $a, b \in \mathbb{Z}$, that $a \equiv b \pmod{n}$ iff n divides $b - a$.

Reflexive   Let $a \in \mathbb{Z}$. Then $a - a = 0$ and any n divides 0, since $n \cdot 0 = 0$. Thus, $a \equiv a \pmod{n}$.

Symmetric   Let $a, b \in \mathbb{Z}$, and suppose $a \equiv b \pmod{n}$. Then n divides $b - a$, or by definition of "divides," there is some $x \in \mathbb{Z}$ such that $nx = b - a$. To obtain $b \equiv a \pmod{n}$, we must show that n divides $a - b$, but $a - b = -(b - a) = -nx = n(-x)$ and $-x \in \mathbb{Z}$. Thus, n divides $a - b$ and it follows that $b \equiv a \pmod{n}$.

Transitive   Let $a, b, c \in \mathbb{Z}$, and suppose $a \equiv b \pmod{n}$ and $b \equiv c \pmod{n}$. Then n divides $b - a$ and divides $c - b$. From the definition of "divides," we get integers x and y such that $nx = b - a$ and $ny = c - b$. For transitivity we need $a \equiv c \pmod{n}$ or n to divide $c - a$. But $c - a = (c - b) + (b - a) = ny + nx = n(x + y)$. Since $x + y$ is an integer, n divides $c - a$ and $a \equiv c \pmod{n}$.   ∎

DISCUSSION. The differences in the definitions of reflexive, symmetric, and transitive lead to different proof formats. To prove a relation reflexive, we don't get to assume any relationship. With symmetric and transitive, however, we get to assume something to get us started. I have written the proofs of symmetric and transitive so as to reveal the way we might reason out the proof. Here is a proof of symmetry without such motivation:

Let $a, b \in \mathbb{Z}$, and suppose $a \equiv b \pmod{n}$. Then there is some $x \in \mathbb{Z}$ such that $nx = b - a$. Since $-x \in \mathbb{Z}$ and $n(-x) = -(b - a) = a - b$, we see that n divides $a - b$ and $b \equiv a \pmod{n}$.   ∎

While equally correct, this proof may seem magically to invent the key step of using $-x$.

Why is congruence (mod n) important? Modular arithmetic appears in many areas of mathematics and its applications. Instead of considering all infinitely many integers, we can consider the finitely many types of integers (mod n). The smallest interesting value for n, congruence (mod 2), splits integers into two types, odd and even. Such a simple characterization has appeared in everyday life as well as more mathematical situations for thousands of years. Some modern applications need n to be quite large. Public key cryptography requires computers to do exact arithmetic (mod n), where n is the product of two primes each at least one hundred digits long. The practical problems of quickly carrying out precise computations for large n are difficult and, fortunately, do not concern us. Some theoretical questions are much easier, and we consider some of them in Theorem 4.2.2, after illustrating them. (See Gallian, 158–160, for a brief discussion of public key cryptography.)

EXAMPLE 1

Most clocks reduce time to congruences (mod 12), since we use only 12 hours before starting over. Thus, 5 hours after 9 o'clock is 2 o'clock. In effect, "clock arithmetic" concludes that "$5 + 9 = 2$." Again, 29 hours after 9 o'clock would again be 2 o'clock, or "$29 + 9 = 2$." A more mathematical description says $5 \equiv 29 \pmod{12}$, so $5 + 9 = 14$ and $29 + 9 = 38$ are congruent (mod 12) to each other and also to 2. ◇

THEOREM 4.2.2. Suppose $a, b, c, d \in \mathbb{Z}$, and $n \in \mathbb{N}$. If $a \equiv c \pmod{n}$ and $b \equiv d \pmod{n}$, then

i) $a + b \equiv c + d \pmod{n}$,

ii) $a - b \equiv c - d \pmod{n}$, and

iii) $ab \equiv cd \pmod{n}$.

Proof. We'll prove only part iii. See Problem 5 for the other parts. Let $a, b, c, d \in \mathbb{Z}$ and $n \in \mathbb{N}$, and suppose $a \equiv c \pmod{n}$ and $b \equiv d \pmod{n}$. So there are integers s and t such that $c - a = ns$ and $d - b = nt$. To show that $ab \equiv cd \pmod{n}$, we need to show that $cd - ab$ is a multiple of n. Unfortunately, the terms cd and ab seem unrelated. We "mix and match" to find something related to each term. More specifically, we add and subtract the same term bc, which shares one letter with each term. Thus, $cd - ab = cd - bc + bc - ab = c(d - b) + b(c - a)$. Our initial work with the hypotheses gives us helpful substitutions: $c(d - b) + b(c - a) = cnt + bns = n(ct + bs)$. Hence, n divides $cd - ab$ and $ab \equiv cd \pmod{n}$, as desired. ∎

REMARK. Theorem 4.2.2 conspicuously doesn't include division or exponentiation because they don't always hold. Problem 5 asks for counterexamples.

General Equivalence Relations

EXAMPLE 2

The basic trigonometric identity $\sin^2 x + \cos^2 x = 1$ suggests a relation on angles, which are real numbers. For $x, y \in \mathbb{R}$, define xAy iff $\sin^2 x + \cos^2 y = 1$. Prove A is an equivalence relation on $\mathbb{R}$.

Proof. Let $x \in \mathbb{R}$. The trigonometric identity $\sin^2 x + \cos^2 x = 1$ holds for all such x, so A is reflexive: xAx.

For symmetry, let $x, y \in \mathbb{R}$ and suppose xAy; that is, $\sin^2 x + \cos^2 y = 1$. From $\sin^2 x + \cos^2 x = 1$, we see that $\cos^2 y = \cos^2 x$ and, similarly, $\sin^2 y = \sin^2 x$. We can substitute these values to get $\sin^2 y + \cos^2 x = 1$, which means yAx and A is symmetric.

For transitivity, let $x, y, z \in \mathbb{R}$ and suppose xAy and yAz. Then $\sin^2 x + \cos^2 y = 1$ and $\sin^2 y + \cos^2 z = 1$. As with the argument for symmetry, this last equality and $\sin^2 y + \cos^2 y = 1$ give us $\cos^2 y = \cos^2 z$. A substitution gives us $\sin^2 x + \cos^2 z = 1$, or xAz. Thus, A is transitive, and hence is an equivalence relation. ■ ◊

The following property provides a way to derive many equivalence relations, as Examples 3 and 4 illustrate:

LEMMA 4.2.3. Suppose $f : X \to Y$ is a function and define $\sim_f$ on X by $a \sim_f b$ iff $f(a) = f(b)$. Then $\sim_f$ is an equivalence relation.

Proof. See Problem 4. ■

EXAMPLE 3

When we round numbers to the nearest integer (or nearest hundred, etc.), we are creating an equivalence relation. For instance, 234.81 and 235.403 are essentially the same thing as 235, at least as far as rounding is concerned. We can say this more formally by defining a rounding function $r : \mathbb{R} \to \mathbb{Z}$ and using the previous lemma. We use the floor function from Problem 13 of Section 3.1, where $\lfloor x \rfloor$ is the greatest integer less than or equal to x. We need to shift this over by a half to match our rules of rounding: $r(x) = \lfloor x + 0.5 \rfloor$. Thus, $r(0.5) = \lfloor 0.5 + 0.5 \rfloor = \lfloor 1 \rfloor = 1$, but $r(0.49) = \lfloor 0.49 + 0.5 \rfloor = \lfloor 0.99 \rfloor = 0$. In terms of Lemma 4.2.3, we have $0.5 \sim_r 1$ and $0.49 \sim_r 0$. ◊

EXAMPLE 4

From elementary geometry, we know that two triangles are similar, provided that their corresponding angles have the same measure. For instance, a triangle with angles of 30°, 60°, and 90° is similar to any triangle with those same angles. In the previous lemma, we can pick X to be the set of all Euclidean triangles and Y to be $\mathbb{R} \times \mathbb{R} \times \mathbb{R}$. We need to be a little bit careful defining the function, since we want a triangle to be similar to itself regardless of how we list its vertices. For instance, we need $f(\triangle ABC) = f(\triangle BCA)$. Let $f(\triangle ABC)$ be the ordered triple (a, b, c), where a, b, and c are the measures of the angles in nondecreasing order. Thus, (30, 60, 90) is a possible image, but (90, 30, 60) is not. (We say "nondecreasing," rather than the simpler "increasing" because we want (45, 45, 90) to be an acceptable triple.) Then $\sim_f$ corresponds exactly to the relation for similar triangles, and our lemma shows it is an equivalence relation. ◊

Equivalence Classes

An equivalence relation on a set X splits X into subsets called *equivalence classes* that match with the sets that form a partition, as defined in Section 1.6. We'll show this matching in Section 4.3.

DEFINITION. For an equivalence relation R on a set X and $a \in X$, the *equivalence class* of a is $[a] = \{b \in X : aRb\}$. (If we have two or more equivalence relations on the same set, we can use subscripts on equivalence classes to distinguish them.)

EXAMPLE 5

We can split the integers into subsets using congruence (mod 5) by putting two numbers a and b together, provided that $a \equiv b \pmod 5$. Any multiple of 5, say, $5x$, is put together with 0, since 5 divides $5x - 0 = 5x$ and so $0 \equiv 5x \pmod 5$. Then $[0] = \{5x : x \in \mathbb{Z}\}$ is the equivalence class of 0. We could have used 15 or -5 or any other multiple of 5 to name this class, since any two multiples of 5 are congruent (mod 5). That is, $[0] = [15] = [-5]$. There are four other equivalence classes. For instance, $[1] = \{b : 1 \equiv b \pmod 5)\} = \{5x + 1 : x \in \mathbb{Z}\}$. We summarize the equivalence classes for $\equiv \pmod 5$ in Table 1.

TABLE 1.   Equivalence Classes for $\equiv \pmod 5$

$[0]$	$=$	$\{5x : x \in \mathbb{Z}\}$	$= \{\ldots, -10, -5, 0, 5, 10, \ldots\}$
$[1]$	$=$	$\{5x + 1 : x \in \mathbb{Z}\}$	$= \{\ldots, -9, -4, 1, 6, 11, \ldots\}$
$[2]$	$=$	$\{5x + 2 : x \in \mathbb{Z}\}$	$= \{\ldots, -8, -3, 2, 7, 12, \ldots\}$
$[3]$	$=$	$\{5x + 3 : x \in \mathbb{Z}\}$	$= \{\ldots, -7, -2, 3, 8, 13, \ldots\}$
$[4]$	$=$	$\{5x + 4 : x \in \mathbb{Z}\}$	$= \{\ldots, -6, -1, 4, 9, 14, \ldots\}$

Every integer appears in exactly one of these equivalence classes, which is the heart of the idea of a partition.   ◇

EXAMPLE 6

For the relation A in Example 2, which angles x and y satisfy xAy? That is, when is $\sin^2 x + \cos^2 y = 1$? Earlier, we saw that $\cos^2 y = \cos^2 x$ and $\sin^2 y = \sin^2 x$. Equivalently, $\sin y = \pm \sin x$ and $\cos y = \pm \cos x$. Figure 1 illustrates the angles with the same sines and cosines in absolute value. Since adding 360^0 (or 2π) to an angle gives an angle with the same sine and cosine, we see that $[x] = \{x + 360k : k \in \mathbb{Z}\} \cup \{-x + 360k : k \in \mathbb{Z}\} \cup \{180 - x + 360k : k \in \mathbb{Z}\} \cup \{x - 180 + 360k : k \in \mathbb{Z}\}$.

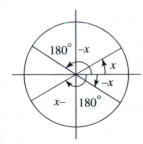

FIGURE 1. Possible angles for y so that $\sin^2 x + \cos^2 y = 1$.   ◇

EXAMPLE 7

The equivalence relation defined from rounding in Example 3 splits $\mathbb{R}$ into intervals for equivalence classes. For instance, $[1] = [0.5, 1.5)$. In general, for $z \in \mathbb{Z}$, we have $[z] = [z - 0.5, z + 0.5)$. Any number in an equivalence class can be used to name the class. For instance, $[\sqrt{2}] = [1] = [0.91]$, since $r(\sqrt{2}) = 1 = r(1) = r(0.91)$.   ◇

EXAMPLE 8

In Section 1.3, we described the rational numbers, which form the set $\mathbb{Q}$, as fractions of the form $\frac{p}{q}$, where $p \in \mathbb{Z}$ and $q \in \mathbb{N}$. Of course, you have known for a long time that two fractions can be equal even if they have different numerators and denominators. In more sophisticated language, we have an equivalence relation on fractions determining when two fractions are equal. Equality of fractions gives rise to equivalence classes. For instance, the fractions equal to $\frac{1}{2}$ have denominators twice as big as their numerators. These form the equivalence class $[\frac{1}{2}] = \{\frac{p}{q} : q = 2p\}$. In general, we know $\frac{p}{q} = \frac{r}{s}$ iff $ps = rq$, so $[\frac{p}{q}] = \{\frac{r}{s} \in \mathbb{Q} : ps = rq\}$. In Problem 12, we define the relation F (for fraction) on $\mathbb{Z} \times \mathbb{N}$ by $(p, q)F(r, s)$ iff $ps = rq$. In formal set theory, rational numbers can be defined to be equivalence classes of ordered pairs from $\mathbb{Z} \times \mathbb{N}$ for the relation F. Problem 12 develops this idea somewhat, showing how arithmetic of fractions and inequalities of fractions fit with equivalence classes. Of course, no one working with fractions thinks of them as equivalence classes. Even more, pedagogically, it is a good thing that grade school students do not have to understand equivalence relations and equivalence classes in order to understand fractions. But it is valuable for college students to realize that the equality of fractions is a sophisticated concept. Further, the language of equivalence relations and equivalence classes enables mathematicians to prove the familiar properties of rational numbers from properties of integers. These properties of integers can be proven from more fundamental properties of sets, although we do not pursue such an exercise here. ◊

PROBLEMS

*1. For each statement that follows, decide whether it is true or (at least sometimes) false. If it is false, explain why. Explain your answer or cite the part of the text supporting it.

   a) To prove symmetry of a relation R, we assume one relationship holds, say, aRb.

   b) To prove reflexivity of a relation R, we assume one relationship holds, say, aRb.

   c) To prove transitivity of a relation R, we assume one relationship holds, say, aRb.

   d) For R an equivalence relation on X and $a \in X$, $[a]$ is a subset of R.

   e) For R an equivalence relation on X and $a \in X$, $[a]$ is a subset of X.

2. On $\mathbb{R}$ define $\equiv (\mathrm{mod}\ \mathbb{Z})$ by $x \equiv y\ (\mathrm{mod}\ \mathbb{Z})$ iff $y - x \in \mathbb{Z}$.

   *a) Prove $\equiv (\mathrm{mod}\ \mathbb{Z})$ is an equivalence relation on $\mathbb{R}$.

   b) Prove for all $x \in \mathbb{R}$, there is some $y \in [0, 1)$ such that $x \equiv y\ (\mathrm{mod}\ \mathbb{Z})$. Hint: Use the floor function $\lfloor x \rfloor$, defined in Problem 13 of Section 3.1. Remark: We can use a circle to represent $\mathbb{R}\ (\mathrm{mod}\ \mathbb{Z})$, as in Figure 2.

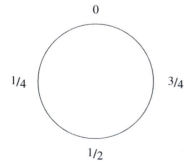

0

1/4

3/4

1/2

FIGURE 2

*c) Give three positive and two negative elements of the equivalence class $[0.2]$ for the relation $\equiv \pmod{\mathbb{Z}}$.

3. The given relations on $\mathbb{R}$ are not equivalence relations. For each relation, state which of the properties—reflexive, symmetric, and transitive—fail. Give counterexamples for the properties that a relation doesn't satisfy and proofs for the properties that a relation satisfies.

   a) Define xAy iff $xy > 0$.

   b) Define xBy iff $xy \geq 0$.

   c) Define xCy iff $x + y = 0$.

   d) Define xDy iff $y - x > 2$.

   e) Define xEy iff $|y - x| \leq 1$.

4. Prove Lemma 4.2.3.

5. ***a)** Prove the rest of Theorem 4.2.2.

   b) Give an example to show that $a \div b$ need not be congruent to $c \div d \pmod{n}$, even if $a \equiv c \pmod{n}$ and $b \equiv d \pmod{n}$.

   c) Repeat part b for a^b and c^d.

6. In Problem 2, we defined $\equiv \pmod{\mathbb{Z}}$ on $\mathbb{R}$. Suppose $a, b, c, d \in \mathbb{R}$, $a \equiv c \pmod{\mathbb{Z}}$, and $b \equiv d \pmod{\mathbb{Z}}$.

   a) Prove $a + b \equiv c + d \pmod{\mathbb{Z}}$,

   b) Prove $a - b \equiv c - d \pmod{\mathbb{Z}}$.

   c) Must $ab \equiv cd \pmod{\mathbb{Z}}$? If so, prove it; if not, give a counterexample.

7. On $\mathbb{R}$, define $\equiv \pmod{\mathbb{Q}}$ by $x \equiv y \pmod{\mathbb{Q}}$ iff $y - x \in \mathbb{Q}$.

   a) Prove $\equiv \pmod{\mathbb{Q}}$ is an equivalence relation on $\mathbb{R}$.

   b) Give three elements in $[\sqrt{2}]$ between 0 and 1.

   c) Suppose $a, b, c, d \in \mathbb{R}$, and $a \equiv c \pmod{\mathbb{Q}}$ and $b \equiv d \pmod{\mathbb{Q}}$. Prove $a + b \equiv c + d \pmod{\mathbb{Q}}$ and $a - b \equiv c - d \pmod{\mathbb{Q}}$.

   d) If $a \equiv c \pmod{\mathbb{Q}}$ and $b \equiv d \pmod{\mathbb{Q}}$, must $ab \equiv cd \pmod{\mathbb{Q}}$? If so, prove it; if not, give a counterexample.
   [Remark: It is very hard to visualize $\mathbb{R} \pmod{\mathbb{Q}}$.]

8. **a)** On $\mathbb{Z}$ with $n \in \mathbb{N}$, prove that $a \equiv b \pmod{n}$ iff the remainder of dividing a by n equals the remainder of dividing b by n.

   b) Use part a to give a different proof of Theorem 4.2.1.

9. Use Lemma 4.2.3 to show that the following relations are equivalence relations:

   a) Congruence of triangles.

   *b) Parallel on the set of all lines in the Euclidean plane. Hints: Describe a line, using an equation. What do parallel lines have in common? Don't forget vertical lines.

   c) Two finite sets A and B have the same size, written $A \sim B$, iff they have the same number of elements. For a finite set X, show that $\sim$ is an equivalence relation on $\mathcal{P}(X)$.

10. Define a relation R on a set X to be *cyclic* iff for all $a, b, c \in X$, if aRb and bRc, then cRa.

    a) Prove if a relation is symmetric and transitive, then it is cyclic.

    b) Prove if a relation is symmetric and cyclic, then it is transitive.

c) Suppose a relation is transitive and cyclic. Must it be symmetric? If so, prove it; if not, give a counterexample.

d) Suppose a relation is reflexive and cyclic. Must it be symmetric? transitive? If so, prove it; if not, give a counterexample.

11. Define $\equiv$ (mod n and k) on $\mathbb{Z}$ by $a \equiv b \pmod{n \text{ and } k}$ iff $a \equiv b \pmod{n}$ and $a \equiv b \pmod{k}$.

a) Prove $\equiv$ (mod n and k) is an equivalence relation on $\mathbb{Z}$.

b) Find a value j so that $\equiv$ (mod j) matches $\equiv$ (mod n and k) and prove your answer.

c) Define $\equiv$ (mod n or k) on $\mathbb{Z}$ by $a \equiv b \pmod{n \text{ or } k}$ iff $a \equiv b \pmod{n}$ or $a \equiv b \pmod{k}$. Is $\equiv$ (mod n or k) an equivalence relation on $\mathbb{Z}$? Prove or disprove.

12. On $\mathbb{Z} \times \mathbb{N}$, define F by $(p, q)F(r, s)$ iff $ps = rq$.

a) Prove F is an equivalence relation on $\mathbb{Z} \times \mathbb{N}$.

*b) Mimicking multiplication of fractions, define $(p, q) \times (t, u) = (pt, qu)$. Suppose $(p, q) F(r, s)$ and $(t, u)F(y, z)$. Prove $((p, q) \times (t, u))F((r, s) \times (y, z))$.

c) Mimicking addition of fractions, define $(p, q) + (t, u) = (pu + tq, qu)$. Suppose $(p, q) F(r, s)$ and $(t, u)F(y, z)$. Prove $((p, q) + (t, u))F((r, s) + (y, z))$.

d) Define inequality on $\mathbb{Z} \times \mathbb{N}$ by $(p, q) \leq (r, s)$ iff $ps \leq rq$. Prove that $\leq$ is reflexive and transitive. (Hint: The denominators are positive.)

e) For $\leq$ as defined in part d, prove that if $(p, q) \leq (r, s)$ and $(r, s) \leq (p, q)$, then $(p, q) F(r, s)$. Give an example to show $\leq$ is not antisymmetric.

f) As you learned long ago, you "can't divide by 0." Define G on $\mathbb{Z} \times \mathbb{Z}$ by $(p, q)G(r, s)$ iff $ps = rq$. Show that G is not an equivalence relation. [Remark: Since G is not an equivalence relation, it is pointless to use it to define addition and multiplication on $\mathbb{Z} \times \mathbb{Z}$.]

13. Suppose R and S are relations on a set X. Define $R \cap S$ by $a(R \cap S)b$ iff aRb and aSb.

a) If R and S are reflexive, must $R \cap S$ be reflexive? If yes, prove it; if not, give a counterexample.

b) If R and S are symmetric, must $R \cap S$ be symmetric? If yes, prove it; if not, give a counterexample.

c) If R and S are transitive, must $R \cap S$ be transitive? If yes, prove it; if not, give a counterexample.

14. Repeat Problem 13, using $R \cup S$ in place of $R \cap S$, where $a (R \cup S) b$ means aRb or aSb.

15. For R a relation on a set X, its inverse relation R^{-1} satisfies xRy iff $yR^{-1}x$.

a) What is the inverse relation of $\leq$ on $\mathbb{R}$?

b) If R is reflexive, must R^{-1} be reflexive? If yes, prove it; if not, give a counterexample.

c) If R is symmetric, must R^{-1} be symmetric? If yes, prove it; if not, give a counterexample.

d) If R is transitive, must R^{-1} be transitive? If yes, prove it; if not, give a counterexample.

16. (Linear algebra) Define two $n \times n$ matrices to be *similar* iff there is an invertible $n \times n$ matrix M such that $MAM^{-1} = B$.

a) For $A = \begin{bmatrix} 1 & 2 \\ 3 & 4 \end{bmatrix}$ and $M = \begin{bmatrix} 2 & 1 \\ 3 & 2 \end{bmatrix}$, find B. What properties do A and B share?

b) Prove similarity of matrices is an equivalence relation on $n \times n$ matrices.

c) Explain why the equivalence class of the matrix $F = \begin{bmatrix} 5 & 0 \\ 0 & 5 \end{bmatrix}$ has only F in it. [Remark: Most equivalence classes for similarity are noticeably more difficult to describe than the one for the matrix in this part.]

17. a) Generalize the definition of composition of functions in Section 3.1 to composition of relations.

b) Let $A = \{a, b, c\}$, $B = \{1, 2, 3, 4\}$, and $C = \{x, y, z\}$. Let the relation R from A to B be the set of ordered pairs $R = \{(a, 2), (b, 1), (b, 3), (c, 4)\}$. Let the relation S from B to C be the set $S = \{(1, x), (1, y), (3, y), (3, z), (4, z)\}$. Use your definition to list the ordered pairs in the composition $S \circ R$.

c) On the set of lines in the Euclidean plane, what is the composition of the relation perpendicular ($\perp$) with itself?

d) In part c, use the relation parallel ($\parallel$) with itself.

e) In part c, compose perpendicular and parallel.

f) Prove that if R and S are reflexive on a set X, then their composition is reflexive.

18. Critique the incorrect "proofs" that follow. Point out reasoning errors and unclear presentation. (If the claim is false, there must be an error in the argument. However, it is not enough to say that the claim is false. You need to find an error in the reasoning.)

a) Claim: If R is an equivalence relation on a set X and $R \subseteq S$, then S is an equivalence relation. "Proof": Suppose R is the relation $\equiv$ (mod 12). For $\equiv$ (mod n) to contain R, all multiples of 12 must be congruent to 0. Thus, n must divide 12. But whichever of the factors 1, 2, 3, 4, or 6 of 12 you pick for n, $\equiv$ (mod n) is an equivalence relation by Theorem 4.2.1. ▲

b) Claim: If $a \equiv b \pmod{n}$ and $a \equiv b \pmod{k}$, then $a \equiv b \pmod{nk}$. "Proof": Suppose $a \equiv b \pmod{n}$ and $a \equiv b \pmod{k}$. Then both n and k divide $b - a$. Thus, nk divides $b - a$, fulfilling the definition of $a \equiv b \pmod{nk}$. ▲

*c) (See Problem 15 for definition of S^{-1}.) Claim: A relation S on X is symmetric iff $S = S^{-1}$. "Proof": Suppose S is symmetric and aSb. Then bSa by symmetry and $bS^{-1}a$ by the definition of S^{-1}. So, bSa iff $bS^{-1}a$, showing $S = S^{-1}$. ▲

19. Critique the questionable "proofs" that follow. Point out reasoning errors and unclear presentation. If the argument is a proof, say so. (If the claim is false, there must be an error in the argument. However, it is not enough to say that the claim is false. You need to find an error in the reasoning.)

*a) Claim: If $a \equiv b \pmod{36}$, then $a \equiv b \pmod{12}$. "Proof": Suppose $a \equiv b \pmod{36}$. By the definition of congruence, 36 divides $b - a$. Further, 12 divides 36, so 12 divides $b - a$, showing $a \equiv b \pmod{12}$. ▲

b) Claim: A symmetric and transitive relation is reflexive. "Proof": Let R be a symmetric and transitive relation on a set X. Let a and b be any two elements of X, and suppose aRb. By symmetry we have bRa. From aRb and bRa, transitivity gives us aRa (and bRb). Hence, R is reflexive. ▲

c) Claim: The relation S on $\mathbb{R} \times \mathbb{R}$ is symmetric, where $(q, r)S(t, u)$ iff $q + r = t + u$. "Proof": For any q and r in $\mathbb{R}$, we have $q + r = r + q$ so $(q, r)S(r, q)$, showing symmetry. ▲

REFERENCE

GALLIAN, J. 2002. *Contemporary abstract algebra*, 5th ed. Boston: Houghton Mifflin.

4.3 PARTITIONS AND EQUIVALENCE RELATIONS

"The essence of mathematics lies in its freedom."

—Georg Cantor (1845–1918)

The main goal of this section is to show that each equivalence relation on a set X gives a unique partition of that set and, conversely, each partition gives a unique equivalence relation. In the language of functions, there is a one-to-one function from the set of equivalence relations on X onto the set of partitions of X. We start by restating the definition of a partition from Section 1.6.

DEFINITION.   A *partition* of a set B is a family of sets $\{S_i : i \in I\}$ such that

i) each S_i is nonempty and

ii) for each $b \in B$, there is a unique S_i such that $b \in S_i$.

THEOREM 4.3.1.   For any equivalence relation R on a set X, the family of its equivalence classes $\{[a] : a \in X\}$ is a partition of X.

Proof.   For an equivalence relation R and $a \in X$, the equivalence class $[a] = \{b \in X : aRb\}$ is clearly a subset of X. For any a, from aRa, we find $a \in [a]$, so $[a] \neq \emptyset$ and a is in some class. Thus the reflexive property of an equivalence relation gives us part i and the existence in part ii of the definition of a partition.

We need both symmetry and transitivity to show the uniqueness of the classes. Let c be any element of X and suppose c is in both $[a]$ and $[b]$. That is, aRc and bRc. By symmetry, cRb and by transitivity, aRb. Thus, $b \in [a]$. But we need more: $[b] \subseteq [a]$ and, conversely, $[a] \subseteq [b]$. We'll show the first inclusion. Let $d \in [b]$. So bRd, and we already have aRb. Transitivity gives aRd and so $d \in [a]$, showing $[b] \subseteq [a]$ as needed. The other inclusion is similar. Hence, $[a] = [b]$, showing uniqueness. Thus, R is a partition. ∎

THEOREM 4.3.2.   Given a partition $\{B_i : i \in I\}$ of a set X, define the relation A on X by sAt iff there is some $i \in I$ such that $s, t \in B_i$. Then A is an equivalence relation on X. Further, the equivalence classes of A are the sets B_i.

Proof.   See Problem 8. ∎

The two previous theorems reveal that equivalence relations and partitions are two aspects of the same mathematical idea. Why should we want to have both ideas? Consider briefly the familiar geometry relation "parallel," which is certainly a useful equivalence relation. In addition, we benefit from the partition of lines into sets with the same slope (and the separate set of vertical lines).

The most useful equivalence relations (or partitions) allow us to work with representatives of an equivalence class, instead of the entire class, or with the relation. For instance, consider adding fractions with different denominators. To find the sum $\frac{1}{6} + \frac{1}{4}$, grade school students will convert each fraction to an equivalent one with a common denominator, obtaining $\frac{2}{12} + \frac{3}{12} = \frac{5}{12}$. Of course, they have no idea that there is an equivalence relation relating $\frac{1}{6}$ with $\frac{2}{12}$ or think of these fractions as in the same subset of a partition. They simply switch from one representative to another, more convenient one. Similarly, in the previous section we considered arithmetic (mod n). In effect, we saw how to do "clock arithmetic" on the partition of $\mathbb{Z}$ into n classes. However, we don't need to work with equivalence classes there. Rather, Theorem 4.2.2 allows us to add, subtract and multiply representatives from the classes. In general, it is useful to shift between the whole equivalence class and representative members.

PROBLEMS

1. The collection of sets $P = \{\,\{1\},\ \{2, 3, 5, 7\},\ \{4, 6, 9, 10\},\ \{8\}\,\}$ is a partition.

   *a) What is the set for which P is a partition?

   b) Theorem 4.3.2 defines the equivalence relation derived from P in a rather artificial way. Give a more natural description of this relation.

   c) Extend your description of the relation in part b to all natural numbers in $\mathbb{N}$.

   d) Give the corresponding partition for the set of integers from 1 to 20.

*2. Prove that the family of sets $\{B_i : i \in \mathbb{Z}\}$, where $B_i = [i, i + 1)$, is a partition of $\mathbb{R}$. Describe the corresponding equivalence relation by using Lemma 4.2.3 and an appropriate function.

3. Prove that the family of sets $\{C_r : r \in [0, \infty)\}$, where $C_r = \{(x, y) : x^2 + y^2 = r^2\}$, is a partition of $\mathbb{R} \times \mathbb{R}$. Describe the corresponding equivalence relation by using Lemma 4.2.3 and an appropriate function.

4. On the set $C^1(\mathbb{R})$ of continuously differentiable real functions, define the class of a function f to be $[f] = \{g \in C^1(\mathbb{R}) : \text{There is some } k \in \mathbb{R} \text{ such that for all } x \in \mathbb{R},\ g(x) = f(x) + k\}$.

   *a) Give two functions in the class of h, where $h(x) = x^2 + \sin(x)$.

   b) Prove that $\{[f] : f \in C^1(\mathbb{R})\}$ is a partition of $C^1(\mathbb{R})$.

   c) Use derivatives to describe the corresponding equivalence relation.

5. On $\mathbb{N}$, define the relation S by aSb iff ab is a square.

   *a) Give three numbers related to 6 by S.

   b) Prove that S is an equivalence relation.

   c) Describe $[1]_S$ and $[3]_S$. Use the equivalence class $[1]_S$ to describe a generic equivalence class $[n]_S$.

6. For $n \in \mathbb{N}$, define R on $\mathbb{Z}$ by aRb iff $a^2 \equiv b^2 \pmod{n}$.

   a) Prove that R is an equivalence relation on $\mathbb{Z}$.

   b) Describe the partition of $\mathbb{Z}$ determined by R when $n = 8$.

7. Prove the following strengthened version of Theorem 4.2.3: Suppose P is a partition on a set Y and $f : X \rightarrow Y$ is a function. Define the relation F on X by aFb iff $f(a)$ and $f(b)$ are in the same set of the partition P. Then F is an equivalence relation on X. Explain why this is a strengthening of Theorem 4.2.3.

8. Prove Theorem 4.3.2.

9. Suppose that $\{S_i : i \in I\}$ is a partition of a set Y and X is a nonempty subset of Y. Prove that $\{X \cap S_i : i \in I \text{ and } X \cap S_i \neq \emptyset\}$ is a partition of X.

10. Suppose $\{A_i : i \in I\}$ and $\{B_k : k \in K\}$ are each a partition of a set X.

    a) Define $C_{i,k} = A_i \cap B_k$. Is $\{C_{i,k} : i \in I \text{ and } k \in K\}$ a partition of X? If so, prove it; if not give a counterexample.

    b) Repeat part a with $D_{i,k} = A_i \cup B_k$ in place of $C_{i,k}$.

11. Suppose $\{A_i : i \in I\}$ is a partition of a set X and $\{B_k : k \in K\}$ is a partition of a set Y. For $i \in I$ and $k \in K$, define $C_{i,k} = A_i \times B_k$, a subset of $X \times Y$. Is $\{C_{i,k} : i \in I \text{ and } k \in K\}$ a partition of $X \times Y$? If so, prove it; if not give a counterexample.

12. Critique the incorrect "proofs" that follow. Point out all reasoning errors and unclear presentation. (If the claim is false, there must be an error in the argument. However, it is not enough to say that the claim is false. You need to find an error in the reasoning.)

    *a) Claim: For an equivalence relation R on X and $x, y, z \in X$, if $x \notin [y]$ and $y \notin [z]$, then $x \notin [z]$. "Proof": From the givens, we can say "not yRx" and "not zRy." By symmetry, we can switch the order of these two statements. By transitivity, we get "not zRx," which gives us $x \notin [z]$. ▲

    b) Claim: For a collection of subsets $\{B_i : i \in I\}$ of a set X, the B_i are pairwise disjoint iff $\bigcap_{i \in I} B_i = \emptyset$. "Proof": Suppose $\{B_i : i \in I\}$ is a collection of subsets of X. First, suppose the B_i are pairwise disjoint. Then $\bigcap_{i \in I} B_i \subseteq B_1 \cap B_2 = \emptyset$. Now suppose $\bigcap_{i \in I} B_i = \emptyset$ and consider any two of the sets, B_i and B_j. Then $B_1 \cap B_2 \subseteq \bigcap_{i \in I} B_i = \emptyset$. ▲

    c) Suppose $\{A_i : i \in I\}$ is a partition of a set X with more than one set. For each $i \in I$, let $C_i = \overline{A_i}$, the complement of A_i in X. Claim: $\{C_i : i \in I\}$ is a partition of X. "Proof": Given any C_k, we first show it is nonempty. Since each A_i is nonempty, we can pick some $x \in A_k$ for $i \neq k$. Then $x \in C_i$, showing it is nonempty. Next, we show that the union of the C_i is all of X. Let $x \in X$. For any given i, either $x \in A_i$ or $x \notin A_i$. In the second case, $x \in C_i$. In the first case, by disjointness, $x \notin A_k$ for $k \neq i$. Thus, $x \in C_k$. So $X \subseteq \bigcup_{i \in I} C_i$. Further, for each i, $C_i \subseteq X$, so their union is also a subset of X, showing equality. Finally, we must show disjointness. For $i \neq k$, we know $A_i \cap A_k = \emptyset$. Then de Morgan's Laws give us $C_i \cap C_k = \overline{A_i} \cap \overline{A_j} = \overline{A_i \cup A_k} = \overline{X} = \emptyset$. ▲

13. Critique the questionable "proofs" that follow. Point out reasoning errors and unclear presentation. If the argument is a proof, say so. (If the claim is false, there must be an error in the argument. However, it is not enough to say that the claim is false. You need to find an error in the reasoning.)

    a) Claim: For an equivalence relation R on X and $x, y \in X$, $x \in [y]$ iff $y \in [x]$. "Proof": $x \in [y]$ iff yRx iff xRy iff $y \in [x]$. ▲

    *b) Suppose $\{A_i : i \in I\}$ is a partition of a set X and $\{B_k : k \in k\}$ is a partition of a set Y. Claim: $W = \{A_i : i \in I\} \cup \{B_k : k \in k\}$ is a partition of $X \cup Y$. "Proof": Since each A_i and each B_j are nonempty, each set in W is nonempty. Similarly, the union of the A_i gives all of X and the union of the B_j gives all of Y. So the union of every set in W gives all of $X \cup Y$. In the same manner, all the A_i are disjoint and all the B_j are disjoint, so any two sets in W are disjoint. This fulfills the three conditions for a partition, so W is a partition of $X \cup Y$. ▲

    c) For any subsets A and B of a set X, the collection of sets $\{A \cap B, A \cap \overline{B}, \overline{A} \cap B, \overline{A} \cap \overline{B}\}$ is a partition of X. "Proof": For any $x \in X$, exactly one of the two propositions $x \in A$ or $x \in \overline{A}$ is true. Similarly, there are two options for x and B. Thus, there are four disjoint possibilities for x and A and B together, which are listed in the collection of sets. Consequently, we have a disjoint collection whose union is X, showing it is a partition. ▲

4.4 PARTIAL ORDERS

With several examples in Section 4.1, we defined a partial order to be a reflexive, transitive, and antisymmetric relation. We won't prove these motivating relations ($|$ on $\mathbb{N}$, $\subseteq$ and $\leq$) to be partial orders, each for a different reason. You already proved "divides" on $\mathbb{N}$ was a partial order in Problem 7 of Section 2.3, although we hadn't yet given names for these properties. The definitions of "subset" and "set equality" readily give the reflexive and antisymmetric properties for $\subseteq$, and the transitive property was shown in Problem 7a of Section 2.5. The relation $\leq$ is so fundamental that either we take its properties as axioms or we need to make other explicit axioms about numbers. Problem 17 explores this second option. Instead, in this section, we consider different aspects of general partial orders.

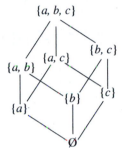

FIGURE 1. Hasse diagram for $\subseteq$ on $\mathcal{P}(\{a, b, c\})$.

Hasse Diagrams

Example 4 of Section 4.1 used graphs with vertices and arrows to represent relations. If we know a relation $\sqsubseteq$ is a partial order, eliminating many of the arrows in such a graph gives a clearer picture of the relation. Hasse diagrams, as in Figures 1, 2, and 3, use the properties of a partial order to minimize the edges needed to represent it. First, there is no need for loops, since, by reflexivity, we already know any element is related to itself. Second, when $x \sqsubseteq y$, we place x below y. Thus, we don't need arrows to indicate the direction, since antisymmetry assures us the relation only goes one way—up. Finally, from transitivity, there is no need to include an edge from x to z if we already have edges from x to y and from y to z. We also eliminate edges between elements with longer upward paths of edges connecting them.

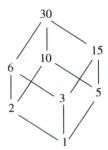

FIGURE 2. Hasse diagram for "divides" on D_{30}, the positive divisors of 30.

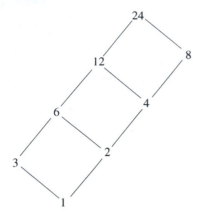

FIGURE 3. Hasse diagram for "divides" on D_{24}, the positive divisors of 24.

EXERCISE 1. Draw the Hasse diagram for the relation divides on D_{128}, the positive divisors of $128 = 2^7$.

EXERCISE 2. Compare and contrast the Hasse diagrams for Figures 1, 2, and 3 and Exercise 1. We will investigate when two partial orders are similar or different in the final subsection of this chapter.

Hasse diagrams appear in many contexts, including family trees and evolutionary trees, portraying the lineage of people or species. See Figure 4.

Properties of Partial Orders

The partial orders in Figures 1, 2, 3, and 4 show branching in the ordering. The most familiar partial order, $\leq$ on $\mathbb{R}$, doesn't display such branching. Rather, the order of real numbers matches the ordering of points on a line. We use the term *linear* for partial orders like $\leq$.

DEFINITION. A partial order $\trianglelefteq$ on a set X is a *linear order* (or a *total order*) iff for all $a, b \in X$, we have $a \trianglelefteq b$ or $b \trianglelefteq a$.

Different parts of mathematics focus on different aspects of order, leading to a profusion of definitions. Even the differences in the linear order $\leq$ on $\mathbb{N}$, $\mathbb{Z}$, $\mathbb{Q}$, and $\mathbb{R}$ have led mathematicians to make distinctions. Analysts investigate upper bounds and least upper bounds of subsets of $\mathbb{R}$. Algebraists are more interested in maximal

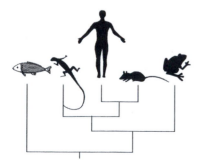

FIGURE 4. Evolutionary tree. (Reprinted with permission from *Science* 310 (11 Nov., 2005), p. 979. Copyright 2005 AAAS.)

elements of things. Number theorists find the successor concept valuable. After the definitions of these and other terms, we will consider examples illustrating their differences and some of their importance.

DEFINITIONS. Let $\trianglelefteq$ be a partial order on a set X.

- For a subset S of X, an element u of X is an *upper bound* of S iff for all $s \in S$, $s \trianglelefteq u$.
- For a subset S of X, an element l of X is a *lower bound* of S iff for all $s \in S$, $l \trianglelefteq s$.
- A subset S of X is *bounded* iff S has an upper bound and a lower bound.
- For a subset S of X, an element b of X is a *least upper bound* of S iff b is an upper bound of S and for every upper bound u of S, $b \trianglelefteq u$.
- An element $g \in X$ is a *greatest element* of X iff for all $x \in X$, $x \trianglelefteq g$.
- An element $m \in X$ is a *maximal element* of X iff for all $x \in X$ if $m \trianglelefteq x$, then $m = x$.
- Given $a \in X$, a *successor* of a is an element b such that $a \trianglelefteq b$ and for all $c \in X$, if $a \trianglelefteq c$ and $c \trianglelefteq b$, then $a = c$ or $c = b$.
- A linear order $\trianglelefteq$ on a set X is *dense* iff for all $a, b \in X$, if $a \neq b$ and $a \trianglelefteq b$, then there is $c \in X$ distinct from a and b such that $a \trianglelefteq c$ and $c \trianglelefteq b$.
- The definitions of *greatest lower bound, least element, minimal element,* and *predecessor* are left to the reader.

EXAMPLE 1

The linear order $\leq$ has no greatest element or maximal element in $\mathbb{N}$, $\mathbb{Z}$, $\mathbb{Q}$, or $\mathbb{R}$. The set $\{x \in \mathbb{Q} : x^2 < 2\}$ has many upper bounds, such as 3 and 17.2, but has no least upper bound in $\mathbb{Q}$. It does have a least upper bound in $\mathbb{R}$, namely, $\sqrt{2}$, which Problem 7 of Section 2.2 showed is not a rational number. In $\mathbb{N}$ and $\mathbb{Z}$, each element n has a successor, namely, $n + 1$. However, neither $\mathbb{Q}$ nor $\mathbb{R}$ have successors to elements. The lack of successors is closely related to the fact that their orders are dense, which we now show. For any two distinct rationals or reals a and b, let $c = \frac{a+b}{2}$, which is strictly between them, fulfilling the definition of dense. Problem 4 asks you to show that the concepts of successor and dense are essentially opposites. Problem 3 considers least upper bounds. $\diamondsuit$

EXAMPLE 2

Figure 5 shows the Hasse diagram for the relation "divides" on the set $\{1, 2, 3, \ldots, 8\}$. This set has no greatest element under the relation "divides." The numbers 5, 6, 7, and 8 are all maximal elements. Problem 5 explores maximal and greatest elements.

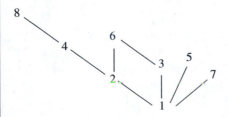

FIGURE 5. Hasse diagram for "divides" on $\{1, 2, 3, 4, 5, 6, 7, 8\}$. $\diamondsuit$

"Almost" Partial Orders

Several relations, such as $<$ and "divides" on $\mathbb{Z}$, have some, but not all, the properties of partial orders. Each is closely related to a partial order.

"Less than" versus "Less than or Equal to" When we consider the usual order of numbers in $\mathbb{R}$, we use $<$ as often as $\leq$. Partial orders correspond to $\leq$. We could generalize $<$, getting what are called *strict partial orders*. However, as Theorem 4.4.1 shows, partial orders and strict partial orders pair up so nicely there is little value in having both general concepts. Of course, the difference between $\leq$ and $<$ is the phrase "or equal to." We generalize $<$ in Theorem 4.4.1.

DEFINITION. A relation $\prec$ on a set X is *irreflexive* iff for all $x \in X$, it is not the case that $(x \prec x)$, abbreviated $x \not\prec x$. Equivalently, no x is related to itself by $\prec$.

THEOREM 4.4.1. Let $\preceq$ be a partial order on a set X and define $\prec$ on X by $x \prec y$ iff $x \preceq y$ and $x \neq y$. Then $\prec$ is irreflexive and transitive. Conversely, suppose $\sqsubset$ is an irreflexive, transitive relation on a set X. Define $\sqsubseteq$ on X by $x \sqsubseteq y$ iff $x \sqsubset y$ or $x = y$. Then $\sqsubseteq$ is a partial order on X.

Proof. See Problem 7. ∎

Partial Orders and Equivalence Relations In Section 4.1, we noted that on $\mathbb{Z}$ the relation "divides" is not antisymmetric, although it is reflexive and transitive. While it is not a partial order, it is close—only the intrusion of minus signs upsets the ordering. For instance, not only does 3 divide 12, but -3 divides 12, and both 3 and -3 divide -12 as well as each other. If we clump each pair x and $-x$ together, we reduce $\mathbb{Z}$ to $\mathbb{N}$, where "divides" is a partial ordering. In other contexts, we can also benefit from such clumping, or more mathematically, from equivalence classes. While $\subseteq$ is the most important partial order on $\mathcal{P}(X)$, there is another fairly natural relation on $\mathcal{P}(X)$, at least if X is finite. For subsets A and B of X, define $A \preceq B$ iff the number of elements in A is less than or equal to the number of elements in B. Because the relation $\preceq$ is built from the partial order $\leq$, it is natural to expect that $\preceq$ is also a partial order. It is not quite; for instance, $\{1, 2, 3\} \preceq \{2, 4, 6\}$ and $\{2, 4, 6\} \preceq \{1, 2, 3\}$, but these sets are not equal, just equal in size. Thus, antisymmetry fails, although reflexivity and transitivity hold, just as with "divides" on $\mathbb{Z}$. The construction of Theorem 4.4.2 shows how to take any reflexive and transitive relation and obtain a closely related partial order. The key is to create equivalence classes. For the example of "divides" on $\mathbb{Z}$, a generic equivalence class is $[x] = \{x, -x\}$. For $\preceq$ on $\mathcal{P}(X)$, an equivalence class contains all subsets of the same size.

THEOREM 4.4.2. Suppose $\preceq$ is reflexive and transitive on a set X. Define $\approx$ on X by $a \approx b$ iff $a \preceq b$ and $b \preceq a$. Then $\approx$ is an equivalence relation on X. Let $X_\approx = \{[x]_\approx : x \in X\}$ be the set of equivalence classes of $\approx$, and define $\preceq$ on $X_\approx$ by $[a]_\approx \preceq [b]_\approx$ iff there are $x \in [a]_\approx$ and $y \in [b]_\approx$ such that $x \preceq y$. Then $\preceq$ is a partial order on $X_\approx$.

Proof. Problem 9a shows that $\approx$ is an equivalence relation. Thus, the set $X_\approx$ of the equivalence classes of $\approx$ makes sense, as does the relation $\preceq$. For clarity, we will write $[x]$ for the equivalence class $[x]_\approx$ and so on. For any $[x]$ in $X_\approx$, we know $x \preceq x$ and $x \in [x]_\approx$, so $[x] \preceq [x]$, showing reflexivity. For transitivity, let $[x]$, $[y]$, and $[z]$ be any equivalence classes and assume $[x] \preceq [y]$ and $[y] \preceq [z]$. By definition of $\preceq$, there are elements $p \in [x]$, $q, r \in [y]$, and $s \in [z]$ such that $p \preceq q$ and $r \preceq s$. Now $q, r \in [y]$ means $y \approx q$ and $y \approx r$. In turn, the definition of $\approx$ gives us $q \preceq y$ and $y \preceq r$. Thus, we have $p \preceq q$, $q \preceq y$, $y \preceq r$, and $r \preceq s$. Since $\preceq$ is transitive, we have $p \preceq s$, showing $[x] \preceq [z]$, finishing transitivity. Problem 9b shows that $\preceq$ is antisymmetric. ∎

Order Isomorphisms

You may have noticed that the Hasse diagrams in Figures 1 and 2 look essentially the same, but they each differ from the Hasse diagram in Figure 3. What do we mean by "essentially the same"? Certainly, the sets need to be the same size, but that isn't enough, since all three diagrams have eight elements. The structure of the partial orders must match up in some way. Isomorphisms are special bijections that match structure as well as elements of sets. (The word "isomorphism" has Greek roots: *iso* means "equal" or "same," and *morph* means "form" or "shape." So, an isomorphism maps something to something else with the same form.)

DEFINITION. Suppose $\preceq$ is a partial order on a set X and $\sqsubseteq$ is a partial order on Y. A function $f : X \to Y$ is an *order isomorphism* from X to Y iff f is a bijection and for all $a, b \in X$, $a \preceq b$ iff $f(a) \sqsubseteq f(b)$.

EXAMPLE 3

Show that the partial order of Figure 1 is isomorphic to the one in Figure 2, but the one in Figure 2 is not isomorphic to the one in Figure 3.

Solution The function f given by Table 1 converts the Hasse diagram in Figure 1 to the diagram in Figure 2, giving us an isomorphism. This function can be found by matching corresponding points on the figures. It would be tedious to check every pair of subsets with their corresponding divisors to see whether $A \subseteq B$ always matches with $f(A)|f(B)$. Instead, we use a mathematical connection. The singletons correspond to the prime divisors: $\{a\}$ to 2, $\{b\}$ to 3, and $\{c\}$ to 5. Then union in $\mathcal{P}(\{a, b, c\})$ corresponds to multiplication in D_{30}. For instance, $\{a\} \cup \{c\} = \{a, c\}$ corresponds to $2 \cdot 5 = 10$.

TABLE 1. Isomorphism

x	$\emptyset$	$\{a\}$	$\{b\}$	$\{c\}$	$\{a, b\}$	$\{a, c\}$	$\{b, c\}$	$\{a, b, c\}$
$f(x)$	1	2	3	5	6	10	15	30

There are many bijections from D_{30} to D_{24}. In fact, by Theorem 3.2.5, there are $8! = 40{,}320$ such bijections. How do we show that none of them can be an order isomorphism? We find some property where the partial orders differ. Note that the three elements 2, 3, and 5 in D_{30} are unrelated to each other—none divides the other two because they are primes. However, in D_{24}, for any trio of numbers, the smallest divides one of the other two. So, no function from D_{30} to D_{24} can be an order isomorphism. ◇

The theorem that follows shows that the subset relation is, in a reasonable sense, the most general type of partial order. Given a partial order on a set X, the trick is to find an appropriate collection of subsets so that $\subseteq$ matches the given order.

THEOREM 4.4.3. Let $\preceq$ be any partial order on a set X. Then there is an order isomorphism from X with $\preceq$ to some subset of $\mathcal{P}(X)$ with the relation $\subseteq$.

Proof. Let X be any set and for $x \in X$, define $A_x = \{y \in X : y \preceq x\}$. That is, we match each element to the set of elements "less than or equal" to the element. Define $f : X \to \{A_x : x \in X\}$ by $f(x) = A_x$. Claims: f is a bijection and for all $x, y \in X$, $x \preceq y$ iff $A_x \subseteq A_y$. See Problem 10 for these claims. ∎

The concept of an isomorphism appears in many areas of modern mathematics. We will consider it again in Chapters 6 and 7.

PROBLEMS

*1. For each statement that follows, decide whether it is true or (at least sometimes) false. Explain your answer or cite the part of the text supporting it.

a) A linear order is a partial order.

b) A partial order is a linear order.

c) A maximal element is a greatest element.

d) A greatest element is a maximal element.

e) Every set with an upper bound has a least upper bound.

f) Every set with a least upper bound has a greatest lower bound.

g) Every relation is either reflexive or irreflexive.

2. Define a relation $\sqsubseteq$ on the set $\{Rock, Paper, Scissors\}$ based on the children's game. Which of the properties of a partial order does your relation satisfy? Which fail? Explain.

3. *a) Suppose X is partially ordered by $\trianglelefteq$ and a subset S of X has at least one least upper bound. Show that S has exactly one least upper bound.

b) An axiom of $\mathbb{R}$ asserts, "Every nonempty bounded subset of $\mathbb{R}$ has a least upper bound in $\mathbb{R}$." Explain why the word "nonempty" is essential in the definition.

c) Does the axiom in part b hold if we replace $\mathbb{R}$ by $\mathbb{N}$? $\mathbb{Z}$? $\mathbb{Q}$? Explain.

d) Assume the axiom in part b and prove every nonempty bounded subset of $\mathbb{R}$ has a greatest lower bound. Hint: For a subset A of $\mathbb{R}$, consider $-A = \{-a : a \in A\}$.

e) Explain why every nonempty subset of $\mathbb{N}$ has a greatest lower bound. Relate your answer to the well ordered property of $\mathbb{N}$, given in Section 2.4.

f) Suppose $\sqsubseteq$ is a partial order on a finite set X and every subset of X has an upper bound under $\sqsubseteq$. Must every subset have a least upper bound under $\sqsubseteq$? If so, prove it; if not give a counterexample.

4. Suppose $\trianglelefteq$ is a linear order on a set X. Show that $\trianglelefteq$ is dense on X iff no element of X has a successor.

5. *a) If a partial order on X has a greatest element g, prove g is a maximal element.

   b) If a partial order has just one maximal element, must that element be the greatest element of X? Prove your answer.

   c) Prove every partial order on a finite set X has at least one maximal element.

   d) If a linear order on X has a maximal element, prove that maximal element is the greatest element.

6. Suppose R is a partial order on a set X.

   ***a)** If Y is a subset of X, prove that $R \cap (Y \times Y)$ is a partial order on Y. (For instance, R could be $\leq$ on $X = \mathbb{R}$ and Y could be $\mathbb{Q}$. Then $R \cap (\mathbb{Q} \times \mathbb{Q})$ is just $\leq$ on $\mathbb{Q}$.)

   b) Define R^{-1} on X by $aR^{-1}b$ iff bRa. Prove R^{-1} is a partial order on X.

   c) If R is linear, prove that R^{-1} and $R \cap (Y \times Y)$ are linear.

   ***d)** If R is dense, must R^{-1} and $R \cap (Y \times Y)$ be dense? For each one, if so, prove it; if not, give a counterexample.

7. Prove Theorem 4.4.1.

8. a) If R and S are antisymmetric relations on a set X, must $S \cap R$ be antisymmetric? If yes, prove it; if not, give a counterexample.

   b) Repeat part a for $S \cup R$.

9. a) Show that $\approx$ in Theorem 4.4.2 is an equivalence relation.

   b) Show that $\preceq$ in Theorem 4.4.2 is antisymmetric.

10. a) In Theorem 4.4.3, show that f is a bijection.

   b) In Theorem 4.4.3, show that for all $x, y \in X$, $x \preceq y$ iff $A_x \subseteq A_y$.

   c) In Theorem 4.4.3, could we have made an order isomorphism matching x with $B_x = \{y \in X : y \neq x$ and $y \preceq x\}$? Prove or give a counterexample.

11. On $\mathbb{R} \times \mathbb{R}$, define $\preceq$ by $(a, b) \preceq (p, q)$ iff $a < p$ or $(a = p$ and $b \leq q)$.

   a) Prove $\preceq$ is a partial order on $\mathbb{R} \times \mathbb{R}$.

   b) Prove $\preceq$ is linear.

   c) Prove $\preceq$ is dense.

   d) Find an example of a nonempty subset A of $\mathbb{R} \times \mathbb{R}$ such that A is bounded for $\preceq$, but A has no least upper bound.

12. Let $\mathbb{F}$ be the set of all functions from $\mathbb{R}$ to $\mathbb{R}$. On $\mathbb{F}$, define $\leq$ by $f \leq g$ iff for all $x \in \mathbb{R}$, $f(x) \leq g(x)$.

   a) Prove $\leq$ is a partial order on $\mathbb{F}$.

   b) Show that $\leq$ is not linear.

   c) Even though $\leq$ is not linear, show that it fulfills the idea of being dense. That is, given any two different functions f and g, with $f \leq g$, show that there is a distinct third function h such that $f \leq h$ and $h \leq g$.

13. Suppose $\sqsubseteq$ and $\preceq$ are partial orders on a set X.

    a) Prove $\trianglelefteq$ is a partial order on X, where $a \trianglelefteq b$ iff $a \sqsubseteq b$ and $a \preceq b$.

    b) Is $\trianglelefteq$ a partial order on X, where $a \trianglelefteq b$ iff $a \sqsubseteq b$ or $a \preceq b$? If so, prove it; if not, state which properties can fail and give a counterexample for each failed property.

    c) Is the composition of $\sqsubseteq$ and $\preceq$ a partial order on X? If so, prove it; if not, state which properties can fail and give a counterexample for each failed property.

14. Suppose $\sqsubseteq$ is a partial order on a set X and $\preceq$ is a partial order on a set Y.

    a) Prove $\trianglelefteq$ is a partial order on $X \times Y$, where $(a, b) \trianglelefteq (c, d)$ iff $a \sqsubseteq c$ and $b \preceq d$.

    b) Prove $\trianglelefteq$ is a partial order on $X \times Y$, where $(a, b) \trianglelefteq (c, d)$ iff $a \neq c$ or $(a = c$ and $b \preceq d)$.

    c) Suppose both $\sqsubseteq$ and $\preceq$ are linear orders and X and Y each have at least two elements. Determine which of the relations in parts a and b is a linear order on $X \times Y$. Prove your answer.

    d) Suppose both $\sqsubseteq$ and $\preceq$ are dense. For the order in part c that is linear, determine whether it is also dense. Prove your answer.

15. Suppose R is a symmetric and antisymmetric relation on a set X. Prove R is transitive.

16. A set X is well ordered by a partial order $\preceq$ iff, for all nonempty subsets A of X, A has a least element in A. If a set is well ordered, prove it is linearly ordered.

17. To define $\leq$ on $\mathbb{R}$ and show some of its properties, we start with axioms for a subset $\mathbb{P}$ of $\mathbb{R}$, which will turn out to be the positive numbers.

    Axiom 1. For all $x \in \mathbb{R}$, $x \in \mathbb{P}$ or $-x \in \mathbb{P}$ or $x = 0$.

    Axiom 2. For all $x \in \mathbb{R}$, if $x \in \mathbb{P}$, then $-x \notin \mathbb{P}$.

    Axiom 3. For all $x, y \in \mathbb{P}$, $x + y \in \mathbb{P}$ and $xy \in \mathbb{P}$.

    Define $x \leq y$ iff $y - x \in \mathbb{P}$ or $y = x$.

    a) Prove $\leq$ is a partial order on $\mathbb{R}$.

    *b) Prove $\leq$ is a linear order on $\mathbb{R}$.

    c) Prove for all $x, y, z \in \mathbb{R}$, if $x \leq y$, then $x + z \leq y + z$.

    d) Prove for all $x, y \in \mathbb{R}$ and $z \in \mathbb{P}$, if $x \leq y$, then $xz \leq yz$.

    Note that if we drop the conclusion $xy \in \mathbb{P}$ from Axiom 3, the set of negative numbers satisfies the other axioms (and part d would fail). The following parts should convince you that $\mathbb{P}$ really is the set of positive numbers:

    e) Prove $1 \in \mathbb{P}$. Hint: Use a proof by contradiction.

    f) Prove $\mathbb{N} \subseteq \mathbb{P}$. Hint: Use part e.

    g) Prove if $x \in \mathbb{P}$ and $-y \in \mathbb{P}$, then $xy \notin \mathbb{P}$.

    h) Prove for all $q \in \mathbb{N}$, $1/q \in \mathbb{P}$.

    i) Use induction to prove all positive rationals are in $\mathbb{P}$. That is, for all $p, q \in \mathbb{N}$, prove $\frac{p}{q} \in \mathbb{P}$.

    j) Prove for all $x \in \mathbb{R}$ that if $x \neq 0$, then $x^2 \in \mathbb{P}$. Explain why this property shows that $\mathbb{P}$ contains all positive real numbers.

18. There are just two nonisomorphic partial orders on a set with two elements: Either one element is "bigger" than the other, or else they are unrelated.

    *a) Find the number of nonisomorphic partial orders on a set with three elements. Draw Hasse diagrams for them.

b) Find the number of nonisomorphic partial orders on a set with four elements. Draw Hasse diagrams for them.

19. a) Define $\trianglelefteq$ on the set $\{0, 1, 2\} \times \{0, 1, 2\}$ by $(a, b) \trianglelefteq (c, d)$ iff $a \leq c$ and $b \leq d$. (Problem 14a shows $\trianglelefteq$ is a partial order.) Find and prove an order isomorphism from $\{0, 1, 2\} \times \{0, 1, 2\}$ with $\trianglelefteq$ to D_{36}, the positive divisors of 36 with the relation "divides."

b) Change the set $\{0, 1, 2\} \times \{0, 1, 2\}$ of part a to $\{0, 1\} \times \{0, 1, 2, 3\}$ and find a number n so that there is an order isomorphism from $\{0, 1\} \times \{0, 1, 2, 3\}$ to D_n, the set of all positive divisors of an integer n. Find and prove the order isomorphism.

c) Generalize part b for any set $\{0, 1, \ldots, j\} \times \{0, 1, \ldots, k\}$.

20. Let D_n be the set of all positive divisors of an integer n. Give conditions on n and k so that there is an isomorphism from D_n to D_k, where the partial order for each is "divides." Define and prove the isomorphism.

21. Find $k \in \mathbb{N}$ such that D_k with "divides" is isomorphic to $\mathcal{P}(\{1, \ldots, n\})$ with $\subseteq$. Define the isomorphism. [Remark: In light of Theorem 4.4.3, this problem shows that "divides" is also the most general type of relation on a finite set. However, Theorem 4.4.3 applies to infinite sets as well.]

22. Given a partial ordering $\sqsubseteq$ on a set X and $a, b \in X$ with $a \sqsubseteq b$, define the *interval* $[a, b] = \{c : a \sqsubseteq c \text{ and } c \sqsubseteq b\}$. Warning: For partial orders that are not linear, these intervals look unusual.

a) For the relation "divides" on D_{144}, the positive divisors of 144, find $[3, 72]$. Draw the Hasse diagram of $[3, 72]$. Compare with the Hasse diagram of D_{24}.

b) In $\mathbb{N}$, suppose a divides b. Prove there is an isomorphism from $D_{b/a}$ to the subset $[a, b]$ of D_b.

c) In $\mathcal{P}(X)$ with $\subseteq$ and $A \subseteq B \subseteq X$, prove there is an isomorphism from $[A, B]$ to $\mathcal{P}(B - A)$.

23. Critique the incorrect "proofs" that follow. Point out reasoning errors and unclear presentation. (If the claim is false, there must be an error in the argument. However, it is not enough to say that the claim is false. You need to find an error in the reasoning.)

*a) Claim: A maximal element of a partially ordered set is a greatest element. "Proof": Suppose m is a maximal element of X with partial ordering $\trianglelefteq$. For $x \in X$, either $x \trianglelefteq m$ or $m \trianglelefteq x$. In the first case, m already satisfies the property of being the greatest element. The second case contradicts the definition of "maximal," so m is the greatest element of X. ▲

b) Claim: If $\sqsubset$ is irreflexive and transitive on X and $a \sqsubseteq b$ iff $a \sqsubset b$ or $a = b$, then $\sqsubseteq$ is a partial order on X. "Proof": We suppose the givens. For reflexivity, with any $a \in X$, we have $a = a$, so $a \sqsubseteq a$. For transitivity, suppose $a \sqsubseteq b$ and $b \sqsubseteq c$. Now $\sqsubseteq$ splits into $\sqsubset$ and $=$. So we have $a \sqsubset b$ and $b \sqsubset c$ or $a = b$ and $b = c$. Both $\sqsubset$ and $=$ are transitive, so either way we have $a \sqsubset c$ or $a = c$ and so $a \sqsubseteq c$. For antisymmetry, let $a, b \in X$, and suppose $a \sqsubseteq b$ and $b \sqsubseteq a$. Because $\sqsubset$ is irreflexive, we can't have $a \sqsubset b$ and $b \sqsubset a$. So we must have $a = b$ and $b = a$, which is plenty to show antisymmetric. ▲

c) Claim: A relation that is symmetric and antisymmetric is reflexive. "Proof": Suppose R is symmetric and antisymmetric on X, and suppose aRb. By symmetry, bRa. By antisymmetry, we have $a = b$. By substituting a for b in the given aRb, we have aRa. Since this holds for any a and b, R must be reflexive. ▲

   d) Claim: The interval $[0, 1]$ with the usual order $\leq$ is well ordered. (See Problem 16 for the definition of "well ordered.") "Proof": By Problem 6, parts a and c, we know $\leq$ is a linear order on $[0, 1]$. Further, every subset of $[0, 1]$ is bounded below by 0 and above by 1. So, every nonempty subset is bounded. By Problem 3d, every nonempty subset has a greatest lower bound. But such a greatest lower bound fulfills the definition of a least element for the subset, showing that $[0, 1]$ fulfills the definition of "well defined." ▲

24. Critique the questionable "proofs" that follow. Point out reasoning errors and unclear presentation. If the argument is a proof, say so. (If the claim is false, there must be an error in the argument. However, it is not enough to say that the claim is false. You need to find an error in the reasoning.)

   *a) Claim: For the partial order $\leq$, every proper subset of $\mathbb{R}$ either has a lower bound or an upper bound. "Proof": For any interval (a, b) or $[a, b]$, if a is a number, then a is a lower bound. Similarly, if b is a number, it is an upper bound. Otherwise, we have the interval $(-\infty, \infty) = \mathbb{R}$, which is not a proper subset. ▲

   b) Claim: An irreflexive relation is antisymmetric. "Proof": Suppose $\prec$ is irreflexive on X and $c, d \in X$ satisfy $c \prec d$ and $d \prec c$. Together these relations give us $c \prec c$, which would contradict irreflexive. Hence, antisymmetric is vacuously true. ▲

   c) Claim: The only relation on a set X that is reflexive, symmetric, and antisymmetric is $=$. "Proof": Let R be any relation on X that has all three properties. For $p, q \in X$, we must show $p = q$ iff pRq. The first direction is just the reflexive property of R. Now suppose pRq. By symmetry, qRp. By antisymmetry, $p = q$. ▲

DEFINITIONS FROM CHAPTER 4

Section 4.1

- A *(binary) relation* R from a set X to a set Y is a subset of $X \times Y$. A subset of $X \times X$ is called a *relation on X*. We generally write aRb, rather than the more formal $(a, b) \in R$, to indicate that a and b are related by R.

- A relation R on a set X is *reflexive* iff for all $x \in X$, xRx.

   A relation R on a set X is *symmetric* iff for all $x, y \in X$, if xRy, then yRx.

   A relation R on a set X is *transitive* iff for all $x, y, z \in X$, if xRy and yRz, then xRz.

   A relation R on a set X is *antisymmetric* iff for all $x, y \in X$, if xRy and yRx, then $x = y$.

- A relation R on a set X is an *equivalence relation* iff R is reflexive, symmetric, and transitive.

- A relation R on a set X is a *partial order* (or *partial ordering*) iff R is reflexive, transitive, and antisymmetric.

Section 4.2

- For an equivalence relation R on a set X and $a \in X$, the *equivalence class* of a is $[a] = \{b \in X : aRb\}$. (If we have two or more equivalence relations on the same set, we can use subscripts on equivalence classes to distinguish them.)

Section 4.3

- A *partition* of a set B is a family of sets $\{S_i : i \in I\}$ such that
  - (i) each S_i is nonempty and
  - (ii) for each $b \in B$, there is a unique S_i such that $b \in S_i$.

Section 4.4

- A partial order $\trianglelefteq$ on a set X is a *linear order* (or a *total order*) iff for all $a, b \in X$, we have $a \trianglelefteq b$ or $b \trianglelefteq a$.

- Let $\trianglelefteq$ be a partial order on a set X.
  - For a subset S of X, an element u of X is an *upper bound* of S iff for all $s \in S$, $s \trianglelefteq u$.
  - For a subset S of X, an element l of X is a *lower bound* of S iff for all $s \in S$, $l \trianglelefteq s$.
  - A subset S of X is *bounded* iff S has an upper bound and a lower bound.
  - For a subset S of X, an element b of X is a *least upper bound* of S iff b is an upper bound of S and for every upper bound u of S, $b \trianglelefteq u$.
  - An element $g \in X$ is a *greatest element* of X iff for all $x \in X$, $x \trianglelefteq g$.
  - An element $m \in X$ is a *maximal element* of X iff for all $x \in X$ if $m \trianglelefteq x$, then $m = x$.
  - Given $a \in X$, a *successor* of a is an element b such that $a \trianglelefteq b$ and for all $c \in X$, if $a \trianglelefteq c$ and $c \trianglelefteq b$, then $a = c$ or $c = b$.
  - A linear order $\trianglelefteq$ on a set X is *dense* iff for all $a, b \in X$, if $a \neq b$ and $a \trianglelefteq b$, then there is $c \in X$ distinct from a and b such that $a \trianglelefteq c$ and $c \trianglelefteq b$.

- A relation $\prec$ on a set X is *irreflexive* iff for all $x \in X$, it is not the case that $(x \prec x)$, abbreviated $x \nprec x$. Equivalently, no x is related to itself by $\prec$.

- Suppose $\preceq$ is a partial order on a set X and $\sqsubseteq$ is a partial order on Y. A function $f : X \to Y$ is an *order isomorphism* from X to Y iff f is a bijection and for all $a, b \in X$, $a \preceq b$ iff $f(a) \sqsubseteq f(b)$.

PART II

CHAPTER 5

INFINITE SETS

"The infinite! No other question has ever moved so profoundly
the spirit of man."

—David Hilbert (1862–1943)

THE CONCEPT OF the infinite has intrigued and perplexed people for thousands of years. Over the centuries, authorities like Aristotle (384–322 B.C.) and Thomas Aquinas (1225–1274) argued that (with the exception of God) nothing was actually infinite, but some things could be potentially infinite, a distinction discussed in Section 1.4. Following such admonitions, mathematicians before the nineteenth century frequently used potential infinities, but avoided actual infinities. They thought of an induction proof, for instance, as a potentially infinite process. The work of Georg Cantor (1845–1918) transformed this situation. Cantor's investigations of the foundations of the real numbers and analysis led him to work with actual infinities. He proved results on the different sizes of infinite sets and other topics that some mathematicians resisted initially. By the time of Cantor's death, mathematicians widely recognized the importance of his work. His results, which form much of Sections 5.2 and 5.3, still delight and astound.

5.1 THE SIZES OF SETS

One of the most basic properties of a set is how many elements it has. For millennia, people have used numbers for finite collections. For infinite sets, it is initially unclear that we can be more precise than "not finite," the literal meaning of infinite. A closer look at the finite case suggests a way to proceed. For a child to determine there are, say, five toys in a box, she or he counts them. In more mathematical language, the child builds a one-to-one function from the toys onto the initial set of numbers {1, 2, 3, 4, 5}. (Luckily, children don't need to understand or use this terminology in order to count.) Cantor (and some others before him in a more informal way) simply extended this key observation to general sets. We use the term *cardinality* to refer to this sense of the size of a set. We need some preliminary results before it makes sense to treat an infinite cardinality as a number, so we delay that idea until later in this section.

DEFINITION. Two sets A and B have the *same cardinality*, denoted $A \sim B$, iff there is a one-to-one and onto function $f : A \to B$.

EXAMPLE 1

Show $\mathbb{N} \sim \mathbb{S}$, where $\mathbb{S} = \{n^2 : n \in \mathbb{N}\}$, the squares of the positive integers.

Proof. On $\mathbb{N}$ define $f(x) = x^2$. Each $x \in \mathbb{N}$ has exactly one image, namely, x^2, which is in $\mathbb{S}$; so, f is a function from $\mathbb{N}$ into $\mathbb{S}$. For one-to-one, let $v, w \in \mathbb{N}$, and assume $f(v) = f(w)$. Thus, $v^2 = w^2$. Since both v and w are positive, $v = w$. For onto, let $y \in \mathbb{S}$. Since y is the square of a positive integer, $\sqrt{y} \in \mathbb{N}$ and $(\sqrt{y})^2 = y$. ∎

Over 200 years before Cantor, Galileo discussed the preceding bijection without the terminology and with a more intuitive explanation. As this matching shows, a proper subset of a set can sometimes have the same cardinality as the entire set. Such a situation seemed to contradict common sense, as expressed in Euclid's fifth common notion, "The whole is greater than the part." Galileo thought of this matching as a paradox, rather than the start of an exploration of the actual infinite. ◊

EXAMPLE 2

The function $f : \mathbb{R} \to (-\frac{\pi}{2}, \frac{\pi}{2})$ given by $f(x) = \arctan(x)$ and illustrated in Figure 1 is a bijection, since it has an inverse function, $f^{-1}(x) = \tan(x)$. Thus, $\mathbb{R} \sim (-\frac{\pi}{2}, \frac{\pi}{2})$. It may be surprising that a nonempty finite interval has the same cardinality as the entire, infinitely long real line. Since the finite interval $(-\frac{\pi}{2}, \frac{\pi}{2})$ has infinitely many points, we have two competing senses of infinity: length and size. Cantor's concept of cardinality elucidated one of these, the size of a set. The even more modern subject of measure theory gave a solid foundation for the concept of length, including infinitely long lines.

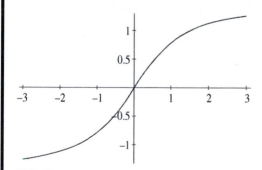

FIGURE 1. $f(x) = \arctan(x)$. ◊

At this point, it would be natural to try to show that $\sim$ is an equivalence relation. However, equivalence relations need to have a set on which they act. There is no such appropriate set here because Russell's Paradox from Section 1.4 makes it impossible to have a universal "set of all sets." Our first theorem evades this problem.

THEOREM 5.1.1. Let A, B, and C be any sets. Then

    i) (reflexive) $A \sim A$.
    ii) (symmetric) If $A \sim B$, then $B \sim A$.
    iii) (transitive) If $A \sim B$ and $B \sim C$, then $A \sim C$.

Proof. See Problem 2. ∎

    The relation $\sim$ tells us when sets have the same size, but we expect the idea of size to tell us more than just which things have the same size. We next compare sets to decide when one set is smaller than another. For instance, a subset of a set surely ought to have a cardinality "less than or equal to" the entire set. We modify the definition of $\sim$ to compare sets of possibly different sizes. Theorems 5.1.2 and 5.1.3 give us some of the properties we expect in comparing sizes.

DEFINITION. Given two sets A and B, $A \lesssim B$ iff there is a function $f : A \to B$ that is one-to-one. If $A \lesssim B$, but they do not have the same cardinality, we say the cardinality of A is *strictly smaller than* the cardinality of B.

THEOREM 5.1.2. For any sets A and B, if $A \subseteq B$, then $A \lesssim B$.

Proof. See Problem 3. ∎

    The following theorem tells us $\lesssim$ has two of the three properties for a partial order:

THEOREM 5.1.3. Let A, B, and C be any sets. Then

    i) (reflexive) $A \lesssim A$.
    ii) (transitive) If $A \lesssim B$ and $B \lesssim C$, then $A \lesssim C$.

Proof. See Problem 4. ∎

    Is $\lesssim$ antisymmetric? No. Example 1 gives a one-to-one function from $\mathbb{N}$ to $\mathbb{S}$, and the function g given by $g(x) = x$ is one-to-one from $\mathbb{S}$ to $\mathbb{N}$. So, $\mathbb{N} \lesssim \mathbb{S}$ and $\mathbb{S} \lesssim \mathbb{N}$. Yet these sets are clearly not equal. Rather, they are equal in size. Cantor conjectured that this weakened sense of antisymmetry holds in general. The theorem asserting this fact is named for the mathematicians who independently proved it in 1894 and 1896. We state this useful theorem next, but postpone its proof until Section 5.2.

THE SCHRÖDER–BERNSTEIN THEOREM (5.2.2). For any sets A and B, if $A \lesssim B$ and $B \lesssim A$, then $A \sim B$.

    We expect even more of an ordering of the sizes of sets. In Section 4.4, we defined a partial order $\sqsubseteq$ to be a linear order iff for all elements a and b, we have $a \sqsubseteq b$ or $b \sqsubseteq a$. Surely if $\lesssim$ reflects ordering of the size of sets, it should be linear.

To the surprise of many mathematicians, the elementary axioms of set theory were insufficient to prove linearity. (See Problem 7 in Section 5.4.) The Axiom of Choice, discussed in the optional Section 5.4, is needed to prove this proposition, as well as a number of other results.

While proofs involving the Axiom of Choice are not essential in an introductory proof course, all mathematics majors should know certain facts about this axiom. First of all, it can't be proven from the other elementary axioms of set theory. Secondly, almost all mathematicians accept it and use it whenever it is needed. Thirdly, it is needed for many important results in mathematics, but also leads to some puzzling results. Finally, the most used proposition logically equivalent to the Axiom of Choice is called Zorn's Lemma (even though it is really an axiom, not a theorem).

The phrase "has the same cardinality" for $\sim$ suggests that the cardinality of a set is a number, the set's size. For finite sets, we naturally use the appropriate integer in the definition that follows. (The set $\mathbb{N} \cap [1, n]$ is just a formal way of writing the set $\{1, \ldots, n\}$.) Cantor's choice of the cardinalities of infinite sets needs some explanation. Following Cantor, we reserve the first letter of the Hebrew alphabet, $\aleph$, pronounced "aleph," for infinite cardinal numbers. Since the "smallest" infinite set is $\mathbb{N}$, Cantor called its size $\aleph_0$. (Theorem 5.4.1 makes precise what we mean by the smallest infinite size.) Cantor couldn't determine which aleph corresponded to the size of the real numbers, so he used c, for "continuum," for the size of the reals. In Section 5.3, we'll give his proof that $\aleph_0$ is strictly smaller than c; that is, $\mathbb{R}$ is "uncountable," as defined next.

DEFINITION. We write $|A|$ for the *cardinality* of the set A.

$$|\varnothing| = 0. \quad |\mathbb{N} \cap [1, n]| = n. \quad |\mathbb{N}| = \aleph_0. \quad |\mathbb{R}| = c.$$

A set is *finite* iff its cardinality is an integer. A set X is *countably infinite* (or *denumerable*) iff $X \sim \mathbb{N}$. A set is *countable* iff it is finite or countably infinite; otherwise, it is *uncountable*.

With our definition of cardinality, we can rewrite the results of Examples 1 and 2 as $|\mathbb{S}| = |\mathbb{N}| = \aleph_0$ and $|(-\frac{\pi}{2}, \frac{\pi}{2})| = |\mathbb{R}| = c$, respectively. In the next two sections, we will determine the cardinalities of many sets, such as $\mathbb{Q}, \mathbb{N} \times \mathbb{N}, \mathcal{P}(\mathbb{N})$, and $\mathbb{R} \times \mathbb{R}$. We define one more type of set now, the set of all functions from one set to another set. The exponential notation is inspired by Theorem 3.2.4.

DEFINITION. For nonempty sets A and B, the set of all functions from A to B is B^A.

THEOREM 5.1.4. If A and B are finite nonempty sets, then $|B^A| = |B|^{|A|}$.

Proof. This theorem simply restates Theorem 3.2.4. ∎

Nonmathematicians, to the distress of mathematicians, often use "infinity" as a synonym of "unimaginably large." Such sloppiness does a disservice to the wonder of the infinite. The familiarity of the functions in Examples 1 and 2 might also hide the surprising nature of what they tell us. Students are encouraged to spend some time thinking about those examples and Problems 7 and 8. It is hoped you will get a feel from them about how much bigger infinite sets must be than any finite set, no matter how unimaginably big.

PROBLEMS

*1. For each statement that follows, decide whether it is true or (at least sometimes) false. Explain your answer or cite the part of the text supporting it.

a) $\sim$ is an equivalence relation.

b) $\lesssim$ is a partial order.

c) $\mathbb{N} \lesssim \mathbb{R}$.

d) Every finite set is countable.

e) Every countable set is finite.

f) In Example 1, $|\mathbb{S}| = \aleph_0$.

g) If a set is not denumerable, it is uncountable.

h) There is an uncountable set.

i) The smallest infinite cardinality is $\aleph_0$.

2. Prove Theorem 5.1.1.

3. Prove Theorem 5.1.2.

4. Prove Theorem 5.1.3.

5. For the intervals in parts a through f, give explicit one-to-one onto functions showing the equivalences. For parts g, h, and i, use the Schröder–Bernstein Theorem, which will require you to find two explicit one-to-one functions for each part. Prove that your functions are bijections or one-to-one, as appropriate.

*a) Show that $[1, 3] \sim [2, 8]$.

b) Show that $(-1, 3.5)$ and $(\sqrt{2}, 4)$ have the same cardinality.

c) Show that $[a, b] \sim [c, d]$, where $a < b$ and $c < d$.

d) Show that $[0, 1) \sim [0, \infty)$.

e) Show that $[a, b) \sim [0, \infty)$, for $a < b$.

f) Show that $(0, 1) \sim \mathbb{R}$ (referred to in Section 5.3).

g) Show that $[1, 5]$ and $(2, 7.5)$ have the same cardinality.

h) Show that any closed interval $[a, b]$ and the corresponding open interval (a, b) have the same cardinality, where $a < b$.

i) State and prove a proposition generalizing the previous parts.

6. Let E be the set of positive even integers and D be the set of positive odd integers. For each part to follow, give an explicit function and show that it is a bijection:

a) $D \sim E$.

b) $\mathbb{N} \sim E$.

c) $\mathbb{N} \sim D$.

d) $\mathbb{N} \sim \mathbb{Z}$. (Hint: Start with the inverse functions for parts b and c (referred to in Section 5.2).

7. This problem considers a paradox Galileo discussed concerning infinite sets.

   *a) Let C_1 be the set of points on a circle of radius 1 and C_2 be the set of points on a circle of radius 2 with the same center. Prove that $C_1 \sim C_2$. (Hint: Use a picture.)

   b) Galileo thought of the circles as wheels locked together. When the larger wheel rolls one revolution along a road, it covers a length 4π. Now imagine what happens to the smaller wheel while the larger one makes its revolution. The smaller wheel would also cover a length of 4π along an imaginary road at its height and make one revolution. But its circumference is only 2π. How is this possible?

8. The *Hilbert Hotel* is an imaginary hotel with a countable infinity of rooms numbered consecutively with all of the natural numbers. The Hilbert Hotel is never really full: If every room has a guest staying in it and a new guest arrives, the manager can ask each current guest to shift to the next higher room, leaving room number 1 for the new guest.

   a) Suppose that 100 more guests arrive. Give an explicit way for the hotel manager to accommodate all the current guests and these new guests without any two guests having to share a room.

   b) Suppose that a countable infinity of new guests arrive. Give an explicit way for the hotel manager to accommodate all the current guests and these new guests without any two guests having to share a room.

   c) Suppose that, of the original guests, a countable infinity were so important that the manager does not want to ever move them, regardless of how many more guests arrive. Also, assume that there is a countable infinity of less important guests. Explain how the manager can leave all of the important guests in peace and still accommodate a countable infinity of new guests.

9. For any sets A and B, show that $|A \times B| = |B \times A|$.

10. For any sets A and B, show that if A and B have the same cardinality, then $\mathcal{P}(A)$ and $\mathcal{P}(B)$ have the same cardinality.

11. Suppose for sets A, B, C, and D that $|A| = |B|$ and $|C| = |D|$ (referred to in Sections 5.2 and 5.3).

    a) Prove that $|A \times C| = |B \times D|$.

    b) Give a counterexample to show that $|A \cap C|$ need not equal $|B \cap D|$.

    c) Repeat part b, using $A - C$ and $B - D$.

    d) Repeat part b, using $A \cup C$ and $B \cup D$.

    e) If $A \cap C = \emptyset = B \cap D$, prove that $A \cup C \sim B \cup D$.

    f) Suppose that $|A| = k$ and $|C| = n$, finite cardinalities, and $A \cap C = \emptyset$. What is $|A \cup C|$? Prove your answer.

    g) Note that $[0, 1] \cap (1, 2] = \emptyset$. What are the cardinalities of $[0, 1]$, of $(1, 2]$, and of their union?

12. In the definition of B^A, we required A and B to be nonempty sets. This problem explores extending this definition to include the empty set. (Hint: $\emptyset$ and $\{\emptyset\}$ differ.)

    a) What would $\emptyset^A$ mean for A a nonempty set? Does the size of this set match the value of 0^a, for $a \neq 0$?

    b) What would $B^\emptyset$ mean for B a nonempty set? Does the size of this set match the value of b^0, for $b \neq 0$?

    c) What would $\emptyset^\emptyset$ mean? What is $|\emptyset^\emptyset|$? (*Note:* 0^0 is generally not defined because of the ambiguity of 0^a and b^0 when $a = b = 0$.)

13. ***a)** For any set A, prove that $A^{\{1,2\}} \sim A \times A$.

    b) For any set A, prove that $A^{\{1,2,3\}} \sim A \times A \times A$.

    c) Generalize parts a and b. Prove your generalization.

14. For any set A, prove that $\{0, 1\}^A \sim \mathcal{P}(A)$ (referred to in Section 5.3).

15. Suppose that A, B, C, and D are sets with $A \sim B$ and $C \sim D$.

    a) Prove that $C^A \sim D^A$. (Hint: Use composition.)

    b) Prove that $C^A \sim C^B$.

    c) Prove that $C^A \sim D^B$ (referred to in Section 5.3).

16. Critique the incorrect "proofs" that follow. Point out reasoning errors and unclear presentation. (If the claim is false, there must be an error in the argument. However, it is not enough to say that the claim is false. You need to find an error in the reasoning.)

    *a) Claim: If A is a proper subset of B, then the cardinality of A is strictly smaller than the cardinality of B. "Proof": Suppose that $A \subseteq B$ and $b \in B$, but $b \notin A$. The function $f : A \to B$, given by $f(a) = a$, is clearly one-to-one, so $A \lesssim B$. But just as clearly, there is no $a \in A$ such that $f(a) = b$, so this function is not onto. Hence, the cardinality of A is less than the cardinality of B. ▲

    b) Claim: $\mathbb{N}$ is finite. "Proof": We use induction. By definition, $|\emptyset| = 0$, an integer, so $\emptyset$ is finite. Suppose that any set with $n = k$ elements is finite. Consider a set X with cardinality $k + 1$. Since k is an integer, $k + 1$ is an integer and X is also finite. By the Principle of Mathematical Induction, this holds for all of $\mathbb{N}$, showing that $\mathbb{N}$ is finite. ▲

    c) Claim: (the Schröder–Bernstein Theorem) If $A \lesssim B$ and $B \lesssim A$, then $A \sim B$. "Proof": Suppose that A and B are sets satisfying $A \lesssim B$ and $B \lesssim A$. Then there are one-to-one functions $f : A \to B$ and $g : B \to A$. By Theorem 3.2.2, $g \circ f$ is a one-to-one function from A to A. But $A = A$, so the domain and range are equal and $g \circ f$ is onto. By Problem 10a of Section 3.2, g is onto as well as one-to-one. Hence, $B \sim A$. Then Theorem 5.1.1 gives us $A \sim B$. ▲

17. Critique the questionable "proofs" that follow. Point out reasoning errors and unclear presentation. If the argument is a proof, say so. (If the claim is false, there must be an error in the argument. However, it is not enough to say that the claim is false. You need to find an error in the reasoning.)

    a) Claim: If X is an infinite set and $y \notin X$, then $X \sim X \cup \{y\}$. "Proof": Suppose that X is infinite. Either it is countable or uncountable. Case 1. If X is countably infinite, there is a one-to-one onto function $f : X \to \mathbb{N}$. Now extend f by defining $f(y) = 0$. Then $f : X \cup \{y\} \to \mathbb{N} \cup \{0\}$ is clearly a bijection. The function $g : \mathbb{N} \cup \{0\} \to \mathbb{N}$ given by $g(n) = n + 1$ is also a bijection. So, $X \cup \{y\} \sim \mathbb{N}$. Case 2. If X is uncountable, then $X \sim \mathbb{R}$. From Problem 5f, $\mathbb{R} \sim (0, 1)$. Similar to the first case, we have $X \cup \{y\} \sim (0, 1]$. By part h of Problem 5, we have $(0, 1] \sim (0, 1)$. So, $X \cup \{y\} \sim (0, 1) \sim \mathbb{R} \sim X$. ▲

    *b) Claim: For any set A, $\emptyset \lesssim A$. "Proof": The empty set is vacuously a one-to-one function from $\emptyset$ to A. ▲

    c) Claim: Any set has a strictly smaller cardinality than its power set. "Proof": Let A be any set. For a contradiction, suppose $f : A \to \mathcal{P}(A)$ were a bijection. The power set $\mathcal{P}(A)$ includes the empty set plus all of the singletons $\{a\}$, where $a \in A$. So, some element x_1 has to have $f(x_1) = \emptyset$. In turn, we need some x_2 so that $f(x_2) = \{x_1\}$, and so on. Thus, A is infinite and we still won't have any element to map to $\{x_1, x_2\}$. So, f is not onto. Contradiction. ▲

5.2   COUNTABLE SETS

"Mathematics has been called the science of the infinite."

—Hermann Weyl (1885–1955)

Cantor's investigations went far beyond the examples of the previous section, which were well known before him. We concentrate in this section on results about countable sets, ones whose elements can be paired with some or all of the elements of $\mathbb{N}$. Even though the counting numbers are very familiar, Cantor found some surprises among the countable sets. He proved Theorems 5.2.1, 5.2.3, and 5.2.4, which follow:

THEOREM 5.2.1. $\mathbb{N} \times \mathbb{N} \sim \mathbb{N}$.

Proof.   We need to find a bijection from $\mathbb{N} \times \mathbb{N}$ to $\mathbb{N}$. The arrows in Figure 1 give the essential idea of this mapping and may well already convince you. Since we are matching a two-dimensional array with a one-dimensional one—a rather unsettling notion—we will turn this picture into an explicit function. We start with one key pattern: The images of the ordered pairs $(n, 1)$ in Figure 1 are the numbers $1, 3, 6, 10, \ldots$ Conveniently, these values match the sum of the first n natural numbers: $f(n, 1) = \sum_{i=1}^{n} i$. Now, Example 1 of Section 2.4 gives us the formula for this sum, $(n + 1)n/2$. So we want $f(n, 1) = (n + 1)n/2$.

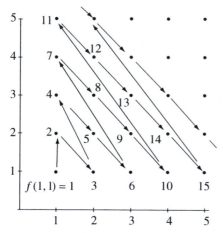

FIGURE 1. $f : \mathbb{N} \times \mathbb{N} \to \mathbb{N}$.

Next we need to extend f to other ordered pairs (x, y). Let's start this process with an example. In Figure 1 consider the diagonal with the ordered pairs $(1, 4)$, $(2, 3)$, $(3, 2)$, and $(4, 1)$ and their images 7, 8, 9, and 10, respectively. Of course, we already have the formula $f(4, 1) = (4 + 1)4/2 = 10$. The other images decrease as we go up the diagonal, which suggests that we can write these images as $(4 + 1)4/2 - [\text{something}]$. The bigger the second coordinate y is, the more we subtract. For instance, when $y = 2$, we subtract 1. If the second coordinate is y, we subtract $y - 1$. So if (x, y) is any of these four ordered pairs, we have $f(x, y) = (4 + 1)4/2 - (y - 1)$. Now we need to modify the expression $(4 + 1)4/2$ so it is in terms of the coordinates x and y. Note that for each of these ordered pairs, the sum of their coordinates is 5: $1 + 4 = 2 + 3 = 3 + 2 = 4 + 1 = 5$. So we can rewrite $(4 + 1)4/2$ as $(x + y)(x + y - 1)/2$. Then the formula for ordered pairs on this diagonal is $f(x, y) = (x + y)(x + y - 1)/2 - (y - 1)$.

In general, consider the ordered pairs (x, y) on the diagonal $y = -x + n + 1$ ending in $(n, 1)$. As with the example, their images must be less than $(n + 1)n/2$, written as $(n + 1)n/2 - $ [something]. Since for every pair (x, y) on this diagonal, $x + y = n + 1$, we get $f(x, y) = (n + 1)n/2 - $ [something] $= (x + y)(x + y - 1)/2 - $ [something]. As in the example, [something] $= (y - 1)$. Thus, the general formula is $f(x, y) = (x + y)(x + y - 1)/2 - (y - 1)$. This clearly gives us a function from $\mathbb{N} \times \mathbb{N}$ to $\mathbb{N}$. Furthermore, the increase from $f(n, 1) = (n + 1)n/2 = n^2/2 + n/2$ to the next such number $f(n + 1, 1) = (n + 2)(n + 1)/2 = n^2/2 + 3n/2 + 1$ is $n + 1$, which is the number of ordered pairs on the next diagonal from $(1, n + 1)$ to $(n + 1, 1)$. Since our pattern starts with $f(1, 1) = 1$, every number in $\mathbb{N}$ is an image exactly once. Thus f is one-to-one and onto and $\mathbb{N} \times \mathbb{N} \sim \mathbb{N}$. ∎

Unfortunately, not every pair of sets can be so readily matched with an explicit bijection. However, the Schröder–Bernstein Theorem provides a handy way to substitute an easier-to-find pair of one-to-one functions for a bijection. Since the proof is rather subtle, we illustrate its setup with an example before we give the proof.

THEOREM 5.2.2. (Schröder–Bernstein Theorem). For any sets A and B, if $A \lesssim B$ and $B \lesssim A$, then $A \sim B$.

EXAMPLE 1

Let $A = \{a_i : i \in \mathbb{Z}\}$ and $B = \{b_j : j \in \mathbb{Z}\}$. Consider $f : A \to B$ and $g : B \to A$, given by $f(a_i) = b_{2i}$ and $g(b_j) = a_{3j}$. (See Figure 2.) Both f and g are one-to-one, but not onto. We need to construct a one-to-one function h from A onto B built from the functions f and g. (There is a very simple one-to-one onto function, namely, $k(a_i) = b_i$; but k doesn't illustrate the proof.) The solid arrows going down match elements a_i in A with their images $f(a_i)$ in B. Similarly,

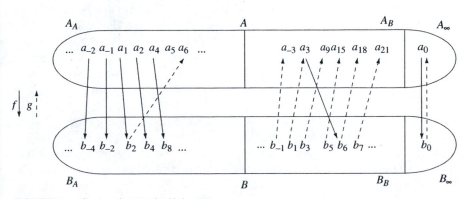

FIGURE 2. $f(a_x) = b_{2x}$ and $g(b_y) = a_{3y}$.

a dashed arrow matches b_j in B with its image $g(b_j)$ in A. Every element in either set has an arrow leading from it to the other set because both f and g are functions. However, the idea of the proof is to go backwards along the arrows. Because these functions are one-to-one, there is at most one arrow pointing toward any number. For instance, a_{54} has a dashed arrow pointing to it, coming from b_{18}, since $g(b_{18}) = a_{3 \cdot 18} = a_{54}$. In turn, b_{18} has a solid arrow pointing to it from a_9, which has a dashed arrow pointing to it from b_3. But b_3 has no prior arrow. We say that the *ancestors* of a_{54} are b_{18}, a_9, and b_3. We call b_3 the *first ancestor* of a_{54} because it has no other ancestors. (It is also the first ancestor of b_{18}, of a_9, and of itself.) Similarly, a_1 is a first ancestor of b_2, a_6, b_{12}, and so on. The elements a_0 and b_0 behave differently from the others, since $f(a_0) = b_0$ and $g(b_0) = a_0$. We can go "back" forever between these two, meaning that neither has a first ancestor. We split the elements of A into three parts, starting with those whose first ancestor is in A. We call that set A_A. Similarly, A_B contains those elements of A whose first ancestors are in B, and A_∞ are those elements with no first ancestor. We do the same with B to get B_A, B_B, and B_∞. The proof generalizes this situation before constructing the new function h.   ◇

Proof.   From the hypotheses $A \lesssim B$ and $B \lesssim A$, we have one-to-one functions $f : A \to B$ and $g : B \to A$. We can't assume that either f or g is onto. The idea of the proof is to partition A into three parts A_A, A_B, and A_∞, and similarly, B into B_A, B_B, and B_∞, so that on each pair of matching parts one (or both) of the functions is onto as well as one-to-one.

Define the idea of an *ancestor* of an element recursively. Any element is an ancestor of itself. If $f(x) = y$, we say that x is an ancestor of y. Similarly, if $g(v) = w$, we say that v is an ancestor of w. Further, every ancestor of v is an ancestor of $g(v) = w$ and every ancestor of x is an ancestor of $f(x) = y$. An element $a \in A$ is a *first ancestor of* $z \in A$ (or $z \in B$) iff a is an ancestor of z and there is no $b \in B$ such that $g(b) = a$. Similarly, an element $b \in B$ is a *first ancestor of* $z \in A$ (or $z \in B$) iff b is an ancestor of z and there is no $a \in A$ such that $f(a) = b$.

Because f is one-to-one, for $x \in A$, the ancestors of $f(x)$ are all the ancestors of x together with $f(x)$. Similarly, for $v \in B$, the ancestors of $g(v)$ are all the ancestors of v together with $g(v)$. We can turn these observations around to see that as we go backwards from the set of ancestors of $f(x)$ to the set of ancestors of x, we lose exactly one element, and similarly from the ancestors of $g(v)$ to those of v. Thus, if an element has finitely many ancestors, we can take them away one at a time to get back to the element's unique first ancestor. (This statement can be made more rigorous by induction, but at the cost of clarity.) Thus, for any element a in A, one of three things happens: Its first ancestor is in A or it is in B or it has infinitely many ancestors. In the first case, we say that $a \in A_A$. In the second case, $a \in A_B$; and in the third case, $a \in A_\infty$. Similarly, the elements of B are split into three sets: B_A if the first ancestor is in A, B_B if the first ancestor is in B, and B_∞ if there is no first ancestor.

Consider the claim that $f : A_A \to B_A$ is one-to-one and onto. First, all of f is one-to-one, so this part of f is as well. Next, every element y of B_A has to have exactly one $x \in A$ with $f(x) = y$ because the first ancestor of y is in A. (The first ancestor of y need not be x. Rather, x and y have the same first ancestor.) But then, f is onto B_A, by $f(x) = y$.

Similarly, $g : B_B \to A_B$ is one-to-one and onto. Thus, $g^{-1} : A_B \to B_B$ is one-to-one and onto. Finally, consider $f : A_\infty \to B_\infty$. As before, because f is one-to-one on all of A, it is one-to-one on A_∞. For onto, let $y \in B_\infty$. By definition, y has infinitely many ancestors, which means it has to have some x satisfying $f(x) = y$. But this is just what we need for onto. Now we can patch these together to give the explicit bijection, which we'll call h:

$$h(x) = \begin{cases} f(x) & \text{if } x \in A_A \\ g^{-1}(x) & \text{if } x \in A_B \\ f(x) & \text{if } x \in A_\infty \end{cases}.$$

∎

THEOREM 5.2.3. $\mathbb{N} \sim \mathbb{Q}$.

Proof. The first step toward using the Schröder–Bernstein Theorem is easy: $f(x) = \frac{x}{1}$ is one-to-one from $\mathbb{N}$ to $\mathbb{Q}$. For the other direction, we show that $\mathbb{Q} \lesssim \mathbb{Z} \times \mathbb{N} \lesssim \mathbb{N} \times \mathbb{N} \lesssim \mathbb{N}$ and use the transitivity of $\lesssim$ from Theorem 5.1.3. A rational number $\frac{p}{q}$ has $p \in \mathbb{Z}$ and $q \in \mathbb{N}$. It seems natural to map $\frac{p}{q}$ to (p, q). Unfortunately, fractions can have different representations. For instance, $\frac{1}{2} = \frac{3}{6}$, but $(1, 2) \neq (3, 6)$. We resolve this potential problem by defining $g : \mathbb{Q} \to \mathbb{Z} \times \mathbb{N}$ by stipulating $g(x) = (a, b)$, where $x = \frac{a}{b}$ and $\frac{a}{b}$ is in lowest terms. Thus, g is uniquely defined and so is a function. One-to-one follows directly from the definition. Next, from Problem 6d of Section 5.1, we know that $\mathbb{Z} \sim \mathbb{N}$; so, from Problem 11a of Section 5.1, $\mathbb{Z} \times \mathbb{N} \sim \mathbb{N} \times \mathbb{N}$. Finally, the function h from Theorem 5.2.1 gives us $\mathbb{N} \times \mathbb{N} \lesssim \mathbb{N}$. This completes the string of "inequalities," yielding $\mathbb{Q} \lesssim \mathbb{N}$. Now the Schröder–Bernstein Theorem completes our work for us. ∎

THEOREM 5.2.4. Every subset of a countable set is countable.

Proof. The general proof needs to start with a countable set B and a subset C of B. We start with the special case of $B = \mathbb{N}$ because the definition of countable uses $\mathbb{N}$ and some of its finite subsets.

We first show that every subset A of $\mathbb{N}$ is countable. If $A = \emptyset$, it is certainly countable. So suppose $A \neq \emptyset$. We need to build a bijection g from A to all of $\mathbb{N}$ or to some initial segment $\{1, 2, \ldots, n\}$ so we can tell its cardinality. The key is the well ordering of $\mathbb{N}$, assuring us that A has a least element, say, a_1. Not surprisingly, we define $g(a_1) = 1$. Consider $A_1 = A - \{a_1\}$. If $A_1 = \emptyset$, we're done. Otherwise, we can repeat the previous step: Map the least element of A_1—say, a_2—to the next available number, 2, and form the smaller set $A_2 = A_1 - \{a_2\}$. We define the sets A_{n+1} recursively: $A_{n+1} = A_n - \{a_{n+1}\}$, where a_{n+1} is the least element of A_n. Of course, $g(a_n) = n$. The function g is one-to-one because we always shift up to the next available integer for the next image.

One of two things happens in $\mathbb{N}$. Case 1. We run out of elements in A after a finite number of steps, and so there is some n such that $A \sim \{1, 2, \ldots, n\}$. Case 2. We don't run out of elements. Then g is onto and $A \sim \mathbb{N}$. Either way, A is countable.

Now let B be any countable set and C be any subset of B. We already dealt with the case $C = \emptyset$; so, assume that $C \neq \emptyset$, and thus $B \neq \emptyset$. Because B is countable, we have a bijection f from B either onto all of $\mathbb{N}$ or onto $\{1, 2, \ldots, n\}$. Then the restriction of f from the subset C maps to a subset $f[C]$ of $\mathbb{N}$ or of $\{1, 2, \ldots, n\}$, which still makes $f[C]$ a subset of $\mathbb{N}$. (In Problem 5 of Section 3.2, we called this restriction $f|_C$.) The previous paragraph showed the countability of $f[C]$. By the transitivity of $\sim$ (Theorem 5.1.1), B is also countable. ∎

In the previous proof, the well ordering of $\mathbb{N}$ gave us a rule to select elements one at a time. Thus, even though the preceding proof is quite general, in every possible situation there is a way to proceed. The statement of the theorem that follows may not seem any more general. However, the proof requires infinitely many arbitrary choices. Such a situation transcends the elementary axioms of set theory, relying on the Axiom of Choice. We therefore defer the proof until Section 5.4, although we use the result in the problems.

THEOREM 5.2.5. If I is a countable index set and for each $i \in I$ the set A_i is countable, then $\bigcup_{i \in I} A_i$ is countable. In short, the countable union of countable sets is countable.

EXAMPLE 2

Let $\mathbb{N}_1 = \mathbb{N}$, $\mathbb{N}_2 = \{\sqrt{n} : n \in \mathbb{N}\}$, and, for $k \in \mathbb{N}$, $\mathbb{N}_k = \{\sqrt[k]{n} : n \in \mathbb{N}\}$. Then each $\mathbb{N}_k$ is countable. It is curious that $\mathbb{N}_1 \subseteq \mathbb{N}_2 \subseteq \mathbb{N}_4$ and so on. Even though these sets seem to be getting larger, Theorem 5.2.5 assures us that the set of all of them together, $\bigcup_{k \in \mathbb{N}} \mathbb{N}_k$, is still countable. These numbers are some of the *algebraic* numbers, so called because they satisfy algebraic equations, such as $x^k = n$. Problem 6 considers the set of all algebraic numbers. ◇

PROBLEMS

1. a) Prove that the set of integer multiples of three has the same cardinality as $\mathbb{Z}$.

   b) Prove that $\{x \in \mathbb{Z} : x \equiv 3 \pmod 4\} \sim \mathbb{Z}$.

   c) Prove for $0 \leq b < n$ and $b, n \in \mathbb{N}$, that $\{x \in \mathbb{Z} : x \equiv b \pmod n\} \sim \mathbb{Z}$.

2. a) Prove that $\mathbb{Z} \times \mathbb{Z}$ and $\mathbb{Z}$ have the same cardinality.

   b) Prove that $\mathbb{Q} \times \mathbb{Q}$ and $\mathbb{N}$ have the same cardinality.

*3. For any nonzero rational number, prove that there is a countable infinity of distinct ways of writing it as a fraction. (For instance, $\frac{1}{2}$ and $\frac{3}{6}$ are distinct ways of writing the same rational number.)

4. For a set A and $n \in \mathbb{N}$, by A^n we mean $A \times A \times \ldots \times A$ (n times). More formally, $A^1 = A$ and $A^{n+1} = A^n \times A$ (referred to in Section 5.3).

   *a) Prove that $\mathbb{N}^3 \sim \mathbb{N}$.

   b) Prove for all $n \in \mathbb{N}$ that $\mathbb{N}^n \sim \mathbb{N}$.

   c) Prove for all $n \in \mathbb{N}$ that $\mathbb{Q}^n \sim \mathbb{Q}$.

   d) Prove that for all $n \in \mathbb{N}$, if A is countable, then A^n is countable.

5. **a)** How many real numbers strictly between 0 and 1 can be written by using one decimal place?

  b) How many real numbers strictly between 0 and 1 can be written by using at most n decimal places, where n is some element of $\mathbb{N}$?

  c) How many real numbers strictly between 0 and 1 can be written by using a finite number of decimal places? Prove your answer, assuming your answers to parts a and b.

  d) How many real numbers can be written by using a finite number of decimal places? Prove your answer, assuming your answer to part c.

  e) How many real numbers strictly between 0 and 1 can be written by using at most n decimals followed by an infinitely repeating pattern that repeats after at most k decimal places? Prove your answer, assuming your answer to part b. (For instance, 0.1234567567567... uses $n = 4$ decimals followed by the infinitely repeating pattern 567567567... with $k = 3$ decimal places.)

  f) Relate the set described in part e with $\mathbb{Q}$.

6. An nth *degree rational polynomial* is an expression of the form $a_n x^n + a_{n-1} x^{n-1} + \cdots + a_2 x^2 + a_1 x + a_0$, where all $a_i \in \mathbb{Q}$ and $a_n \neq 0$. An *algebraic number* is a real or complex number r that is a root of some rational polynomial for some finite degree. That is, $a_n r^n + a_{n-1} r^{n-1} + \cdots + a_2 r^2 + a_1 r + a_0 = 0$, for appropriate a_i and n. (For instance, $\sqrt[3]{2}$ is algebraic because it is a root of $x^3 - 2 = 0$.) Assume the fact that an nth degree rational polynomial has at most n complex roots (referred to in Problem 7 of Section 5.3).

  *a) For all $n \in \mathbb{N}$, prove that the set of all rational polynomials of degree n is countable.

  b) Prove that the set of all rational polynomials of any finite degree is countable.

  c) Prove that the set of all algebraic numbers is countable (proven by Cantor in 1874).
  Remark. Numbers that are not algebraic are called *transcendental*. For instance, e and π were proven transcendental in 1873 and 1882, respectively.

7. **a)** Find the number of conceivable words (meaningful or not) of length n from an alphabet of 26 letters.

  b) Prove that the set of all conceivable English words of finite length is countable.

  c) Prove that the set of all possible "words," using an alphabet with a countable number of letters, is countable.

  d) A proof (or a novel or any other communication in a language) is an ordered set of words. If a language has a countable alphabet, prove that there are a countable number of proofs.

  e) For a proof to be comprehensible, a person must be able to at least read it. Estimate how many symbols long the longest comprehensible proof could be, where a symbol can be either a letter or a math symbol. Use this estimate to estimate the maximum (finite) number of comprehensible proofs. Explain why the actual number of comprehensible proofs would be far smaller.

  f) What do your answers in parts d and e tell you about proofs?

8. Let S_n be the set of all subsets of $\mathbb{N}$ that have size n. For instance, $\{3, 7\} \in S_2$ and $\{1, 4, 9\} \in S_3$.

  *a) Prove that S_2 is countable. (Hint: Relate S_2 to $\mathbb{N} \times \mathbb{N}$.)

  b) Prove for all $n \in \mathbb{N}$ that S_n is countable.

  c) Prove that the set of all finite subsets of $\mathbb{N}$ is countable.

9. Define a subset B of a set A to be *co-finite* iff $A - B$ is finite. (For instance, $\{n \in \mathbb{N} : n^2 - 3n > 0\} = \{4, 5, 6, \ldots\}$ is co-finite in $\mathbb{N}$, since its complement is $\{1, 2, 3\}$, a finite set.) Similarly, B is *co-countable* in A iff $A - B$ is countable.

   a) Prove that the set of finite subsets of a set has the same cardinality as the set of co-finite sets.

   b) Prove that the set of countable subsets of a set has the same cardinality as the set of co-countable sets.

   c) Prove that the set of co-finite subsets of $\mathbb{Q}$ is countable.

10. *a) Use Theorem 5.2.2 to give an alternative proof of Theorem 5.2.1 by finding an explicit one-to-one function from $\mathbb{N} \times \mathbb{N}$ into $\mathbb{N}$. (Note that Theorem 5.2.2 does not depend on Theorem 5.2.1, so this is a legitimate proof.)

   b) Use the function $f : \mathbb{N} \times \mathbb{N} \to \mathbb{N}$ given by $f(x, y) = 2^{x-1}(2y - 1)$ to give an alternative proof of Theorem 5.2.1.

11. Use Theorems 5.2.1, 5.2.2, and 5.2.4 to prove that the following special case of Theorem 5.2.5, does not depend on the Axiom of Choice. Suppose that I is a finite index set and the sets A_i for $i \in I$ are pairwise disjoint and countable. Then $\bigcup_{i \in I} A_i$ is countable.

12. a) Given two sets B and C, prove that B and $C - B$ are disjoint and $B \cup C = B \cup (C - B)$.

   b) Use part a and Problem 11 to prove that a finite union of countable sets is countable.

13. (Adapted from Wilmott.) A magical gnome can accomplish each succeeding task in half the time as the previous one. To the left of the gnome, there is a countable infinity of balls numbered consecutively with each element of $\mathbb{N}$. At one minute to midnight, the gnome transfers the two lowest-numbered balls (1 and 2) at his left to in front of him, and then the lowest numbered ball (1) in front of him to his right. At one-half minute to midnight, he moves the two lowest-numbered balls (now 3 and 4) at his left to in front of him, and the lowest numbered ball (now 2) in front of him to his right. He completes countably infinitely many such transfers by midnight.
   Critique the following arguments:

   a) The pile in front of him has one ball (numbered 2) after the first set of moves, two balls (numbered 3 and 4) after the second, three balls after the third set of moves, and so on. Therefore, there are infinitely many balls in front of him at midnight.

   b) Ball number 1 is at his right after the first set of moves, ball number 2 is at his right after the second set of moves, etc. Therefore, at midnight every ball will be to his right and there will be no balls in front of him.
   To learn more about this gnome, see Wilmott.

14. Critique the questionable "proofs" that follow. Point out reasoning errors and unclear presentation. If the argument is a proof, say so. (If the claim is false, there must be an error in the argument. However, it is not enough to say that the claim is false. You need to find an error in the reasoning.)

   *a) Claim: If $A \cup B$ is countably infinite, then A and B are infinite. "Proof": We'll show the contrapositive. So, assume that A and B are finite—say, A has n elements and B has k elements. Then $A \cup B$ has $n + k$ elements, which is finite. Since we have shown the contrapositive, the original claim is proven. ▲

   *b) Let $\mathbb{Q}^+ = \{\frac{p}{q} : p, q \in \mathbb{N}\}$. Claim: $\mathbb{Q}^+ \sim \mathbb{N}$. "Proof": Consider Figure 3, which gives a two-dimensional array for $\mathbb{Q}^+$ with repeats. Now start listing the fractions $\frac{p}{q}$, starting with the one satisfying $p + q = 2$, followed by those with $p + q = 3$, etc., skipping over any repeats. This gives a bijection from $\mathbb{Q}^+$ to $\mathbb{N}$. ▲

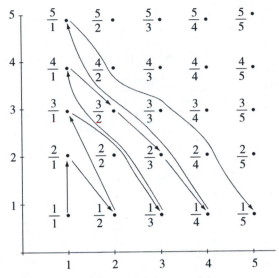

FIGURE 3. $f\left(\frac{1}{1}\right) = 1$, $f\left(\frac{2}{1}\right) = 2$, $f\left(\frac{1}{2}\right) = 3$, etc. We skip $\frac{2}{2}, \frac{4}{2}, \frac{3}{3}$, etc., which equal earlier fractions.

c) Claim: $\mathbb{N} \sim (0, 1)$. "Proof": The function $f : \mathbb{N} \rightarrow (0, 1)$, given by $f(n) = \frac{1}{n+1}$, is one-to-one. So, $\mathbb{N} \lesssim (0, 1)$. For the other direction, consider the following listing of the numbers between 0 and 1: First come the nine one-place numbers 0.1 to 0.9. Then come to the 99 two-place decimals 0.01 to 0.99, followed by the three-place numbers 0.001 to 0.999, etc. Since each piece of the list is finite, from Theorem 5.2.5 we have a countably infinite list. Now, every real number r between 0 and 1 has some decimal expansion $r = 0.a_1 a_2 a_3 \ldots$. For any natural number n, we can find an approximation r_n with the first n decimal places agreeing with r. Since our list is infinite, we just take the limit $\lim_{n \to \infty} r_n = r$. So, each r is in the list. While the list has repeats, for instance, $0.1 = 0.10$, each r from $(0, 1)$ appears in the list, so $(0, 1) \lesssim \mathbb{N}$. Now use the Schröder–Bernstein Theorem. ▲

d) Claim: The denumerable union of denumerable sets is denumerable. "Proof": A denumerable set is countable, since it is countably infinite. By Theorem 5.2.5, the denumerable union of denumerable sets is countable. Since the first of these infinitely many sets already has infinitely many elements in it, the union has infinitely many elements. Hence, the union is infinite and countable, and so is denumerable. ▲

e) Claim: The finite union of finite sets is finite. "Proof": Suppose the finitely many finite sets $A_1, A_2 \ldots, A_n$ have $k_1, k_2 \ldots, k_n$ elements, respectively. Then $\bigcup_{i=1}^{n} A_i$ has at most $\sum_{i=1}^{n} k_i$ elements, which is a finite number. (If the A_i are not disjoint sets, the number of elements is even smaller.) Therefore, the finite union of finite sets is finite. ▲

REFERENCE

WILMOTT, R. 1977. "The Gnome and the Pearl of Wisdom: A Fable," *Mathematics Magazine* 50, no. 3:141–43.

5.3 UNCOUNTABLE SETS

"The most astonishing product of mathematical thought, one of the most beautiful realizations of human activity in the domain of the purely intelligible."

—David Hilbert, on Cantor's work (1862–1943)

It is easy to see that $\mathbb{N} \lesssim \mathbb{R}$ from the identity map $i_{\mathbb{N}}(x) = x$. But in Section 5.1, we asserted more: $\mathbb{R}$ is uncountable. How can we show that there can be no bijection between $\mathbb{R}$ and $\mathbb{N}$? After all, we can hardly look through all the infinitely many possible functions. Cantor published a clever proof by contradiction in 1874, now called the Cantor Diagonal argument. We start with this proof before presenting other arguments, all due to Cantor, about uncountable sets. The realm of uncountable sets is, as Hilbert said, astonishing. For instance, Theorem 5.3.2 implies that there are infinitely many sizes of uncountable sets. Even Cantor was moved to say, in 1877, "I see it, but I do not believe it" about his proofs of Theorem 5.3.4 and its generalizations (see Problem 8).

THEOREM 5.3.1. The interval $(0, 1)$ and $\mathbb{R}$ are uncountable.

Proof. For a contradiction, assume that $(0, 1)$ were countable. Then we could match the numbers $1, 2, 3, \ldots$ with the elements of $(0, 1)$. We will show that any such listing always misses some real number (indeed, many). That will contradict the assumption of countability. Consider any list of reals in $(0, 1)$ or, more accurately, any function $f : \mathbb{N} \to (0, 1)$. In the middle of Table 1 is a general list, and on the right is an example.

TABLE 1

$f(1)$	$0.\underline{b_{11}}b_{12}b_{13}b_{14}\ldots$	$0.\underline{1}570\ldots$
$f(2)$	$0.b_{21}\underline{b_{22}}b_{23}b_{24}\ldots$	$0.0\underline{4}46\ldots$
$f(3)$	$0.b_{31}b_{32}\underline{b_{33}}b_{34}\ldots$	$0.30\underline{7}9\ldots$
$f(4)$	$0.b_{41}b_{42}b_{43}\underline{b_{44}}\ldots$	$0.713\underline{4}\ldots$
$\vdots$	$\vdots$	$\vdots$

We now build our missing number through a two-step process. First, we take the underlined diagonal digits to form a number b. For the general list, $b = 0.b_{11}b_{22}b_{33}b_{44}\ldots$, and for the example, $b = 0.1474\ldots$. This number might be in our list, since it agrees with the nth number of the list in the nth place. Now we switch digits. Replace each 4 in b with a 5 and each other digit with a 4 to get a new number c. In the example, $c = 0.4545\ldots$. We write $c = 0.\overline{b_{11}}\ \overline{b_{22}}\ \overline{b_{33}}\ \overline{b_{44}}\ldots$ in the general situation, where the bar indicates the changed digit. The number c differs from the first number in the first decimal place, from the second number in the second place, and so on. Hence, c is not in the list. But we had assumed that $(0, 1)$ was countable and so its elements could be listed. This is our contradiction, and $(0, 1)$ is uncountable. Finally, Problem 5f from Section 5.1 shows that $(0, 1) \sim \mathbb{R}$, so $\mathbb{R}$ is also uncountable. ∎

The key in the previous proof was to construct a generic missing number by switching digits along the "diagonal." Cantor recast this idea of switching something to find a proof of Theorem 5.3.2, showing that any set has a smaller cardinality than its power set, the set of all of its subsets. This proof cleverly constructs a generic subset that can't be an image. This proof also has a similar flavor to the argument leading to Russell's Paradox, discussed in Section 1.4.

THEOREM 5.3.2. For any set A, its power set $\mathcal{P}(A)$ has a strictly larger cardinality than A has.

Proof. From Example 4 of Section 3.1, for any set A, the function $f : A \rightarrow \mathcal{P}(A)$ given by $f(a) = \{a\}$ is one-to-one, so $A \lesssim \mathcal{P}(A)$. Suppose g is any function from A to $\mathcal{P}(A)$. We need to show that g can't possibly be onto to show there are no bijections between A and $\mathcal{P}(A)$. That is, we need to find some subset C such that for all $y \in A$, we have $g(y) \neq C$. Define $C = \{x \in A : x \notin g(x)\}$. Let $y \in A$. Since $g(y)$ is a subset of A, we could have either $y \in g(y)$ or $y \notin g(y)$. In the first case, $y \in g(y)$ ensures that $y \notin C$, so $C \neq g(y)$. Similarly, in the second case, $y \notin g(y)$ implies that $y \in C$, and again $C \neq g(y)$. Hence, for all $y \in A$, $g(y) \neq C$. Therefore, no function from a set A can be onto all of $\mathcal{P}(A)$. ∎

Cantor realized that Theorem 5.3.2 gave infinitely many different sizes of infinity. We simply take successive power sets: $\mathbb{N}$, $\mathcal{P}(\mathbb{N})$, $\mathcal{P}(\mathcal{P}(\mathbb{N}))$, $\mathcal{P}(\mathcal{P}(\mathcal{P}(\mathbb{N})))$, and so on. (See Problem 5 for a proof.) All but the first of these are uncountable sets. It is natural to ask, "How many different sizes of infinity are there?" More accurately, we would like to know the cardinality of the different sizes of infinity. Unfortunately, there can be no set of all cardinal numbers, just as there can be no universal set. In more advanced set theory, one can prove that there are uncountably many different infinite cardinalities.

Coming down a little from the heady realm of the previous paragraph, we can ask how the size of $\mathbb{R}$, the real numbers, compares with the sizes in the string of power sets in the previous paragraph. The next theorem, Theorem 5.3.3, which answers that question, uses base two notation for real numbers between 0 and 1. We pause to provide a short explanation of base two.

Base Two Notation In base ten, the decimal $0.71828\ldots$ means $\frac{7}{10^1} + \frac{1}{10^2} + \frac{8}{10^3} + \frac{2}{10^4} + \frac{8}{10^5} + \ldots$. In base two, we use powers of 2 in the denominators and restrict the numerators to either 0 or 1. Thus, $0.101101_2 = \frac{1}{2} + \frac{0}{4} + \frac{1}{8} + \frac{1}{16} + \frac{0}{32} + \frac{1}{64} = \frac{45}{64}$. Fact: Every real number between 0 and 1 can be represented as a base two decimal $0.b_1 b_2 b_3 b_4 \ldots_2$. (The subscript $_2$ reminds us what the base is. A proof of this fact would distract us here, but the statement is exactly analogous to the corresponding fact in base ten. See Theorem 8.4.6.) As with ordinary base ten decimals, some values have two representations. In base two, one representation has infinitely repeating zeros and the other has infinitely repeating ones. For instance, $0.11000\ldots_2 = 0.10111\ldots_2$.

THEOREM 5.3.3. $\mathbb{R} \sim \mathcal{P}(\mathbb{N})$.

Proof. We first use Problem 5f from Section 5.1 to replace $\mathbb{R}$ with $(0, 1)$, which has the same cardinality. Also, we use base two representations without infinitely repeating ones. Now we'll use the Schröder–Bernstein Theorem. Define $f : (0, 1) \to \mathcal{P}(\mathbb{N})$ by $f(0.b_1b_2b_3b_4\ldots_2) = \{i \in \mathbb{N} : b_i = 1 \text{ in } 0.b_1b_2b_3b_4\ldots_2\}$. (The function f gives the places with a 1. For instance, $f(0.101_2) = \{1, 3\}$ and $f(0.010101\ldots_2) = \{2, 4, 6, \ldots\}$, the set of positive even integers.) Because we have eliminated double representations, f is a function. Further, f is one-to-one: Suppose that $f(v) = f(w)$. The subsets $f(v)$ and $f(w)$ simply tell us which decimal places of v and w are 1. Since v and w have ones in exactly the same places, they are equal. Hence, f is one-to-one and $(0, 1) \lesssim \mathcal{P}(\mathbb{N})$. Unfortunately, f is not onto, because we had to eliminate infinitely repeating ones.

We'll return to the more familiar base ten for the next function. Define $g : \mathcal{P}(\mathbb{N}) \to (0, 1)$ by $g(A) = 0.a_1a_2a_3a_4\ldots$, where $a_i = 4$ if $i \in A$ and $a_i = 5$ if $i \notin A$. (For instance, $g(\{2, 3, 5, 7\}) = 0.54454545555\ldots$.) Because we avoided the double representation risk with zeros and nines, g is a one-to-one function. Hence, $\mathcal{P}(\mathbb{N}) \lesssim (0, 1)$. The Schröder–Bernstein Theorem now gives us the result. ∎

Dimension is a very different concept from cardinality. Theorem 5.2.1 showed this difference in the discrete case. ("Discrete" means "separate," just as the integers are separated from one another.) Most students find the following theorem for the continuous case more surprising than the discrete version, just as Cantor did:

THEOREM 5.3.4. $\mathbb{R} \sim \mathbb{R} \times \mathbb{R}$.

Proof. We invoke the Schröder–Bernstein Theorem yet again. The equation $f(x) = (x, 0)$ clearly gives a one-to-one function from $\mathbb{R}$ into $\mathbb{R} \times \mathbb{R}$. The other direction requires a more creative function. As before, it is easier to work with the finite interval $(0, 1)$ rather than all of $\mathbb{R}$. Problem 11a from Section 5.1 allows us similarly to replace $\mathbb{R} \times \mathbb{R}$ with $(0, 1) \times (0, 1)$. Define $g : (0, 1) \times (0, 1) \to (0, 1)$ by $g(0.a_1a_2a_3a_4\ldots, 0.b_1b_2b_3b_4\ldots) = 0.a_1b_1a_2b_2a_3b_3a_4b_4\ldots$, where we do not allow repeating nines. For instance, $g(0.1234, 0.5678) = 0.15263748$. Eliminating repeating nines prevents double representations and so ensures that g is a function. Further, the image cannot have repeating nines, so different inputs have to give different images. Thus, g is one-to-one. The Schröder–Bernstein Theorem finishes our proof. ∎

Rather than prove more results about the cardinality of specific sets, let's summarize what we know about familiar sets from the text and previous problems. If a set A has cardinality α, we can write 2^α for the cardinality of $\mathcal{P}(A)$ by Problem 14 of

Section 5.1. We have the following results:

Countably Infinite sets (cardinality $\aleph_0$): $\mathbb{N}$, $\mathbb{Z}$, $\mathbb{Q}$, $\mathbb{N} \times \mathbb{N}$

Uncountable sets:

(cardinality $c = 2^{\aleph_0}$): $\mathbb{R}$, $[a, b]$, (a, b), $\mathbb{R} \times \mathbb{R}$, and $\mathcal{P}(\mathbb{N})$

(cardinality 2^c): $\mathcal{P}(\mathcal{P}(\mathbb{N}))$, $\mathcal{P}(\mathbb{R})$

There are other sets of interest in undergraduate mathematics, some of which we'll consider in the problems. The set of all real sequences, $\mathbb{R}^{\mathbb{N}}$, has cardinality c. (Recall from calculus that a sequence is an ordered list of numbers, such as $\{1, \frac{1}{2}, \frac{1}{3}, \ldots\} = \{\frac{1}{n}\}_{n \in \mathbb{N}}$. In effect, it is a function from $\mathbb{N}$ to $\mathbb{R}$.) The set of all real functions, $\mathbb{R}^{\mathbb{R}}$, has cardinality 2^c, while the set of all continuous real functions has the smaller cardinality c.

The Arithmetic of Infinite Cardinals

We can state results about cardinality more uniformly by using an arithmetic of cardinals. Set theory constructions allow us to extend sums, products, and powers from finite numbers to infinite cardinalities. From Problem 8 of Section 1.4, addition corresponds to disjoint union. Similarly, Problem 2 of Section 1.6 links multiplication with Cartesian products. Theorem 3.2.4 shows the size of the set of all functions from one finite set to another. These facts motivate the following definitions, due to Cantor.

DEFINITION. If A and B are sets with $|A| = \alpha$ and $|B| = \beta$, then $|A \times B| = \alpha \cdot \beta$, $|A^B| = \alpha^\beta$; and if $A \cap B = \emptyset$, then $|A \cup B| = \alpha + \beta$.

REMARK. It is conceivable that these definitions might not make sense. For instance, if there were sets where $|A| = |C|$ and $|B| = |D|$, but $|A \times B| \neq |C \times D|$, we would have two different values for $\alpha \cdot \beta$. Fortunately, Problems 11a, 11e, and 15c of Section 5.1 already addressed this potential problem for multiplication, addition, and exponentiation, respectively.

THEOREM 5.3.5. Let n be a finite cardinal with $n \geq 2$. Then

$$\aleph_0 = n + \aleph_0 = \aleph_0 + \aleph_0 = n \cdot \aleph_0 = \aleph_0 \cdot \aleph_0 = (\aleph_0)^n,$$
$$c = n + c = \aleph_0 + c = c + c = n \cdot c = \aleph_0 \cdot c = c \cdot c = c^n = c^{\aleph_0},$$
$$c = n^{\aleph_0} = \aleph_0^{\aleph_0} = c^{\aleph_0}, \text{ and}$$
$$2^c = n^c = \aleph_0^c = c^c.$$

Proof. It suffices to find one example for each equality, since we can use bijections to convert that example to any other example involving sets of the same cardinality. Further, because of the Schröder–Bernstein Theorem, we can work with inequalities. Problem 9 asks you to find one-to-one functions, one each from a set of the preceding cardinalities to a set of the cardinality to the right of it. (For instance, from a set with cardinality $\aleph_0$ to one with cardinality $n + \aleph_0$.) Next, we need to link the last cardinality of a row back to the first one. Problem 4 of Section 5.2 shows $\aleph_0 = (\aleph_0)^n$. Problems 8d and 12 of this section show that $c = c^{\aleph_0}$ and $2^c = c^c$, respectively. ∎

The Continuum Hypothesis

Are there other sizes of infinity between $\aleph_0$ and $c = 2^{\aleph_0}$? More generally, are there sizes of infinity between the cardinality α of a set and the cardinality 2^α of its power set? In spite of considerable effort, Cantor was unable to answer these questions, although he conjectured each answer was "No." His conjectures, stated next, are called the *continuum hypothesis* and the *generalized continuum hypothesis*. Other mathematicians also struggled with these conjectures over many years. In 1963, Paul Cohen proved that these questions cannot be answered on the basis of the elementary axioms of set theory. Cohen won the Fields medal, the highest honor in mathematics, for this work. Even if we add the Axiom of Choice, discussed in Section 5.4, these questions cannot be answered. Thus, we can take these hypotheses as additional axioms or take their negations as axioms. That is, in some versions of set theory there are "in-between" sizes of infinity, and in others there aren't. Fortunately, such differences affect relatively few areas of mathematics, restricted mainly to advanced set theory and logic.

CONTINUUM HYPOTHESIS. For any set A, if $\mathbb{N} \lesssim A \lesssim \mathbb{R}$, then either $A \sim \mathbb{N}$ or $A \sim \mathbb{R}$. In short, $c = \aleph_1$.

GENERALIZED CONTINUUM HYPOTHESIS. For any sets A and B, if $B \lesssim A \lesssim P(B)$, then either $A \sim B$ or $A \sim P(B)$. In short, $2^{\aleph_n} = \aleph_{n+1}$.

PROBLEMS

1. Determine the cardinality of each set:

   a) $\mathbb{Q}^4$

   *b) $\mathbb{Q}^{\mathbb{Q}}$

   c) $\mathbb{Q}^{\mathbb{R}}$

   d) $\mathbb{R}^{\mathbb{Q}}$

   *e) $\mathbb{N} \cap [1, \pi]$

   f) $\mathbb{N} \cup [1, \pi]$

   g) $\mathbb{N} - [1, \pi]$

   h) $[1, \pi] - \mathbb{N}$

   i) $P(\mathbb{Z}) \times P(\mathbb{Z})$

   j) $P(P(\mathbb{Z}))$

   *k) $P(\mathbb{Z})^{P(\mathbb{Z})}$

   l) $P(P(P(P(\mathbb{Z}))))$

2. Show that the complex numbers, $\mathbb{C}$, have cardinality c.

*3. Let $B = \{(x, y) : x^2 + y^2 \leq 1$ and $x, y \in \mathbb{R}\}$, the unit ball, and $S = \{(x, y) : x \in [-1, 1]$ and $y \in [-1, 1]\}$, a square. Prove that B and S have the same cardinality.

4. a) Without Theorem 5.3.5, prove that $P(\mathbb{R})$ has strictly greater cardinality than $P(\mathbb{N})$.

   b) Generalize part a.

5. Use induction to prove that there are infinitely many sizes of infinity.

6. Modify the argument in Theorem 5.3.2 to give an alternative proof that there can be no universal set.

7. a) Suppose A is a countable subset of an uncountable set B. Prove by contradiction that $B - A$ is uncountable.

   b) Prove that the set of irrational real numbers is uncountable (referred to in Section 8.1).

   c) Prove that the set of transcendental numbers is uncountable. (See Problem 6 from Section 5.2.)

8. Generalize the proof of Theorem 5.3.4 for parts a and b.

    a) Show that $\mathbb{R} \times \mathbb{R} \times \mathbb{R} \sim \mathbb{R}$.

    b) Show for all $n \in \mathbb{N}$ that $|\mathbb{R}^n| = c$. For a set X, recall that X^n means $X \times X \times \ldots \times X$, n times.

    c) Devise a way to encode any countably infinite sequence of real numbers as one real number.

    d) Use part c to show that $|\mathbb{R}^{\mathbb{N}}| = c$.

*9. For each adjacent pair of cardinal numbers in Theorem 5.3.5, give an example of two sets A and B so that A is the size of the first cardinal, B is the size of the second cardinal, and there is an (easy-to-find) one-to-one function from A to B.

10. a) Let $D = \{x \in (0, 1) : \text{The decimal expansion of } x \text{ has only odd digits}\}$. For instance, $.355917 \in D$. Prove that D is uncountable.

    b) Let $E = \{x \in (0, 1) : \text{The decimal expansion of } x \text{ has only even digits}\}$. Prove that $E \sim D$.

    c) Describe the elements of $(0, 1) - (D \cup E)$ and find its cardinality.

11. Let α and β be infinite cardinals with $\alpha \leq \beta$. Make a conjecture about the cardinality of $\alpha + \beta, \alpha \cdot \beta, \alpha^\beta$, and β^α.

*12. Show that $\mathcal{P}(\mathbb{R}) \sim \mathbb{R}^{\mathbb{R}}$, where $\mathbb{R}^{\mathbb{R}}$ is the set of all functions from $\mathbb{R}$ to itself. (Hints: Find an explicit map to show $\mathcal{P}(\mathbb{R}) \lesssim \mathbb{R}^{\mathbb{R}}$. For the other direction, note that each function $f : \mathbb{R} \to \mathbb{R}$ is a subset of $\mathbb{R} \times \mathbb{R}$ and use Theorem 5.3.4.)

13. This problem considers the idea of subtraction involving infinite cardinal numbers, based on the following provisional definition: For sets A and B with $A \subseteq B$, $|A| = \alpha$, and $|B| = \beta$, if $\alpha < \beta$, then $\beta - \alpha = |B - A|$.

    a) If α is finite and $\beta = \aleph_0$, what is $\beta - \alpha$? Prove your answer. (Hint: Use a proof by contradiction.)

    b) If α is countable and $\beta = c$, what is $\beta - \alpha$? Prove your answer.

    c) Give examples of countably infinite sets A, B, C, and D such that $A \subseteq B, C \subseteq D$ and $|B - A| \neq |D - C|$. What possible cardinalities can $|B - A|$ be?

    d) Give examples of sets A, B, C, and D, each with cardinality c, such that $A \subseteq B, C \subseteq D$, and $|B - A| \neq |D - C|$.

    e) Use parts c and d to explain why the definition of $\beta - \alpha$ requires $\alpha < \beta$, rather than $\alpha \leq \beta$.

14. Let $\mathbb{P}_0 = \mathbb{N}$ and define $\mathbb{P}_n$ recursively by $\mathbb{P}_{n+1} = \mathcal{P}(\mathbb{P}_n)$.

    a) Show that $\mathbb{S}_0 = \bigcup_{n \in \mathbb{N}} \mathbb{P}_n$ has cardinality strictly greater than any $\mathbb{P}_n$.

    b) Show that there are infinitely many sizes of infinity larger than $\mathbb{S}_0$.

    c) Generalize parts a and b.

15. This problem proves that the set B of all bijections from $\mathbb{N}$ to $\mathbb{N}$ has cardinality c. Split $\mathbb{N}$ into the family of two-element subsets $A_n = \{2n - 1, 2n\}$ for $n \in \mathbb{N}$. Thus, $A_1 = \{1, 2\}$, $A_2 = \{3, 4\}$, etc.

    a) For each $n \in \mathbb{N}$, show that there are two bijections $g_n : A_n \to A_n$ and $h_n : A_n \to A_n$.

    *b) For $S \subseteq \mathbb{N}$, define $f_S : \mathbb{N} \to \mathbb{N}$ by $f_s(2n) = \begin{cases} g_n(2n) & \text{if } n \in S \\ h_n(2n) & \text{if } n \notin S \end{cases}$ and similarly for $f_S(2n - 1)$. Prove that f_S is a bijection on $\mathbb{N}$.

    c) Use part b to show that $\mathcal{P}(\mathbb{N}) \precsim B$.

    d) Finish proving that $|B| = c$.

16. Prove that the set of all bijections from $\mathbb{R}$ to $\mathbb{R}$ has cardinality 2^c. (Hint: See Problem 15.)

17. We know $\mathbb{Q}$ is a countably infinite subset of the uncountable set $\mathbb{R}$. This problem shows that $\mathbb{R} - \mathbb{Q}$, the set of irrationals, has cardinality c. (The Continuum Hypothesis and part a would show this result, but we don't need the Continuum Hypothesis here.)

    a) Show that $\mathbb{R} - \mathbb{Q} \precsim \mathbb{R}$. Explain why $\mathbb{R} - \mathbb{Q}$ must be uncountable.

    b) Describe the elements of $\mathbb{Q}$ in terms of decimal representation. Explain why this means that irrationals are exactly the reals with infinite, nonrepeating decimals.

    c) Give an example of a specific irrational number between 0 and 1. Let its decimal representation be $0.a_1a_2a_3\ldots$.

    d) Consider the set S of numbers with decimal representations $0.a_1b_1a_2b_2a_3b_3\ldots$, where each b_i is either 4 or 5 and the a_i are the digits from the number in part c. Explain why each such number must be irrational.

    e) Show that there is a one-to-one function from $\mathcal{P}(\mathbb{N})$ onto S. (Hint: In what set are the subscripts for b_i?)

    f) Finish the proof that $\mathbb{R} - \mathbb{Q} \sim \mathbb{R}$.

18. Critique the questionable "proofs" that follow. Point out reasoning errors and unclear presentation. If the argument is a proof, say so. (If the claim is false, there must be an error in the argument. However, it is not enough to say that the claim is false. You need to find an error in the reasoning.)

    *a) Claim: All infinite sets have the same cardinality. "Proof": Let $A = \{a_1, a_2, a_3, \ldots\}$ and $B = \{b_1, b_2, b_3, \ldots\}$ be any infinite sets. Clearly, we can match A and B by $f(a_i) = b_i$, which is one-to-one and onto. So, $A \sim B$. ▲

    *b) Claim: $\mathbb{N} \sim \mathbb{N} \times \mathbb{N}$. "Proof": Clearly, $f : \mathbb{N} \to \mathbb{N} \times \mathbb{N}$ given by $f(n) = (n, 1)$ is one-to-one, so $\mathbb{N} \precsim \mathbb{N} \times \mathbb{N}$. Let $a_1a_2 \ldots a_n$ and $b_1b_2 \ldots b_n$ be elements of $\mathbb{N}$, written in base ten. (For instance, 537 and 682.) Following the idea in the proof of Theorem 5.3.4, we map $(a_1a_2 \ldots a_n, b_1b_2 \ldots b_n)$ to $a_1b_1a_2b_2 \ldots a_nb_n$ in $\mathbb{N}$. (For the previous example, we'd get 563872.) If $a_1b_1a_2b_2 \ldots a_nb_n = c_1d_1c_2d_2 \ldots c_nd_n$, then we have $a_1a_2 \ldots a_n = c_1c_2 \ldots c_n$ and $b_1b_2 \ldots b_n = d_1d_2 \ldots d_n$, so this mapping is one-to-one. Thus, $\mathbb{N} \times \mathbb{N} \precsim \mathbb{N}$. The Schröder–Bernstein Theorem finishes the proof. ▲

    c) By $|X| < |Y|$, we mean that the cardinality of X is strictly smaller than the cardinality of Y. For all $n \in \mathbb{N}$, suppose that A_n is a set and these sets satisfy if $k < n$, then $|A_k| < |A_n|$. Claim: For all $n \in \mathbb{N}$, $|A_n| < |\bigcup_{n \in \mathbb{N}} A_i|$. "Proof": Since $A_n \subseteq \bigcup_{n \in \mathbb{N}} A_i$, by Theorem 5.1.2, $A_n \precsim \bigcup_{n \in \mathbb{N}} A_i$. Further, $|A_n| < |A_{n+1}| \leq |\bigcup_{n \in \mathbb{N}} A_i,|$, so $|A_n| < |\bigcup_{n \in \mathbb{N}} A_i|$. ▲

    d) Claim: $|\mathbb{R} - \mathbb{Q}| = c$. "Proof": We know $\mathbb{Q}$ is a countably infinite subset of the uncountable set $\mathbb{R}$. Then, by Problem 7, $\mathbb{R} - \mathbb{Q}$ is uncountable, and hence, $|\mathbb{R} - \mathbb{Q}| = c$. ▲

    e) Claim: A set is uncountable iff it has a proper uncountable subset. "Proof": First, $\emptyset$ is countable, so we may assume that all sets here are nonempty. ($\Rightarrow$) Suppose that A is uncountable, and for a contradiction, suppose that no proper subset is uncountable. Since $A \neq \emptyset$, there is $B \subseteq A$ with $B \neq \emptyset$. Then B and $A - B$ must be countable. But $A = B \cup (A - B)$ is a countable union of countable sets and so is countable. Contradiction. ($\Leftarrow$) Suppose now A has a proper uncountable subset B. This statement would contradict Theorem 5.2.4, unless A is uncountable. ▲

f) Claim: $\mathbb{Q} \sim \mathbb{R} - \mathbb{Q}$. "Proof": We will show that between any two irrationals there is a rational and that this implies that there are at least as many rationals as irrationals, or $\mathbb{R} - \mathbb{Q} \lesssim \mathbb{Q}$. Then we'll give a one-to-one function for the other direction and invoke the Schröder–Bernstein Theorem. Let a and b be irrationals with $a < b$. Then at some place in their decimal expansions, a and b differ. Without loss of generality, we may suppose that the nth decimal is the first place in which they differ and the nth decimal of a is less than the nth decimal of b. Consider the rational number b' that agrees with b up through the nth decimal place and then terminates. Thus, $a < b' < b$, as claimed. For $b \in \mathbb{R} - \mathbb{Q}$ and $b > 0$, we have $0 < b' < b$ and $b' \in \mathbb{Q}$. So we can map $f(b) = b'$. For $-b < 0$, we can use $f(-b) = -b'$ in the same way to finish defining our function from $\mathbb{R} - \mathbb{Q}$ one-to-one to $\mathbb{Q}$. Thus, $\mathbb{R} - \mathbb{Q} \lesssim \mathbb{Q}$. For the other direction, we know from Example 3 of Section 2.2 that a rational plus an irrational is irrational and, from Problem 7 of that section, that $\sqrt{2} \in \mathbb{R} - \mathbb{Q}$. So, $g : \mathbb{Q} \to \mathbb{R} - \mathbb{Q}$ given by $g(q) = q + \sqrt{2}$ is a one-to-one function. Thus, $\mathbb{Q} \lesssim \mathbb{R} - \mathbb{Q}$. By our friend the Schröder–Bernstein Theorem, $\mathbb{Q} \sim \mathbb{R} - \mathbb{Q}$. ▲

5.4   THE AXIOM OF CHOICE AND ITS EQUIVALENTS

(Optional)

"The moving power of mathematical invention is not reasoning, but imagination."

—Augustus De Morgan (1806–1871)

Ernst Zermelo included the Axiom of Choice in his 1908 paper giving the first axiomatization of set theory. This axiom engendered quite a bit of discussion and discomfort over the next 60 years. Although mathematicians used it to prove many beautiful results, its totally nonconstructive nature was hard for many to accept. In addition, there are counterintuitive results following directly from this axiom. In 1963, Paul Cohen proved that the Axiom of Choice was, indeed, an axiom: It couldn't be proven from the other axioms of set theory. By the 1970s, mathematicians had overwhelmingly decided that the axiom's usefulness outweighed any philosophical drawbacks. For a discussion of some of the philosophical issues, see the subsection on Intuitionism and Constructivism in Section 9.2.

I don't expect students in a first proof course to be able to write proofs using the Axiom of Choice. Still, my own experience convinces me that students heading to graduate school need access to a helpful exposition on proofs involving the Axiom of Choice and Zorn's Lemma. In my first week of graduate school (in 1975), I asked my algebra professor if Zorn's Lemma were needed in the just-finished sketch of a proof. He glared at me, saying only, "Of course." I learned not to ask him "dumb questions," but I was grateful that my college professors had prepared me adequately. Most of the problems in this section are designed to help students develop proofs involving the Axiom of Choice or Zorn's Lemma. I hope students bound for graduate school will find these problems helpful.

Bertrand Russell (1872–1970) gave an intuitive explanation of why the Axiom of Choice is sometimes needed. Imagine first that you are asked to select one shoe from each of infinitely many pairs of shoes. You can use the simple rule of picking the left shoe from each pair. We don't need any fancy set theory for this situation. However, if you need to select one sock from each of infinitely many pairs of socks, you'd find the situation inherently different. The socks are interchangeable, so there is no clear rule describing which sock to pick. With infinitely many choices to make, the elementary axioms of set theory provide no way to accomplish this task. The Axiom of Choice simply asserts that there is such a rule. In Russell's illustration, each A_i in the following axiom is a pair of socks and $f(i)$ is the sock chosen from that pair.

AXIOM OF CHOICE. Given any family of nonempty sets $\{A_i : i \in I\}$, there is a function $f : I \to \bigcup_{i \in I} A_i$ such that $f(i) \in A_i$.

THEOREM 5.2.5. If I is a countable index set and for each $i \in I$ the set A_i is countable, then $\bigcup_{i \in I} A_i$ is countable. In short, the countable union of countable sets is countable.

TABLE 1. Listing the Elements of Each A_i

$\vdots$	$\vdots$	$\vdots$	$\vdots$	$\vdots$	$\vdots$	
A_3	a_{31}	a_{32}	a_{33}	a_{34}	a_{35}	$\cdots$
A_2	a_{21}	a_{22}	a_{23}			
A_1	a_{11}	a_{12}	a_{13}	a_{14}	a_{15}	$\cdots$

Before we start the proof, let's look at a naïve attempt, both to convey the general idea and to point out where we'll need the Axiom of Choice. Because each A_i is countable, we can list its elements, whether there are finitely or infinitely many. Table 1 illustrates this idea using $\mathbb{N}$ for the index set. (A_2 happens to be a finite set.) The representation of $\mathbb{N} \times \mathbb{N}$ in Table 2 should make it seem as though we could simply assign the elements of the different A_i to corresponding rows of $\mathbb{N} \times \mathbb{N}$. (I switched the coordinates in $\mathbb{N} \times \mathbb{N}$ to make the match clearer.) We don't need to fill in each row, since the part filled in is a subset of $\mathbb{N} \times \mathbb{N}$. Recall that $\mathbb{N} \times \mathbb{N}$ is countable by Theorem 5.2.1. Then, by Theorem 5.2.4, every subset of it is also countable. It would seem that we are done. There are three problems for the general theorem, one minor and the others major. First of all, the index set of the theorem is I, not $\mathbb{N}$, but we

TABLE 2 $\mathbb{N} \times \mathbb{N}$.

$\vdots$	$\vdots$	$\vdots$	$\vdots$	$\vdots$	
(3, 1)	(3, 2)	(3, 3)	(3, 4)	(3, 5)	$\cdots$
(2, 1)	(2, 2)	(2, 3)			
(1, 1)	(1, 2)	(1, 3)	(1, 4)	(1, 5)	$\cdots$

could easily overcome this problem. Secondly, in general, we would need infinitely many functions to list the elements of the A_i, one function for each of the infinitely many sets. While we can obtain any one function, we need a simple application of the Axiom of Choice to get all of them at once. Finally, the elements of one of the A_i can be elements of another set A_j. This last problem is fatal to this simplistic approach, because we no longer have a function from the union of the A_i into $\mathbb{N} \times \mathbb{N}$. A short argument and the Axiom of Choice will supply us with an alternative function.

Proof. For the family $\{A_i : i \in I\}$, suppose that I is countable and each A_i is countable. We need to find a one-to-one function $f : \bigcup_{i \in I} A_i \to \mathbb{N} \times \mathbb{N}$. We will first make a relation R from $\mathbb{A} = \bigcup_{i \in I} A_i$ to $\mathbb{N} \times \mathbb{N}$, following the idea from Tables 1 and 2. Then we use the Axiom of Choice to "trim" the relation so that it becomes the desired function.

Because I is countable, there is $g : I \to \mathbb{N}$ that is one-to-one. The Axiom of Choice provides countably many one-to-one functions $h_i : A_i \to \mathbb{N}$ for all the $i \in I$ at once. For $a \in \mathbb{A}$ and $(x, y) \in \mathbb{N} \times \mathbb{N}$, we say that $aR(x, y)$ iff there is some $i \in I$ such that $a \in A_i$ and $g(i) = x$ and $h_i(a) = y$. (In terms of the tables, suppose that $a_{12} = a_{31}$. Then we would have $a_{12}R(1, 2)$ and $a_{12}R(3, 1)$.) Every a in the union $\mathbb{A}$ is related to at least one ordered pair, because a is in at least one A_i.

We next eliminate duplicate ordered pairs by forming a new collection of sets. For $a \in \mathbb{A}$, let $B_a = \{(x, y) \in \mathbb{N} \times \mathbb{N} : aR(x, y)\}$. (In the example in the previous brackets, $B_{a_{12}} = \{(1, 2), (3, 1)\}$.) Each B_a is nonempty, because each a is related to at least one ordered pair. The Axiom of Choice now delivers a function $f : \mathbb{A} \to \bigcup_{a \in \mathbb{A}} B_a$ so that $f(a) \in B_a$. Our next job is to ensure that f is one-to-one. Let $a, b \in \mathbb{A}$, and suppose that $f(a) = f(b)$. Thus, for some $(x, y) \in \mathbb{N} \times \mathbb{N}$, we have $f(a) = f(b) = (x, y)$. Then a and b are in some A_i together (the A_i with $g(i) = x$). Furthermore, $h_i(a) = y = h_i(b)$. But h_i is one-to-one, so $a = b$. We have just proved that $\mathbb{A} \lesssim \mathbb{N} \times \mathbb{N}$. In fact, f is a bijection from $\mathbb{A}$ to its range $f[\mathbb{A}]$, a subset of $\mathbb{N} \times \mathbb{N}$. Theorem 5.2.4 guarantees that this subset is countable, so $\mathbb{A} = \bigcup_{i \in I} A_i$ is countable. ∎

The results that follow were discussed in Section 5.1. The proof of the first one uses the Axiom of Choice, while the second proof uses Zorn's Lemma, discussed in the next subsection. Theorem 5.4.1 assures us that $\aleph_0$ is the smallest size of infinity.

THEOREM 5.4.1. Every infinite set has a countably infinite subset.

Proof. See Problem 3. ∎

THEOREM 5.4.2. For any sets X and Y, we have $X \lesssim Y$ or $Y \lesssim X$.

Proof. See Problem 7. ∎

Sometimes, Theorem 5.4.2 is restated as the *trichotomy law*, saying either one set has a smaller cardinality, a larger cardinality, or the same cardinality as another set. This is analogous to the situation with ordinary real numbers: $x < y$ or $x > y$ or $x = y$.

Zorn's Lemma

Outside of set theory, working mathematicians often find Zorn's Lemma more useful than the Axiom of Choice. The two propositions are logically equivalent, although we will not prove this fact. (See Jech, 10–11.) Zorn's Lemma asserts the existence of a maximal element in certain kinds of partially ordered sets. We need one additional definition beyond those of Section 4.4. A *chain*, as defined next, is a subset for which the partial order is a linear order. For instance, in $\mathcal{P}(\{1, 2, 3, 4, 5, 6\})$ with the partial order $\subseteq$, one chain is $\{\{2, 3\}, \{1, 2, 3\}, \{1, 2, 3, 4, 5\}\}$.

DEFINITION. In a set X with a partial order $\preceq$, a *chain* is a subset C such that for all $a, b \in C$, we have $a \preceq b$ or $b \preceq a$.

Zorn's Lemma   In a partially ordered set X, if every chain has an upper bound, then X has a maximal element.

The details of a proof using Zorn's Lemma often risk overwhelming the role Zorn's Lemma plays in the proof. Our example of a proof uses familiar terms from linear algebra, although the result does not usually appear in a typical undergraduate linear algebra course.

Bases in Linear Algebra   In linear algebra, you saw how to find a basis for a finite dimensional vector space. For instance, the set $\{(1, 0, 0), (0, 1, 0), (0, 0, 1)\}$ is the usual basis of $\mathbb{R}^3$. Some infinite dimensional vector spaces have natural bases; consider the space of all polynomials with the basis $\{1, x, x^2, x^3, \dots\}$. Others, such as the set of all continuous real functions, have no obvious choice of a basis. We'll use Zorn's Lemma to show that every vector space has a basis. Before the theorem and proof, we include several definitions and an example from linear algebra.

DEFINITIONS. A set of vectors $W = \{\overrightarrow{w_i} : i \in I\}$ from a vector space V over $\mathbb{R}$ is *linearly independent* iff for all finite subsets $\{\overrightarrow{w_1}, \overrightarrow{w_2}, \dots, \overrightarrow{w_n}\}$ from W and all scalars $\alpha_1, \alpha_2, \dots, \alpha_n \in \mathbb{R}$, if $\alpha_1\overrightarrow{w_1} + \alpha_2\overrightarrow{w_2} + \cdots + \alpha_n\overrightarrow{w_n} = \overrightarrow{0}$, then each $\alpha_i = 0$. A set of vectors $W = \{\overrightarrow{w_i} : i \in I\}$ from a vector space V *spans* V iff for all $\overrightarrow{v} \in V$, there is a finite subset $\{\overrightarrow{w_1}, \overrightarrow{w_2}, \dots, \overrightarrow{w_n}\}$ from W and scalars $\beta_1, \beta_2, \dots, \beta_n \in \mathbb{R}$ such that $\beta_1\overrightarrow{w_1} + \beta_2\overrightarrow{w_2} + \cdots + \beta_n\overrightarrow{w_n} = \overrightarrow{v}$. A set of vectors W from a vector space V is a *basis* of V iff it is linearly independent and spans V.

EXAMPLE 1

The set $\{(1, 0, 0), (0, 1, 0)\}$ is linearly independent, but doesn't span $\mathbb{R}^3$, since, for instance, $(1, 2, 3)$ is not a linear combination of these two vectors. The set $\{(1, 0, 0), (0, 1, 0), (0, 0, 1), (1, 2, 3)\}$ spans $\mathbb{R}^3$, but is not linearly independent, since $1(1, 0, 0) + 2(0, 1, 0) + 3(0, 0, 1) + (-1)(1, 2, 3) = (0, 0, 0)$. $\Diamond$

THEOREM 5.4.3. Every vector space has a basis.

DISCUSSION. To use Zorn's Lemma, we first give a partially ordered set. This will be the collection of all sets of independent vectors of the vector space. The partial order will be the familiar relation $\subseteq$ for subsets. Next we find an upper bound of a general chain and invoke Zorn's Lemma. The maximal element guaranteed by this lemma will be our candidate for a basis, but we will need to show that the candidate really is a basis. You will notice that Zorn's Lemma appears in just one line, although the whole proof is built around it.

Proof. Let V be any vector space and let $\mathbb{L}$ be the collection of all sets of linearly independent vectors. Then $\subseteq$ is a partial order on $\mathbb{L}$. To use Zorn's Lemma, let $C = \{C_i : i \in I\}$ be a chain from $\mathbb{L}$. That is, each C_i is a set of independent vectors from V, and for any two of these sets C_i and C_j, we have $C_i \subseteq C_j$ or $C_j \subseteq C_i$. Our candidate for the upper bound of C is $H = \bigcup_{i \in I} C_i$, the union of all of the vectors from these independent sets. This is certainly a set of vectors. Is H independent? Let $\{\vec{h_1}, \vec{h_2}, \ldots, \vec{h_n}\}$ be any finite subset of H and suppose for scalars $\alpha_i \in \mathbb{R}$ we have $\alpha_1 \vec{h_1} + \alpha_2 \vec{h_2} + \cdots + \alpha_n \vec{h_n} = \vec{0}$. Each $\vec{h_1}$ is in some C_i. Since there are only finitely many of these sets and they are linearly ordered, one of them, say, C_m, is the biggest one. Then for each i we have $\vec{h_i} \in C_m$. Since C_m is linearly independent, each $\alpha_i = 0$. This shows that all of H is linearly independent.

We have fulfilled the hypothesis of Zorn's Lemma, which now gives us a maximal element in $\mathbb{L}$, which we'll call B. We claim that B is a basis of V. First of all, $B \in \mathbb{L}$, so B is linearly independent. Does B span V? For a contradiction, suppose that $\vec{v} \in V$ is not spanned by B. Then, for any finite subset $\{\vec{b_1}, \vec{b_2}, \ldots, \vec{b_n}\}$ of B and all scalars $\beta_i \in \mathbb{R}$, we have $\beta_1 \vec{b_1} + \beta_2 \vec{b_2} + \cdots + \beta_n \vec{b_n} \neq \vec{v}$. Because the scalars β_i were general, we can say that for any nonzero scalar $\alpha \in \mathbb{R}$, we have $\beta_1 \vec{b_1} + \beta_2 \vec{b_2} + \cdots + \beta_n \vec{b_n} \neq -\alpha \vec{v}$. Shift $\alpha \vec{v}$ to the other side to get $\beta_1 \vec{b_1} + \beta_2 \vec{b_2} + \cdots + \beta_n \vec{b_n} + \alpha \vec{v} \neq \vec{0}$. Then $B \cup \{\vec{v}\}$ would be a set of independent vectors and even bigger than B. But B was maximal, a contradiction. Hence, B spans and so is a basis. ∎

DISCUSSION. The chains of interest are infinite. The word "finite" in the definition of linearly independent allows us to reduce the indefinitely large set of vectors H to a manageable size. Similarly, we need to deal only with finitely many vectors in B to show spanning.

Well Ordering Principle

A third proposition, "Every set can be well ordered," is equivalent to the Axiom of Choice. This statement is less intuitive than Zorn's Lemma and also less used outside of set theory. In addition, its use often involves something called transfinite induction. We leave such considerations to more advanced texts. (See Halmos, 66–73.)

Paradoxes

While the Axiom of Choice is essential for many important results, and in Russell's example sounds completely innocuous, it also implies a number of counterintuitive propositions. It is worthwhile for undergraduates to have an intuitive grasp of one of these results, the Banach–Tarski paradox. In the 1830s, W. Bolyai and P. Gerwien showed that, given any two polygons with the same area, one of them could be cut into a finite number of smaller polygons that could be reassembled into the other polygon. (Reassembling allows sliding and/or turning the smaller polygons, similar to the game Tangram.) In 1900, Max Dehn proved that the corresponding three-dimensional decomposition for polyhedra was not always possible. However, in 1924, Stefan Banach and Alfred Tarski used the Axiom of Choice to show that any three-dimensional ball could be "decomposed" into a finite number of subsets that could be reassembled as two balls of the same radius. This seems close to saying that the familiar idea of volume is meaningless. However, the subsets involved are totally unimaginable and nonconstructive—certainly nothing that could be physically carved from an actual ball. In the language of measure theory, these subsets are "nonmeasurable." (The existence of nonmeasurable sets, which depends on the Axiom of Choice, was already discomforting.) Fortunately, measure theory, developed in the early twentieth century, gives us a way to have a meaningful theory of volume and still use the Axiom of Choice. Nevertheless, the Banach–Tarski paradox and other paradoxes made many mathematicians in the first half of the twentieth century at best uneasy with the Axiom of Choice.

Historical Remarks

Some proofs by Cantor and others in the nineteenth century used the Axiom of Choice or the Well Ordering Principle, and sometimes the use wasn't recognized. The arguments were generally uncritically accepted at the time. (Russell's example of choosing one sock from each of infinitely many pairs illustrates how innocent some of the uses can appear.) Part of the reason that Ernst Zermelo published his axioms of set theory in 1908 was to respond to the objections some mathematicians had to his explicit use of the Axiom of Choice in an earlier paper. Zermelo's work helped mathematicians realize how often this axiom, or propositions equivalent to it, were used in proofs.

The nonconstructive nature of the axiom disturbed many. The Banach–Tarski paradox and others like it made many of these mathematicians oppose its use. In 1938, Kurt Gödel published a result showing that the Axiom of Choice and the Continuum Hypothesis, discussed in Section 5.3, were consistent with the elementary axioms of set theory. That is, if the elementary axioms were free from contradiction, adding these two axioms would not lead to a contradiction. This result reassured mathematicians to some degree. Paul Cohen's proof of the independence of the Axiom of Choice and the Continuum Hypothesis eliminated the hope that the Axiom of Choice and the Continuum Hypothesis were theorems. They needed to be assumed or denied. The now-wide acceptance of the Axiom of Choice probably reflects the practical approach of most mathematicians. If we need it to get the theorems we want, we'll accept it as an axiom and live with the paradoxes that come with it.

Zorn's Lemma came into the discussion in fits and starts. Felix Hausdorff seems to be the first to publish, in 1909, the use of such a maximal principle. Several others published papers, using similar principles, before Zorn published his in 1935. Apparently, all failed to notice others doing similar things. However, Zorn's timing was impeccable, even if unintended. Many mathematicians were ready to make use of such an axiom. Zorn did not call it a lemma, which is a misnomer, since it functions as an axiom, not something to prove. Nevertheless, the name quickly became standard and its use widespread. Thus, Zorn's Lemma is not a lemma, nor was it first done by Zorn.

For more information about the history of the Axiom of Choice and related topics, see Moore.

PROBLEMS

1. For each family of sets, give, if possible, an explicit rule to pick one element from each set. (If you can give such a rule, you don't need the Axiom of Choice.)

   *a) $\{[a, b] : a, b \in \mathbb{R} \text{ and } a < b\}$.

   b) The family of all open intervals in $\mathbb{R}$.

   *c) All nonempty subsets of $\mathbb{R}$.

   d) All finite nonempty subsets of $\mathbb{R}$.

   e) All countable nonempty subsets of $\mathbb{R}$.

   f) All nonempty subsets of $\mathbb{Z}$.

   g) $\{P_n : n \in \mathbb{N}\}$, where P_n is the set of all polynomials of degree n.

   h) $\{F_{r,s} : r, s \in \mathbb{R}\}$, where $F_{r,s}$ is the set of all functions $f : \mathbb{R} \to \mathbb{R}$ such that $f(r) = s$.

   i) The family of all bases for the vector space $\mathbb{R}^3$.

2. If possible, find a maximal element for each chain, given the set and the partial order.

   *a) Set: $\mathbb{R}$; partial order: $\leq$; chain: $\{r^2 : r \in [1, 7)\}$.

   *b) Set: $\mathbb{R}$; partial order: $\leq$; chain: $\{r^2 : r \in \mathbb{N}\}$.

   c) Set: $\mathcal{P}(\mathbb{R})$, partial order: $\subseteq$; chain: $\{[\frac{1}{n}, n + 1] : n \in \mathbb{N}\}$.

   d) Set: $\mathbb{R} \times \mathbb{R}$; partial order: $\preceq$, where $(a, b) \preceq (c, d)$ iff $a \leq c$ and $b \leq d$; chain: $\{(n, n^2) : n \in \mathbb{N}\}$.

   e) Set: $\mathbb{R} \times \mathbb{R}$; partial order: $\sqsubseteq$, where $(a, b) \sqsubseteq (c, d)$ iff $a < c$ or $(a = c \text{ and } b \leq d)$; chain: $\{(1 - \frac{1}{n}, n^2) : n \in \mathbb{N}\}$.

*3. Use the following idea and the Axiom of Choice to give a proof of Theorem 5.4.1: If a set A_i is infinite, there is an element $a_i \in A_i$ and $A_i - \{a_i\}$ is infinite.

4. To compare sizes of sets in Section 5.1, we defined $A \lesssim B$ iff there is $f : A \to B$ that is one-to-one. For nonempty sets, define the alternative comparison $A \gtrsim B$ iff there is $g : B \to A$ that is onto. Use parts b, c, and d to prove that these definitions are logically equivalent.

   a) Use Figure 1 to explain why these two definitions should be equivalent for finite sets.

   b) Suppose that $f : A \to B$ is one-to-one and $v \in A$. Use the idea of Figure 1 to define $g : B \to A$ from f and prove that g is a function and onto.

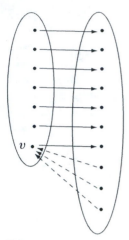

FIGURE 1

c) Suppose that $g : B \to A$ is onto. Show for all $a \in A$ that $g^{-1}[\{a\}]$ is a nonempty subset of B. Also, show that if $s \neq t$ in A, then $g^{-1}[\{s\}] \cap g^{-1}[\{t\}] = \emptyset$. (Hint: See Section 3.3.)

d) Use the Axiom of Choice and part c to define a function $f : A \to B$ and prove that your function is one-to-one.

5. Let $f : X \to Y$ be any function. Prove that $f[X] \lesssim X$.

6. Let $\{S_n : n \in \mathbb{N}\}$ be a family of nonempty pairwise disjoint sets. That is, for all $k, n \in \mathbb{N}$, if $k \neq n$, then $S_k \cap S_n = \emptyset$. Use the Axiom of Choice to show that $\bigcup_{n \in \mathbb{N}} S_n$ has a countably infinite subset. *Note:* Theorem 5.2.5 assures you only a countable subset.

7. Use the given outline to prove that for all sets X and Y, $X \lesssim Y$ or $Y \lesssim X$ (Theorem 5.4.2); (referred to in Section 5.1).

   Let $S = \{(A, B, f) : A \subseteq X, B \subseteq Y$ and $f : A \to B$ is a bijection$\}$. Define $\sqsubseteq$ on S by $(A, B, f) \sqsubseteq (C, D, g)$ iff $A \subseteq C, B \subseteq D$ and for all $a \in A$, $f(a) = g(a)$.

   Example. $X = [0, \infty), Y = \mathbb{R}, A = [1, 2], B = [1, 4], C = [1, 3], D = [1, 9]$, and $f(x) = g(x) = x^2$. Then $A \subseteq C, B \subseteq D$, and $f = g$ on A. *Note:* There is no promise that we can extend this function to all of X and get a bijection.

   a) Prove that $\sqsubseteq$ is a partial order on S.

   b) For a chain $C = \{(A_i, B_i, f_i) : i \in I\}$, define $f : \bigcup_{i \in I} A_i \to \bigcup_{i \in I} B_i$ based on the f_i and show that f is a function, one-to-one, and onto.

   c) Use Zorn's Lemma to get a maximal element in S, say, (T, W, h).

   d) Prove by contradiction that $T = X$ or $W = Y$.

   e) If $T = X$, show that $X \lesssim Y$. If $W = Y$, show that $Y \lesssim X$.

8. Use the given outline to derive the Well Ordering Principle: Any nonempty set can be well ordered. For X a nonempty set, let $S = \{(W, \sqsubseteq) : W \subseteq X$ and $\sqsubseteq$ is a well ordering of $W\}$. That is, for every $V \subseteq W$, if $V \subseteq \emptyset$, then there is a unique $v \in V$ such that for all $x \in V, v \sqsubseteq x$. Define $\preceq$ on S by $(W, \sqsubseteq) \preceq (W^*, \sqsubseteq^*)$ iff $W \subseteq W^*$ and $\sqsubseteq^*$ "extends" $\sqsubseteq$. That is, for all $a, b \in W$, and $c \in W^*$, we have $(a \sqsubseteq b$ iff $a \sqsubseteq^* b)$ and (if $c \notin W$, then $a \sqsubseteq^* c$). In effect, W is the initial part of W^*.

a) Prove that $\preceq$ is a partial order on S.

b) For a chain $C = \{(W_i, \sqsubseteq_i) : i \in I\}$, in order to find an upper bound, give a candidate for an order $\sqsubseteq_Y$ on the set $\bigcup_{i \in I} W_i = Y$.

c) Show that $\sqsubseteq_Y$ is a partial order.

d) To show that $\sqsubseteq_Y$ well orders Y, let T be any nonempty subset of Y. There is some $x \in T$. Why is x in some W_i? Consider $T_i = T \cap W_i$. Why must T_i have a least element, say, t_i? Why must t_i be the least element of T?

e) Show that $(Y, \sqsubseteq_Y)$ is an upper bound of the chain C.

f) Use Zorn's Lemma to obtain a maximal element of S, say, $(W, \sqsubseteq)$.

g) Prove by contradiction that $W = X$.

REFERENCES

HALMOS, P. 1974. *Naive set theory*. New York: Springer-Verlag.

JECH, T. 1973. *The axiom of choice*. New York: American Elsevier.

MOORE, G. 1982. *Zermelo's axiom of choice: Its origins, development and influence*. New York: Springer-Verlag.

DEFINITIONS FROM CHAPTER 5

Section 5.1

■ Two sets A and B have the *same cardinality*, denoted $A \sim B$, iff there is a one-to-one and onto function $f : A \rightarrow B$.

■ Given two sets A and B, $A \precsim B$ iff there is a function $f : A \rightarrow B$ that is one-to-one. If $A \precsim B$, but they do not have the same cardinality, we say that the cardinality of A is *strictly smaller than* the cardinality of B.

■ We write $|A|$ for the *cardinality* of the set A.
$$|\emptyset| = 0. \quad |\mathbb{N} \cap [1, n]| = n. \quad |\mathbb{N}| = \aleph_0. \quad |\mathbb{R}| = c.$$
A set is *finite* iff its cardinality is an integer. A set X is *countably infinite* (or *denumerable*) iff $X \sim \mathbb{N}$. A set is *countable* iff it is finite or countably infinite; otherwise, it is *uncountable*.

■ For nonempty sets A and B, the set of all functions from A to B is B^A.

■ If A and B are sets with $|A| = \alpha$ and $|B| = \beta$, then $|A \times B| = \alpha \cdot \beta$, $|A^B| = \alpha^\beta$ and, if $A \cap B = \emptyset$, then $|A \cup B| = \alpha + \beta$.

Section 5.4

■ In a set X with a partial order $\preceq$, a *chain* is a subset C such that for all $a, b \in C$, we have $a \preceq b$ or $b \preceq a$.

■ A set of vectors $W = \{\vec{w_i} : i \in I\}$ from a vector space V over $\mathbb{R}$ is *linearly independent* iff for all finite subsets $\{\vec{w_1}, \vec{w_2}, \ldots, \vec{w_n}\}$ from W and all scalars

$\alpha_1, \alpha_2, \ldots, \alpha_n \in \mathbb{R}$, if $\alpha_1 \overrightarrow{w_1} + \alpha_2 \overrightarrow{w_2} + \cdots + \alpha_n \overrightarrow{w_n} = \overrightarrow{0}$, then each $\alpha_i = 0$. A set of vectors $W = \{\overrightarrow{w_i} : i \in I\}$ from a vector space V *spans* V iff for all $v \in V$, there is a finite subset $\{\overrightarrow{w_1}, \overrightarrow{w_2}, \ldots, \overrightarrow{w_n}\}$ from W and scalars $\beta_1, \beta_2, \ldots, \beta_n \in \mathbb{R}$ such that $\beta_1 \overrightarrow{w_1} + \beta_2 \overrightarrow{w_2} + \cdots + \beta_n \overrightarrow{w_n} = \overrightarrow{v}$. A set of vectors W from a vector space V is a *basis* of V iff both it is linearly independent and it spans V.

CHAPTER 6

INTRODUCTION TO DISCRETE MATHEMATICS

"Mathematics is the music of reason."

—James Sylvester (1814–1897)

THE **EXPLOSIVE RISE** of computing has greatly increased the importance of discrete mathematics, the mathematics underlying computer science. ("Discrete" means separate and is in contrast to "continuous." Computers, with their pixels and on/off electronic structure, are inherently discrete. In contrast, calculus is a part of continuous mathematics.) While computers simply compute, discrete mathematics is full of intriguing ideas. To extend Sylvester's image in the preceding quote, computers are instruments on which we can play the music of discrete mathematics.

Discrete mathematics splits into two main areas: graph theory and combinatorics. Each of these names needs some explanation. A graph in graph theory is a set of points connected with some edges, as in Figure 1, a very different object from the graphs of calculus. By now, the use of the term *graph* for both ideas is firmly established, and students and mathematicians have to live with the confusion. Graph theory allows us to represent a variety of situations visually and then reason about them. Combinatorics takes counting to new heights. A preliminary description of it is the art of efficient counting and the search for patterns underlying such counting. Problems and techniques now part of these areas have been around a long time, although the subjects themselves are twentieth-century creations.

6.1 GRAPH THEORY

Each part of Figure 1 quickly conveys information, even if the subjects vary greatly.

DEFINITION. A *graph G* is a finite nonempty set *V* of *vertices* and a set *E* of *edges*, where an edge is a subset of *V* with two elements. We abbreviate the edge

213

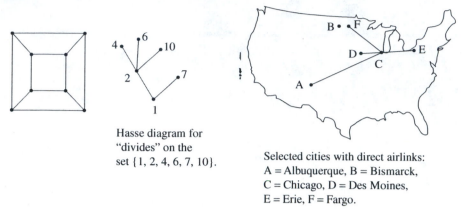

Hasse diagram for "divides" on the set {1, 2, 4, 6, 7, 10}.

Selected cities with direct airlinks:
A = Albuquerque, B = Bismarck,
C = Chicago, D = Des Moines,
E = Erie, F = Fargo.

FIGURE 1

$\{a, b\}$ by ab. If needed, we identify the sets V and E of a particular graph G by writing $V(G)$ and $E(G)$.

In terms of the definition, a graph has only a little more structure than a set. In fact, one can think of a graph as a symmetric relation on the set of vertices, since the edge ab equals the edge ba. However, the drawings of graphs in Figure 1 suggest many properties to consider. We will consider only a few, starting with perhaps the easiest: the number of edges at a vertex, called the *degree* of a vertex.

DEFINITION. For $v \in V$ in a graph, the *degree* of a vertex v is the number of $x \in V$ such that vx is an edge. A graph is *regular* iff all vertices have the same degree.

EXAMPLE 1

The first graph in Figure 1 is regular, since each of the eight vertices has degree 3. The degrees in the second graph range from 1 to 4, and those of the third graph range from 0 to 4. ◇

THEOREM 6.1.1. In a graph the number of vertices of odd degree is even.

Proof. Consider summing the degrees of all the vertices of a graph. Each edge of the graph will contribute two to that sum, one for each vertex that is part of the edge. Thus, the sum of the degrees must be even. For the sum of a collection of integers to be even, there must be an even number of odd numbers. That is, there is an even number of vertices of odd degree. ∎

The graphs in Figure 1 suggest other, interrelated properties.

DEFINITION. A *trail* from vertex v_1 to vertex v_n is a set of vertices $\{v_i : 1 \le i \le n\}$ such that if $1 \le i < i + 1 \le n$, then $v_i v_{i+1}$ is an edge of the graph, and no edge is used twice. A *circuit* is a trail from v_1 to v_n with $v_1 = v_n$ and $n \ge 2$. A graph is *connected* iff for each pair of distinct vertices, there is a trail from one to the other. A *tree* is a connected graph with no circuits.

EXAMPLE 2

Any two vertices in the first graph of Figure 1 are connected by a variety of trails. Some trails in this graph form circuits leading back to where they start. The second graph is also connected, but none of its trails form circuits. Because of the pattern made by graphs like this second graph, these graphs are called trees. The third graph is not connected because no trail connects Bismarck to the other cities. ◇

REMARKS. By definition, the set $\{v_1\}$ is a trail, but not a circuit. Problem 10 asks you to show that a circuit has to have at least three distinct vertices.

THEOREM 6.1.2. A graph is a tree iff for any two distinct vertices there is exactly one trail from one to the other.

Proof. See Problem 21. ∎

Euler Trails and Circuits

The beginnings of graph theory come from a puzzle posed by the people of Königsberg, now the town of Kaliningrad, in Russia. The town had seven bridges across the Pregel River connecting four different areas of the city. The people wondered if it were possible to take a walk, crossing each bridge exactly once, preferably returning to the start. In 1736, Leonard Euler (1707–1783) showed the impossibility of such a walk, establishing some of the basic ideas of graph theory in the process. He recast the situation, shown in Figure 2, as finding a circuit (or even a trail) that included every edge of the graph. (Technically, the figure is a *multigraph*, defined at the end of this section.) The edges represent the bridges, and the vertices represent the four areas of the city.

Consider an area other than the starting or ending place. Each time one enters this area, one crosses a bridge, and then one has to cross another bridge to leave it, using up an even number of bridges connected to that area. Thus, each such area needs to have an even number of bridges in the trail or circuit, one pair for each entry and exit. In graph theory terms, all degrees, except possibly the start and end, have to be even. Since all four vertices in the graph of Figure 2 have odd degree, Euler realized that no trail (let alone a circuit) crossing each bridge just once was possible. Euler also analyzed what happens at the start and the end. If the start is the same as the end, then the same reasoning applies to it as to other places, so all vertices must have even degree. If the start differs from the end, then the start has one bridge for leaving,

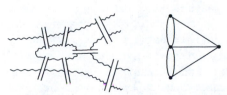

FIGURE 2. The seven bridges of Königsberg and a graphical representation.

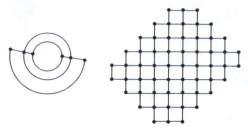

FIGURE 3. Two graphs with Euler paths.

unmatched with any entering bridge, forcing an odd number of bridges. Similarly, the end has an odd number of bridges.

DEFINITION. An *Euler trail* of a graph is a trail that uses every edge exactly once. If the starting and ending vertices are the same, the Euler trail is an *Euler circuit*.

What we call Euler trails and the way to find them have appeared independently in several cultures. For instance, children of the Shongo people of the Congo have for generations challenged one another with designs such as the second graph in Figure 3 (see Zaslavsky, 105–109). Both graphs in Figure 3 have Euler trails.

THEOREM 6.1.3. (Euler).

(i) A connected graph has an Euler circuit iff each vertex has even degree.

(ii) A connected graph has an Euler trail iff the graph has at most two vertices of odd degree.

Proof.

(i) ($\Rightarrow$) See Problem 14.

($\Leftarrow$) For the other direction, we need an algorithm, a method to find such a circuit. See Epp, 628–629, for a careful proof. See Problem 15 for an exploration of this idea.

(ii) See Problem 16. ∎

Isomorphism

While the first graph of Figure 4 (repeated from Figure 1) faithfully represents the edges of a cube, we are more familiar with the second drawing in Figure 4. The two graphs are in some sense equivalent, even though in the second graph some edges cross one another, while none cross in the first graph. When are two graphs sufficiently alike for the purposes of graph theory? The definition of a graph involves only the set of vertices and the set of edges, so a matching of these sets from the graphs suffices. We can explicitly match the vertices and edges of the two graphs in Figure 4 by using the function f given by $f(v_i) = w_i$. For instance, in the first graph of Figure 4, the edges from the vertex v_1 are v_1v_2, v_1v_4, and v_1v_5. In the second graph, the edges from

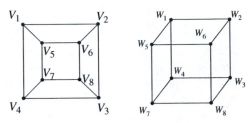

FIGURE 4. Isomorphic graphs representing a cube.

w_1 match: w_1w_2, w_1w_4, and w_1w_5. All of the edges match, as well as the vertices. The following definition generalizes this matching.

DEFINITION. Two graphs G and H are *isomorphic* iff there is a bijection $f : V(G) \to V(H)$ such that v_iv_j is an edge of G iff $f(v_i)f(v_j)$ is an edge of H. The function is called an *isomorphism*.

EXAMPLE 3

Show that the graphs in Figure 5 are not isomorphic to the graphs in Figure 4 or to each other.

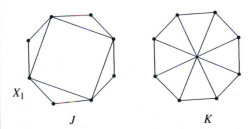

FIGURE 5. Graphs nonisomorphic to each other or to the graphs of Figure 4.

Solution   There are many possible bijections between the vertices of the graph J of Figure 5 and the vertices of any of the other graphs, since they each have eight vertices. However, no bijection can also make the edges match, even though all of the graphs have 12 edges. In particular, in J some of the vertices have four edges, whereas all vertices of the other graphs have three edges. Thus, wherever we map X_1 from J, its image would have to have four vertices with edges connecting to it. Since none of the other graphs have such a vertex, there can be no isomorphism.

In general, to show that two graphs are not isomorphic, we need to find some property involving vertices and edges where they differ.

All the vertices of the graph K have degree 3, but even so, K is not isomorphic to the graphs of Figure 4. Consider the number of circuits made with four edges, starting at any point. For the cube (either graph of Figure 4) there are three circuits, corresponding to the three squares meeting at a vertex of a cube. But any vertex of K has only two circuits made with four edges.   ◇

Variations on Graphs

Different applications have led to modifications of edges in a variety of ways, all based visually. A *directed graph* uses ordered pairs to designate directed edges (drawn as arrows). A *graph with loops* allows an edge from a vertex to itself. (We used directed graphs with loops to represent relations in Chapter 4.) A *multigraph* permits more than one edge between two vertices, as in Figure 2. A *weighted* graph assigns a number to each edge, such as a distance. We'll consider weighted graphs in the next section.

Historical Remarks

Graph theory became a recognized discipline in the twentieth century. Before then, various mathematicians investigated problems that didn't fit into already recognized branches of mathematics. One problem of note led Euler in 1758 to publish a formula linking the number of vertices, edges, and faces in a polyhedron, or in modern terms, a *planar* graph. This formula has had wide-ranging importance in geometry and topology, as well as graph theory. In graph theory, map coloring built on Euler's formula and in turn spurred much new mathematics. After a 120-year effort, this work culminated in 1976 with the computer-assisted proof of the Four Color Map Theorem (see Problem 23).

PROBLEMS

*1. For each statement that follows, decide whether it is true or (at least sometimes) false. Explain your answer or cite the part of the text supporting it.

   a) Graph theory studies the same type of graphs as calculus does.

   b) A triangle can be thought of as a graph in which every vertex has degree 2.

   c) Some trees are trails.

   d) All trails are trees.

   e) Two graphs with the same number of vertices and edges are always isomorphic.

2. Find an Euler trail, if one exists, for each graph in Figure 6.

3. If possible, draw a graph satisfying each given condition. If not possible, explain why not.

   *a) A connected graph with six vertices and eight edges.

FIGURE 6

  b) A graph with no circuits, seven vertices, and five edges.

  *c) A connected graph with ten vertices and seven edges.

  d) A graph that is not connected, and has seven vertices and at least 11 edges.

  e) A graph with six vertices and 17 edges.

  f) A graph with seven vertices, and vertices with every degree from 1 to 6.

4. We consider regular graphs of degree 2.

  a) Draw a regular graph of degree 2 with five vertices. Explain why all such graphs are isomorphic.

  *b) Draw two nonisomorphic regular graphs of degree 2, each with six vertices.

  c) Draw as many nonisomorphic graphs as you can, all regular, of degree 2 with eight vertices.

  d) Draw as many nonisomorphic graphs as you can, all regular, of degree 2 with 11 vertices.

  e) Describe the possible kinds of regular graphs of degree 2 with n vertices.

5. Suppose that a graph has six vertices.

  *a) What is the maximum number of edges?

  b) If the graph is regular of degree 3, how many edges are there?

  c) If the graph is regular of degree 4, how many edges are there?

6. Suppose that a graph has n vertices.

  a) What is the maximum number of edges?

  b) If the graph is regular of degree d, how many edges are there? Are there any values of n and d for which no regular graph of degree d exists on n vertices? If so, give a relationship involving n and d and explain why these values have no such regular graphs.

  c) If it has no vertex of degree 0, what is the minimum number of edges?

  d) Justify your answers for parts a, b, and c.

7. Make a conjecture about the number of edges in a tree with n vertices.

8. For each part, prove it or give a counterexample.

  a) If a graph is connected, it has no vertex of degree 0.

  b) If a graph has no vertex of degree 0, it is connected.

9. Given a graph G, its *complement* $\overline{G}$ has the same vertices as G and exactly those edges that are not edges of G (referred to in Problem 15 of Section 6.2).

  a) Draw the complement of the first graph in Figure 1.

  b) If G has $|V|$ vertices and $|E|$ edges, how many edges does $\overline{G}$ have?

  c) If G is regular, must $\overline{G}$ be regular? Prove or give a counterexample.

  d) If G is connected, must $\overline{G}$ be connected? Prove or give a counterexample.

10. We verify that a circuit has to have at least three different vertices. Suppose that $\{v_1, \ldots, v_n\}$ is a circuit.

  a) Explain why the definition prohibits $n = 2$.

  b) Explain why the definition prohibits $n = 3$.

  c) Explain why a circuit has to have at least three different vertices in it.

11. The *complete graph* on n vertices, called K_n, has all possible edges on the n vertices. (See the first graph in Figure 7.) The *complete bipartite graph* $K_{n,k}$ has $n + k$ vertices divided into two subsets of sizes n and k, and its edges consist of all possible edges from vertices in the subset with n vertices to vertices in the subset with k vertices. (See the second graph in Figure 7.)

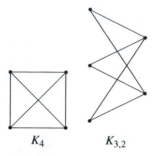

$K_4 \qquad\qquad K_{3,2}$

FIGURE 7

   a) Draw K_5, $K_{4,2}$, and $K_{3,3}$.

   ***b)** Find the degree of each vertex and the total number of edges in K_n. Prove your answers.

   c) Find the degree of each kind of vertex and the total number of edges in $K_{n,k}$. Prove your answers.

   d) For which n and k does $K_{n,k}$ have an Euler trail? Justify your answer.

   e) Define the *complete tripartite graph* $K_{n,k,j}$ and draw $K_{2,3,2}$.

   f) Find the degree of each kind of vertex and the total number of edges in $K_{n,k,j}$. Prove your answers.

   g) For which n, k, and j does $K_{n,k,j}$ have an Euler trail? Justify your answer.

12. **a)** Find a graph isomorphic to its complement. (Problem 9 defines complement.)

   b) Prove: No graph with six vertices can be isomorphic to its complement.

   c) For which number of vertices will your proof for part b hold? Justify your answer.

13. A *cubic* graph is a regular graph of degree 3.

   a) Draw cubic graphs with four, six, and eight vertices.

   b) Can cubic graphs have an odd number of vertices? Give an example or prove that there are no odd cubic graphs.

   c) For all $2n \geq 6$, describe a cubic graph with $2n$ vertices.

14. Recast the discussion before Theorem 6.1.3 as a proof of the ($\Leftarrow$) direction of part i.

15. Suppose that G is a connected graph with, say, w vertices and e edges, that each vertex has even degree, and that $w \geq 3$.

   a) Explain why we can always find some circuit, even if the initial one is not an Euler circuit. Suppose that this circuit has $c > 0$ edges.

   b) If the circuit of part a is not an Euler circuit, explain why there must be some vertex v in that circuit that has at least two edges not in the circuit.

c) Explain how you can make a circuit starting and ending at v by using, say, d edges not in the original circuit.

d) Explain how to join the circuits of a and c together to make a bigger circuit. How many edges does the new circuit have?

e) If the enlarged circuit of part d is not an Euler circuit, what can we do next?

16. Use part i of Theorem 6.1.3 to prove part ii. (Hint: How could you modify a graph with two vertices of odd degree to get a graph for part i?)

17. In a graph G, suppose that there is a trail from vertex a to vertex b and a trail from vertex a to vertex c. Prove that there is a trail from b to c in G (referred to in Problem 20 of Section 6.2).

*18. Prove that $\sim$ is an equivalence relation on V, where V is the set of vertices of a nonempty graph and $v \sim w$ iff there is a trail from v to w.

19. Suppose that G is a graph with n vertices and $n \geq 2$. Show that there are two vertices of G with the same degree.

20. Describe the minimum number of vertices in a graph with n edges, $n \geq 1$. For instance, a graph with four or five edges must have at least four vertices.

21. In a graph G, suppose that in the trail $\{v_1, \ldots, v_n\}$ the vertex v appears twice, say, $v_i = v = v_j$ and $i < j$ (referred to in Section 6.2).

a) Prove that $\{v_i, \ldots, v_j\}$ is a circuit.

b) Prove that there are at least two trails in G from v_{j-1} to v.

c) Prove that if $\{v_0, \ldots, v_n\}$ is any circuit in a graph G and $v_i \neq v_k$ in this circuit, then there are at least two trails in G from v_i to v_k.

d) Prove Theorem 6.1.2.

22. a) Four married couples meet at a party, and some of the people shake hands. No individual shakes hands with her or his spouse (or with herself or himself). One woman notices that of the seven other people, everyone shook a different number of hands. How many hands did that woman shake? Explain.

b) Repeat part a with five couples.

c) Generalize to n couples.

23. A *planar graph* is a graph that can be drawn on a plane with no edges crossing one another. (The first graphs in Figures 4 and 5 are planar.) For such graphs, the concept of a *face* is readily seen, although an exact definition is somewhat complicated. Mathematicians count the area surrounding the graph as a face so that the number of faces corresponds to related polyhedra. Thus, the first graph in Figure 4 has 6 faces, along with 8 vertices and 12 edges.

a) Make drawings to show that the following graphs are planar: K_4, $K_{2,4}$, and $K_{2,2,2}$. See Problem 11 for these graphs. (Fact: K_5 and $K_{3,3}$ are not planar. You are welcome to verify this fact.)

b) List the number of vertices, edges, and faces of the graphs in part a, and graph J of Figure 5.

c) Find a numerical relationship between the numbers of vertices, edges, and faces of a planar graph. (You will probably need to draw more graphs than those for the earlier parts.)

d) For each graph you drew for parts a and c, color its faces with the fewest number of colors so that faces sharing an edge are colored differently.

e) Color a map of South America with the fewest number of colors so that countries with a common boundary have different colors.

f) Color a map of the United States with the fewest number of colors so that states with a common boundary have different colors. (States touching only at a corner, such as Colorado and Arizona, may be the same color.)

For a proof of Euler's formula and related material on map coloring, see Chapter 1 of Beck et al. For more on planar graphs see Bogart, 143–46.

REFERENCES

BECK, A. et al. 2000. *Excursions into mathematics*, Millennium ed. New York: A. K. Peters.
BOGART, K. 1983. *Introductory combinatorics*. Boston: Pitman.
EPP, S. 1995. *Discrete mathematics with applications*, 2nd ed. Boston: PWS.
ZASLAVSKY, C. 1999. *Africa counts: number and pattern in African cultures*, 3rd ed. Chicago: Lawrence Hill Books.

6.2 TREES AND ALGORITHMS

"A fool sees not the same tree that a wise man sees."

—William Blake (1757–1827)

The simplicity of the pictures of mathematical trees can hide the insight they provide; this is a mathematical echo of Blake's quote about living trees. Trees as graphs appear in a variety of applications, some involving large trees and the need for effective algorithms, which we introduce. We start with an historic and continuing application of trees in chemistry before looking at some properties of all trees and then turning to algorithms.

Chemists have used graphs like those in Figure 1 to represent the bonds between atoms (or ions) in a chemical compound. We intentionally take a naïve view of the chemistry, ignoring the three-dimensional structure, the possibility of multiple bonds, cycles, and other considerations. Graphs modeling chemical compounds are connected because all of the atoms of a compound are bonded one to another. Because

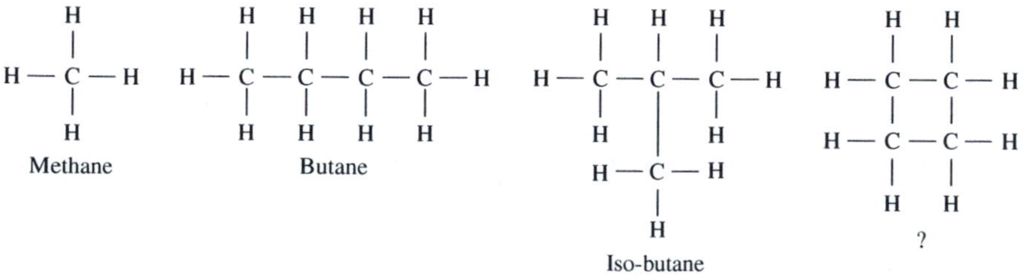

FIGURE 1

we ignore compounds with cycles, such as benzene, all the graphs are trees. An 1857 paper by the mathematician Arthur Cayley (1821–1895) gave a general chemical formula for *saturated hydrocarbons*, compounds made of hydrogen and carbon without multiple bonds and cycles. For our purposes, it suffices to assume that carbon atoms are bonded to four atoms and hydrogen atoms are bonded to one atom.

EXAMPLE 1

Butane is a saturated hydrocarbon with four carbon atoms. Determine the number of hydrogen atoms in butane and different possible trees representing different types of butane.

Solution If the four carbon atoms were separate, they could each attach to four hydrogen atoms, creating four molecules of methane (CH_4). (The first graph in Figure 1 illustrates methane.) That would give us a total of 16 hydrogen atoms. However, we require a connected graph, so the carbon atoms must be attached to one another. Each carbon–carbon bond reduces the potential sites for hydrogens to bond by two, one for each of the bonded carbon atoms. With three carbon–carbon bonds, we see that there are $16 - 3 \cdot 2 = 10$ hydrogen atoms in a molecule of butane, whose chemical compound is thus C_4H_{10}. The second and third graphs of Figure 1 illustrate the two forms of butane. These forms are not isomorphic from a graph theory point of view. They also have slightly different chemical properties and are called *isomers* in chemistry. (If we allowed cycles, the fourth graph of Figure 1 would be mathematically possible, although not saturated and chemically very artificial.) Cayley considered the number of isomers of saturated hydrocarbons in an 1875 paper. ◇

We defined trees as connected graphs without any circuits. These two conditions force all trees on a given number of vertices to have the same number of edges, the minimum number leaving the graph connected.

THEOREM 6.2.1. If a tree has at least one vertex, then $|E| = |V| - 1$.

Proof. We use strong induction on the number of vertices. When $|V| = 1$, no edges are possible, so $|E| = |V| - 1 = 0$. Now suppose that for all trees with between 1 and k vertices, we have $|E| = |V| - 1$. Consider a tree T with $k + 1$ vertices and select a particular edge vw. (See Figure 2.) Our idea is to "split" T into two smaller trees, one on each "side" of vw. Then we count the edges of these smaller trees to count the edges of T. We also split the proof into two claims.

Define $V_v = \{t \in V(T) : \text{The trail from } v \text{ to } t \text{ in } T \text{ does not include } vw\}$ and $V_w = \{t \in V(T): \text{The trail from } w \text{ to } t \text{ in } T \text{ does not include } vw\}$. We want to make graphs from these sets of vertices, so we let G_v be the graph with vertices in V_v and

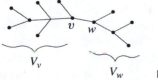

FIGURE 2

edges in E_v, the subset of $E(T)$ that include only vertices from V_v. Define G_w and E_w similarly.

Claim 1: $v \in V_v$, $w \in V_w$, and both G_v and G_w are trees. Proof: Since $\{v\}$ and $\{w\}$ are trails, we have the first two parts of this claim. From the definition of V_v, everything in it is connected with v. By Problem 17 of Section 6.1, any two vertices of V_v are connected by a trail. Further, since T had no circuits, G_v with fewer edges has no circuits. Thus, G_v is a tree. The same holds for G_w.

Claim 2: V_v and V_w are disjoint and $V_v \cup V_w = V(T)$. Proof: Let x be any vertex of T. By Theorem 6.1.2, there is a unique trail from x to v in T. Suppose first that this trail doesn't have vw in it. Then $x \in V_v$, and we can make a trail from x to w by extending the trail $\{x, \ldots, v\}$ to $\{x, \ldots, v, w\}$. Since trails are unique in T, this shows that $x \notin V_w$, so V_v and V_w are disjoint. Now suppose that the trail from x to v in T does include the edge vw. Since that edge is used just once, it must be the very last edge, going from w to v. (Otherwise, v would appear twice in the trail and there would be a circuit, contradicting Problem 21 of Section 6.1.) Thus, using the trail from x to v, but leaving out the final v, we find that $x \in V_w$. Hence, every point x is in exactly one of the sets V_v or V_w, as required.

These claims have succeeded in splitting T into two smaller trees. By the induction hypothesis, $|E_v| = |V_v| - 1$ and $|E_w| = |V_w| - 1$. Additionally, $E(T) = E_v \cup E_w \cup \{vw\}$. So, $|E(T)| = |E_v| + |E_w| + 1 = |V_v| - 1 + |V_w| - 1 + 1 = |V_v| + |V_w| - 1 = |V(T)| - 1$, as desired. By the Principle of Strong Induction, the theorem holds for all trees with at least one vertex. ∎

DEFINITION. A *leaf* of a tree is a vertex of degree 1.

THEOREM 6.2.2. Every tree with at least one edge has at least two leaves.

Proof. From our hypothesis and Theorem 6.2.1, $|E| = |V| - 1$. From the proof of Theorem 6.1.1, the sum of the degrees of all vertices is $2|E| = 2|V| - 2$. If every vertex had degree 2 or more, the sum of the degrees would be at least $2|V|$, which is bigger. Hence, at least one vertex has degree less than 2. Since a tree is connected, each vertex has degree at least 1. So, we have at least one leaf. If there were just one leaf, the sum of all the degrees would be at least $1 + 2(|V| - 1) = 2|V| - 1$, which is too big. Thus, there are at least two leaves. ∎

Algorithms

Theorem 6.2.2 assures us that a tree with edges has leaves, but it gives no hint of how to find them in an actual tree. From the drawing of a small tree, we can readily find the leaves. But if there were hundreds or thousands of vertices and no picture, how would we find one leaf, let alone two? Unsurprisingly, we'd want to use a computer for such a large problem, but computers need to follow step-by-step procedures, called *algorithms*. The strong connection between discrete mathematics and computers makes the development and analysis of algorithms an increasingly important aspect of discrete mathematics. The idea for the next algorithm is simple:

From any vertex, say, v in Figure 2, follow any trail in a tree to its "end," which will have to be a leaf.

ALGORITHM.

1. Pick any vertex (V_1).
2. Find all edges involving this vertex.
3. If there is just one edge, this vertex is a leaf and we're done.
4. Otherwise, pick one of the edges (say, $V_1 V_2$) and use it to get a new vertex (V_2).
5. Go back to step 2, using the new vertex in place of the previous one. In future steps, do not use any previous edge or vertex.

Since the tree is finite, we will eventually run out of new vertices as we build our trail $\{V_1, V_2, V_3, \ldots\}$. The last vertex we find must be a leaf—otherwise, it would have degree greater than 1 and we could have gone another step in the algorithm.

This leaf-finding algorithm is hardly impressive or particularly fast to implement, but the preceding paragraph should convince you that it will always work.

Our next algorithm tackles the harder task of finding a kind of tree needed in some applications. Recall that in a weighted graph, each edge comes with a positive number or weight. We start with a connected graph with weights on each edge and seek a tree connecting all the vertices and having the minimum total weight. Such a tree is called a *minimal spanning tree*. For instance, the weights might give the cost of a direct connection between two vertices, and we want the cheapest way to ensure that all vertices are connected. Figure 3 gives a weighted graph and two possible spanning trees. There can be a huge number of spanning trees, so finding a minimal one might not be easy. (Indeed, a complete graph on n vertices has n^{n-2} spanning trees.) Fortunately, the next, "greedy" algorithm, developed in 1956 by Joseph Kruskal (1929–), as well as others, always gives a minimal spanning tree. The term *greedy* in computer science describes an algorithm that makes the minimal choice at each step. Luckily, the greedy choice here leads to the overall minimal cost. Figure 4 shows the

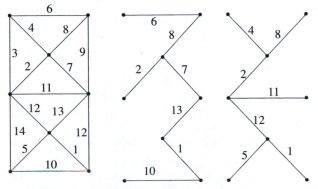

FIGURE 3. A weighted graph and two spanning trees, neither minimal.

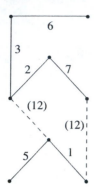

FIGURE 4.   Minimal spanning tree. (Either edge of weight 12 can be used.)

options for a minimal spanning tree resulting from Kruskal's algorithm for the graph of Figure 3. Because two edges have the same cost, there are two minimal spanning trees for this example.

KRUSKAL'S ALGORITHM.   Given a connected weighted graph,

1. Select an edge with the minimum weight in the original graph.
2. If the selected edges form a connected graph including all the vertices, they form a minimal spanning tree.
3. Otherwise, among all edges that do not make a cycle with the set of selected edges, select one with minimal weight.
4. Return to step 2.

Kruskal did more than find the algorithm; he proved that it always gave a minimal spanning tree. One part of that proof is easy: showing that the algorithm gives a tree. Suppose that there are n vertices in the original connected graph. Problem 13 assures us that $n - 1$ edges of that graph which do not form any circuits must form a tree. (And any more edges will form a circuit.) Hence, we know that we will get a tree and how many times we need to return to step 2 to get the tree. But how do we know that the tree we get is minimal?

CLAIM.   Kruskal's algorithm gives a minimal spanning tree.

Proof.   Suppose that the graph has $n + 1$ vertices and so every spanning tree has n edges. Among all the finitely many spanning trees, one has to have the minimum weight. Say that tree T has weight W as the sum of its edge weights and W is the minimum possible. Let T_K be (one of) the tree(s) from Kruskal's algorithm, and let W_K be the sum of all the weights of the edges in T_K.

For a contradiction, suppose that $W < W_K$. We'll get our contradiction by finding another tree T' with a weight smaller than W, the minimum weight. Some of the edges of T and T_K have to differ, since $W \neq W_K$. List the edges of T in increasing order of weight: $E(T) = \{e_1, e_2, \ldots, e_d, \ldots, e_n\}$ and suppose that e_d is the smallest one that differs from the edges in T_K. That is, in applying Kruskal's algorithm, we get the edges $e_1, e_2, \ldots, e_{d-1}$ and then a different edge, e'_d, whose weight is less than e_d.

Consider the graph G with the edges of T together with e'_d. Since this graph has $n + 1$ edges, it is no longer a tree and has a circuit C that includes e'_d. Further, the edges $e_1, e_2, \ldots e_{d-1}, e'_d$ are from the tree T_K, so they don't form a circuit. Hence, the circuit involves some other edge of T, say, e_T, whose weight is at least as large as e_d and thus larger than e'_d.

Form a new graph T' from T by replacing e_T with e'_d. That is, $E(T') = E(T) \cup \{e'_d\} - \{e_T\}$. (Equivalently, we get T' from G by dropping the edge e_T.) Claim: T' is a spanning tree with a smaller weight than T. First of all, replacing e_T with e'_d decreases the total weight from W. In addition, T' has n edges, the right number of edges for a tree. By Problem 14, we need only show that T' is connected. For any vertices u and v, there was a trail between them in G, since G contained the tree T. If this trail didn't involve the edge e_T, then the trail would be in T', and u and v would be connected in T'. If the trail did involve e_T, replace e_T with the rest of the circuit C in G, which is in T'. While this new set of edges may contain repetitions, Problem 17 of Section 6.1 shows that u and v are connected in T', which is all we need. We have proven our claim and achieved a contradiction. ∎

A word of caution regarding algorithms is in order: Many significant applications have no known algorithm providing an exact solution in a reasonable time. One of the most famous, the Traveling Salesman Problem, seems at first quite similar to minimal spanning trees:

The Traveling Salesman Problem   In this hypothetical situation, a salesman needs to visit each city once and then return home, covering the minimum total distance. In graph theory terms, we have a connected weighted graph and we want to find a minimum circuit going to each vertex exactly once. The substitution of a circuit for a tree may not seem a major change in the problem, but the greedy approach Kruskal took generally fails here. Moreover, in spite of prodigious efforts, no one has found an efficient algorithm that always gives the minimum. (I use the term *efficient* here to replace a sophisticated idea.) Indeed, you could win one of the million-dollar Millennium prizes by finding such an algorithm! In fairness, it should be said that most computer scientists think no such efficient algorithm exists for this problem or a host of other related problems, termed *NP complete*. For more information on the Millennium prizes, this problem, and related matters, see Devlin, 105–129.

PROBLEMS

1. **a)** Draw a tree with one vertex of degree 2 and one vertex of degree 3 and no vertices of higher degree. How many leaves does the tree have?
   b) Repeat part a with a tree having two vertices of degree 2 and three of degree 3.
   c) Repeat part a with a tree having three vertices of degree 2 and four of degree 3.

*d) Make a conjecture about the number of leaves in a tree with i vertices of degree 2 and j vertices of degree 3.

e) Repeat part a with a tree having three vertices of degree 2 and four of degree 4.

f) Make a conjecture about the number of leaves in a tree with i vertices of degree 2 and k vertices of degree 4.

g) Make a conjecture about the number of leaves in a tree with i vertices of degree 2, j vertices of degree 3, and k vertices of degree 4.

2. a) The complete graph K_6 has 15 edges, exactly three times as many edges as a tree with six vertices. In a drawing, show how to split the edges of K_6 into three subsets so that each subset gives a tree on the six vertices.

b) In a drawing, show how to split the edges of K_8 into four subsets so that each subset gives a tree on the eight vertices.

c) Generalize parts a and b to complete graphs on more vertices.

3. In a *binary tree*, every vertex has degree 1 or degree 3 and there is at least one edge. (Remark: Most texts require binary trees to be "rooted," a concept we will not consider.)

a) Draw several binary trees, each with a different numbers of vertices.

b) Make a conjecture about the number of vertices in a binary tree.

c) Make a conjecture relating the number of vertices to the number of leaves in a binary tree.

d) Prove your conjectures.

4. *a) Pentane is a saturated hydrocarbon with five carbon atoms. Determine the number of hydrogen atoms in pentane.

b) Repeat part a for octane, which has eight carbon atoms.

c) How many hydrogen atoms are in a saturated hydrocarbon with n carbon atoms? Justify your formula.

d) Draw possible configurations (isomers) for pentane.

e) Draw possible configurations for hexane, with six carbon atoms.

   (Remark: Cayley's count of isomers for saturated hydrocarbons and related material can be found at the website http://www.cs.uwaterloo.ca/journals/JIS/cayley. html.)

5. Mathematicians are free from the constraints of real chemistry. Consider an imaginary chemical D that bonds to five atoms, and define *saturated hydroDs* as compounds made from hydrogen and D atoms with no multiple bonds or cycles.

*a) If a saturated hydroD has four D atoms, how many hydrogen atoms does it have?

b) Repeat part a if there are five D atoms.

c) Find a general formula for the number of hydrogen atoms in a saturated hydroD with n D atoms. Justify your formula.

d) Suppose that the imaginary chemical E bonds with k atoms. Define a *saturated hydroE* and give a formula for the number of hydrogen atoms in a saturated hydroE having n E atoms. Justify your formula.

6. For each graph in Figure 5, apply Kruskal's algorithm to find a minimum spanning tree.

7. A *maximal spanning tree* is a tree with edges from a connected weighted graph that maximizes the total weight of the edges.

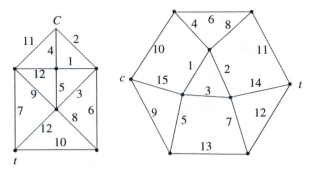

FIGURE 5

    a) Explain how to modify Kruskal's algorithm to find a maximal spanning tree.

    b) Use your algorithm to find a maximal spanning tree for each graph in Figure 5.

    c) Prove that your algorithm gives a maximal spanning tree.

8. Prim's algorithm also finds a minimal spanning tree:

    i) Start with one vertex t_1 of the original graph in the set T and the remaining vertices in S.

    ii) Among all edges connecting a vertex in T with a vertex in S, find an edge with minimal weight, say, ts, with $s \in S$.

    iii) Add both s and the edge ts to T and delete s from S.

    iv) Return to step ii until every vertex is in T.

        a) Use Prim's algorithm to find a minimal spanning tree for each graph in Figure 5. Use vertex t as the original vertex. Are these the same trees given by Kruskal's algorithm in Problem 6?

        b) Show that at each stage the vertices and edges in T form a connected graph.

        c) Show for a graph with n vertices that Prim's algorithm will give a tree after $n - 1$ repetitions of steps ii and iii.

            (Vojtech Jarnik in 1930 and Robert Prim in 1957 developed this algorithm independently of each other. The proof that this tree is minimal is somewhat involved. See Epp, 691.)

9. For each graph in Figure 5, find a reasonable candidate for a minimum circuit to solve the Traveling Salesman problem.

10. A *breadth first search* in a connected (unweighted) graph is an algorithm to find the shortest trails from a starting vertex A to all the other vertices. The search first includes all edges from A. For each edge AB already included so far, the search next includes any edge BC that doesn't form a circuit with already included edges. This continues until all vertices have a trail connecting them to A.

    a) Explain why a breadth first search gives a spanning tree.

    b) For each graph in Figure 6, do a breadth first search starting with the vertex marked A.

    c) Show for each vertex X that a breadth first search always finds a shortest trail from A to X.

11. A highly centralized organization might need to build connections between vertices so as to minimize the weights on the trails from its center to each other vertex. Call a graph

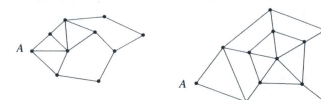

FIGURE 6

satisfying these conditions a *minimal centered graph* with center C. (A breadth first search is a special case of this type of graph, where the weight of each edge is 1.)

a) Explain why a minimal centered graph will be a spanning tree. Give an example to show that the graph need not be a minimal spanning tree as in Kruskal's algorithm.

b) For each graph in Figure 5, suppose that C is the center and find a minimal centered graph with center C.

c) Describe an algorithm to find a minimal centered graph with center C.

12. Suppose that we allowed trees to have infinitely many vertices.

　*a) Design an infinite tree with no leaves.

　b) Design an infinite tree with exactly one leaf.

*13. Suppose that a graph has n vertices, $n - 1$ edges, and no circuits. Prove that it is a tree.

14. Suppose that a connected graph with n vertices has $n - 1$ edges. Prove that it is a tree.

15. Find all trees (up to isomorphism) whose complements are also trees. Justify your answer. (See Problem 9 of Section 6.1 for the complement of a graph.)

16. It would seem reasonable that a "forest" would be a collection of trees. Graph theorists define a *forest* to be a graph with no circuits.

a) Draw four different forests, each with eight vertices, but with differing numbers of edges.

b) Explain why every forest, as defined here, is a collection of trees.

c) How are the number of vertices, edges, and trees in a forest related? Prove your answer.

17. Suppose that the vertices of a tree have only three possible degrees, 1, 2, or k, with $k \geq 3$. If there are j vertices of degree 2 and n vertices of degree k, how many vertices of degree 1 are there? Prove your answer.

18. The *diameter* of a tree is the number of edges in the longest trail in the tree.

a) Explain why for every n greater than 2, there is a tree with n vertices and diameter 2.

b) If every vertex of a tree is either 1 or 2 and the tree has n vertices, what is the diameter?

c) Find the maximum number of vertices in a binary tree with diameter 2. (See Problem 3 for binary trees.)

d) Repeat part c with diameter 4.

e) Repeat part c with diameter 6.

f) Repeat part c with diameter 8.

g) Describe a way to determine the maximum number of vertices in a binary tree with diameter $2n$.

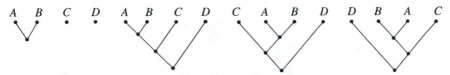

FIGURE 7. The start of an evolutionary tree and three equivalent representations of the completed tree.

19. An algorithm to build an evolutionary tree starts with n current species as the leaves at the top. Then we add a vertex for an hypothesized "common ancestor" of the two most closely related species, together with edges from this ancestor to the two current species. Now we consider these two current species and their common ancestor as a leaf to reduce the problem to $n - 1$ leaves, and we continue the process until we get to a common ancestor of all the species. Figure 7 illustrates this process, as well as several equivalent representations of a final tree. (We avoid the very difficult biological problem of how to determine how species are related to one another.) The number of possible nonequivalent evolutionary trees grows rapidly with the number of current species, which we illustrate in this problem.

a) Find the number of nonequivalent evolutionary trees with three leaves, A, B, and C.

b) Find the number of nonequivalent evolutionary trees with four leaves, A, B, C, and D. (Hint: There are two basic shapes to the evolutionary trees with four leaves; see Figures 7 and 8.) Count how many ways there are to letter each basic shape.

FIGURE 8. The second basic shape of an evolutionary tree with four leaves.

c) Find the number of nonequivalent evolutionary trees with five leaves, A, B, C, D, and E. (Even with only three basic shapes, as illustrated in Figure 9, there are too many ways to write out all of them in a reasonable time.)

d) Form a conjecture for the number of nonequivalent evolutionary trees on n species. (Hint: Factor your answers for the previous parts.)

20. Critique the following argument:

Claim: A tree has one more vertex than it has edges. "Proof": To start the induction proof, note that a tree with one vertex has no edges, so it satisfies the claim. Now suppose that every tree with n vertices has $n - 1$ edges. To show the induction step with $n + 1$, vertices, we can add a vertex v to a tree with n vertices and one edge from v to any of the

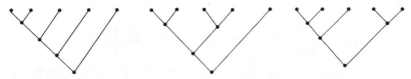

FIGURE 9. The three basic shapes of evolutionary tress with five leaves.

already existing vertices, say, *w*, to make a new graph. Clearly, this new graph has $n + 1$ vertices and *n* edges, satisfying the desired relationship. So, we need only ensure that the new graph is still a tree. By Problem 17 of Section 6.1, the new graph is connected. And Problem 14 here shows that it is a tree. By PMI, the claim holds. ▲

REFERENCES

DEVLIN, K. 2002. *The Millennium problems: The seven greatest unsolved mathematical puzzles of our time*. New York: Basic Books.

EPP, S. 1995. *Discrete mathematics with applications*, 2nd ed. Boston: PWS.

6.3   COUNTING PRINCIPLES I

"Never underestimate results that count something."

—John Fraleigh (1930–)

Before people could write, they already asked the question, "How many?" They noted recurring patterns, enabling them to answer this basic question without tediously counting the objects one at a time. Addition and multiplication, for instance, were well developed over 4000 years ago, by the time of the oldest mathematics texts we have. We'll elaborate on these obvious techniques to find other counting patterns. Before reading the next two sections entirely, you are encouraged to try to solve on your own the upcoming four problems, which motivate general ideas that we'll develop. We'll use these general approaches to solve the following four problems previewed here, and others in this section and the next section, Counting Principles II:

EXAMPLE 1

When I was younger, license plates had numbers up to six digits long. Now Minnesota license plates have three letters followed by a number up to three digits long. How many different license plates are there with each method? Does the Minnesota method provide enough different license plates for California? ◊

EXAMPLE 2

Each fall, sports writers are asked weekly to order their choices for the top 20 Division I football teams among the 119 teams. How many different lists could there be if the sports writers choose the teams at random? After the first two weeks, suppose that 30 of the teams are undefeated. How many different lists of the top 20 teams from those 30 would there be then? ◊

EXAMPLE 1

From Section 6.4. Three women and two men in an office have to share a menial daily task. The boss claims that each day two of them are assigned randomly to this task. Suppose that the women feel that they are assigned more frequently than they would be if the assignments were truly random. How can they determine what would be fair? ◊

EXAMPLE 2

From Section 6.4. How many ways can 18 children choose from nine peaches, six pears, and three plums so that each child gets one fruit?   ◇

Frequently, we have a succession of choices, each with a certain number of options, and the number of options for any one choice is not influenced by the other choices made. For instance, suppose I have three clean shirts—red, green, and blue—and I have two acceptable pairs of pants—jeans and slacks. If I'm willing to wear any shirt with any pair of pants, how many ways can I choose to appear? Figure 1 gives a tree explicitly listing all of the ways. The number of ways, six, also comes from the product $3 \cdot 2$ of the number of options for each choice. The multiplication principle merely states this result in general and is much handier than drawing a tree each time.

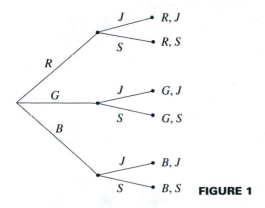

FIGURE 1

THEOREM 6.3.1. (Multiplication Principle). If there are n ordered choices and the ith choice has k_i options, regardless of the other choices, then there are $k_1 \cdot k_2 \cdot \ldots \cdot k_n = \prod_{i=1}^{n} k_i$ ways to make all of the choices.

Proof. See Problem 15.   ∎

EXAMPLE 1

When the author was younger, license plates had numbers up to six digits long. Now Minnesota license plates have three letters followed by a number up to three digits long. How many different license plates are there with each method? Does the Minnesota method provide enough different license plates for California?

Solution   There are 10 digits and 26 letters. By the multiplication principle, there could have been $10^6 = 1,000,000$ old-style license plates with six or fewer digits. (A number with fewer digits can be thought of as starting with blanks, which act like zeros. We don't use initial zeros. You might reasonably argue that there is no license plate numbered 0, in which case there would be 999, 999 old-style license plates.) Now, there are $26^3 10^3 = 17,576,000$ possible Minnesota license plates with three letters followed by a number up to three digits long.

(As before, omitting the number 0 gives $26^3 999 = 17,558,424$ possible plates. We likewise will ignore various other issues, such as the difficulty of distinguishing the letter O from the number 0 or the need to avoid certain letter combinations.) But even $17,576,000$, while ample for Minnesota, is not enough for California, which had over $35,000,000$ people in 2004. $\Diamond$

In Example 1, we say the selection is done *with replacement*, since the same number or letter can be used more than once. Example 2 illustrates selection without replacement, but still uses the multiplication principle.

EXAMPLE 2

Each fall, sports writers are asked weekly to order their choices for the top 20 Division I football teams among the 119 teams. How many different lists could there be if the sports writers choose the teams at random? After the first two weeks, suppose that 30 of the teams are undefeated. How many different lists of the top 20 teams from those 30 would there be then?

Solution There are clearly 119 choices for the first-place team. Once that team is chosen, 118 candidates remain for the second place. While which options remain depends on the previous choice, the number of options is the same regardless of which team is chosen first. Thus, the multiplication principle is still valid. The same reasoning applies at each choice, with each new choice decreasing the remaining options by one. Hence, there are $119 \cdot 118 \cdot 117 \cdot \ldots \cdot 100 = \prod_{i=1}^{20} (119 + 1 - i) = 5.97 \times 10^{40}$ ways to choose the top 20 teams at random. If we limit ourselves to the 30 teams who won their first two games, we get $30 \cdot 29 \cdot \ldots \cdot 11 = \prod_{i=1}^{20} (30 - 1 + i) = 7.31 \times 10^{25}$, still a huge number, but trillions of times smaller. $\Diamond$

The counting pattern in Example 2 relates closely to the idea of a permutation in Chapter 3. Recall that a permutation is a one-to-one function from a set to itself, and from Theorem 3.2.5, there are $n!$ permutations if the set has n elements. Suppose in Example 2 that there were only 20 teams, numbered 1 to 20. Ordering them gives one of the 20! permutations of these numbers. The orderings in Example 2 have the same flavor, but simply don't order all of the teams. This sort of situation appears so often that it has its own name and notation, or actually two common notations:

DEFINITION. The number of ways to list k objects in order from a set of n objects is the number of *permutations of n objects k at a time*, denoted $_nP_k$, or $P(n, k)$.

THEOREM 6.3.2. If $k, n \in \mathbb{N}$ and $k \leq n$, then $_nP_k = \prod_{i=1}^{k} (n + 1 - i) = \frac{n!}{(n-k)!}$.
An argument similar to the solution in Example 2 is at least as convincing as a formal induction proof and explains the idea behind the proof better.

Proof. There are n options for the first position. Once the first position is chosen, there are $n - 1$ options left for the second position. In general, each time another position is chosen, there is one fewer option for the next choice. The total number of choices will be the product of k decreasing terms, starting with n times $(n - 1)$. The hardest part of this expression is to determine the last term. We can rewrite the first term as $n = n + 1 - 1$ and the second term as $n - 1 = n + 1 - 2$. In general, the ith term is $n + 1 - i$ and the last term is the kth one, which is $n + 1 - k$. Thus, $_nP_k = n(n-1)(n-2)\ldots(n+1-k) = \prod_{i=1}^{k}(n+1-i)$, giving the first equality. Since $\frac{n!}{(n-k)!} = \frac{n(n-1)\ldots(n+1-k)(n-k)\ldots1}{(n-k)(n-k-1)\ldots1}$, by canceling the common terms from $(n-k)$ down to 1, we see that this formula equals the other one in the theorem. ∎

Since there is a multiplication principle, you could reasonably expect an "addition principle" describing when to add various options together. We worked out a more general situation in Example 2 of Section 2.1 when we considered the size of the union of two sets. If set A has $|A|$ elements and B has $|B|$ elements, we saw that $|A \cup B| = |A| + |B| - |A \cap B|$. This equation reduces to addition when $|A \cap B| = 0$, that is, when the sets have nothing in common. In short, we have already proven an addition principle:

ADDITION PRINCIPLE. If A and B are finite disjoint sets, then $|A \cup B| = |A| + |B|$.

EXAMPLE 3

License plates alternate letters and numbers to make it easier for people to remember them.

a) Modify Example 1 to allow the three letters to either precede or follow the numbers on license plates. How many possible license plates are there in this situation?
b) Suppose that in an effort to get more possible license plates, we allow the six characters to be in three pairs, with each pair being both letters or both numbers, but not all six characters being all letters or all numbers. How many possible license plates could there be?

Solution

a) The option of the letters preceding the numbers is disjoint from the option where the letters follow the numbers. So we can count these options separately and add. From Example 1, each option gives 17, 576, 000 possibilities, or a total of 35, 152, 000 possible plates.
b) See Problem 8. (Answer: 157, 372, 800.) ◇

Students sometimes have difficulty distinguishing when to add and when to multiply options. Common sense, familiar examples, and practice provide better guides than any set of rules would. So this text will defer to the exercises, rather than give such a guideline.

Counting problems often use the language of probability, which we discuss very briefly. In our naïve approach, we assume that every option is equally likely.

DEFINITION. If every element of a finite set T is equally likely, then the *probability* of a subset S is $P(S) = \frac{|S|}{|T|}$. (Recall that $|X|$ is the number of elements in the set X.)

In everyday language, this definition gives the probability of success (being in S) as the number of ways of succeeding divided by the total number of possibilities.

EXAMPLE 4

a) In four rolls of a fair die, what is the probability that all the outcomes are different?

b) What is the probability that at least one six appears in four rolls of a fair die?

Solution

a) By definition of a fair die, each face is equally likely to appear on each roll. Thus, the denominator of the probability is $6^4 = 1296$. The numerator is $_6P_4 = 360$. So, the probability is $\frac{360}{1296}$, or approximately 0.2778.

b) The denominator is still $6^4 = 1296$. For the numerator, it is easier first to count the ways that *no* sixes appear, which is $5^4 = 625$. So, there are $6^4 - 5^4 = 671$ ways for at least one six to appear, giving a probability of $\frac{671}{1296}$, or approximately 0.5177. Around 1650, the gambler Chevalier de Méré bet people he'd be able to roll a six within four rolls and used the results to make money, until people caught on. Problem 13 considers the altered bet he then made.   ◇

PROBLEMS

1. A combination for a lock consists of three numbers between 1 and 39, with the requirement that adjacent numbers be different. How many possible combinations are there?

2. *a) How many five-digit numbers are there where the digits in the odd places are odd and the digits in the even places are even?

   b) Repeat part a, but prohibit the repetition of digits.

3. a) How many arrangements of the eight letters A to H have the letter A before the letter B?

   ***b)** How many arrangements of the eight letters A to H have the letters BAG together in that order?

   c) How many arrangements of the eight letters A to H have the letters $FACE$ together in some order?

   d) How many arrangements of the eight letters A to H have the letters D and H separate from each other?

4. At an ice cream shop, one can buy a single-dip cone, a double-dip cone or a triple-dip cone, choosing from any of the 12 flavors of ice cream.

   a) How many different orders of a cone are there with no repetition of flavors in a cone?

   b) How many different orders of a cone are there, allowing repetition of flavors in a cone?

5. Suppose in making license plates with up to six digits, we allow one or more 0s at the start. For instance, 000234, 0234, and 234 would give three different license plates. How many license plates could there be now?

6. A certain calculator will display (up to) 12 digits for a number, along with a decimal point and, if needed, a minus sign. Ignore alternate notation, such as scientific notation, for parts a and b.

   ***a)** If the first digit is not 0, how many different numbers can the calculator display?

b) If we allow a number to start with one or more 0s after an initial decimal place, how many different numbers can the calculator display?

c) In scientific notation, a specific calculator represents a number in the form $x.yEEz$, where x is a positive or negative nonzero digit, y is a series of up to 11 digits, and z is an integer between -999 and 999. How many different numbers can this calculator display in scientific notation?

7. A student has four math books, four novels, and four other books to arrange on a shelf.

   a) How many ways can the books be arranged if the math books are first, followed by the novels?

   b) How many ways can the books be arranged if books of the same type are together?

8. Finish Example 3.

9. There are five (distinguishable) girls and four (distinguishable) boys. How many ways can they sit in a row of nine chairs under the following conditions?

   a) (no extra conditions.)

   b) All the girls are on the left.

   c) All the boys sit together, and all the girls sit together.

   d) All the boys sit together.

   e) No two girls sit together.

   f) No two boys sit together.

10. There are six bills in a wallet, each worth a different amount: \$1, \$2, \$5, \$10, \$20, and \$50.

    a) What is the probability that one bill chosen randomly will be enough to pay for an \$18 book?

    b) If the first bill isn't enough to pay for the book, what is the probability that drawing a second one will suffice?

    c) What is the probability that two bills chosen randomly will be enough for the book?

11. In a lottery, people choose four digits in order. Find the probability of winning each of the following prizes:

    a) Grand prize: The four digits you selected match the computer-chosen ones in order.

    b) Second prize: The four digits you selected match the computer-chosen ones, but in a different order. (Assume that you chose four different digits.)

    c) Third prize: Three of the four digits you selected match exactly three of the ones the computer chose, in any order. (Assume that you chose four different digits.)

    d) Repeat part b, assuming that you chose three different digits.

    e) Repeat part c, assuming that you chose three different digits.

12. *****a)** How many distinguishable ways are there to arrange 12 knights sitting around the Round Table?

    b) If the knights sit at random, what is the probability that knights A and B are together?

    c) If the knights sit at random, what is the probability that knights A, B, C, and D are evenly spaced around the table?

    d) If there are six knights and six ladies sitting randomly around the table, what is the probability that no two knights are together?

*13. It is said that, after people refused to bet Chevalier de Méré that he wouldn't roll a six in four rolls (as in Example 4), he started betting on rolling a pair of sixes in 24 rolls of a pair of dice. Unfortunately, he had done his math wrong and so lost money overall on these bets. What really was his probability of winning? What is the smallest number of rolls of a pair of fair dice to get the probability of rolling a pair of sixes to be over 0.5? (The Chevalier asked the noted mathematician Blaise Pascal (1623–1662) for help, which collaboration was a key spur to the development of the theory of probability.)

14. The Sieve of Eratosthenes is a (very inefficient, but ancient) way to find prime numbers. Write down the numbers from 2 on, as far as desired. Then cross out all multiples of 2 greater than 2 (roughly half of the numbers). Next, cross out multiples of 3 greater than 3 (roughly one-third of the remaining numbers). Continue in the same way: The first uncrossed number is the next prime, and all of its multiples are not, so they should be crossed out.

    a) After the multiples of 2 and 3 are crossed out, approximately what percentage of numbers remains?

    b) After the multiples of 2, 3, and 5 are crossed out, approximately what percentage of numbers remains?

    c) After the multiples of 2, 3, 5, and 7 are crossed out, approximately what percentage of numbers remains?

    d) Describe the general process to determine the approximate percentage of numbers remaining after multiples of the first n primes are crossed out.

15. Prove the Multiplication Principle (Theorem 6.3.1).

16. Give an induction proof of Theorem 6.3.2.

17. Prove for all $n \in \mathbb{N}$ with $n \geq 2$ that $_{n+1}P_3 = n^3 - n$.

18. a) Prove for all $n \in \mathbb{N}$ with $n \geq 3$ that $_{n+1}P_3 - {_nP_3} = 3{_nP_2}$.

    b) Find a formula for $_{n+1}P_4 - {_nP_4}$ similar to part a and prove your formula.

    c) Generalize part b to find and prove a formula for $_{n+1}P_k - {_nP_k}$, for $k < n$.

19. a) All of the students in a certain calculus class are also taking at least one of the following classes: English 111 (abbreviated E), Computer Science 120 (C), and History 133 (H). Suppose that 12 students are in E; 15 are in C; 10 are in H; 5 are in both E and C; 2 are in both E and H; 4 are in both C and H; and 1 student is taking E, C, and H, in addition to calculus. How many students are in the calculus class?

    b) Draw a Venn diagram for three generic sets A, B, and C and use it to explain why the following formula is correct for all finite sets A, B, and C:

$$|A \cup B \cup C| = |A| + |B| + |C| - (|A \cap B| + |A \cap C| + |B \cap C|) + |A \cap B \cap C|.$$

    c) For finite sets A, B, C, and D, devise a formula for $|A \cup B \cup C \cup D|$ in terms of the sizes of these four sets and their various intersections. Justify your formula.

    d) Describe the general pattern for the formula for the size of the union n finite sets in terms of their sizes and the sizes of their intersections. (This general pattern is called *inclusion–exclusion*.)

20. (The Birthday Problem.) This problem develops a way to find the probability that at least two people from a set of n unrelated people share a birthday. Assume that all dates are equally likely and that there are 365 days in a year.

a) What is the probability that two people chosen at random have different birthdays?

b) What is the probability that three people chosen at random have different birthdays? (Hint: We assume that the first two birthdays have to differ, before worrying about the third birthday.)

c) Describe how to generalize part b to find the probability that n people chosen at random have different birthdays.

d) State how to convert the probability in part c to the probability that at least two of the people share a birthday.

e) Find the smallest number of people for which the probability in part d is greater than 0.5.

21. This problem seeks a formula for the number of ways one can pair n players in a round of a tournament, where we make as many pairs as possible. Let $P(n)$ be the number of ways if there are n players.

a) Explain why $P(1) = 1$ and $P(2) = 1$.

b) Find $P(3)$ and $P(4)$.

c) Find $P(5)$ and $P(6)$. Describe how these values relate to one of the values in part b.

d) Find a general formula or pattern for $P(n)$ and prove it.

6.4 COUNTING PRINCIPLES II

"Mathematics is being lazy.
Mathematics is letting the principles do the work for you so that you do not have to do the work for yourself."

—George Pólya (1887–1985)

The problems in Section 6.3 specified the order in which things were chosen. We turn to situations where the choices don't involve order.

EXAMPLE 1

Three women and two men in an office have to share a menial daily task. The boss claims that each day two of them are assigned randomly to this task. Suppose that the women feel that they are assigned more frequently than they would be if the assignments were truly random. How can they determine what would be fair?

Solution Because the number of employees is small, we explicitly list in Table 1 all of the ten possibilities, providing one way to solve this problem. Of course, we desire a general approach. First of all, how many total ways are there to choose a pair of people from a set of five? If the choices were in order, we'd have $5 \cdot 4 = 20$ ways. Any particular pair $\{x, y\}$ would

TABLE 1. The 10 Possible Pairs from $\{W_1, W_1, W_3, M_1, M_2\}$, Where W_i are Women and M_j are Men

$W_1 W_2$	$W_1 W_3$	$W_2 W_3$	$W_1 M_1$	$W_1 M_2$	$W_2 M_1$	$W_2 M_2$	$W_3 M_1$	$W_3 M_2$	$M_1 M_2$

have 2 ways to be chosen in those 20: first x and then y or y followed by x. To compensate for the double counting, we divide by 2 to get $\frac{20}{2} = 10$ total ways. We need to compare the ways involving women with the total number of ways. Thus 10 is the denominator of the probability and we next determine the numerator.

The wording of the example is intentionally a bit vague about what we count as a "success" to go in the numerator. This vagueness allows us to look at two options. (I try to avoid such vagueness in the problem sets.) Let's look first at how many ways two women can be chosen from three women. We find the number of ways by reasoning as we did to choose two from five. Thus, there are $(3 \cdot 2)/2 = 3$ ways to choose two women from among three. In comparison, there is only one way to choose two men from two men: Choose them both. (The formula also works: $(2 \cdot 1)/2 = 1$.) Hence, in a fair, random process, we would expect two women to be chosen $\frac{3}{10} = 30\%$ of the time, three times as often as two men would be chosen, which is $\frac{1}{10} = 10\%$. Such a difference might at first seem rather surprising, since the number of men and women differ only by one.

Let's modify the question: How likely is it for at least one women to be chosen? This outcome is the complement of two men being chosen, so the probability must be $\frac{9}{10} = 90\%$. Again, this value might seem surprisingly high. Still, the probability that at least one man is chosen is $\frac{7}{10} = 70\%$, which might seem equally surprising. We need mathematics to sort out apparent discrepancies from actual ones. ◇

When the order doesn't matter, choosing k objects from n options amounts to picking a subset of size k from a set of size n. As with permutations, this situation occurs sufficiently often to have acquired its own names (*combinations* or *binomial coefficients*) and notations.

DEFINITION. The number of subsets of size k from a set of size n is called the number of *combinations of n items k at a time* and is denoted $\binom{n}{k}$, or $_nC_k$, or $C(n, k)$. (Mathematicians read $\binom{n}{k}$ as "n choose k.")

THEOREM 6.4.1. If $k, n \in \mathbb{Z}$ and $0 \le k \le n$, then $\binom{n}{k} = \frac{n!}{k!(n-k)!}$.

Proof. First recall that $0! = 1$, so the fraction $\frac{n!}{k!(n-k)!}$ is always defined. Let's do the case $k = 0$ separately. The only subset of size $k = 0$ is $\emptyset$, so $\binom{n}{0} = 1$. Further, $\frac{n!}{0!n!} = 1$, showing that this case holds.

For $k > 0$, the desired formula can be rewritten $\frac{n!}{k!(n-k)!} = \frac{n!}{(n-k)!} \cdot \frac{1}{k!} = {_nP_k} \cdot \frac{1}{k!}$, which suggests a strategy generalizing the reasoning of Example 1. We know that there are $_nP_k$ ways to choose with order k elements from a set of n elements. Those k elements form a subset of size k. Any other order of those same k elements will yield the same subset, and by Theorem 6.3.2, there are $k!$ such orderings, or permutations. Thus, we counted each subset $k!$ times when we used $_nP_k$. Hence, $\binom{n}{k} = {_nP_k} \cdot \frac{1}{k!} = \frac{n!}{k!(n-k)!}$, as claimed. ■

The name *binomial coefficient* for combinations comes from the following theorem:

THEOREM 6.4.2. (The Binomial Theorem). The coefficient of $x^k y^{n-k}$ in $(x + y)^n$ is $\binom{n}{k}$. That is, $(x + y)^n = \sum_{k=0}^{n} \binom{n}{k} x^k y^{n-k}$.

A formal induction proof would likely hide the entire idea of this theorem. The proof following the discussion of the coefficients links the coefficients with the counting of subsets. First let's multiply out several cases of $(x + y)^n$, as in Table 2.

TABLE 2

$$(x+y)^1 = 1x + 1y$$
$$(x+y)^2 = 1x^2 + 2xy + 1y^2$$
$$(x+y)^3 = 1x^3 + 3x^2 y + 3xy^2 + 1y^3$$
$$(x+y)^4 = 1x^4 + 4x^3 y + 6x^2 y^2 + 4xy^3 + 1y^4$$

The coefficients in Table 2 form the rows of numbers in Table 3 now called Pascal's Triangle, even though the pattern was known in several cultures before Pascal independently found the pattern. The first row of Pascal's Triangle corresponds to the boring initial case $(x + y)^0 = 1$. Problems 15 and 16 ask you to prove various properties about the values $\binom{n}{k}$. The most familiar of these properties tells us that the sum of two adjacent coefficients is the coefficient below and between them in Pascal's Triangle. For instance, $\binom{4}{1} + \binom{4}{2} = 4 + 6 = 10 = \binom{5}{2}$.

Proof of Theorem 6.4.2. We need to show that the coefficient of $x^k y^{n-k}$ in $(x + y)^k$ is $\binom{n}{k}$, the number of subsets of size k in a set of size n. To keep track of individual terms, let's label the n factors of $(x + y)^n$, giving us $(x_1 + y_1)$ $(x_2 + y_2) \ldots (x_n + y_n)$.

TABLE 3 Pascal's Triangle

			1		
		1		1	
	1		2		1
1		3		3		1

1   4   6   4   1

1   5   10   10   5   1

First consider the case $n = 2$: $(x_1 + y_1)(x_2 + y_2) = x_1x_2 + x_1y_2 + y_1x_2 + y_1y_2$. Each term has one letter with the subscript 1 and one with the subscript 2. If we multiply this expression by $(x_3 + y_3)$, each new term will have one letter with each subscript and we get every possible arrangement.

In general, to get the term $x^k y^{n-k}$ in the product $(x + y)^n$, we need to select k of the x_i, forcing the remaining $n - k$ terms to be y_j. (For instance, we could select the terms in boldface in $(\mathbf{x_1} + y_1)(x_2 + \mathbf{y_2})(x_3 + \mathbf{y_3})(\mathbf{x_4} + y_4)(x_5 + \mathbf{y_5})$ to get $\mathbf{x_1 y_2 y_3 x_4 y_5}$, which is one term contributing to $x^2 y^3$ in $(x + y)^5$.) The subset of subscripts of the x_i we chose has k elements out of the n possible choices. (The preceding particular instance has $\{1, 4\}$ for the subset of subscripts for x.) Since any equally sized subset gives us a term $x^k y^{n-k}$, we see there are indeed $\binom{n}{k}$ ways to get $x^k y^{n-k}$.   ∎

While permutations and combinations are the basic counting tools, variations abound. Example 2 exhibits just one possibility, where we shift from two types of objects (men and women in Example 1) to three types (peaches, pears, and plums). We'll consider another variation in Example 3 and leave still others for the problems.

EXAMPLE 2

How many ways can 18 children choose from nine peaches, six pears, and three plums so that each child gets one fruit?

Solution   Even though the fruits are not ordered, let's first pretend that they are and then sort out the overcounting. Such reasoning is a useful approach to a number of problems. There are 18 fruits and 18 children, so if both sets were ordered, there would be 18! permutations of the fruit, each giving a distribution of the fruit. But if two children with peaches switch their fruits, these two permutations would give the same distribution. In fact, any permutation of the 9 peaches, together with any permutation of the 6 pears and any permutation of the 3 plums gives the same distribution of fruit. Thus, the permutations way overcount the different distributions. Indeed, there are $9! \cdot 6! \cdot 3!$ rearrangements of the fruit so that the peaches still go to the same children, the pears to the same children, and the plums to the same children. Hence, there are $\frac{18!}{9!6!3!} = 4,084,080$ distributions.

◊

The idea behind Example 2 is very similar to combinations. Combinatorists call the corresponding numbers *multinomial coefficients*. (Problem 17 relates these numbers with multinomials, justifying the name.)

DEFINITION.   The value $\binom{n}{k_1, k_2, \ldots, k_j} = \frac{n!}{k_1! k_2! \ldots k_j!}$ is a *multinomial coefficient*, provided that each k_i is a nonnegative integer and $\sum_{i=1}^{j} k_i = n$.

EXAMPLE 3

In a study on variable reinforcement, a psychologist has 13 identical treats to give out to five (distinguishable) rats.

a) How many ways can the treats be distributed among the five rats?
b) How many ways can the treats be distributed if each rat gets at least one treat?
c) How many ways can the treats be distributed if each rat gets at least two treats?

Solution a) Imagine a machine that starts at the first cage and at each step either gives a treat or moves to the next cage. There will be $13 + 4$ steps: one for each of the 13 treats (T) and the 4 moves (M). A particular distribution is determined by where the 4 moves come. For instance, *TTMMTTTMTTMTTTTT* would correspond to the first rat getting 2 treats, the second one none, the third rat 3 treats, the fourth one 2, and the last rat the remaining 6 treats. In total, there are $\binom{17}{4} = 2380$ possible ways of placing 4 Ms in 17 places.

  b, c) See Problem 10.   ◇

Historical Remarks

General notation for explicit formulas such as $_nP_k$ and $\binom{n}{k}$ is more recent than their earliest use and even proofs of them. A text from India written before the year 600 gives a problem whose solution is $\binom{16}{4} = 1820$. The text gives this number, which seems too big to find by listing the possibilities. By 1200, what we call Pascal's Triangle appears in the work of the Muslim mathematician Ibn Mun'im. The Jewish mathematician Levi Ben Gerson (1288–1344) gives induction proofs of the formulas for $_nP_k$ and $\binom{n}{k}$, although the formula and proofs are written in words, rather than symbols.

PROBLEMS

1. A Brownie troop has seven girls. Each weekly meeting starts with a short ceremony involving three girls. A school year has 36 weeks. Can the troop's leader choose the girls for the ceremony so that no threesome ever does the ceremony twice during the school year?

2. A contestant chooses at random two balls from an urn containing five (identical) red balls, four (identical) blue balls, and three (identical) green balls. (Probability problems frequently feature urns, which are large vases, because contestants have to reach into them to choose a ball, but their necks are narrow enough that the contestants can't see which color a ball is.)

    a) How many ways can two balls be chosen?
    *b) What is the probability that both balls are red?
    c) What is the probability that the balls are red or green?
    d) What is the probability that both balls are red or both balls are green?
    e) What is the probability that one ball is red and one is green?

3. The first of two urns contains five red balls and five blue balls. The second urn contains six red balls and three blue ones. For parts a, b, and c, a contestant draws three balls from each urn.

   a) What is the probability that all six balls are red?

   b) What is the probability that the balls from the first urn are red and the balls from the second urn are blue?

   c) What is the probability that one ball from the first urn is red and two from the second urn are red?

   d) The contestant draws one ball from the first urn and, without looking at it, drops it into the second urn. After the balls in the second urn are mixed, the contestant draws one ball from the second urn. What is the probability that it is red?

4. ***a)** Students can pick any 6 questions from 11 on a test. How many ways are there of choosing the questions?

   b) The 11 questions are split into 7 problems and 4 proofs. If the students must do four problems and two proofs, how many ways are there of choosing the questions?

   c) In part b, suppose that two of the problems are linked: They must both be done or neither done. Now how many ways are there of choosing the questions?

5. In an experiment, 30 lab mice are split randomly into two groups of 15, the treatment group getting a new drug and the control group getting the standard drug.

   a) How many ways are there of splitting the mice into groups?

   b) Suppose that 10 of the mice have an unknown gene reducing the effectiveness of the new drug. What is the probability that all 10 are in the treatment group?

   c) In part b, what is the probability that five of the mice with that gene are in the treatment group?

6. **a)** How many different ways can a sample of 10 lightbulbs be chosen at random from a package of 100 bulbs?

   b) Suppose that 20 of the lightbulbs in the package are defective. What is the probability that none of the defective bulbs are chosen in the sample of 10 bulbs?

   c) What is the probability that 1 of the defective bulbs is chosen in the sample of 10 bulbs?

   d) What is the probability that 2 of the defective bulbs are chosen in the sample of 10 bulbs?

   e) What is the probability that 10 of the defective bulbs are chosen in the sample of 10 bulbs?

7. Find the number of distinct arrangements of the letters in each of these words:

   *a) algebra   **b)** parallel   **c)** Mississippi

8. Two teams, each with five runners, race. If the placements at the end were random, what is the probability that one team would have the top three finishers?

9. ***a)** In Figure 1, how many ways can you trace a trail from point A to B, staying on the grid and always moving right or up?

   b) Repeat part a for the points A and C.

   c) Repeat part a for the points A and D.

   d) Generalize the earlier parts: If there are j horizontal steps and k vertical steps from X to Y on a grid, how many different trails are there? Justify your answer.

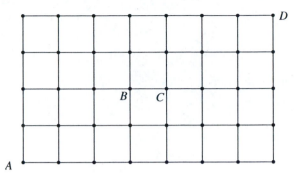

FIGURE 1

   e) How many trails from A to D go through B?

   f) How many trails from A to D never go through B?

   g) How many trails from A to D go through B or C?

10. Finish Example 3.

11. A standard deck of cards has 52 cards split into four *suits* (♣, ♢, ♡, ♠), each with 13 *denominations*: 2, 3, . . . 10, J, Q, K, A, listed in order.

      A poker hand consists of five cards. A hand is termed *one pair* iff it has two cards of one denomination and one each of three other denominations. A hand is termed *two pairs* iff it has two cards of one denomination, two of another denomination, and one of a third denomination. A hand is termed *three of a kind* iff it has three cards of one denomination and one each of two other denominations. A hand is termed a *full house* iff it has three cards of one denomination and two of another denomination. A hand is termed *four of a kind* iff it has four cards of one denomination and one of another denomination. A hand is termed *a straight* iff it has one card of each of five denominations that are consecutive—for instance, 8, 9, 10, J, Q. A hand is termed *a flush* iff it has five cards from the same suit. A hand is termed *a straight flush* iff it is a flush and a straight.

   *a) Find the number of possible poker hands.

   *b) Find the probability of getting one pair.

   c) Find the probability of getting two pair.

   d) Find the probability of getting three of a kind.

   e) Find the probability of getting a full house.

   f) Find the probability of getting four of a kind.

   g) Find the probability of getting a straight flush.

   h) Find the probability of getting a flush that is not a straight flush.

   i) Find the probability of getting a straight that is not a straight flush.

   j) A *bust* is a hand that fits none of the preceding categories. Find the probability of getting a bust.

12. A bridge hand consists of 13 cards from a standard deck. (See Problem 11.) One consideration in bridge is the number of cards in the four suits, called the *distribution* of the hand. For instance a $4 - 3 - 3 - 3$ distribution has 4 cards from one suit and 3 from each of the others.

FIGURE 2

a) Find the number of possible bridge hands.

b) Find the probability of a $4 - 3 - 3 - 3$ distribution.

c) Find the probability of a $4 - 4 - 3 - 2$ distribution.

d) Find the probability of a $4 - 4 - 4 - 1$ distribution.

e) Find the probability of a $5 - 3 - 3 - 2$ distribution.

f) Find the probability of a $10 - 2 - 1 - 0$ distribution.

13. The arrangements of atoms in a chemical molecule can alter the molecular properties. How many geometrically distinguishable ways can five hydrogen atoms (H), two chlorine atoms (Cl), and one hydroxyl group (OH) be put on the eight sites along the chain of three carbon atoms (C) in Figure 2? Explain.

14. A mathematical biology class has 6 math majors and 12 biology majors enrolled. How many ways can they be split into six project groups, each with two biology majors and one math major?

15. Use the definition of $\binom{n}{k}$ or theorems from this section to prove the following properties:

   *a) For $0 \le k \le n$, $\binom{n}{k} = \binom{n}{n-k}$.

   b) For $0 \le k \le n$, $\binom{n}{k} + \binom{n}{k+1} = \binom{n+1}{k+1}$.

   c) $\sum_{k=0}^{n} \binom{n}{k} = 2^n$.

16. For a given row of Pascal's Triangle, the sum of the terms in the even places equals the sum of the terms in the odd places. For instance, in the row 1 6 15 20 15 6 1, the even places give $6 + 20 + 6 = 32$ and the odd places give $1 + 15 + 15 + 1 = 32$. Prove that this property holds for any row.

17. **a)** Use multinomial coefficients to generalize the Binomial Theorem (Theorem 6.4.2) for the case $(x + y + z)^n$.

   Remark: In "binomial," the "bi" stands for "two," so a binomial $(x + y)$ has two terms. A multinomial has multiple terms.

   b) Generalize the proof of Theorem 6.4.2 to fit part a.

   c) State the Multinomial Theorem for the case $(w + x + y + z)^n$.

DEFINITIONS FROM CHAPTER 6

Section 6.1

- A *graph G* is a finite nonempty set V of *vertices* and a set E of *edges*, where an edge is a subset of V with two elements. We abbreviate the edge $\{a, b\}$ by ab. If needed, we identify the sets V and E of a particular graph G by writing $V(G)$ and $E(G)$.

- For $v \in V$ in a graph, the *degree* of a vertex v is the number of $x \in V$ such that vx is an edge. A graph is *regular* iff all vertices have the same degree.

- A *trail* from vertex v_1 to vertex v_n is a set of vertices $\{v_i : 1 \le i \le n\}$ such that if $1 \le i < i+1 \le n$, then $v_i v_{i+1}$ is an edge of the graph, and no edge is used twice. A *circuit* is a trail from v_1 to v_n, with $v_1 = v_n$ and $n \ge 2$. A graph is *connected* iff for each pair of distinct vertices there is a trail from one to the other. A *tree* is a connected graph with no circuits.

- An *Euler trail* of a graph is a trail that uses every edge exactly once. If the starting and ending vertices are the same, the Euler trail is an *Euler circuit*.

- Two graphs G and H are *isomorphic* iff there is a bijection $f : V(G) \to V(H)$ such that $v_i v_j$ is an edge of G iff $f(v_i) f(v_j)$ is an edge of H. The function is called an *isomorphism*.

Section 6.2

- A *leaf* of a tree is a vertex of degree 1.

Section 6.3

- The number of ways to list k objects in order from a set of n objects is the number of *permutations of n objects k at a time*, denoted $_nP_k$ or $P(n, k)$.

- If every element of a finite set T is equally likely, the *probability* of a subset S is $P(S) = \frac{|S|}{|T|}$. (Recall that $|X|$ is the number of elements in the set X.)

Section 6.4

- The number of subsets of size k from a set of size n is called the number of *combinations of n items k at a time* and denoted $\binom{n}{k}$, or $_nC_k$, or $C(n, k)$. (Mathematicians read $\binom{n}{k}$ as "n choose k.")

- The value $\binom{n}{k_1,k_2,\dots,k_j} = \frac{n!}{k_1!k_2!\dots k_j!}$ is a *multinomial coefficient*, provided that each k_i is a nonnegative integer and $\sum_{i=1}^{j} k_i = n$.

CHAPTER 7

INTRODUCTION TO ABSTRACT ALGEBRA

"... [A]lgebra is the intellectual instrument which has been created for rendering clear the quantitative aspects of the world."
—Alfred Whitehead (1861–1947)

HISTORICALLY, **ALGEBRA HAS** its roots in solving equations. Indeed, "algebra" comes from the Arabic word "al jabr," meaning "the restoring," part of the process of manipulating an equation to solve it. Like so many other areas of mathematics, algebra changed radically in the nineteenth century. Algebraists realized that the processes they used in the context of numbers worked similarly in many settings. For instance, the process of solving equations in structures now called groups and fields mimics the process for solving simple equations in high school. They began investigating the properties underlying different systems and proved theorems based only on the properties, not the particular systems. This shift led to an emphasis on structure and abstraction characteristic of much of modern mathematics, which followed the lead of algebra. This chapter will give a taste of the abstract structural approach, as well as examples of different kinds of systems.

7.1 OPERATIONS AND PROPERTIES

Operations

The operations of addition and multiplication on positive integers and fractions have been used in many cultures over thousands of years. Basically, an operation simply turns two inputs, such as 3 and 7, into an output, say, their sum or product. Formally, this process means that an operation is a type of function. While subtraction of positive numbers is also ancient, it doesn't qualify as an operation, since the difference of two positive numbers might not be positive. As recently as 400 years ago, some mathematicians found the notion of negative numbers perplexing. Since then the advantages of making subtraction an operation overcame any hesitation people had about negative numbers. However, the inclusion of negative numbers has required grade school teachers to explain the meaning of these numbers. An even harder job for teachers is to explain why a negative times a negative should mean anything, let alone be positive. Whatever explanation you may have heard in grade school, the mathematical reason that a negative times a negative is positive is structural. We'll give the proof in Section 7.3.

DEFINITION. A *(binary) operation* on a set A is a function from $A \times A$ to A.

EXAMPLE 1

For any set S, union, intersection, and difference are operations on $\mathcal{P}(S)$. $\Diamond$

EXAMPLE 2

On any subset of real numbers, the maximum and minimum of two numbers are operations. The average of two numbers is an operation on $\mathbb{Q}$ or $\mathbb{R}$, but not on $\mathbb{Z}$, since the average of two integers, while a number, is not always an integer. Division is not an operation on any of these three sets because we can't divide by 0. $\Diamond$

EXAMPLE 3

The greatest common divisor (gcd) and the least common multiple (lcm) are operations on $\mathbb{N}$. $\Diamond$

EXAMPLE 4

From Theorem 4.2.2, addition, subtraction, and multiplication (mod n) are operations on $\mathbb{Z}_n = \{0, 1, \ldots, n - 1\}$, the set of remainders after dividing by n. $\Diamond$

EXAMPLE 5

Let S_n be the set of all permutations on the set $\{1, 2, \ldots n\}$. By Problem 11 b of Section 3.2, composition is an operation on S_n. $\Diamond$

EXAMPLE 6

Addition and multiplication of $n \times n$ matrices and addition of vectors are operations in linear algebra. Scalar multiplication and inner products are not operations, because they involve elements from different sets, the vector space and the set of scalars. $\Diamond$

EXAMPLE 7

We can make new operations from others. On the Cartesian product $\mathbb{Z}_2 \times \mathbb{Z}_2 = \{(0, 0), (1, 0), (0, 1), (1, 1)\}$ we can use "component-wise" addition, similar to the addition of vectors. That is, we add in $\mathbb{Z}_2 \times \mathbb{Z}_2$ by adding the values (mod 2) in the first position separately from adding the values (mod 2) in the second position. Table 1 gives the operation $\oplus$ explicitly. To determine from the table a particular "sum," say, $x \oplus y$, we find the x entry in the leftmost

TABLE 1. Addition in $\mathbb{Z}_2 \times \mathbb{Z}_2$.

$\oplus$	(0,0)	(1,0)	(0,1)	(1,1)
(0,0)	(0,0)	(1,0)	(0,1)	(1,1)
(1,0)	(1,0)	(0,0)	(1,1)	(0,1)
(0,1)	(0,1)	(1,1)	(0,0)	(1,0)
(1,1)	(1,1)	(0,1)	(1,0)	(0,0)

column and the y entry in the top row. Then we find the entry inside the table for the x row and y column, which is the value $x \oplus y$. For instance, $(0, 1) \oplus (1, 1) = (1, 0)$. Such tables are called Cayley tables after their inventor Arthur Cayley (1821–1895). In general, Cartesian products with component-wise operations occur frequently in algebra and are called *direct products*. ◊

Properties of Operations

The structural approach focuses on properties of operations rather than particular operations. Of course, the properties we choose to study are ones that appear in a number of particular operations, especially familiar ones. Most of the properties in the following definition may already be familiar to you:

DEFINITION. Let $*$ be an operation on a set S. Then,

> $*$ is *associative* iff for all $a, b, c \in S$, $(a * b) * c = a * (b * c)$;
>
> $*$ is *commutative* iff for all $a, b \in S$, $a * b = b * a$;
>
> $*$ has an *identity* iff there is some $e \in S$ such that for all $a \in S$, $a * e = a = e * a$.
>
> If $*$ has an identity e, then $*$ has *inverses* iff for all $a \in S$, there is some $b \in S$ such that $a * b = e = b * a$.
>
> $*$ is *idempotent* iff for all $a \in S$, $a * a = a \in S$.
>
> $*$ is *closed* on the subset T of S iff for all $a, b \in T$, $a * b \in T$.
>
> $*$ *distributes* over an operation $+$ iff for all $a, b, c \in S$, $a * (b + c) = (a * b) + (a * c)$ and $(b + c) * a = (b * a) + (c * a)$.

Table 2 shows which of the first five of these properties hold for some familiar operations. In this table $\mathbb{Q}^+$ refers to the set of positive rational numbers and $\mathbb{Z}_2 \times \mathbb{Z}_2$ and $\oplus$ come from Example 7. The letter Y stands for "yes," and the letter N stands for "no."

The first four sets and operations in the table have several key properties in common. They are all examples of *groups*, a fundamental algebraic structure we'll investigate in Section 7.2. Because of their common properties, we can solve certain types of equations in all of them. The ninth entry in the table lacks inverses and so fails to be a group. That failure is connected to our inability to solve an equation like $3 \cdot x = 7$ in $\mathbb{Z}$.

TABLE 2.   Properties of Selected Operations

Operation	Assoc	Com	Iden	Inv	Idemp
$+$ on $\mathbb{Z}$ (1)	Y	Y	Y	Y	N
$\cdot$ on $\mathbb{Q}^+$ (2)	Y	Y	Y	Y	N
$\oplus$ on $\mathbb{Z}_2 \times \mathbb{Z}_2$ (3)	Y	Y	Y	Y	N
$\circ$ on S_n (4)	Y	N	Y	Y	N
$\cup$ on $\mathcal{P}(A)$ (5)	Y	Y	Y	N	Y
min on $\mathbb{R}$ (6)	Y	Y	N	N	Y
gcd on $\mathbb{N}$ (7)	Y	Y	N	N	Y
$-$ on $\mathbb{R}$ (8)	N	N	N	N	N
$\cdot$ on $\mathbb{Z}$ (9)	Y	Y	Y	N	N

The fifth, sixth, and seventh operations in Table 2 have a different cluster of properties in common and are examples of another type of algebraic structure, a *semi-lattice*, discussed in Section 7.4.

Often, we use pairs of operations: addition and multiplication or union and intersection. We'll discuss algebraic structures involving two operations in Sections 7.3 and 7.4. Distributivity plays a crucial role in the interactions of operations.

The property of closure plays a different role than the other properties of an operation, since it applies to subsets. We noted earlier that subtraction is not an operation in $\mathbb{N}$, although it is in $\mathbb{Z}$. That is, subtraction is not closed in the subset $\mathbb{N}$. As another example, addition and multiplication in $\mathbb{Z}$ are not closed in $\{1, 2, 3, 4\}$, although they are closed in $\mathbb{N}$ as well as in the set of even integers. This difference suggests that some subsets are algebraically more fruitful to study than others. We'll investigate such "subsystems" in the later sections.

PROBLEMS

*1. For each statement that follows, decide whether it is true or (at least sometimes) false. Explain your answer or cite the part of the text supporting it.

   a) Subtraction is an operation on $\mathbb{Q}$.

   b) Division is an operation on $\mathbb{N}$.

   c) Division is an operation on $\mathbb{Q}^+$.

   d) It is possible for an operation to be idempotent and have an identity.

   e) Addition distributes over multiplication in $\mathbb{Z}$.

   f) Multiplication distributes over addition in $\mathbb{R}$.

2. For the operations given by the table in each part, determine which of the properties of commutative, identity, inverses, and idempotent the operation has. For each property that fails for an operation, give a counterexample. Verify that each of these operations is not associative by giving a counterexample.

*a)

*	a	b	c
a	a	c	b
b	c	b	a
c	b	a	c

b)

⊙	a	b	c	d	e
a	e	d	b	c	a
b	c	e	d	a	b
c	d	a	e	b	c
d	b	c	a	e	d
e	a	b	c	d	e

c)

⊙	a	b	c	d	e
a	a	d	a	e	c
b	c	b	b	a	d
c	a	b	c	d	e
d	b	c	d	d	b
e	d	a	e	c	e

3. High school algebra often makes use of the property of *cancellation*. An operation $*$ on a set X satisfies *cancellation* iff for all $a, b, c \in X$, if $a * b = a * c$, then $b = c$, and similarly, if $b * a = c * a$, then $b = c$. Which of the operations discussed in this section and in Problem 2 satisfies cancellation? (You may omit composition on S_n, which we will consider in Section 7.2.)

4. An operation $*$ on a set X satisfies *solvability* iff for all $a, b \in X$, there is $x \in X$ (a *solution*) such that $a * x = b$, and similarly, there is $y \in X$ such that $y * a = b$. Which of the operations discussed in this section and in Problem 2 satisfy solvability? (You may omit composition on S_n, which we will consider in Section 7.2.)

5. *a) Does intersection distribute over set subtraction on $\mathcal{P}(\mathbb{N})$? Explain.

   b) Repeat part a with union and set subtraction. Explain.

6. a) Consider a variety of examples to determine whether the greatest common divisor (gcd) distributes over the least common multiple (lcm) on $\mathbb{N}$. That is, does $\gcd(a, \operatorname{lcm}(b, c)) = \operatorname{lcm}(\gcd(a, b), \gcd(a, c))$?

   b) Repeat part a to determine whether lcm distributes over gcd.

7. a) On $\mathbb{R}$, show that the operation of the maximum of two numbers distributes over the minimum. That is, for all $a, b, c \in \mathbb{R}$, show that $\max(a, \min(b, c)) = \min(\max(a, b), \max(a, c))$. (Hint: Examine the possible orderings of a, b, and c.)

   b) Show that the minimum distributes over the maximum.

   c) Does addition distribute over the maximum? Justify your answer.

   *d) Does the maximum distribute over addition? Justify your answer.

   e) Does multiplication distribute over the maximum? Justify your answer.

   f) Does the maximum distribute over multiplication? Justify your answer.

8. High school algebra talks about cancellation for multiplication, even though it isn't strictly correct: $0 \cdot 7 = 0 \cdot 3$, but $7 \neq 3$. However, if we exclude 0, then cancellation does work for multiplication in $\mathbb{R}$ or any of its subsets. We say that a set X with additive identity 0 has *multiplicative cancellation* iff for all $a, b, c \in X$, if $a \neq 0$ and $a \cdot b = a \cdot c$, then $b = c$, and similarly, if $b \cdot a = c \cdot a$, then $b = c$.

   a) Verify that $\mathbb{Z}_5 = \{0, 1, 2, 3, 4\}$ with addition and multiplication (mod 5) has multiplicative cancellation by writing out the Cayley table for its multiplication.

   b) Does $\mathbb{Z}_4$ have multiplicative cancellation?

   c) Determine for various values of n whether $\mathbb{Z}_n$ has multiplicative cancellation. Make a conjecture about which $\mathbb{Z}_n$ in general have multiplicative cancellation.

9. The Cayley table of an operation $*$ on a set X forms a *Latin square* iff every element of X appears exactly once in each row and column of the table (referred to in Section 7.2).

a) Which of the tables in Problem 2 are Latin squares?

b) Fill in the following Cayley table so that it forms a Latin square (this problem is similar to a Sudoku puzzle, but there are no subsquares):

*	e	a	b	c	d	f
e	e	a	b	c	d	f
a	a		e	d		
b	b		a		c	
c	c			e	b	
d	d	c				
f	f				e	

10. a) Prove that if the Cayley table of an operation $*$ is a Latin square, then $*$ has cancellation. (See Problems 3 and 9.)

   b) Prove that if the Cayley table of an operation $*$ is a Latin square, then $*$ has solvability. (See Problem 4.)

   c) Find an example of an operation on an infinite set that has cancellation, but its Cayley table is not a Latin square.

   d) Prove that if an operation has cancellation and solvability, then its Cayley table is a Latin square.

11. Which of the following subsets of $\mathbb{Q}$ have closure for the operation of the average of two numbers?

   *a) $\mathbb{Z}$

   b) $\mathbb{Q}^+$

   c) E, the even integers

   d) $\mathbb{Q} \cap [-3, 2]$

   *e) $\{13\}$

   f) $\mathbb{Q}_E = \{\frac{p}{q} : q \text{ is even}, \frac{p}{q} \text{ in lowest terms}\}$

12. Suppose that a set X has n elements. Justify your answer to each part.

   a) Determine the number of operations on X.

   b) Suppose that $e \in X$. How many operations on X have e as an identity?

   c) How many commutative operations on X are there?

   d) How many commutative operations on X have e as an identity?

   e) How many idempotent commutative operations does X have?

13. Critique the questionable "proofs" that follow. Point out reasoning errors and unclear presentation. If the argument is a proof, say so. (If the claim is false, there must be an error in the argument. However, it is not enough to say that the claim is false. You need to find an error in the reasoning.)

   *a) Claim: Subtraction has an identity and inverses on $\mathbb{Z}$. "Proof": Clearly, 0 is the identity for subtraction because for all $x \in \mathbb{Z}$, we have $x - 0 = x$. Similarly, each element x is its own inverse, since $x - x = 0$. ▲

   b) Claim: If an operation $*$ on a set X has an identity and inverses, then for all $a, b \in X$, $a * b * a^{-1} = b$. "Proof": Suppose that $*$ has an identity e and inverses in X and $a, b \in X$. Now $a * a^{-1} = e$, the identity, and $b * e = b$. So $a * b * a^{-1} = b * e = b$. ▲

   c) Claim: If an operation $*$ on a set X has an identity, the identity is unique. "Proof": Suppose that for a contradiction, e and e' are both identities and $e \neq e'$. Then $e = e * e' = e'$. Contradiction. ▲

d) If an operation $*$ on X is commutative, then it is associative. "Proof": Consider $a * (b * a) = a * (a * b)$ by commutativity for the operation inside the parentheses. By commuting outside the parentheses, we get $(a * b) * a$, which gives us associativity: $a * (b * a) = (a * b) * a$. The general case is similar. ▲

7.2 GROUPS

"Wherever groups disclosed themselves, or could be introduced, simplicity crystallized out of comparative chaos."

—Eric Bell

The concept of a group emerged slowly over nearly a hundred years, starting in the late 1700s as mathematicians explored a variety of examples. Although the most familiar groups are the sets $\mathbb{Z}$, $\mathbb{Q}$, $\mathbb{R}$, and $\mathbb{C}$ under the operation of addition, those examples didn't lead mathematicians to define the abstract concept of a group. Rather, work with sets of permutations and sets of symmetries of geometric shapes made clear the need for a general term. We will consider such groups later in the section once we build some familiarity with groups.

DEFINITION. A *group* is a set G with an operation $*$ such that $*$ is associative and has an identity and inverses. More formally,

i) for all $a, b, c \in G$, $a * (b * c) = (a * b) * c$,

ii) there is an *identity* $e \in G$ such that for all $a \in G$, $a * e = a = e * a$, and

iii) for all $a \in G$, there is an *inverse* $a^\# \in G$ such that $a * a^\# = e = a^\# * a$.

If in addition, for all $a, b \in G$, $a * b = b * a$, the group is *commutative* (or *Abelian*).

REMARK. Once we have shown inverses are unique in Corollary 7.2.2, we will write a^{-1} for the inverse of a.

EXAMPLE 1

Each of the sets $\mathbb{Z}$, $\mathbb{Q}$, $\mathbb{R}$, and $\mathbb{C}$ forms a commutative group with the operation of addition. Even though multiplication is an operation with identity 1 for each of these sets, they don't form groups with multiplication, because 0 doesn't have a multiplicative inverse. We denote the nonzero elements of a set X by X^*. Then $\mathbb{Q}^*$, $\mathbb{R}^*$, and $\mathbb{C}^*$ each forms a commutative group with the operation of multiplication. Also, $\mathbb{Q}^+$ and $\mathbb{R}^+$ form commutative groups with the operation of multiplication, where the $^+$ indicates just the positive elements of the set. ◇

EXAMPLE 2

Consider the vector space $\mathbb{R}^3 = \{(x_1, x_2, x_3) : x_i \in \mathbb{R}\}$ with addition defined "component-wise": $(x_1, x_2, x_3) + (y_1, y_2, y_3) = (x_1 + y_1, x_2 + y_2, x_3 + y_3)$. This operation is associative (and commutative), basically because each component is. The identity is the vector $(0, 0, 0)$, and the inverse of the vector (x_1, x_2, x_3) is the vector $(-x_1, -x_2, -x_3)$. The definition of a vector space implies that the vectors always form a commutative group under vector addition. ◇

Associativity eliminates the need for cumbersome parentheses when we operate on a string of elements. In ordinary arithmetic, we simply write $12 + 34 + 56 + 78 = 180$ without saying which pair of numbers is added first. We know that $(12 + 34) + (56 + 78)$ is the same as $((12 + 34) + 56) + 78$, and similarly for other arrangements. Ordinary arithmetic is also commutative, which means that we can add the numbers in any order. Some groups are not commutative, such as the ones in Example 4 and Problem 5 in this section. So, don't assume that you can change the order of elements.

Properties of Groups

Historically, algebra developed from the search to solve equations. Let's start with two of the easiest equations to solve: $x + 5 = 3$ and $5y = 3$. Your training has you "solve for the unknown" by shifting everything else to the "other side." That is, $x = 3 - 5 = -2$ and $y = 3/5 = 0.6$. Of course, those answers are correct, but why does the process always work? And are there any other answers? High school students may not wonder about such questions, satisfied with a successful method to get an answer. However, part of the power of mathematics lies in the guarantee of the method and answer. Our first proof gives that guarantee, based on the algebraic properties of a group.

THEOREM 7.2.1. For any group G, and any $a, b \in G$, there are unique $x, y \in G$ such that $x * a = b$ and $a * y = b$.

Proof. To show existence, we must first find a candidate for a solution to each equation and then show that our candidate satisfies the equation. Our previous experience in algebra suggests that we try shifting the a to the other side of the equation to get $b * a^\#$, where $a^\#$ is an inverse of a. Now we check whether this choice works. Pick $x = b * a^\#$. Then $x * a = (b * a^\#) * a$. Now we use the group properties to simplify this expression. Associativity shifts parentheses, so $(b * a^\#) * a = b * (a^\# * a)$. From the definition of an inverse, we reduce $b * (a^\# * a)$ to $b * e$. Finally, the definition of an identity e yields $b * e = b$. Hence, $x * a = b$, as desired.

It is tempting for the second part of the theorem to pick $y = b * a^\#$ as well, but the properties of a group don't lead us anywhere with this choice. Problem 7 suggests a more successful alternative.

We start the proof of uniqueness for $x * a = b$ by assuming two solutions, say, $v * a = b$ and $w * a = b$. We use the properties of a group to convert $v * a = w * a$ to $v = w$. Since $*$ is an operation (a function) and $v * a$ and $w * a$ are the same element, $(v * a) * a^\# = (w * a) * a^\#$. Apply associativity to both sides to obtain $v * (a * a^\#) = w * (a * a^\#)$. The definitions of inverses and identity simplify this equation to $v * e = w * e$ and finally to $v = w$. So, $x * a = b$ has a unique solution. Problem 7 completes the proof. ∎

We often call the uniqueness part of Theorem 7.2.1 *cancellation*, defined next. (From the preceding $v * a = w * a$, we cancel the a to deduce $v = w$.)

DEFINITION.   An operation $*$ on a set X has *cancellation* iff for all $a, b, c \in X$, if $a * b = a * c$, then $b = c$, and if $b * a = c * a$, then $b = c$.

If the group is not commutative, we can't cancel an element if it appears on different "sides." For instance, from $v * a = a * w$, we can't always conclude that $v = w$. See Example 4, for instance. The uniqueness in Theorem 7.2.1 carries over to our first corollary, allowing us to use a^{-1} for the inverse of a henceforth. The second corollary basically restates the theorem in a different way. In terms of Problem 9 of Section 7.1, Corollary 7.2.3 tells us that the Cayley table of a group is a Latin square.

COROLLARY 7.2.2.   In a group, the identity is unique and the inverse of an element is unique.

Proof.   See Problem 8.   ■

COROLLARY 7.2.3.   Every element of a group appears exactly once in each row and column of its Cayley table.

Proof.   See Problem 8.   ■

EXAMPLE 3

The set $\mathbb{Z}_n = \{0, 1, 2, \ldots, n - 1\}$ is a group under addition (mod n). Theorem 4.2.2 shows that this addition is an operation. Not surprisingly, 0 is the identity, just as it is in $\mathbb{Z}$. The inverse of x is $n - x$ for $x \neq 0$, since $x + (n - x) = n \equiv 0 \pmod n$. The inverse of 0 is 0. We will skip the uninformative proof of associativity. Table 1 shows a typical Cayley table for a $\mathbb{Z}_n$, in this case $\mathbb{Z}_6$.

TABLE 1.   Addition in $\mathbb{Z}_6$

+	0	1	2	3	4	5
0	0	1	2	3	4	5
1	1	2	3	4	5	0
2	2	3	4	5	0	1
3	3	4	5	0	1	2
4	4	5	0	1	2	3
5	5	0	1	2	3	4

$\Diamond$

We turn briefly to groups of permutations. In Section 3.1, we defined a permutation to be a bijection from a set to itself. After the main problem set, we discuss symmetry groups, which are related to permutation groups and an important type of group that arises in geometry and other areas.

THEOREM 7.2.4. The set S_X of all permutations on a set X forms a group under composition.

Proof. See Problem 11 of Section 3.2. ∎

EXAMPLE 4

Recall that S_n is the set of all permutations of the set $\{1, 2, \ldots, n\}$. From Theorem 3.2.5, we know that S_n has $n!$ elements. Table 2 gives the Cayley table for S_3 with the operation of composition. We abbreviate the names of the permutations in a way that reminds us of what they do. For instance, we write (12) for the permutation that switches 1 and 2 and leaves 3 fixed. Similarly, (123) represents the permutation taking 1 to 2 and 2 to 3 and 3 back to 1. For ease, we denote the permutation leaving every number fixed by (1), which is the identity.

TABLE 2. The Cayley table of S_3

∘	(1)	(123)	(132)	(12)	(13)	(23)
(1)	(1)	(123)	(132)	(12)	(13)	(23)
(123)	(123)	(132)	(1)	(13)	(23)	(12)
(132)	(132)	(1)	(123)	(23)	(12)	(13)
(12)	(12)	(23)	(13)	(1)	(132)	(123)
(13)	(13)	(12)	(23)	(123)	(1)	(132)
(23)	(23)	(13)	(12)	(132)	(123)	(1)

The operation of composition often appears "backwards" to students, since $f \circ g$ is the function they obtain by performing g first and then f. That is, $f \circ g(x) = f(g(x))$. For instance, we can determine $(123) \circ (12)$ by seeing what happens to each number. (12) takes 1 to 2, and (123) takes 2 to 3; so we know the composition takes 1 to 3. Similarly, (12) takes 2 to 1, which goes back to 2 under (123). Finally, (12) leaves 3 fixed, which then goes to 1 by (123). So we see that $(123) \circ (12) = (13)$.

The elements (1), (12), (13), and (23) are their own inverses, and the other two elements, (123), and (132), are inverses of each other. This group is not commutative, illustrated by the compositions $(12) \circ (13) = (132)$ and $(13) \circ (12) = (123)$. While cancellation holds here as in any group, a modification does not: $(123) \circ (12) = (12) \circ (132)$, but we can't cancel the (12), since $(123) \neq (132)$. To cancel a term a, it has to appear on the same side of the operation: $a * b = a * c$, not on opposite sides, as here.

It would be tedious to do every entry in the table, let alone check for associativity. We depend instead on Theorem 7.2.4, which assures us that S_n is a group. Then we need to check only enough entries to force the rest by using Corollary 7.2.3. ◊

Subgroups

Two familiar groups, $\mathbb{Z}$ and $\mathbb{Q}$, use the same operation of addition as $\mathbb{R}$ and are subsets of $\mathbb{R}$. We call $\mathbb{Z}$ and $\mathbb{Q}$ *subgroups of* $\mathbb{R}$.

DEFINITION. A nonempty subset H of a group G is a *subgroup* of G iff H is a group using the same operation as G.

EXAMPLE 5

Even though $\mathbb{Q}^+$, the positive rationals under multiplication, is a group and a subset of $\mathbb{R}$, it is not a subgroup of $\mathbb{R}$, because the operation for $\mathbb{R}$ is addition, not multiplication. The set of odd integers is not a subgroup under addition, because the addition isn't closed on this set. That is, the sum of two odd integers is not odd. While $\mathbb{N}$ has closure for addition, it fails to include the identity or inverses, so it is not a subgroup of $\mathbb{R}$. $\Diamond$

EXAMPLE 6

In $\mathbb{Z}_6$, the sets $\{0, 2, 4\}$, $\{0, 3\}$, and $\{0\}$ are subgroups, as is the whole set. (See Table 1.) $\Diamond$

The first step in showing that a subset is a subgroup is to verify closure; that is, for a and b in the subset, we must show that $a * b$ is in the subset as well. (Because $*$ is an operation, $a * b$ is certainly in the whole set.) We don't need to verify associativity, since it already works for all elements of the group, including those of the subset. We do need to verify that the identity is in the subset as well as the inverse of each element of the subset.

Cosets and Lagrange's Theorem

We split the integers quite naturally into the evens and the odds. The even integers form a subgroup $2\mathbb{Z} = \{2n : n \in \mathbb{Z}\}$, but the odds do not. Indeed, the mathematical definition of an odd integer, something of the form $1 + 2n$, suggests that we describe odd integers in terms of even integers. A natural name for the set of odd integers $1 + 2\mathbb{Z}$ matches the language of cosets, to follow.

DEFINITION. For $a \in G$ and H a subgroup of G, the *coset* $a * H = \{a * h : h \in H\}$.

REMARK. A course in algebra studies the structure of groups much more deeply and so distinguishes "left" cosets from "right" cosets, a distinction we don't need. Our cosets are technically left cosets. Problem 19 defines right cosets.

EXAMPLE 7

The subgroup $\{0, 3\} = T$ of $\mathbb{Z}_6$ has cosets $1 + T = \{1 + 0, 1 + 3\} = \{1, 4\} = 4 + T$ and $2 + T = \{2, 5\} = 5 + T$, in addition to $T = 0 + T = 3 + T$. $\Diamond$

EXAMPLE 8

Consider the subgroup $J = \{(1), (123), (132)\}$ of S_3, the group in Example 4. From Table 2, J has two cosets: $(12) \circ J = \{(12) \circ (1), (12) \circ (123), (12) \circ (132)\} = \{(12), (23), (13)\}$ and $J = (1) \circ J = (123) \circ J = (132) \circ J$. Note that $(12) \circ J = (13) \circ J = (23) \circ J$. $\Diamond$

As Examples 7 and 8 illustrate, a subgroup nicely organizes the elements of a finite group into equally sized cosets, which we show in Lemma 7.2.5. This lemma leads directly to Theorem 7.2.6 (Lagrange's Theorem), which has been described as one of the two most important elementary theorems about groups.

LEMMA 7.2.5. For any group G, elements a and b, and subgroup H,

   i) $a \in a * H$,

   ii) either $a * H \cap b * H = \emptyset$ or $a * H = b * H$, and

   iii) if G is finite, then $|a * H| = |b * H|$.

Proof.

   i) $a = a * e \in a * H$ because $e \in H$.

   ii) If $a * H \cap b * H = \emptyset$, we're done; so suppose that $c \in a * H \cap b * H$. That is, there are $h_1, h_2 \in H$ such that $a * h_1 = c = b * h_2$. Then $a = b * h_2 * h_1^{-1} \in b * H$. But we need more; we need $a * H \subseteq b * H$ and the other inclusion $b * H \subseteq a * H$. For the first inclusion, let $a * h$ be a general element of $a * H$. Then $a * h = (b * h_2 * h_1^{-1}) * h = b * (h_2 * h_1^{-1} * h) \in b * H$. See Problem 17 for the other inclusion.

   iii) Define $f : a * H \rightarrow b * H$ by $f(a * h) = b * h$. Problem 17 asks you to prove that f is a one-to-one onto function. ∎

THEOREM 7.2.6. (Lagrange's Theorem). If G is a finite group and H is a subgroup of G, then $|H|$ divides $|G|$.

Proof. Let G be a finite group and H any subgroup of G. First we will show that the cosets $g * H$ form a partition of G. (See Section 1.6 for the definition of a partition.) From part i of Lemma 7.2.5, $g \in g * H$, so each $g * H$ is nonempty and $G = \bigcup_{g \in G} g * H$, fulfilling two of the three conditions of a partition. Part ii of the Lemma is just the remaining condition for a partition. Now we use the third part of the Lemma to compare the sizes of H and G. Since G is finite, there are a finite number of cosets, say, $g_1 * H, g_2 * H, \ldots, g_k * H$. Also, these cosets are disjoint, so $|G| = |g_1 * H| + |g_2 * H| + \ldots + |g_k * H|$. Part iii of Lemma 7.2.5 tells us that all the terms on the right side are the same size as $|e * H| = |H|$. That is, $|G| = k|H|$, fulfilling the claim of Lagrange's Theorem. ∎

Finite groups, such as $\mathbb{Z}_n$ and S_n, have fascinated algebraists for decades. The goal of classifying all possible finite groups appears incredibly difficult, in spite of prodigious efforts. Even the classification of the "building blocks" of finite groups, the so-called finite simple groups, required over 10,000 journal pages to print the proofs. Dozens of mathematicians labored for decades to complete this classification, announced in 1981. (See Gallian, 413–424.) Unfortunately, it remains unclear how to determine all the possible ways to build groups from these building blocks.

PROBLEMS

*1. For each statement that follows, decide whether it is true or (at least sometimes) false. Explain your answer or cite the part of the text supporting it.

   a) All groups are commutative.

b) If a subset of a group is closed under the group operation, then the subset forms a subgroup.

c) The cosets of a subgroup form a partition of the group.

d) A subgroup of a group is a coset of the group.

2. For each set and operation, determine whether they form a group. For those that are not groups, give a counterexample for each property of a group that fails.

   *a) $\mathbb{N}$ with the operation of the maximum of two numbers.

   *b) $\{\frac{p}{q} : p$ and q are odd integers$\}$ with the operation of multiplication.

   c) $\mathbb{R}$ with the operation of the average of two numbers.

   d) $\mathbb{N} \cup \{0\}$ with the operation $a \ominus b = |a - b|$.

   e) For which of the previous sets and operations can we solve the equation $a * x = b$?

   f) For which of the previous sets and operations does cancellation hold?

*3. Prove that the set $\{3^x : x \in \mathbb{Z}\}$ is a group with the operation of multiplication.

4. Define the operation $\oplus$ on $\mathbb{Z}$ by $x \oplus y = x + y - 1$. Prove that $\mathbb{Z}$ is a group with this operation.

5. (Linear Algebra) Let $SL(2, \mathbb{R}) = \left\{ \begin{bmatrix} a & b \\ c & d \end{bmatrix} : a, b, c, d \in \mathbb{R} \text{ and } ad - bc = 1 \right\}$, the set of matrices with determinant 1. Show that $SL(2, \mathbb{R})$ is a group with the operation of matrix multiplication. Show that $SL(2, \mathbb{R})$ is not commutative. (This group is called the *special linear group*.)

6. Suppose that G is a group with operation $*$ and K is a group with operation $\circledast$. Define $\odot$ on $G \times K$ by $(g, k) \odot (h, j) = (g * h, k \circledast j)$. Prove that $G \times K$ is a group with the operation $\odot$. (This group generalizes Example 7 from Section 7.1. Referred to in Sections 7.3 and 7.5.)

7. **a)** In the proof of Theorem 7.2.1, show that $y = a^{\#} * b$ satisfies $a * y = b$.

   b) Prove uniqueness for $a * y = b$.

   c) Give a solution x for $a * x * b = c$. Is the solution unique? Prove your answer.

8. Prove Corollaries 7.2.2 and 7.2.3, using the following suggestions:

   *a) Prove the uniqueness of the identity with a good choice of b in Theorem 7.2.1.

   b) Prove the uniqueness of inverses with a good choice of b in Theorem 7.2.1.

   c) Relate the rows and columns of the Cayley table to solving equations in Theorem 7.2.1 for Corollary 7.2.3.

9. Prove the following for any group G:

   a) The inverse of a^{-1} is a. (That is, $(a^{-1})^{-1} = a$.)

   b) The inverse of $a * b$ is $b^{-1} * a^{-1}$.

   c) Find the inverse of $a * (b * c)$ and prove that your answer is correct.

   d) If $a * b * a * b = e$, then $b * a * b * a = e$.

10. **a)** Explain why $\{0, 1, 2, 3\}$ is not a subgroup of $\mathbb{Z}_8 = \{0, 1, 2, 3, 4, 5, 6, 7\}$.

    b) However, $\mathbb{Z}_8$ does have a subgroup with four elements. List the elements in that subgroup.

    c) Find a subgroup of $\mathbb{Z}_8$ with two elements.

    d) Find all other subgroups of $\mathbb{Z}_8$.

11. a) Show that the set of odd integers is a subgroup of the group in Problem 4.

    b) Show that the set of even integers is not a subgroup of the group in Problem 4. Is it a coset of the set of odd integers? Prove your answer.

12. a) List all subsets of $\mathbb{Z}_9$ that form subgroups.

    b) Repeat part a with $\mathbb{Z}_{10}$.

    *c) Repeat part a with $\mathbb{Z}_{12}$.

    d) Make a conjecture based on parts a, b, and c.

13. *a) List all the subgroups of S_3. (See Table 2.)

    b) List all subgroups of $\mathbb{Z}_2 \times \mathbb{Z}_2$. (See Table 1, Section 7.1.)

    c) List all elements of $\mathbb{Z}_3 \times \mathbb{Z}_3$, using the definition in Problem 6.

    d) List all subgroups of $\mathbb{Z}_3 \times \mathbb{Z}_3$.

    e) Does your conjecture in Problem 12d hold with the groups in this problem?

14. Show that $A = \left\{ \begin{bmatrix} a & b \\ 0 & 1/a \end{bmatrix} : a, b \in \mathbb{R} \text{ and } a \neq 0 \right\}$ is a subgroup of $SL(2, \mathbb{R})$. (See Problem 5 for $SL(2, \mathbb{R})$.)

15. Describe the cosets in $\mathbb{Z}$ of the subgroup $5\mathbb{Z} = \{5x : x \in \mathbb{Z}\}$. Relate the cosets of $5\mathbb{Z}$ to the elements of $\mathbb{Z}_5$.

16. Consider the group $G \times K$ as defined in Problem 6.

    a) Prove that $\overline{G} = \{(g, e_K) : g \in G\}$ is a subgroup, where e_K is the identity of K.

    b) Prove that $\overline{K} = \{(e_G, k) : k \in K\}$ is a subgroup, where e_G is the identity of G.

    c) If $G = K$, prove that $\overline{D} = \{(x, x) : x \in G\}$ is a subgroup.

    d) For $(a, b) \in G \times K$, describe the elements in the coset $(a, b) \odot \overline{G}$ for part a.

    e) Repeat part d for $\overline{D}$.

17. a) Finish the proof of part ii of Lemma 7.2.5.

    b) Show that f is one-to-one and onto in part iii of Lemma 7.2.5.

18. Let $a \in G$ and G be a group with identity e.

    a) Prove that $\{e, a\}$ is a subgroup iff $a = a^{-1}$.

    b) If $a \neq a^{-1}$, what conditions guarantee that $\{e, a, a^{-1}\}$ is a subgroup? Prove your answer.

19. *a) List the elements in each coset of the subgroup $\{(1), (12)\}$ in S_3. See Example 4.

    b) Define the *right coset* for $a \in G$ and H a subgroup of G by $H * a = \{h * a : h \in H\}$. List the elements in each right coset of $\{(1), (12)\}$. Note: Some of these right cosets differ from the (left) cosets of $\{(1), (12)\}$.

    c) Restate and prove Lemma 7.2.5 for right cosets.

20. a) Suppose that H and K are subgroups of the group G. Prove that $H \cap K$ is a subgroup of G.

    b) In part a, must $H \cup K$ be a subgroup? Prove it or give a counterexample.

21. Critique the questionable "proofs" that follow. Point out reasoning errors and unclear presentation. If the argument is a proof, say so. (If the claim is false, there must be an error in the argument. However, it is not enough to say that the claim is false. You need to find an error in the reasoning.)

a) Claim: In a group G, $a^2 * b^2 = (a * b)^2$, where x^2 means $x * x$. "Proof": Clearly, $a * b = a * b$. Now square both sides to get $a^2 * b^2 = (a * b)^2$. ▲

*b) Claim: In a group G, if $a^2 * b^2 = (a * b)^2$, then $a * b = b * a$, where x^2 means $x * x$. "Proof": If we write out $a^2 * b^2 = (a * b)^2$, we get $a * a * b * b = a * b * a * b$. Cancel the a on the left of each side and the b on the right of each side to get $a * b = b * a$. ▲

c) Suppose that a finite group G has $2n$ elements. Then there are two elements of G that are their own inverses. "Proof": First of all, the identity e is its own inverse, since $e * e = e$. For a contradiction, suppose that all other elements differ from their inverses. Then by Problem 9a, we can pair a and a^{-1} for all the nonidentity elements, giving an even number of nonidentity elements. But G has an even number of elements, and there is only one identity by Corollary 7.2.2. Contradiction. So some other element besides the identity is its own inverse. ▲

d) (Converse of Lagrange's Theorem.) Suppose that G is a finite group with n elements and k is a positive integer dividing n. Then there is a subgroup of G with k elements. "Proof": We'll show the contrapositive. Suppose that k does not divide n, the number of elements in a finite group G. Let H be any subgroup of G. By Lagrange's Theorem, $|H|$ divides $|G|$, so $k \neq |H|$. ▲

Groups in Geometry

Figure 1 shows the square $ABCD$ and two of the permutations of its corners that make the square land exactly on itself. We call these permutations *symmetries* of the square. (Not every permutation of the corners is a symmetry: The permutation switching A and B and leaving C and D fixed twists the square and so fails to be a symmetry.) The study of symmetry has deeply enriched our understanding of areas from crystallography to archeology. (See Sibley, Chapter 5, for more information.) For our formal definition, we use the concept of the distance $d(a, b)$ between two points a and b.

DEFINITION. Suppose a set X has a distance d on it. A *symmetry* σ of X is a permutation of X that preserves distance. That is, for all $a, b \in X$, $d(a, b) = d(\sigma(a), \sigma(b))$.

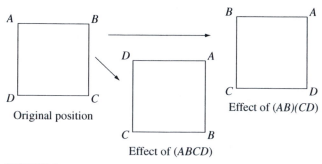

Original position

Effect of $(ABCD)$

Effect of $(AB)(CD)$

FIGURE 1

THEOREM 7.2.7. The set $Sym(X)$ of all symmetries on a set X is a subgroup of S_X.

Proof. First we show closure. Let σ and τ be symmetries of X. Then $\sigma \circ \tau$ is a permutation of X by Problem 11 of Section 3.2. Further, for all $a, b \in X$, we have

$$d(\sigma \circ \tau(a), \sigma \circ \tau(b)) = d(\sigma(\tau(a)), \sigma(\tau(b)))$$
$$= d((\tau(a)), (\tau(b)) \quad \text{because } \sigma \text{ is a symmetry}$$
$$= d(a, b) \quad \text{because } \tau \text{ is a symmetry.}$$

The identity permutation given by $\iota(a) = a$ certainly preserves distance: $d(\iota(a), \iota(b)) = d(a, b)$.

Finally, given a symmetry σ, we know it already has an inverse σ^{-1}, which is a permutation. We need to show that this inverse is also a symmetry. To do so, we apply σ to the points $\sigma^{-1}(a)$ and $\sigma^{-1}(b)$ to show that σ^{-1} also preserves distance:

$$d(\sigma^{-1}(a), \sigma^{-1}(b)) = d(\sigma(\sigma^{-1}(a)), \sigma(\sigma^{-1}(b))) = d(\iota(a), \iota(b)) = d(a, b). \quad \blacksquare$$

We investigate $Sym(ABCD)$, the 8 symmetries of the square in Problem 22. Following the notation of Example 4, $Sym(ABCD) = \{(A), (ABCD), (AC)(BD), (ADCB), (AC), (BD), (AB)(CD), (AD)(BC)\}$.

PROBLEMS (CONTINUED)

22. a) Draw arrangements of the corners of the square to illustrate what the different symmetries do to the square.

   b) Describe each of the symmetries of the square in geometric terms. (Options include a rotation of a certain angle and a mirror reflection over a particular line. What angle of rotation does the identity have?)

   c) Complete the accompanying Cayley table for $Sym(ABCD)$. (Hint: Start with the row and column for the identity. Next, determine the compositions involving two rotations. Determine enough other compositions to fill in the table, using Corollary 7.2.3.)

$\circ$	(A)	(ABCD)	(AC)(BD)	(ADCB)	(AC)	(BD)	(AB)(CD)	(AD)(BC)
(A)		(ABCD)						
(ABCD)			(ADCB)		(AD)(BC)	(AB)(CD)		
(AC)(BD)			(ABCD)				(AD)(BC)	
(ADCB)			(ABCD)	(AC)(BD)			(BD)	
(AC)				(AD)(BC)	(A)			(ADCB)
(BD)								
(AB)(CD)				(AC)		(ABCD)		
(AD)(BC)	(AD)(BC)		(AB)(CD)					

   d) List the inverse of each element.

   e) Show that $Symm(ABCD)$ is not commutative.

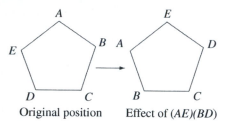

Original position   Effect of (AE)(BD)

FIGURE 2

    f) Verify that the set of rotations form a subgroup of *Symm(ABCD)*.

    g) Find a subgroup with four elements differing from the subgroup of part e.

    h) Determine the number of subgroups with two elements.

23. Figure 2 gives a regular pentagon and one of its symmetries.

    a) List the elements of *Symm(ABCDE)*, the symmetries of a regular pentagon.

    b) Describe each of the symmetries of a regular pentagon in geometric terms. (Options include a rotation of a certain angle and a mirror reflection over a particular line.)

    c) Explain why the set of rotations forms a subgroup of *Symm(ABCDE)*.

    d) List the inverse of each element of *Symm(ABCDE)*.

    e) Show that *Symm(ABCDE)* is not commutative.

24. a) Determine the number of rotations and the number of mirror reflections that are symmetries of a regular *n*-gon. Explain your answer.

    b) Explain why the set of rotations forms a subgroup of the symmetries of a regular *n*-gon.

REFERENCES

GALLIAN, J. 2002. *Contemporary abstract algebra*, 5th ed. Boston: Houghton Mifflin.
SIBLEY, T. 1998. *The geometric viewpoint: A survey of geometries*, Boston: Addison Wesley.

7.3 RINGS AND FIELDS

Ordinary arithmetic, as well as other familiar algebraic systems, has both addition and multiplication. The interaction of these operations deserves special consideration. In particular, the distributivity of multiplication over addition underlies some important and familiar arithmetic facts. In the last half of the nineteenth century, mathematicians generalized the integers to structures they called *rings* and the rationals and reals to structures they called *fields*. A ring is a group for one operation with one property for the second operation and distributivity linking the two operations. Fields have more properties, making it possible to add, subtract, multiply, divide, and solve equations in ways quite similar to ordinary arithmetic. To distinguish the additive identity and the multiplicative identity in a field or a ring, we call a multiplicative identity a *unity*.

DEFINITION. A *ring R* is a set with two operations + and · so that

+ is associative;   · is associative;

+ is commutative;   · distributes over +.

+ has an identity 0;

+ has inverses, written $-x$;

DEFINITION. A *field F* is a set with two operations + and · so that

+ is associative;   · is associative;

+ is commutative;   · is commutative;

+ has an identity 0;   · has a unity 1, where $1 \neq 0$;

+ has inverses, written $-x$;   · has inverses for non-zero elements, written x^{-1};

                                          · distributes over +.

EXAMPLE 1

The sets $\mathbb{Z}$, $\mathbb{Q}$, $\mathbb{R}$, and $\mathbb{C}$, with the usual addition and multiplication, are rings. The last three are fields, but $\mathbb{Z}$ is not.   ◇

EXAMPLE 2

The set $P(\mathbb{R})$ of all polynomials in the variable x over the reals $\mathbb{R}$ (or any ring) is a ring with the familiar addition and multiplication of polynomials. Thus, $(3x^2 - 2x + 1) \cdot (4x + 5) = 12x^3 + 7x^2 - 6x + 5$ in the ring $P(\mathbb{R})$, as well as in $P(\mathbb{Z})$, $P(\mathbb{Q})$, and $P(\mathbb{C})$. Polynomial rings are not fields, since a general polynomial such as $x^2 - 3x + 1$ doesn't have a multiplicative inverse that is still a polynomial.   ◇

EXAMPLE 3

The set $\mathbb{M}_n(\mathbb{R})$ of $n \times n$ matrices with entries from the reals $\mathbb{R}$ (or any ring) is a ring with the familiar addition and multiplication of matrices. Matrix multiplication is usually not commutative. The importance of these rings is one good reason the commutative property of multiplication is not assumed in the definition of a ring. When $n \geq 2$, a matrix ring is not a field, because the matrices with zero determinant have no multiplicative inverses.   ◇

EXAMPLE 4

The set $\mathbb{Z}_n$ with addition and multiplication (mod n) is a ring. Some of these rings are fields, something we'll investigate in this section. (See Example 6 and Problems 14 and 15.)   ◇

EXAMPLE 5

The set $2\mathbb{Z}$ of even integers is a ring with the usual addition and multiplication, but doesn't have a unity.   ◇

We can use all the properties of groups from Section 7.2 for the additive operation of a ring. Some familiar additional properties of numbers hold in any ring, as our first theorem shows. These properties all depend on distributivity, the only structural link between addition and multiplication.

THEOREM 7.3.1. Let R be a ring and $a, b \in R$. Then

i) $a \cdot 0 = 0 = 0 \cdot a$;

ii) $a \cdot (-b) = -(a \cdot b) = (-a) \cdot b$;

iii) $(-a) \cdot (-b) = a \cdot b$.

iv) If R has a unity (1), then $(-1) \cdot (-1) = 1$.

Proof. Let $a, b \in R$. First, $0 = 0 + 0$, since 0 is the additive identity. Thus, $a \cdot 0 = a \cdot (0 + 0) = a \cdot 0 + a \cdot 0$, using distributivity. Adding 0 yields $0 + a \cdot 0 = a \cdot 0 + a \cdot 0$ and canceling $a \cdot 0$ gives $0 = a \cdot 0$. The other equality for part

i) is similar.

ii) The first issue in this part is to keep track of the notation: $-b$ is the additive inverse of b, $-(a \cdot b)$ is the additive inverse of $a \cdot b$, and $-a$ is the additive inverse of a. See Problem 4 for the proof.

iii) See Problem 4.

iv) See Problem 4. ∎

Parts iii and iv of Theorem 7.3.1 contain the structural heart of the rule "A negative times a negative is positive." This phrase is thus a consequence of algebraic properties, not an arbitrary rule imposed by grade school teachers or the quirk of numbers. Similarly, the rule "You can't divide by zero" comes from part i of this theorem. The equation $a \div b = c$ involving division simply means that $c \cdot b = a$. Division by zero would correspond to the equation $a \div 0 = c$, or equivalently, $c \cdot 0 = a$. In any ring, part i guarantees that $c \cdot 0 = 0$, so the only possible such division is $0 \div 0$. However, then we could pick multiple values for c, and $\div$ would not be determined. Thus, we can't divide by zero in any ring (except the ring with just one element, 0). The structural approach to algebra has given great insight and has shaped modern mathematics. Teachers should remember, however, that elementary school children learning arithmetic may not be developmentally ready to think structurally. See Problems 2 and 3 for other instances of arithmetic rules that come from structural properties.

EXAMPLE 6

Use Tables 1 and 2, giving multiplication in $\mathbb{Z}_5$ and $\mathbb{Z}_6$, to investigate which ring is a field.

Solution For any n, the multiplication (mod n) is associative and commutative, distributes over addition (mod n), and has 1 as a unity. Also, addition forms a commutative group. Thus, any $\mathbb{Z}_n$ satisfies all but one of the properties of a field. The remaining property is the existence of multiplicative inverses. For $\mathbb{Z}_5$, we see that each nonzero element has such an inverse: $1 \cdot 1 = 1$, $2 \cdot 3 = 1 = 3 \cdot 2$, and $4 \cdot 4 = 1$. However, 2, 3, and 4 fail to have multiplicative inverses in $\mathbb{Z}_6$, so $\mathbb{Z}_6$ is not a field. In addition, there are some 0 entries outside of the zero row and column

TABLE 1. · in $\mathbb{Z}_5$

·	0	1	2	3	4
0	0	0	0	0	0
1	0	1	2	3	4
2	0	2	4	1	3
3	0	3	1	4	2
4	0	4	3	2	1

TABLE 2. · in $\mathbb{Z}_6$

·	0	1	2	3	4	5
0	0	0	0	0	0	0
1	0	1	2	3	4	5
2	0	2	4	0	2	4
3	0	3	0	3	0	3
4	0	4	2	0	4	2
5	0	5	4	3	2	1

of the table for $\mathbb{Z}_6$. Example 8 indicates how such zeros can create havoc in solving equations. Problems 14 and 15 ask you to prove $\mathbb{Z}_n$ is a field iff n is a prime. ◇

Rings can have subrings, just as groups can have subgroups. Indeed, a subring, as defined in Problem 7, is a subgroup. Problem 12 asks you to extend Lagrange's Theorem to finite rings.

Historically, abstract algebra grew from the study of solving equations. Theorem 7.2.1 showed that a group was enough to solve an additive equation like $a + x = b$. With two operations, we can write the first degree equation $a \cdot x + b = 0$. Problem 5 shows that in a field, such an equation has a unique solution, provided that $a \neq 0$. High school algebra goes on to consider higher degree equations. While we don't need to set first degree expressions equal to zero in order to solve them, factoring higher degree equations depends on setting the equation equal to zero. The special role of zero in the next theorem is the key to factoring in any field, as Corollary 7.3.3 makes clear. For ease, we write $x - a$ as a shorthand for $x + (-a)$.

THEOREM 7.3.2. If F is a field, $a, b \in F$, and $a \cdot b = 0$, then $a = 0$ or $b = 0$.

Proof. If $a = 0$, we're done. So, suppose that $a \neq 0$ and $a \cdot b = 0$. Since F is a field and $a \neq 0$, a has a multiplicative inverse a^{-1}. Then $a \cdot b = 0$ becomes $a^{-1} \cdot (a \cdot b) = a^{-1} \cdot 0 = 0$. The properties of associativity, inverses, and unity give $b = 0$. ∎

COROLLARY 7.3.3. For $a, b, x \in F$, where F is a field, x is a solution of $(x - a) \cdot (x - b) = 0$ iff $x = a$ or $x = b$.

Proof. See Problem 11. ∎

Corollary 7.3.3 and its generalization in Problem 11 show that if you can factor a polynomial over a field at all, there is only one way to do so. This statement is a weaker version of the following important theorem, whose proof goes beyond this text:

THEOREM 7.3.4. An nth degree polynomial with coefficients from a field has at most n roots in that field.

Proof. See Gallian, 289. ∎

EXAMPLE 7

For the field $\mathbb{Q}$, the equations $x^2 - 2 = 0$ and $x^2 + 1 = 0$ can't be factored and have no roots. When we expand the field to $\mathbb{R}$, the first equation factors to $(x - \sqrt{2})(x + \sqrt{2}) = 0$, giving roots $\pm\sqrt{2}$. The second equation has no roots in $\mathbb{R}$, but it does in $\mathbb{C}$, the complex numbers. There we have $x^2 + 1 = (x + i)(x - i) = 0$, giving the roots $\pm i$, the square roots of -1. Thus, the existence of roots depends significantly on the field. $\Diamond$

EXAMPLE 8

Consider $(x - 2)(x - 3) = 0$ in $\mathbb{Z}_6$. From the given factoring, 2 and 3 are roots of this equation. Let's multiply the factors (mod 6) to get $(x - 2)(x - 3) = x^2 - 5x + 6 \equiv x^2 + x$ (mod 6). But $x^2 + x = 0$ factors easily into $x(x + 1) = 0$, giving two more roots $x = 0$ and $x = -1 \equiv 5$ (mod 6). Thus, this second degree equation manages to have four different roots in $\mathbb{Z}_6$. This doesn't contradict Corollary 7.3.3, because $\mathbb{Z}_6$ is not a field. $\Diamond$

Historical Remarks

Solving polynomial equations has engaged mathematicians for thousands of years. Over that time, people's understanding of numbers has expanded from positive integers to include rational numbers, irrational numbers, and complex numbers. The oldest existing mathematics texts from 4000 years ago show a mastery of solving word problems involving first degree equations. By 1600 B.C., the ancient Babylonians had a verbal version of what we call the quadratic formula, giving an explicit way to solve second degree equations. While the Greeks gave proofs of geometric versions of the quadratic formula, progress in solving higher degree equations algebraically had to wait for the European Renaissance. In 1545, Gerolamo Cardano (1501–1576) published a book showing how to solve any third- or fourth-degree polynomial equation, on the basis of work of several mathematicians. Since negative numbers were not recognized, there were numerous cases; for instance, $x^3 + ax = b$ differed from $x^3 + b = ax$. The method of solving these equations generally involved complex numbers, even when the answer was a real number. Gradually, negative and complex numbers became more accepted, culminating in the proof in 1799 by Carl Friedrich Gauss (1777–1855) of the Fundamental Theorem of Algebra: Every n^{th}-degree polynomial with coefficients in $\mathbb{C}$ has all n (possibly repeated) roots in $\mathbb{C}$. This result shows that there is no algebraic need to expand our number system beyond the complex numbers. At the same time, various mathematicians were struggling to solve the general fifth-degree equation. An analysis of the permutation of roots led both to the development of groups and to a stunning discovery: There could not be a general formula for the roots of a fifth-degree equation. After Niels Abel (1802–1829) first proved this result, Evariste Galois (1811–1832) explored the general context of the solvability of equations, leading to a profound link between groups and what we now call fields.

PROBLEMS

*1. For each statement that follows, decide whether it is true or (at least sometimes) false. Explain your answer or cite the part of the text supporting it.

   a) Every ring is a field.

b) Every field is a ring.

c) Every ring is a group.

d) In a field, every second-degree equation has exactly two roots.

e) In a ring, a second-degree equation has at most two roots.

f) In every ring, a negative times a positive is negative.

2. Without a calculator, many of us would do the following work to find $321 \cdot 213$:

$$
\begin{array}{r}
3\ 2\ 1 \\
\times\ 2\ 1\ 3 \\
\hline
9\ 6\ 3 \\
3\ 2\ 1 \\
6\ 4\ 2 \\
\hline
6\ 8\ 3\ 7\ 3
\end{array}
$$

a) Use distributivity and place value to explain why this procedure is legitimate. (Ignore the idea of carrying.)

b) In base 10, the number 34 means $10 \cdot 3 + 4$. Use distributivity to rewrite $(10a + b)$ $(10c + d)$. Then explain how this answer corresponds to the handwritten procedure for the product of two two-digit numbers.

*3. To square a number ending in a 5, say, $X5$, as a child you may have learned to multiply $X(X + 1)$ and put a 25 after it. For instance, $35^2 = 1225$ "because" $3 \cdot (3 + 1) = 3 \cdot 4 = 12$. Use distributivity and place value to justify this rule.

4. *a) Prove part ii of Theorem 7.3.1.

   b) Use part ii to prove part iii of Theorem 7.3.1.

   c) Use part iii to prove part iv of Theorem 7.3.1.

5. Prove that we can solve first degree equations in a field. That is, for F a field and any $a, b \in F$, if $a \neq 0$, there is a unique $x \in F$ such that $a \cdot x + b = 0$.

6. **a)** On a commutative group G with the operation $+$, for all $a, b \in G$, define $a \otimes b = 0$, the identity of the group. Prove that G with $+$ and $\otimes$ is a ring.

   b) On $\mathbb{Q}$, define an unusual multiplication $\odot$ by $a \odot b = 7ab$. Prove that $\mathbb{Q}$ with ordinary addition and $\odot$ is a ring.

   c) Does the ring in part b have a unity? Is it a field? Prove your answers.

7. Given a ring R, we define a subset S to be a *subring* of R iff S is a ring, using the same operations as R.

   a) Give examples of four subrings of $\mathbb{Z}$.

   b) Explain why we don't need to show associativity, distributivity, or commutativity when we prove that a subset of a ring is a subring.

   c) Show that the *Gaussian integers* $\mathbb{Z}[i] = \{a + bi : a, b \in \mathbb{Z}\}$ is a subring of $\mathbb{C}$, the complex numbers.

   d) If R is a ring and $a \in R$, show that the set $\{a \cdot r : r \in R\}$ is a subring of R.

8. Which of the following sets are subrings of $M_2(\mathbb{R})$, the set of 2×2 real matrices?

a) $\mathbb{D} = \left\{ \begin{bmatrix} a & 0 \\ 0 & d \end{bmatrix} : a, d \in \mathbb{R} \right\}$.

b) $\mathbb{U} = \left\{ \begin{bmatrix} 1 & b \\ 0 & 1 \end{bmatrix} : b \in \mathbb{R} \right\}$.

c) $\mathbb{W} = \left\{ \begin{bmatrix} a & b \\ -b & a \end{bmatrix} : a, b \in \mathbb{R} \right\}$.

d) Which of the subrings in parts a, b, and c is a field? Justify your conclusion.

9. Let $\mathbb{Q}[\sqrt{2}] = \{a + b\sqrt{2} : a, b \in \mathbb{Q}\}$. Prove that $\mathbb{Q}[\sqrt{2}]$ with ordinary addition and multiplication is a field (referred to in Problem 18 of Section 7.5).

10. Given two commutative rings $(R, +, \cdot)$ and $(S, +, \cdot)$, we can use Problem 6 of Section 7.2 to define a commutative group on $R \times S$ by $(a, b) + (c, d) = (a + c, b + d)$.

*a) Prove that $(a, b) \cdot (c, d) = (a \cdot c, b \cdot d)$ gives a commutative ring on $R \times S$. If R has a unity 1_R and S has a unity 1_S, prove that $R \times S$ has a unity (referred to in Problems 16 and 17 in Section 7.5).

If R and S are the same ring, we can define other multiplications on $R \times S$.

b) Prove that $(a, b) \cdot (c, d) = (a \cdot c + b \cdot d, a \cdot d + b \cdot c)$ gives a commutative ring. If $R = S$ has a unity 1, prove that $R \times S$ has a unity.

c) Prove that $(a, b) \cdot (c, d) = (a \cdot c, a \cdot d + b \cdot c + b \cdot d)$ gives a commutative ring. If $R = S$ has a unity 1, prove that $R \times S$ has a unity.

d) If R and S each have at least two elements, prove that the ring of part a is not a field. (Hint: Find two nonzero elements of $R \times S$ whose product is $(0, 0)$, the identity. Then use Theorem 7.3.2.)

e) If $R = S$ has at least two elements, prove that the ring of part b is not a field.

f) If $R = S$ has at least two elements, prove that the ring of part c is not a field.

g) Let $\overline{S} = \{(0, s) : s \in S\}$. Decide which of the multiplications in parts a, b, and c make $\overline{S}$ a subring.

11. a) Prove Corollary 7.3.3.

b) For all $n \in \mathbb{N}$, show in a field F that x is a solution to $(x - a_1) \cdot (x - a_2) \cdot \ldots \cdot (x - a_n) = 0$ iff $(x = a_1$ or $x = a_2$ or $\ldots$ or $x = a_n)$.

12. Prove Lagrange's Theorem for subrings of finite rings. (See Theorem 7.2.6.)

13. For $a \neq 0$, the Quadratic Formula asserts that the solutions to $ax^2 + bx + c = 0$ are $x = \frac{-b + \sqrt{b^2 - 4ac}}{2a}$ and $x = \frac{-b - \sqrt{b^2 - 4ac}}{2a}$. We investigate to what extent this formula holds in any field F. We first reinterpret this formula, starting by rewriting a fraction $\frac{p}{q}$ as $p \cdot q^{-1}$. Also, 2 and 4 are shorthand for $1 + 1$ and $1 + 1 + 1 + 1$, respectively. The denominator $2a$ requires us to restrict ourselves to fields in which $2 \neq 0$. (There are fields in which $1 + 1 = 0$.)

a) Show that for $d, y \in F$, if $y^2 = d$, then $(-y)^2 = d$ and no element of F other than $\pm y$ satisfies $x^2 = d$. We can thus interpret $\pm\sqrt{d}$ to be $\pm y$.

b) With the previous interpretations and restrictions, show that if there is some $y \in F$ such that $y^2 = b^2 - 4ac$, then both $\frac{-b+\sqrt{b^2-4ac}}{2a}$ and $\frac{-b-\sqrt{b^2-4ac}}{2a}$ satisfy $ax^2 + bx + c = 0$.

14. Use the contrapositive of Corollary 7.3.3 to show that if $n \in \mathbb{N}$, $n > 1$, and n is not a prime number, then $\mathbb{Z}_n$ is not a field.

15. Complete the following steps to prove that if p is a prime number, then $\mathbb{Z}_p$ is a field:

a) For $a \in \mathbb{N}$, if $a < p$ and p is prime, prove that $\gcd(a, p) = 1$.

b) Use Lemma 2.4.4 to show that if p is prime and $\gcd(a, p) = 1$, there is $k \in \mathbb{Z}$ such that $ak \equiv 1 \pmod{p}$.

c) Use the Division Algorithm and parts a and b to show that for p prime, $a \in \mathbb{Z}_p$, and $a \neq 0$, there is $r \in \mathbb{Z}_p$ such that r is the inverse of a.

*16. The cross product of two vectors in $\mathbb{R}^3$ is an operation defined by $(a, b, c) \times (x, y, z) = (bz - cy, cx - az, ay - bx)$. Which of the properties of a field does $\mathbb{R}^3$ with vector addition and cross product satisfy? For those properties that fail, provide a counterexample.

17. Critique the questionable "proofs" that follow. Point out reasoning errors and unclear presentation. If the argument is a proof, say so. (If the claim is false, there must be an error in the argument. However, it is not enough to say that the claim is false. You need to find an error in the reasoning.)

*a) Claim: In $\mathbb{Z}$, $0 = 2$. "Proof": Let $x = 0$. Then $x^2 = x$. Substituting x^2 for the first x in $x - x = 0$, we have $x^2 - x = 0$, which equals x. From $x^2 - x = x$, we have $x^2 - 2x = 0$. Cancel an x to get $x - 2 = 0$, or $x = 2$. But $x = 0$. Therefore, $0 = 2$. ▲

b) Claim: For any b and c in a ring R, $(b - c)^2 = (c - b)^2$, where $x^2 = x \cdot x$. "Proof": For $b, c \in R$, we expand $(b - c)^2$ to get $b^2 - 2bc + c^2$. Similarly, $(c - b)^2 = c^2 - 2bc + b^2$. Since addition is commutative, the two terms are equal. ▲

*c) Claim: For any b and c in a ring R, $(b - c)^2 = (c - b)^2$, where $x^2 = x \cdot x$. "Proof": For $b, c \in R$, we know that $-(b - c) = -b - (-c) = -b + c = c - b$ by commutativity. Then, Theorem 7.3.1, part iii, gives $(c - b)^2 = (-(b - c)) \cdot (-(b - c)) = (b - c) \cdot (b - c) = (b - c)^2$. ▲

d) Claim: In a ring R, if for every $a \in R$, $a \cdot a = a$, then for every $a \in R$, $a + a = 0$. "Proof": For $a \in R$, we have $a \cdot a = a$. Similarly, $(a + a) \cdot (a + a) = a + a$. Use distributivity for the product on the left to get $(a + a) \cdot (a + a) = a \cdot a + a \cdot a + a \cdot a + a \cdot a$, which reduces to $a + a + a + a$. But in $a + a + a + a = a + a$, we can cancel to get $a + a = 0$. ▲

e) Claim: If S is a subring of R and $r \in R$, then $r \cdot S = \{r \cdot s : s \in S\}$ is a subring of R. "Proof": Let $r \in R$ and S a subring of R. For $s, t \in S$, we have $r \cdot s + r \cdot t = r \cdot (s + t) \in r \cdot S$ by distributivity. Similarly, $(r \cdot s) \cdot (r \cdot t) = r \cdot (s \cdot r \cdot t) \in r \cdot S$ by associativity. ▲

REFERENCE

GALLIAN, J. 2002. *Contemporary abstract algebra*, 5th ed. Boston: Houghton Mifflin.

7.4 LATTICES

Lattices give a taste of the variety of algebraic structures beyond those of the last two sections. Although the operations of union and intersection differ greatly from addition and multiplication, they have interesting properties shared with other operations, such as those in Examples 2 and 3. Furthermore, union and intersection are tied to the partial order $\subseteq$, as Example 1 indicates. The connection between paired operations and partial orders, as in Examples 1, 2, and 3, led mathematicians in the twentieth century to study this connection more abstractly. (Section 4.4 discusses partial orders.) The advent of computers and their use of the algebra of logic enhanced the importance of this subject area. For ease, we start with one operation, looking at semi-lattices and their related partial orders before considering lattices, which have two operations. (We call the general operations "meet" ($\sqcap$) and "join" ($\sqcup$), similar to intersection and union, respectively.)

DEFINITION.   A set S with an operation $\sqcap$ is a *semi-lattice* iff for all $a, b, c \in S$,

$\sqcap$ is idempotent: $a \sqcap a = a$,

$\sqcap$ is associative: $a \sqcap (b \sqcap c) = (a \sqcap b) \sqcap c$, and

$\sqcap$ is commutative: $a \sqcap b = b \sqcap a$.

EXAMPLE 1

Intersection and union each satisfy the properties for a semi-lattice on $\mathcal{P}(X)$, the power set of any set X. Problem 13c of Section 2.1 connects these operations with the subset relation. We have for any sets F and G that $F \cap G = F$ iff $F \subseteq G$ iff $F \cup G = G$. $\Diamond$

EXAMPLE 2

On $\mathbb{R}$, the operation of the minimum of two real numbers, $\min(a, b)$, satisfies the three properties of a semi-lattice. Similarly, $\mathbb{R}$ and $\max(a, b)$, the maximum of two numbers, form a semi-lattice. The relation $\leq$ is connected to min and max in the same way that $\subseteq$ is connected to $\cap$ and $\cup$. That is, for all $a, b \in \mathbb{R}$, $\min(a, b) = a$ iff $a \leq b$ iff $\max(a, b) = b$. $\Diamond$

EXAMPLE 3

On $\mathbb{N}$, the greatest common divisor of two numbers, $\gcd(a, b)$, satisfies the three properties of a semi-lattice. Also, the least common multiple, lcm, forms a semi-lattice on $\mathbb{N}$. The relation "divides," written $a|b$, underlies the definitions of both gcd and lcm. By definition, $\gcd(a, b)$ is the largest integer dividing both a and b. In turn, both a and b divide $\text{lcm}(a, b)$, which is the smallest such positive integer. As with the two previous examples, $\gcd(a, b) = a$ iff $a|b$ iff $\text{lcm}(a, b) = b$.

REMARK.   Even though greatest common divisors are defined for negative integers, we need to restrict the least common multiple operation to positive numbers. After all, every negative multiple of ab is a common multiple of a and b and there is no least (most negative) such common multiple. Also, the equation $\gcd(a, a) = a$ fails if a is negative. $\Diamond$

The connection between the operation of a semi-lattice and a partial order is no accident. The properties of a semi-lattice are just strong enough to derive a partial order from the operation.

THEOREM 7.4.1. If S is a semi-lattice with the operation $\sqcap$, then the relation $\sqsubseteq$ given by $a \sqsubseteq b$ iff $a \sqcap b = a$ is a partial order on S.

Proof. We need to show that $\sqsubseteq$ is reflexive, transitive, and antisymmetric. Let $x, y, z \in S$. From idempotency, we have $x \sqcap x = x$, giving by definition $x \sqsubseteq x$, showing reflexivity.

For transitivity, assume that $x \sqsubseteq y$ and $y \sqsubseteq z$. Then $x \sqcap y = x$ and $y \sqcap z = y$. To show that $x \sqsubseteq z$, we use associativity to convert $x \sqcap z$ to x as follows: $x \sqcap z = (x \sqcap y) \sqcap z = x \sqcap (y \sqcap z) = x \sqcap y = x$. Hence, $x \sqsubseteq z$.

See Problem 4 for antisymmetry. ∎

Can we turn the process around and define an operation from a partial order on a set? Not always, as Example 4 illustrates.

EXAMPLE 4

The relation "divides" is a partial order on the set $S = \{2, 4, 6, 9, 18\}$. The definition of Theorem 7.4.1 becomes $a|b$ iff $a \sqcap b = a$. This definition tells us some of the values of $\sqcap$, such as $2 \sqcap 4 = 2$ and $6 \sqcap 18 = 6$; but $2 \sqcap 9$ is not defined, because 2 doesn't divide 9 nor does 9 divide 2. If we expand the set S to $D_{36} = \{1, 2, 3, 4, 6, 9, 12, 18, 36\}$, all positive divisors of 36, then we do have a natural operation: gcd, the greatest common divisor of two numbers. (See Figure 1, which gives the Hasse diagrams for | on D_{36} and for S.)

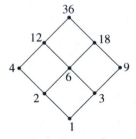
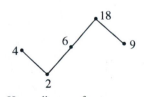

Hasse diagram for
"divides" on D_{36}

Hasse diagram for
"divides" on $\{2, 4, 6, 9, 18\}$

FIGURE 1

While not every partial order gives a semi-lattice, if every pair of elements has a greatest lower bound for the partial order, the greatest lower bound forms a semi-lattice. Similarly, the least upper bound, if it exists, is another natural operation related to the partial order. Examples 1, 2, and 3 illustrate this pairing of two operations and a partial order and help motivate the idea of a lattice. A lattice has two operations, each forming a semi-lattice. However, we can't use just any two such operations. For instance, both max and gcd form semi-lattices on $\mathbb{N}$, but we don't expect them to match. In particular, the partial orders they provide by Theorem 7.4.1 are quite different. In comparison, the partial orders for max and min from that theorem are $\geq$ and $\leq$, respectively, which are the same except for a switching of order. What abstract property corresponds to the desired match of the two operations? The absorptive property, given in the definition of a lattice, accomplishes this purpose, as Theorem 7.4.2 shows. In effect, this theorem says the two operations in a lattice are compatible, since they define the same partial order in Theorem 7.4.1, except for switching the order.

DEFINITION. A *lattice* is a set S with two operations $\sqcap$ and $\sqcup$ such that for all $a, b, c \in S$,

idempotency: $a \sqcup a = a \qquad a \sqcap a = a,$

associativity: $a \sqcup (b \sqcup c) = (a \sqcup b) \sqcup c \qquad a \sqcap (b \sqcap c) = (a \sqcap b) \sqcap c,$

commutativity: $a \sqcup b = b \sqcup a \qquad a \sqcap b = b \sqcap a,$ and

absorptivity: $a \sqcup (a \sqcap b) = a \qquad a \sqcap (a \sqcup b) = a.$

THEOREM 7.4.2. For all $x, y \in S$, where S is a lattice, $y \sqcap x = x$ iff $y \sqcup x = y$.

Proof. ($\Rightarrow$) Substitution of y for a and x for b in the absorptive property gives $y \sqcup (y \sqcap x) = y$. From our assumption that $y \sqcap x = x$, we have $y \sqcup x = y$, as desired. ($\Leftarrow$) This direction is similar. ∎

Just as groups can have subgroups, lattices (and semi-lattices) can have sublattices (and subsemi-lattices). However, Lagrange's Theorem (7.2.6) does not have to hold.

DEFINITION. A subset S of a lattice L is a *sublattice* of L iff S is a lattice with the same operations as L.

EXAMPLE 5

Let $L = [0, 1]$ with the operations of max and min. Then $[0.2, 0.5]$, or any other interval in $[0, 1]$, is a sublattice. ◇

EXAMPLE 6

Let D_{36} be the set of all positive divisors of 36, which is a lattice with the operations of gcd and lcm. In Example 4, we saw that D_{36} has 9 elements. The subset $\{2, 4, 6, 12\}$ is a sublattice with 4 elements. Since 4 does not divide 9, we see that Lagrange's Theorem fails. In addition, there are sublattices with 2 and 6 elements, such as $\{2, 4\}$ and $\{1, 2, 3, 4, 6, 12\}$, even though 2 and 6 do not divide 9. ◇

THEOREM 7.4.3. If $a \in L$ and L is a lattice with partial order $\sqsubseteq$, then $\{x \in L : x \sqsubseteq a\}$ is a sublattice.

Proof. See Problem 5. ∎

Lattices, especially those used in applications, often have more properties. As in Section 7.3, distributivity and identities are valuable properties. The lattice of a power set has these properties and more.

EXAMPLE 7

For any set X, the power set $\mathcal{P}(X)$ with $\cup$ and $\cap$ forms a lattice. Further, each operation distributes over the other and each has an identity. The identity for union is $\emptyset$, which is the minimal element of the lattice: For any subset A, $\emptyset \cup A = A$ and $\emptyset \subseteq A$. Similarly, X is the identity for $\cap$ and is maximal: For any subset A, $X \cap A = A$ and $A \subseteq X$. Even more, subsets of X have *complements* in X, making $\mathcal{P}(X)$ a Boolean algebra, as defined next. $\qquad \diamond$

DEFINITION. A *Boolean algebra* B is a set B with two operations $\cap$ and $\sqcup$ such that, for all $a, b, c \in B$,

idempotency: $a \sqcup a = a \qquad a \cap a = a$,

associativity: $a \sqcup (b \sqcup c) = (a \sqcup b) \sqcup c \qquad a \cap (b \cap c) = (a \cap b) \cap c$,

commutativity: $a \sqcup b = b \sqcup a \qquad a \cap b = b \cap a$,

absorptivity: $a \sqcup (a \cap b) = a \qquad a \cap (a \sqcup b) = a$,

distributivity: $a \sqcup (b \cap c) = (a \sqcup b) \cap (a \sqcup c) \qquad a \cap (b \sqcup c) = (a \cap b) \sqcup (a \cap c)$,

identity: there is $0 \in B$ such that for all $a, 0 \sqcup a = a \qquad$ there is $1 \in B$ such

$\qquad\qquad\qquad\qquad\qquad\qquad\qquad\qquad\qquad\qquad$ that for all $a, 1 \cap a = a$

with $0 \neq 1$, and

complements: for all $a \in B$, there is $a' \in B$ such that $a \sqcup a' = 1$, and $a \cap a' = 0$.

Part of the importance of Boolean algebras in computer science comes from the design of logical circuit boards. Circuit boards treat logic in terms of truth tables, which match the properties of Boolean algebras. The name "Boolean algebra" honors George Boole (1815–1864), who made the many algebraic properties of logic explicit in his work.

EXAMPLE 8

We can think of the logical words "and" ($\wedge$) and "or" ($\vee$) as operations taking two propositions and giving another proposition. For instance, the propositions "It is hot" (P) and "It is raining" (Q) combine to give "It is hot and it is raining," ($P \wedge Q$) as well as "It is hot or it is raining" ($P \vee Q$). The logical word "not" ($\neg$) acts as the complement. Unfortunately, the properties of a Boolean algebra don't immediately match. The proposition "It is hot and it is hot" ($P \wedge P$) doesn't equal the proposition "It is hot" (P), needed for idempotency. Rather, they are logically equivalent. That is, they have equivalent truth tables, or $(P \wedge P) \Leftrightarrow P$ is a tautology. Similarly, we get tautologies for all the other equivalences, such as $(P \wedge Q) \Leftrightarrow (Q \wedge P)$, the commutativity of $\wedge$. If we think of propositions only in terms of their truth tables, we get a Boolean algebra. (In fancier language, the classes of logically equivalent propositions form a Boolean algebra.)

Following Theorem 7.4.1, we can define a partial order on (equivalence classes of) propositions by $P \preceq Q$ iff $(P \wedge Q) \Leftrightarrow P$. This is not a tautology, and we wouldn't expect one, since we don't want every proposition P to be related to every proposition Q. Table 1 gives the truth table for $(P \wedge Q) \Leftrightarrow P$, which is the same as the truth table for $P \Rightarrow Q$, given in

TABLE 1. Truth Table for $(P \wedge Q) \Leftrightarrow P$

P	Q	$P \wedge Q$	$(P \wedge Q) \Leftrightarrow P$
T	T	T	T
T	F	F	F
F	T	F	T
F	F	F	T

TABLE 2. Truth Table for $P \Rightarrow Q$

P	Q	$P \Rightarrow Q$
T	T	T
T	F	F
F	T	T
F	F	T

Table 2. In effect, $\Rightarrow$ is a partial order on classes of logically equivalent propositions. Tautologies are the maximal elements under this ordering, and totally false statements are the minimal elements. ◇

PROBLEMS

*1. For each statement that follows, decide whether it is true or (at least sometimes) false. Explain your answer or cite the part of the text supporting it.

a) A semi-lattice cannot have an identity.

b) We can derive a partial order in a semi-lattice.

c) Two semi-lattices make a lattice. That is, if $(L, \sqcap)$ and $(L, \sqcup)$ are semi-lattices, then $(L, \sqcap, \sqcup)$ is a lattice.

d) A lattice with distributivity and identities is a Boolean algebra.

e) In logic, $P \wedge Q = Q \wedge P$.

2. For each set and operation, determine which of the properties of a semi-lattice fail. For those that fail, give a counterexample.

a) On $\mathbb{Q}$, the average of two numbers.

b) On $\mathbb{N}$, the product of two numbers.

c) On $\mathbb{Z}_3$, the operation Δ given by $a \Delta b = 2(a + b)(\mathrm{mod}\ 3)$. (Hint: Write out the Cayley table.)

*3. Each Hasse diagram in Figure 2 represents a lattice. Determine whether each lattice is distributive and has identities and complements. For each property, if it fails for a given lattice, give a counterexample.

4. Prove antisymmetry in Theorem 7.4.1.

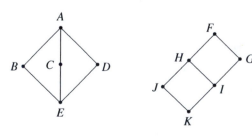
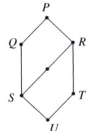

FIGURE 2. Hasse diagrams for three lattices

5. ***a)** Prove Theorem 7.4.3.

   b) For $a \in L$, and L a lattice, prove that $\{b \in L : a \sqsubseteq b\}$ is a sublattice of L.

   c) For $a, c \in L$, and L a lattice, prove that $K = \{b \in L : a \sqsubseteq b \text{ and } b \sqsubseteq c\}$ is a sublattice of L.

   d) Prove that K in part c has an identity for both operations $\sqcup$ and $\sqcap$.

   e) Prove by example that not every sublattice of a lattice has to have the form of one of the sublattices in parts a, b, and c.

6. Prove that the intersection of two sublattices is a sublattice.

7. Given two lattices $(L, \sqcap, \sqcup)$ and $(K, \overline{\wedge}, \underline{\vee})$, we can define operations on their Cartesian product $L \times K$ by $(a, b) \widehat{\sqcap} (c, d) = (a \sqcap c, b \overline{\wedge} d)$ and $(a, b) \widehat{\sqcup} (c, d) = (a \sqcup c, b \underline{\vee} d)$.

   a) Prove that $L \times K$ with these operations is a lattice.

   b) In $L \times K$, define $(a, b) \widehat{\in} (c, d)$ iff $a \sqsubseteq c$ for the partial order in L and $b \lessdot d$ for the partial order of K. Prove that $\widehat{\in}$ is the partial order derived from $\widehat{\sqcap}$ in part a, using Theorem 7.4.1.

   c) For $k \in K$, prove that $L_k = \{(x, k) : x \in L\}$ is a sublattice of $L \times K$.

8. Let D_n be the set of positive divisors of $n \in \mathbb{N}$. Assume, as can be proven, that D_n is a distributive lattice with the operations gcd and lcm.

   *a) For any D_n, find the identity for gcd and the identity for lcm. Prove that your choices are correct.

   b) Determine which of the following values of n make D_n a Boolean algebra: $n = 24$, $n = 26$, $n = 28$, and $n = 30$. (Hint: Draw the Hasse diagrams.) For those that are Boolean algebras, give the complement of each element. For those failing to be Boolean algebras, give an element without a complement.

   c) Make a conjecture about which values of n make D_n a Boolean algebra. Describe the complement of an element in such a D_n.

9. Let B be a Boolean algebra.

   a) Prove that B has a unique identity for $\sqcup$ and a unique identity for $\sqcap$.

   *b) Prove that each $a \in B$ has a unique complement in B.

   c) Prove that for all $a \in B$, $a \neq a'$.

   d) Prove one of De Morgan's laws: For all $a, b \in B$, $(a \sqcup b)' = a' \sqcap b'$. (Hints: Show that $(a' \sqcap b') \sqcup (a \sqcup b) = 1$ and $(a' \sqcap b') \sqcap (a \sqcup b) = 0$, using properties of a Boolean algebra. Then use part b. The other of De Morgan's laws, $(a \sqcap b)' = a' \sqcup b'$, is proven similarly.)

10. Suppose that L and K are Boolean algebras. Is $L \times K$ with the operations of Problem 7 a Boolean algebra? Prove or give a counterexample.

11. **a)** Find an example of a lattice that has an identity for one of the operations $\sqcup$ and $\sqcap$, but has no identity for the other operation.

   b) Find an example of a lattice that has no identity for either of the operations $\sqcup$ or $\sqcap$.

12. Let $S(\mathbb{Z}_{12})$ be the set of all subgroups of $\mathbb{Z}_{12}$.

   a) List the subgroups of $\mathbb{Z}_{12}$.

   b) Is $S(\mathbb{Z}_{12})$ a semi-lattice with the operation $\cap$? Explain.

   c) Is $S(\mathbb{Z}_{12})$ a semi-lattice with the operation $\cup$? Explain.

7.5   HOMOMORPHISMS

"Algebra is generous: she often gives more than is asked for."

—Jean d'Alembert (1717–1783)

Mathematicians developed abstract algebra to generalize and explore individual structures, as we did in earlier sections. But mathematicians benefited even more from the abstract approach, fulfilling d'Alembert's quote. The abstract approach gives a powerful way to study relationships between structures. A *homomorphism* is a structural mapping from one system to another so that the image captures some essential aspects of the original system. Often, the image is simpler than the original structure. Example 1 illustrates this process with modular arithmetic.

EXAMPLE 1

Define $f : \mathbb{Z} \to \mathbb{Z}_{10}$ by $f(x) = k$ iff $x \equiv k \pmod{10}$. For instance, $f(1087) = 7$ and $f(2058) = 8$. Addition and multiplication on the infinitely many elements of $\mathbb{Z}$ are reduced to the much simpler addition and multiplication on the 10 elements of $\mathbb{Z}_{10}$. Theorem 4.2.2 guarantees that the operations in $\mathbb{Z}_{10}$ reflect the corresponding operations in $\mathbb{Z}$. For instance, the last digits in $1087 + 2058 = 3145$ match $7 + 8 \equiv 5 \pmod{10}$, and $1087 \cdot 2058 = 2,237,046$ matches $7 \cdot 8 \equiv 6 \pmod{10}$. In $\mathbb{Z}_{10}$ we keep track of what happens only in the ones column of the corresponding sum or product in $\mathbb{Z}$.   ◇

We can generalize the idea of Example 1 to other algebraic systems. If $a * b = c$ in one system and we map a, b, and c to a', b', and c', respectively, in a related system, we expect $a' \cdot b' = c'$, where $\cdot$ is the corresponding operation in the second system. For the formal definition, we use function notation to rewrite this connection succinctly. We write $(S, *)$ to indicate a set S with an operation $*$. Similarly, $(S, *, \circledast)$ indicates a set S with two operations $*$ and $\circledast$.

DEFINITION.   A *homomorphism* from $(S, *)$ to $(T, \cdot)$ is a function $f : S \to T$ such that, for all $a, b \in S$, $f(a * b) = f(a) \cdot f(b)$. If $(S, *, \circledast)$ and $(T, \cdot, \odot)$ have two operations each, then we also require for all $a, b \in S$, $f(a \circledast b) = f(a) \odot f(b)$. If f is onto T, we say that T is a *homomorphic image* of S. If f is one-to-one and onto, it is an *isomorphism* and S and T are *isomorphic*.

The Greek prefix "homo" means "same," whereas the prefix "iso" means "equal." The root "morph" means "form." Homomorphic groups have similar form, and isomorphic ones have exactly the same form.

EXAMPLE 2

Show that the function $f : \mathbb{Z} \to \mathbb{Z}$ given by $f(x) = 3x$ is a homomorphism of the group $(\mathbb{Z}, +)$ onto the group of multiples of three, $(3\mathbb{Z}, +)$. Show that this function is not a homomorphism from the ring $(\mathbb{Z}, +, \cdot)$ to the ring $(3\mathbb{Z}, +, \cdot)$.

Solution Let $x, y \in \mathbb{Z}$. By distributivity, $f(x + y) = 3(x + y) = 3x + 3y = f(x) + f(y)$. Thus, f is a homomorphism for addition. The reader can show that f is onto $3\mathbb{Z}$. However, f is not a homomorphism for multiplication. For instance, $f(2 \cdot 4) = 3(2 \cdot 4) = 24$, whereas $f(2) \cdot f(4) = 3(2) \cdot 3(4) = 72 \neq 24$. Thus, this function doesn't "preserve" multiplication. ◇

EXAMPLE 3

In my high school days long before calculators, I learned how to convert tedious multiplication problems, such as $4, 321 \cdot 56, 789 = 245, 385, 269$, to easier addition problems using logarithms. I used a table of logarithms (base 10) to find the approximations $\log{(4321)} \doteq 3.64$ and $\log{(56789)} \doteq 4.75$. Then I added: $3.64 + 4.75 = 8.39$. Another consultation of the logarithm table told me $\log{(245, 500, 000)} \doteq 8.39$, giving a reasonable approximation for the product. We can show that this approximating process is valid by showing that $f : \mathbb{R}^+ \to \mathbb{R}$ given by $f(x) = \log{(x)}$ is an isomorphism from $(\mathbb{R}^+, \cdot)$ onto $(\mathbb{R}, +)$. Let $x, y \in \mathbb{R}^+$. From a basic property of logarithms, we have $f(x \cdot y) = \log{(x \cdot y)} = \log{(x)} + \log{(y)} = f(x) + f(y)$, exactly the definition for a homomorphism. (We leave the proof of one-to-one and onto to the reader, although picturing the graph of $y = \ln(x)$ is probably sufficiently convincing.)

Nowadays, statistical packages use programs for fitting lines to data and natural logarithms to fit exponential functions to data. It is easy for a computer to find the best linear fit to data. To find the best exponential curve $y = ce^{kx}$, we first convert the data pairs (s, t) to the pairs $(s, \ln(t))$. Now we have the computer find the best linear fit $Y = mX + b$ to this transformed data. Note that the function $f(x) = \ln(x)$ also transforms ce^{kt} to $\ln(ce^{kt}) = \ln(c) + kt$. This last formula is linear in t with slope k and y-intercept $\ln(c)$, which correspond to the m and the b of the best linear fit to the transformed data. We can use other isomorphisms to fit other types of curves to data. ◇

We turn now to an investigation of the characteristics that a system and any homomorphic image of it must share.

THEOREM 7.5.1. If $(G, *)$ is a group and $f : G \to J$ is a homomorphism onto $(J, \cdot)$, then $(J, \cdot)$ is a group. Also, $f(e_G) = e_J$, where e_G and e_J are the identities of G and J, respectively, and for all $g \in G$, $f(g^{-1}) = f(g)^{-1}$.

Proof. Essentially, the group properties of $(G, *)$ induce the corresponding properties for $(J, \cdot)$. We'll show associativity explicitly. Let $a', b', c' \in J$. Because f is onto, there are $a, b, c \in G$ such that $f(a) = a'$, $f(b) = b'$, and $f(c) = c'$. Then

$$a' \cdot (b' \cdot c') = f(a) \cdot (f(b) \cdot f(c)) \qquad \text{substitution}$$
$$= f(a) \cdot f(b * c) = f(a * (b * c)) \qquad \text{definition of a homomorphism}$$
$$= f((a * b) * c) \qquad \text{associativity in } G$$
$$= f(a * b) \cdot f(c) = (f(a) \cdot f(b)) \cdot f(c) \qquad \text{definition of homomorphism}$$
$$= (a' \cdot b') \cdot c'.$$

See Problem 8 for the rest of this proof. ∎

Theorem 7.5.1 tells us that homomorphisms preserve identities and inverses. Examples 4 and 5 use other properties preserved by homomorphisms to show that two systems can't be related by a homomorphism from one onto the other. In essence, we show that the systems differ in some essential way.

EXAMPLE 4

Show that there is no isomorphism from $\mathbb{Z}_2 \times \mathbb{Z}_2$ onto $\mathbb{Z}_4$, where $\mathbb{Z}_2 \times \mathbb{Z}_2$ is the group defined in Example 7 of Section 7.1.

Solution   Both groups have four elements, so there are many one-to-one functions from $\mathbb{Z}_2 \times \mathbb{Z}_2$ onto $\mathbb{Z}_4$. How can we show that none of these functions could be isomorphisms? We need to find some algebraic property of one of the groups that cannot be matched in the other group. For instance, each element x of $\mathbb{Z}_2 \times \mathbb{Z}_2$ satisfies $x \oplus x = (0, 0)$, the identity.

CLAIM.   This property must hold for all of the elements' images in $\mathbb{Z}_4$. Consider $f(x) + f(x)$, which equals $f(x \oplus x) = f(0, 0)$ by the definition of a homomorphism. From Theorem 7.5.1, $f(0, 0) = 0$. So, $f(x) + f(x) = 0$. In $\mathbb{Z}_4$, the only elements satisfying this condition are 0 and 2. Hence, the four elements of $\mathbb{Z}_2 \times \mathbb{Z}_2$ would have only two places to go, an impossibility for an isomorphism. There are homomorphisms from $\mathbb{Z}_2 \times \mathbb{Z}_2$ into $\mathbb{Z}_4$. For instance, $f(a, b) = 2a$ is a homomorphism into $\mathbb{Z}_4$, but it isn't onto, since neither 1 nor 3 is an image.   ◇

EXAMPLE 5

Show that there is no homomorphism from $(\mathbb{Q}, +)$ onto $(\mathbb{Z}, +)$.

Proof.   For a contradiction, suppose that $f : \mathbb{Q} \to \mathbb{Z}$ were a homomorphism onto $\mathbb{Z}$. Then some $q \in \mathbb{Q}$ must satisfy $f(q) = 1$. Consider $f(q/3)$. We have $f(q) = f(q/3 + q/3 + q/3) = f(q/3) + f(q/3) + f(q/3)$. Since $f(q) = 1$, we must have $3f(q/3) = 1$, but no element of $\mathbb{Z}$ can be a solution of $3z = 1$, giving a contradiction.   ■   ◇

Thus, the rationals and the integers are structurally very different as groups, even though in Section 5.2 we saw that they had the same cardinality.

A homomorphism often maps multiple elements to the same element. The elements going to the same place often share important properties. The next example approaches this idea from solving systems of linear equations in linear algebra. This example motivates the more general situation for all groups in Theorem 7.5.2, where the simplicity of the structure seems to make the proof clearer.

EXAMPLE 6

Consider the two related systems of equations $\begin{cases} 2x - y - z &= -3 \\ 4x - 3y + z &= 1 \end{cases}$ and $\begin{cases} 2x - y - z &= 0 \\ 4x - 3y + z &= 0 \end{cases}$.

We can rewrite these as $A\vec{v} = \vec{b}$ and $A\vec{v} = \vec{0}$, where $A = \begin{bmatrix} 2 & -1 & -1 \\ 4 & -3 & 1 \end{bmatrix}$, $\vec{v} = \begin{bmatrix} x \\ y \\ z \end{bmatrix}$, $\vec{b} = \begin{bmatrix} -3 \\ 1 \end{bmatrix}$, and $\vec{0} = \begin{bmatrix} 0 \\ 0 \end{bmatrix}$. The general solution to the first system is derived from one solution to it, say, $\begin{bmatrix} 1 \\ 2 \\ 3 \end{bmatrix}$, together with all solutions to the second system, $c \begin{bmatrix} 2 \\ 3 \\ 1 \end{bmatrix}$, where c is a constant (scalar). That is, the solutions to the first system are of the form $\begin{bmatrix} 1 \\ 2 \\ 3 \end{bmatrix} + c \begin{bmatrix} 2 \\ 3 \\ 1 \end{bmatrix}$. The set of vectors $K = \left\{ c \begin{bmatrix} 2 \\ 3 \\ 1 \end{bmatrix} : c \in \mathbb{R} \right\}$ is a subspace of the vector space $\mathbb{R}^3$, called the *kernel*.

Recall that vector spaces are groups, and so K is a subgroup. Further, the set of solutions to the first system is a coset of K: $\begin{bmatrix} 1 \\ 2 \\ 3 \end{bmatrix} + K = \left\{ \begin{bmatrix} 1 \\ 2 \\ 3 \end{bmatrix} + c \begin{bmatrix} 2 \\ 3 \\ 1 \end{bmatrix} : c \in \mathbb{R} \right\}$.

We can think of the matrix A as a function, called a linear transformation, from $\mathbb{R}^3$ to $\mathbb{R}^2$. Linear transformations are homomorphisms. Thus, the matrix A takes cosets of K in $\mathbb{R}^3$ to individual vectors in $\mathbb{R}^2$. (The definition of a linear transformation $f : V \to W$ requires that, for all $\vec{x}, \vec{y} \in V$ and all scalars c, $f(\vec{x} + \vec{y}) = f(\vec{x}) + f(\vec{y})$ and $f(c\vec{x}) = cf(\vec{x})$. The first equation is exactly the requirement for a group homomorphism.) $\diamond$

THEOREM 7.5.2. If $f : G \to J$ is a homomorphism from a group $(G, *)$ onto a group $(J, \cdot)$, then

i) $K = \{g \in G : f(g) = e_J\}$ is a subgroup of G, where e_J is the identity of J, and

ii) for all $j \in J$, the pre-image of j, $f^{-1}[\{j\}] = \{g \in G : f(g) = j\}$, is a coset of K.

Proof.

i) Suppose that $g, g' \in K$; that is, $f(g) = e_J = f(g')$. Then $f(g * g') = f(g) \cdot f(g') = e_J \cdot e_J = e_J$. Thus, $g * g' \in K$, giving closure for K. See Problem 9a for the rest of this part. We call K the *kernel* of the homomorphism.

ii) For $j \in J$, there is some $a \in G$ such that $f(a) = j$, because f is onto. The only plausible coset of K is $a * K$, so we need to show that $a * K = \{g \in G : f(g) = j\}$. See Problem 9b for the proof of $a * K \subseteq \{g \in G : f(g) = j\}$.

For the other containment, let $g \in G$ satisfy $f(g) = j$, or equivalently, $f(g) = f(a)$. We need to show that $g = a * k$, for some $k \in K$. By Theorem 7.2.1, we know that the only choice for k is $a^{-1} * g$. But is $a^{-1} * g \in K$? Now $f(a^{-1} * g) = f(a^{-1}) \cdot f(g) = f(a)^{-1} \cdot f(a) = e_J$. Thus, indeed, $a^{-1} * g \in K$ and $g \in a * K$. ∎

The structural nature of Theorem 7.5.2 restricts our ability to create homomorphisms from one group onto another group. Corollary 7.5.3 depends on this theorem as well as Lagrange's Theorem (Theorem 7.2.6).

COROLLARY 7.5.3. If G is a finite group and $f : G \rightarrow J$ is a homomorphism onto J, then $|J|$ divides $|G|$.

Proof. From Theorem 7.5.2, $|J|$ is the number of cosets of K. Further, Lagrange's Theorem assures us that $|J| \cdot |K| = |G|$. Thus, $|J|$ divides $|G|$. ∎

Mathematicians value homomorphisms in part because they help us understand a complicated situation in terms of a simpler, but structurally related, one. Example 7 illustrates this process.

EXAMPLE 7

For $m \neq 0$, the first-degree function $f_{m,b}(x) = mx + b$ is a bijection of $\mathbb{R}$. Problem 10 asks you to show that the set $F = \{f_{m,b} : m, b \in \mathbb{R}$ and $m \neq 0\}$ of all of these functions forms a group under composition. With two parameters (the slope m and the y-intercept b), it is awkward to see how these functions interact or which parameter is more fundamental. In a sense that we will make precise through a homomorphism, the slope matters more than the y-intercept. Let's turn this idea into mathematics. Recall that $\mathbb{R}^*$, the nonzero reals, is a group under multiplication. We can easily show that $h : F \rightarrow \mathbb{R}^*$ given by $h(f_{m,b}) = m$ is a homomorphism: $h(f_{m,b} \circ f_{k,c}) = h(f_{mk,mc+b}) = mk = h(f_{m,b}) \cdot h(f_{k,c})$. The lines with slope 1, which are of the form $f_{1,b}$, map to the identity of $\mathbb{R}^*$ and form a subgroup. The cosets of this subgroup structurally combine parallel lines together. For instance, the line $f_{5,2}$ has in its coset all lines of the form $f_{5,b}$.

While the previous homomorphism shows that it is natural to cluster these functions together by their slopes, we need to ask whether they could be clustered together just as well by their y-intercepts. Maybe we can map F to the group $\mathbb{R}$ with addition to mimic the action of the intercepts. Let's try it. Define $g : F \rightarrow \mathbb{R}$ by $g(f_{m,b}) = b$. Is g a homomorphism? No. For instance, $g(f_{2,1} \circ f_{2,2}) = g(f_{4,5}) = 5$, but $g(f_{2,1}) + g(f_{2,2}) = 1 + 2 = 3$. Even though we can classify lines by their y-intercepts, there is no structural relationship in such clustering. ◇

Mathematicians have used the simplifying aspects of homomorphisms to reduce difficult problems to more manageable ones. In the 1880s, two mathematicians, E. Fedorov and A. Schoenflies, were able independently to classify the 230 possible groups of symmetries corresponding to chemical crystals. These groups are quite complicated and involve infinitely many translations in all three dimensions. Fortunately, there is a homomorphism taking each potential infinite group down to a finite "local" group. Thus, these mathematicians were able to build on J. Hessel's 1830 determination of the 32 possible "local" groups to determine the possible infinite groups.

THEOREM 7.5.4. If $(R, +, \cdot)$ is a ring and $f : R \to S$ is a homomorphism onto $(S, \oplus, \odot)$, then $(S, \oplus, \odot)$ is a ring. If R has a unity 1, then $f(1)$ is the unity of S. If $x \in R$ has an inverse x^{-1}, then $f(x)$ has $f(x^{-1})$ as its inverse.

Proof. See Problem 14. ∎

In spite of Theorem 7.5.4, the homomorphic image of a field need not be a field. We can map every element of a field to 0, which will give a homomorphism. However, a field has to have a unity not equal to its additive identity.

THEOREM 7.5.5. If $(L, \sqcup, \sqcap)$ is a lattice and $f : L \to M$ is a homomorphism onto $(M, \uplus, \cap\!\!\!\cap)$, then $(M, \uplus, \cap\!\!\!\cap)$ is a lattice.

Proof. See Problem 19. ∎

Just as Lagrange's Theorem doesn't hold for lattices, Corollary 7.5.3 fails for some finite lattices, as Problems 21 and 23 illustrate. Example 8 considers a related situation with an infinite lattice.

EXAMPLE 8

A probability is a number from the interval $[0, 1]$. This interval forms a lattice with the operations max and min and the associated partial order $\leq$. Psychologists have found that many people reason with only approximately five categories of likelihood. From a mathematical point of view, people use an informal homomorphism from the infinite system $[0, 1]$ to a simpler system when they deal with uncertainty. The accompanying function gives an approximation of this mapping. Although this uneven collapsing of probabilities into categories is a homomorphism, it may not serve us well in our evaluation of risk. For instance, people may not worry about behaviors that increase their risk of a kind of cancer from 1 in 10,000 to 1 in 1000, since they may classify both risks as unlikely. However, people often emphasize the distinction between a risk of 1 in a million and a risk of 0, even though this is a much smaller difference than the earlier one. Such discrepancies illustrate what some researchers call "probability blindness." (See Piattelli.) ◇

$$
f(x) = \begin{cases}
0 & \text{if } x = 0 & \text{impossible} \\
0.2 & \text{if } 0 < x < 0.4 & \text{unlikely} \\
0.5 & \text{if } 0.4 < x \leq 0.6 & \text{about even odds} \\
0.8 & \text{if } 0.6 \leq x < 1 & \text{likely} \\
1 & \text{if } x = 1 & \text{certain}
\end{cases}
$$

THEOREM 7.5.6. If $f : L \to M$ is a lattice homomorphism, f is onto M, and $m \in M$, then $L_m = \{a \in L : f(a) = m\}$ is a sublattice of L.

Proof. See Problem 20. ∎

PROBLEMS

1. Define $f : \mathbb{Z} \to \{1, i, -1, -i\}$ by $f(z) = i^z$, where $i = \sqrt{-1}$ in the complex numbers. Prove that f is a homomorphism from $(\mathbb{Z}, +)$ onto $(\{1, i, -1, -i\}, \cdot)$.

2. *a) Prove that $f : \mathbb{R}^* \to \mathbb{R}^*$ given by $f(x) = x^2$ is a homomorphism from $(\mathbb{R}^*, \cdot)$ to itself. Is f an isomorphism? Prove your answer.

   b) Repeat part a with the function $g : \mathbb{R}^* \to \mathbb{R}^*$ given by $g(x) = x^3$.

3. *a) Use Tables 1 and 2 in Section 7.2 to give two reasons why the groups $\mathbb{Z}_6$ and S_3 are not isomorphic, even though both have six elements.

   b) Using Problem 6 of Section 7.2, we can name at least three groups with eight elements: $\mathbb{Z}_8$, $\mathbb{Z}_4 \times \mathbb{Z}_2$, and $\mathbb{Z}_2 \times \mathbb{Z}_2 \times \mathbb{Z}_2$. Show that these groups are not isomorphic.

   c) At the end of Section 7.2, we discussed the eight-element group of the symmetries of a square $Sym(ABCD)$. Show that this group differs from each of the groups in part b.

4. Let $\mathbb{P}(\mathbb{R})$ be the group of all polynomials with real coefficients and with the operation of addition, and let $D : \mathbb{P}(\mathbb{R}) \to \mathbb{P}(\mathbb{R})$ be the derivative function. For instance, $D(x^3 - 2x^2 + 3x - 4) = 3x^2 - 4x + 3$.

   a) Use properties from calculus to prove that D is a homomorphism.

   *b) What is the subgroup of polynomials that D maps to the 0 polynomial?

   c) What is the coset of polynomials mapped to $2x^2 - x + 3$?

   d) $\mathbb{P}(\mathbb{R})$ is also a ring. Is D a homomorphism for the ring $\mathbb{P}(\mathbb{R})$ to itself? Justify your answer.

5. a) For $\mathbb{P}(\mathbb{R})$ as in Problem 4, define $I(g)$ to be the integral of g with the constant term of 0. For instance, $I(x^2 - 2x + 1) = x^3/3 - x^2 + x$. Show that $I : \mathbb{P}(\mathbb{R}) \to \mathbb{P}(\mathbb{R})$ is a group homomorphism.

   b) If we consider $\mathbb{P}(\mathbb{R})$ as a ring, is I a homomorphism? Justify your answer.

   c) If we change the constant term in part a to 3, is the new function still a homomorphism? Prove your answer.

6. Show that the groups $(\mathbb{Q}^*, \cdot)$ and $(\mathbb{Q}^+, \cdot)$ are not isomorphic.

7. Let $T = \left\{ \begin{bmatrix} r & s \\ 0 & t \end{bmatrix} : r, s, t \in \mathbb{R} \text{ and } r \neq 0 \text{ and } t \neq 0 \right\}$.

   a) Prove that T is a group under matrix multiplication.

   b) Is $f : T \to \mathbb{R}^*$ given by $f\left(\begin{bmatrix} r & s \\ 0 & t \end{bmatrix} \right) = r$ a homomorphism, where multiplication is the operation in $\mathbb{R}^*$? Prove your answer.

c) Repeat part b, using $g : T \rightarrow \mathbb{R}$ given by $g\left(\begin{bmatrix} r & s \\ 0 & t \end{bmatrix}\right) = s$, where $\mathbb{R}$ has addition for the operation.

d) Repeat part b, using $h : T \rightarrow \mathbb{R}^* \times \mathbb{R}^*$ given by $h\left(\begin{bmatrix} r & s \\ 0 & t \end{bmatrix}\right) = (r, t)$, where $\mathbb{R}^* \times \mathbb{R}^*$ is the group obtained from $\mathbb{R}^*$, using Problem 6 of Section 7.2.

8. Finish the proof of Theorem 7.5.1.

9. a) Finish part i of Theorem 7.5.2.

   b) In Theorem 7.5.2, suppose $f(a)=h$. Show that $a * K \subseteq \{g \in G : f(g) = h\}$. (Hint: If $b \in a * K$, we can write $b = a * k_1$ for some $k_1 \in K$.)

10. In Example 7, prove that F is a group.

11. a) Show that isomorphisms are "reflexive." That is, there is an isomorphism from any algebraic structure onto itself.

    b) Show that isomorphisms are "symmetric." That is, if there is an isomorphism from one algebraic structure X onto another one Y, then there is an isomorphism from Y onto X.

    c) Show that isomorphisms are "transitive." That is, if $f : X \rightarrow Y$ and $g : Y \rightarrow Z$ are isomorphisms, then there is an isomorphism from X onto Z.

12. Let R and S be any rings. Prove that the function $f : R \rightarrow S$ given by $f(r) = 0_S$ is a ring homomorphism, where 0_S is the identity of S.

13. Let $A = \left\{ \begin{bmatrix} a & b \\ -b & a \end{bmatrix} : a, b \in \mathbb{R} \right\}$ and assume that its operations are matrix addition and multiplication. Find an isomorphism from the field $\mathbb{C}$ onto A. Prove that your function is an isomorphism.

14. Suppose that f is a homomorphism from a ring R onto a ring S.

    a) Prove Theorem 7.5.4.

    b) If R is commutative, prove that S is commutative.

    c) If R is a field and S has more than one element, prove that S is a field.

*15. Homomorphisms do not "preserve" every algebraic property. Find a ring R with the property that for all $a, b \in R$, if $a \cdot b = 0$, then $a = 0$ or $b = 0$; a ring S without this property; and a homomorphism from R onto S.

16. Suppose that $R \times S$ is the ring defined in Problem 10a of Section 7.3 from the rings R and S.

    a) Find a homomorphism from $R \times S$ onto R and prove that it is a homomorphism.

    b) Repeat part a from $R \times S$ onto S.

    c) Find an isomorphism from $R \times S$ onto $S \times R$ and prove that it is an isomorphism.

17. *a) Show that the group $(\mathbb{C}, +)$ is isomorphic to the group $(\mathbb{R} \times \mathbb{R}, +)$, as defined in Problem 6 of Section 7.2.

    *b) Show that the ring $(\mathbb{C}, +, \cdot)$ is not isomorphic to the ring $(\mathbb{R} \times \mathbb{R}, +, \cdot)$ with multiplication as defined by Problem 10a of Section 7.3.

   c) Show that the ring $(\mathbb{C}, +, \cdot)$ is not isomorphic to the ring $(\mathbb{R} \times \mathbb{R}, +, \cdot)$ with multiplication as defined by Problem 10b of Section 7.3.

18. Both $\mathbb{Q}$ and $\mathbb{Q}[\sqrt{2}] = \{a + b\sqrt{2} : a, b \in \mathbb{Q}\}$ are fields. (See Problem 9 of Section 7.3.) Show that these fields are not isomorphic.

19. a) Prove Theorem 7.5.5.

   b) Prove that the homomorphic image of a semi-lattice is a semi-lattice.

20. Prove Theorem 7.5.6, using the following steps:

   a) For $a, b \in L_m$, we have $f(a) = m = f(b)$. Show closure: $f(a \sqcup b) = m$ and $f(a \sqcap b) = m$.

   b) Explain why all of the properties of a lattice hold for any subset of it.

21. Let $\mathbb{N}_n$ be the lattice of the set $\{1, 2, \ldots n\}$ with the operations of max and min.

   a) If $k < n$, prove that $f : \mathbb{N}_n \to \mathbb{N}_k$ given by $f(x) = \left\{ \begin{array}{l|ll} x & \text{if} & x \le k \\ \hline k & \text{if} & x > k \end{array} \right\}$ is a homomor-

   phism. (Hint: Consider cases. Note: Since k does not need to divide n, Corollary 7.5.3 need not hold for lattices.)

   b) Prove that the function f in part a will be a homomorphism from $\mathbb{N}$ onto $\mathbb{N}_k$ as lattices.

22. Recall that D_n is the set of all positive divisors of n. It is a lattice with the operations of gcd and lcm.

   *a) Find an isomorphism from D_6 onto D_{10}. (Hint: Draw the Hasse diagram.)

   b) Find an isomorphism from D_4 onto D_{25}.

   c) Find an isomorphism from D_{12} onto D_{45}.

   d) Describe conditions on n and k so that the lattices D_n and D_k are isomorphic.

23. Verify that the mapping $f : D_{12} \to D_6$ given by $f(x) = \left\{ \begin{array}{l|lc} 1 & \text{if} & x = 1 \\ \hline 2 & \text{if} & x = 2 \text{ or } x = 4 \\ \hline 3 & \text{if} & x = 3 \\ \hline 6 & \text{if} & x = 6 \text{ or } x = 12 \end{array} \right\}$ is

a homomorphism for the lattices with the operations of gcd and lcm. Note $|D_6| = 4$ does not divide $|D_{12}| = 6$, showing that Corollary 7.5.3 does not always hold with lattices.

24. Show that $(D_{18}, \gcd, \text{lcm})$ is not isomorphic to any Boolean algebra.

25. Suppose that B is a Boolean algebra with minimal element 0, maximal element 1, and x' the complement of x. Also, suppose that $f : B \to A$ is a lattice homomorphism onto A.

   a) Prove that $f(0)$ is the minimal element of A.

   b) Prove that $f(1)$ is the maximal element of A.

   c) Prove that $f(x')$ is the complement of $f(x)$ in A.

   d) Prove that A is distributive.

      Parts a to d show that A is a Boolean algebra, provided that $f(0) \ne f(1)$.

26. Let B be a Boolean algebra with x' the complement of x. For $f(x) = x'$, prove that $f : B \to B$ is an isomorphism.

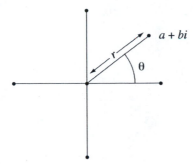

FIGURE 1

27. Complex multiplication, given by $(a + bi)(c + di) = (ac - bd) + (ad + bc)i$, seems, well, complex to some students. Figure 1 suggests that we can write any $a + bi \neq 0$ as $r \cos(\theta) + ri \sin(\theta)$, where r is the length of the vector for the origin to $a + bi$ and θ is the angle this vector makes with the positive x-axis.

   a) Use the formulas for $\cos(\theta + \sigma)$ and $\sin(\theta + \sigma)$ to show that $[r \cos(\theta) + ri \sin(\theta)] \cdot [s \cos(\sigma) + si \sin(\sigma)] = rs \cos(\theta + \sigma) + rsi \sin(\theta + \sigma)$.

   b) Both $\mathbb{C}^*$ and $\mathbb{R}^+$ are groups under multiplication. Prove that $f : \mathbb{C}^* \to \mathbb{R}^+$ given by $f(r \cos(\theta) + ri \sin(\theta)) = r$ is a homomorphism for these groups. Thus, complex multiplication is "like" ordinary multiplication in terms of lengths.

   c) We consider two angles equal if they differ by a multiple of 2π. In effect, we add angles "mod 2π." Let $A = \{\theta : 0 \le \theta < 2\pi\}$ be the group of angles with addition (mod 2π). Show that $g : \mathbb{C}^* \to A$ is a homomorphism converting multiplication to addition (mod 2π). Thus, complex multiplication is "like" addition of angles.

   d) Show that $\mathbb{C}^*$ with multiplication is isomorphic to $\mathbb{R}^+ \times A$, as defined in Problem 6 of Section 7.2. Thus, complex multiplication is just the combining of the operations in $\mathbb{R}^+$ and A.

28. Critique the questionable "proofs" that follow. Point out reasoning errors and unclear presentation. If the argument is a proof, say so. (If the claim is false, there must be an error in the argument. However, it is not enough to say that the claim is false. You need to find an error in the reasoning.)

   *a) Claim: If $f : X \to Y$ is a homomorphism from $(X, *)$ onto $(Y, \circledast)$ and e_Y is the identity of Y and $f(e) = e_Y$, then e is the identity of X. "Proof": Suppose for $e \in X$ that $f(e) = e_Y$. Let $x \in X$. Then $f(x * e) = f(x) \circledast f(e) = f(x) \circledast e_Y = f(x)$. From $f(x * e) = f(x)$, we get $x * e = x$. A similar argument shows that $e * x = x$. Hence, e is the identity of X. ▲

   b) Suppose that $f : G \to J$ is a homomorphism from the group $(G, *)$ onto the group $(J, \circledast)$. Claim: If H is a subgroup of J, then $G_H = \{g \in G : f(g) \in H\}$ is a subgroup of G. "Proof": Let $a, b \in G_H$. That is, $f(a) \in H$ and $f(b) \in H$. Then $f(a * b) = f(a) \circledast f(b) \in H$, showing that $a * b \in G_H$. Hence, G_H is closed. Further, $f(e_G) = e_J \in H$, so the identity of G is in G_H. Next, suppose that $g \in G_H$. That is, $f(g) \in H$. Because H is a subgroup, $f(g)^{-1} = f(g^{-1}) \in H$, showing that $g^{-1} \in G_H$. Finally, any subset of G satisfies associativity. Thus, G_H is a subgroup of G. ▲

   c) Claim: There is a homomorphism from $\mathbb{Z}_6$ onto $\mathbb{Z}_5$. "Proof": Define $f : \mathbb{Z}_6 \to \mathbb{Z}_5$ by $f(x) = 2x \pmod 5$. First we show that f is onto. Clearly, f takes the set $\{0, 1, 2\}$ to $\{0, 2, 4\}$. Also, f takes $\{3, 4, 5\}$ to $\{1, 3, 0\}$ because we are working (mod 5). Hence, f is onto. Let $x, y \in \mathbb{Z}_6$. Then $f(x + y) = 2(x + y)$ (mod 5) $= 2x + 2y$ (mod 5) $= f(x) + f(y)$, showing that f is a homomorphism. ▲

REFERENCE

PIATTELLI, M. 1991. Probability blindness: Neither rational nor capricious. *Bostonia*. March/April, 28–35.

DEFINITIONS FROM CHAPTER 7

Section 7.1

■ A (*binary*) *operation* on a set A is a function from $A \times A$ to A.

■ Let $*$ be an operation on a set S. Then

$*$ is *associative* iff for all $a, b, c \in S, (a * b) * c = a * (b * c)$.

$*$ is *commutative* iff for all $a, b \in S, a * b = b * a$.

$*$ has an *identity* iff there is some $e \in S$ such that for all $a \in S, a * e = a = e * a$.

   If $*$ has an identity e, then $*$ has *inverses* iff for all $a \in S$, there is some $b \in S$ such that $a * b = e = b * a$.

$*$ is *idempotent* iff for all $a \in S, a * a = a \in S$.

$*$ is *closed* on the subset T of S iff for all $a, b \in T, a * b \in T$.

$*$ *distributes* over an operation $+$ iff for all $a, b, c \in S, a * (b + c) = (a * b) + (a * c)$ and $(b + c) * a = (b * a) + (c * a)$.

Section 7.2

■ A *group* is a set G with an operation $*$ such that $*$ is associative and has an identity and inverses. More formally,

   i) for all $a, b, c \in G, a * (b * c) = (a * b) * c$,

   ii) there is an *identity* $e \in G$ such that for all $a \in G, a * e = a = e * a$, and

   iii) for all $a \in G$, there is an *inverse* $a^{\#} \in G$ such that $a * a^{\#} = e = a^{\#} * a$.

   If, in addition, for all $a, b \in G, a * b = b * a$, the group is *commutative* (or *Abelian*).

■ An operation $*$ on a set X has *cancellation* iff for all $a, b, c \in X$, if $a * b = a * c$, then $b = c$ and if $b * a = c * a$, then $b = c$.

■ A nonempty subset H of a group G is a *subgroup* of G iff H is a group using the same operation as G.

■ For $a \in G$ and H a subgroup of G, the *coset* $a * H = \{a * h : h \in H\}$.

■ Define $\odot$ on $G \times K$ by $(g, k) \cdot (h, j) = (g * h, k \circledast j)$, where $*$ is an operation on G and $\circledast$ is an operation on K.

▪ Suppose that a set X has a distance d on it. A *symmetry* σ of X is a permutation of X that preserves distance. That is, for all $a, b \in X$, $d(a, b) = d(\sigma(a), \sigma(b))$.

Section 7.3

▪ A *ring R* is a set with two operations $+$ and $\cdot$ so that

$+$ is associative	$\cdot$ is associative
$+$ is commutative	$\cdot$ distributes over $+$.
$+$ has an identity 0	
$+$ has inverses, written $-x$	

▪ A *field F* is a set with two operations $+$ and $\cdot$ so that

$+$ is associative	$\cdot$ is associative
$+$ is commutative	$\cdot$ is commutative
$+$ has an identity 0	$\cdot$ has a unity 1, where $1 \neq 0$
$+$ has inverses, written $-x$	$\cdot$ has inverses for non-zero elements, written x^{-1}
	$\cdot$ distributes over $+$.

Section 7.4

▪ A set S with an operation $\sqcap$ is a *semi-lattice* iff for all $a, b, c \in S$,

$\sqcap$ is idempotent: $a \sqcap a = a$,

$\sqcap$ is associative: $a \sqcap (b \sqcap c) = (a \sqcap b) \sqcap c$, and

$\sqcap$ is commutative: $a \sqcap b = b \sqcap a$.

▪ A *lattice* is a set S with two operations $\sqcap$ and $\sqcup$ such that for all $a, b, c \in S$,

idempotency: $a \sqcup a = a$ $a \sqcap a = a$,

associativity: $a \sqcup (b \sqcup c) = (a \sqcup b) \sqcup c$ $a \sqcap (b \sqcap c) = (a \sqcap b) \sqcap c$,

commutativity: $a \sqcup b = b \sqcup a$ $a \sqcap b = b \sqcap a$, and

absorptivity: $a \sqcup (a \sqcap b) = a$ $a \sqcap (a \sqcup b) = a$.

▪ A subset S of a lattice L is a *sublattice* of L iff S is a lattice using the same operations as L.

▪ A *Boolean algebra B* is a set B with two operations $\sqcap$ and $\sqcup$ such that for all $a, b, c \in B$,

idempotency: $a \sqcup a = a$ $a \sqcap a = a$,

associativity: $a \sqcup (b \sqcup c) = (a \sqcup b) \sqcup c$ $a \sqcap (b \sqcap c) = (a \sqcap b) \sqcap c$,

commutativity: $a \sqcup b = b \sqcup a$ $a \sqcap b = b \sqcap a$,

absorptivity: $a \sqcup (a \sqcap b) = a$ $a \sqcap (a \sqcup b) = a$,

distributivity: $a \sqcup (b \sqcap c) = (a \sqcup b) \sqcap (a \sqcup c)$ $a \sqcap (b \sqcup c) = (a \sqcap b) \sqcup (a \sqcap c)$,

identity: there is $0 \in B$ such that for all a, $0 \sqcup a = a$ there is $1 \in B$ such that for all a, $1 \sqcap a = a$

with $0 \neq 1$, and

complements: for all $a \in B$, there is $a' \in B$ such that $a \sqcup a' = 1$ and $a \sqcap a' = 0$.

Section 7.5

▪ A *homomorphism* from $(S, *)$ to $(T, \cdot)$ is a function $f : S \to T$ such that for all $a, b \in S$, $f(a * b) = f(a) \cdot f(b)$. If $(S, *, \circledast)$ and $(T, \cdot, \odot)$ have two operations each, then we also require that for all $a, b \in S$, $f(a \circledast b) = f(a) \odot f(b)$. If f is onto T, we say that T is a *homomorphic image* of S. If f is one-to-one and onto, it is an *isomorphism* and S and T are *isomorphic*.

CHAPTER 8

INTRODUCTION TO ANALYSIS

THE MAJOR AREA of analysis grew out of the nineteenth century effort to provide a solid foundation for calculus and related areas. Sir Isaac Newton (1642–1727) and Gottfried Leibniz (1646–1716) deserve credit for finding the key ideas for calculus and fitting them into a unified subject. While the results of Newton, Leibniz, and others were immediately seen as insightful, applicable, and surely correct, the arguments supporting them were often less than satisfactory. Early arguments typically used vague notions of "infinitesimals" (infinitely small numbers), appealing to intuition as much as providing proof. Augustin Cauchy (1789–1857) and a large number of other nineteenth century mathematicians transformed many of those insights into a solid foundation called analysis. While investigating derivatives, integrals, continuity, sequences, and other calculus topics they realized the need to understand the real number system much more profoundly. In turn, their novel insights about $\mathbb{R}$ led to new areas of mathematics. Our short survey of analysis starts with the real numbers before turning to limits, continuous functions, and sequences. In Section 8.3 we will also address the important role that counterexamples play in analysis. In the final section, we dip into dynamical systems. The language of analysis allows us to give an exact definition of the very modern concept of chaos in dynamical systems. The approximations and limits so characteristic of analysis also underlie dynamical systems and chaos.

8.1 REAL NUMBERS, APPROXIMATIONS, AND EXACT VALUES

"Although this may seem a paradox, all exact science is dominated by the idea of approximation."

—Bertrand Russell (1872–1970)

People's numerical intuition generally suffices for the "real world" situations they encounter. (One place it can fail is with Zeno's Paradoxes, discussed after the problems in this section.) However, in the "world of the reals," $\mathbb{R}$, we need a more sophisticated understanding to prove theorems. We'll build our understanding from two familiar and valuable views of real numbers: the points on a line and decimal representations.

We commonly picture a line as a smooth, unbroken straight path along which a point can slide continuously. This image fits well with continuous functions in calculus. We often visualize the variable sliding continuously along the x-axis while a point traces out the graph of the function. But the x-axis is also the set $\mathbb{R}$, made up of numbers. How can we characterize the nature of the entire set from the separate numbers in it? The "density" of $\mathbb{R}$ presents a first difficulty for visualizing it as a set built from individual numbers.

DEFINITION.   A subset W of a partially ordered set X is *dense in X* iff for all $x, y \in X$, if $x < y$, then there is $w \in W$ such that $x < w$ and $w < y$. If X is dense in itself, we say simply that X is *dense*.

The rationals, $\mathbb{Q}$, are already dense in $\mathbb{R}$, even though they are "missing" the many irrational numbers. Indeed, by Problem 7b of Section 5.3, there are uncountably many irrational numbers and only countably many rationals. We strain our intuition to visualize the difference between the number line of all the reals and the string of rational numbers. We need a subtle property to capture formally the unbroken nature of $\mathbb{R}$ as a continuum of individual numbers. Later in this section, we discuss this added condition, called the completeness property.

While our elementary intuition of a line is not sufficiently subtle for analysis, decimal representations have a different shortcoming. Although they are familiar and possess the subtlety of the real numbers, decimal representations are not easy objects to understand or use. In grade school, students already need somehow to digest infinitely repeating decimals, such as $0.333\ldots = \frac{1}{3}$, on the same footing as $0.5 = \frac{1}{2}$ and $0.25 = \frac{1}{4}$. Even more mysteriously, grade school students are told that $0.999\ldots = \frac{3}{3} = 1$. Many calculus students officially "know" this fact, but a large percentage of them don't believe it. Most, not unreasonably, believe that $0.999\ldots$ ought to be ever so slightly less than 1, since $0.9 < 1, 0.99 < 1, 0.999 < 1$, etc. (By the same reasoning, $0.333\ldots$ is ever so slightly less than $\frac{1}{3}$.) Pedagogically speaking, the problem stems from students using fancy things like infinite decimals long before they have the conceptual machinery of a limit and an infinite series to define them. This well reflects the historical order of events, since mathematicians happily developed calculus for 150 years without the concept of a limit. (We will mention only in passing the practical computing problems of infinite decimals. Consider finding the exact value of the square of the infinite nonrepeating decimal $0.919919991\ldots$; this square is a bit more than 0.84625.)

Let's look more closely at whether $0.999\ldots$ equals 1 or is a bit less than 1. If $x < y$, then $y - x > 0$. If we write $y - x$ using decimals, then $y - x > 0$ can be rephrased as "There is some nonzero decimal place in $y - x$." For instance, $1 - 0.99 = 0.01$ and $1 - 0.999999 = 0.000001$. What about $1 - 0.999\ldots$? The difference is less than 0.1, it is less than 0.000001, and it is even less than $0.000000000000000000000000001$. I have had students say that the difference is $0.000\ldots 1$, where they use the ellipsis $\ldots$ to represent infinitely many zeros before that final one. In effect, those students are positing an *infinitesimal*, an infinitely small, but positive, number. There is nothing logically wrong with such a notion. Indeed, in

1960, Abraham Robinson provided a solid mathematical foundation for such a number system, starting a subject area called nonstandard analysis. (See Keisler.) However, as we'll discuss in Example 6, infinitesimals violate the completeness property, which embodies the intuition of the continuity of the real line. Using a nonstandard model of the reals as a foundation for analysis goes way beyond the level of this text. Outside of Example 6, we take the standard approach.

The logically simplest foundation, which we use, simply asserts the properties that the set $\mathbb{R}$ satisfies. The rationals $\mathbb{Q}$ and the reals $\mathbb{R}$ satisfy many familiar fundamental algebraic and inequality properties. In addition, $\mathbb{R}$ satisfies one more key property, the completeness property, which makes calculus and analysis possible. We start with the algebraic properties defining a field—properties allowing us to add, subtract, multiply, and divide, as we have done since grade school. We will use these algebraic properties without mention, focusing on the properties more characteristic for analysis. (For more on fields, see Section 7.3.)

DEFINITION. A *field F* is a set with two operations, written $+$ and $\cdot$, satisfying these properties: For all $x, y, z \in F$,

i) $x + y = y + x$, ii) $x \cdot y = y \cdot x$ (commutativity),

iii) $x + (y + z) = (x + y) + z$, iv) $x \cdot (y \cdot z) = (x \cdot y) \cdot z$ (associativity),

v) $x \cdot (y + z) = (x \cdot y) + (x \cdot z)$, (distributivity),

vi) there is $0 \in F$ such that for all
$\quad x \in F, x + 0 = x = 0 + x$ (additive identity),

vii) there is $1 \in F$ such that $1 \neq 0$ and
$\quad$ for all $x \in F, x \cdot 1 = x = 1 \cdot x$ (multiplicative identity).

viii) for all $x \in F$, there is $y \in F$ such
$\quad$ that $x + y = 0 = y + x$ (additive inverses),

ix) for all $x \in F$, if $x \neq 0$, there is $y \in F$
$\quad$ such that $x \cdot y = 1 = y \cdot x$ (multiplicative inverses).

EXAMPLE 1

The sets $\mathbb{R}$, $\mathbb{Q}$, and $\mathbb{C}$ with addition and multiplication form fields. The integers $\mathbb{Z}$ do not form a field, because property ix of multiplicative inverses doesn't hold in $\mathbb{Z}$. $\Diamond$

Analysis involves more than algebra. In particular, inequalities abound. The next definition adds in the basic properties of $\leq$. As usual, we use $x < y$ as an abbreviation for $x \leq y$ and $x \neq y$.

DEFINITION. An *ordered field F* is a field with a partial order $\leq$ such that for all $x, y, z \in F$,

i) exactly one of the following holds: $x < y$ or $x = y$ or $y < x$,

ii) If $x \leq y$, then $x + z \leq y + z$,

iii) If $x \leq y$ and $0 \leq z$, then $x \cdot z \leq y \cdot z$.

EXAMPLE 2

Both $\mathbb{R}$ and $\mathbb{Q}$ are ordered fields. As a two-dimensional set, $\mathbb{C}$ has no obvious order. Indeed, Problem 6 shows that no possible order on the complex numbers can satisfy the order properties.   ◇

An ordered field must contain at least the rational numbers, which can be built from 1 by the operations of addition, subtraction, multiplication, and inverses of nonzero numbers. For instance, $\frac{2}{3}$ can be written as $(1 + 1) * (1 + 1 + 1)^{-1}$. For the irrational numbers, we need another method. We could start by adding square roots and other roots, but that would not yield all the reals, so we need a different tactic.

Completeness

To distinguish the real numbers from other ordered fields, we need the completeness property. We repeat some definitions from Section 4.4 to prepare for the definitions of the terms "bounded" and "least upper bound," used in the completeness property. Some texts call the least upper bound the *supremum* of a set and its greatest lower bound the *infimum*.

DEFINITION.   Let $\leq$ be a partial order on a set X.

- For a subset S of X, an element u of X is an *upper bound* of S iff for all $s \in S$, $s \leq u$.
- For a subset S of X, an element l of X is a *lower bound* of S iff for all $s \in S$, $l \leq s$.
- A subset S of X is *bounded* iff S has an upper bound and a lower bound.
- For a subset S of X, an element b of X is a *least upper bound* of S iff b is an upper bound of S and for every upper bound u of S, $b \leq u$.
- For a subset S of X, an element g of X is a *greatest lower bound* of S iff g is a lower bound of S and for every lower bound l of S, $l \leq g$.

We often abbreviate least upper bound and greatest lower bound by l.u.b. and g.l.b., respectively.

EXAMPLE 3

Let $S = \{x \in \mathbb{Q} : x^2 < 2\}$. This set is bounded above by 1.5 and below by -1.45, along with many other bounds. In $\mathbb{Q}$, the set S has no least upper bound, since $\sqrt{2}$ is irrational, as shown in Problem 7 of Section 2.2. Of course, in $\mathbb{R}$, the l.u.b. of S is $\sqrt{2}$. This set also has a greatest lower bound in $\mathbb{R}$, but not in $\mathbb{Q}$—namely, $-\sqrt{2}$. In effect, there are "holes" in $\mathbb{Q}$ wherever there is an irrational in $\mathbb{R}$.   ◇

DEFINITION.   A *complete ordered field* is an ordered field satisfying the *completeness property*: Every nonempty bounded set has a least upper bound in the field.

We will simply assume that there is a complete ordered field, namely, $\mathbb{R}$. One can be constructed from the rationals, but that is beyond the level of this text. (See Abbott, 243–249.) Even more, it can be shown that there is essentially just one complete ordered field. (See Hewitt and Stromberg, Section 5.)

THEOREM 8.1.1. If a subset of $\mathbb{R}$ has a least upper bound, it has exactly one least upper bound.

Proof. Suppose that both b and c were least upper bounds. Since b is a l.u.b. and c is an upper bound, we have $b \leq c$. Similarly, $c \leq b$. Antisymmetry then forces $b = c$. ∎

THEOREM 8.1.2. In $\mathbb{R}$, every nonempty bounded set has a unique greatest lower bound in $\mathbb{R}$.

Proof. See Problem 8. ∎

EXAMPLE 4

Consider the bounded set T of finite decimal approximations to 1, where $T = \{0.9, 0.99, 0.999, \ldots\}$. The completeness property assures us of a least upper bound. Certainly, 1 and the infinitely repeating $0.9999\ldots$ are upper bounds. But by Lemma 8.1.1, only one number can be the l.u.b. Of course, we want to claim that $1 = 0.999\ldots$ and that this value is the l.u.b. How do we prove they are equal? Theorem 8.1.3 gives a general method. ◇

EXAMPLE 5

Why does the completeness property assure us that there are no "holes" in $\mathbb{R}$? Suppose, for the sake of argument, that we thought something might be missing—call it μ, say, between 1 and 2. (I specified 1 and 2 just to give us bounded sets.) Split the real interval $[1, 2]$ into two sets A and B, using μ, where $A = \{x \in \mathbb{R} : 1 \leq x \leq 2 \text{ and } x < \mu\}$ and $B = \{y \in \mathbb{R} : 1 \leq y \leq 2 \text{ and } \mu \leq y\}$. We invoke the completeness of $\mathbb{R}$ to obtain a least upper bound a of A and a greatest lower bound b of B.

CLAIM: $a = b$, and they effectively are μ.

Proof. From property i of an ordered field, $b < a$ or $a < b$ or $a = b$. If $b < a$, by Problem 3h, their average $\frac{a+b}{2}$ is between them and must be in either A or B. If it were in A, it would be a greater lower bound of B than b is. Similarly, if it were in B, it would be a lower upper bound of A than a. Either way, we have a contradiction. Now consider the possibility that $a < b$. Again, their average is between them. This time the average would be too big to be in A, implying that $\mu \leq \frac{a+b}{2}$. Similarly, the average would be too small for B, so $\frac{a+b}{2} < \mu$. But these inequalities contradict each other. Hence, $a = b$, which is the dividing point in $\mathbb{R}$ between the sets A and B, filling the supposed hole where μ is. ∎

It is possible that a (and so b) differ from μ by an infinitesimal, something we consider in the next example. ◇

EXAMPLE 6

The complete ordered field $\mathbb{R}$ cannot have infinitesimals.

Proof. Suppose, for a contradiction, that there were a positive infinitesimal λ in $\mathbb{R}$. That is, $\lambda > 0$ and for all $n \in \mathbb{N}$, $\lambda < \frac{1}{n}$. (This condition implies that $\lambda < 0.00\ldots 01 = \frac{1}{10^k}$ for any finite string of $k - 1$ zeroes.) Let I be the set of all positive infinitesimals. Then I is a nonempty set bounded below by 0 and above by 1. From the completeness property, I has a least upper bound, say, β. Is β an infinitesimal? If so, Problem 13a shows that 2β would also be infinitesimal, contradicting the assumption that β is an upper bound. If β is not an infinitesimal, then Problem 13b shows that $\beta/2$ would also fail to be an infinitesimal, again showing that β is not the l.u.b. Either way, there would be no l.u.b., contradicting completeness. Thus, $\mathbb{R}$ has no infinitesimals.   ∎

REMARK.   The preceding reasoning assumes that $\mathbb{N}$ has no infinite numbers in it. This assumption is inherent in the Principle of Mathematical Induction, discussed in Section 2.4.   ◇

Our discussion of 1 and $0.999\ldots$ and Example 6 suggest how to show that two real numbers x and y are equal. We simply show, as in Theorem 8.1.3, that the difference $|y - x|$ is less than every positive real number, which by tradition we denote by ϵ (epsilon). Since 1821, when Cauchy introduced the definition of a limit, mathematicians have used ϵ to represent an arbitrarily small positive real number. In Section 8.2, where we'll need two such small positive quantities, we, like Cauchy, will also use δ (delta). We will postpone the proof of the equality of $0.999\ldots$ and 1 until Section 8.4, where we define infinite series and thus infinite decimals.

DEFINITION.   For all $x \in \mathbb{R}$, $|x| = \left\{ \begin{array}{c|cc} x & \text{if} & 0 \leq x \\ \hline -x & \text{if} & x < 0 \end{array} \right\}$.

See Problem 5 for some inequalities involving absolute value.

THEOREM 8.1.3.   Two real numbers x and y are equal iff for all $\epsilon > 0$, $|y - x| < \epsilon$.

Proof. ($\Rightarrow$) First, suppose that $x = y$. Then $|y - x| = 0$, which is clearly less than any positive ϵ.

($\Leftarrow$) For this direction, we'll show the contrapositive. That is, we'll show that if $x \neq y$, then not (for all $\epsilon > 0$, $|y - x| < \epsilon$). The working negation of the part in parentheses is, "There is $\epsilon > 0$ such that $|y - x| \geq \epsilon$." Suppose that $x \neq y$. Then $|y - x| \neq 0$, and so $|y - x| > 0$. By Example 6 there are no infinitesimals. So the condition for all $n \in \mathbb{N}$, $|y - x| < \frac{1}{n}$, fails. That is, there is some $n \in \mathbb{N}$ such that $|y - x| \geq \frac{1}{n}$. Now pick $\epsilon = \frac{1}{n}$ to finish proving the contrapositive.   ∎

REMARK.   The proof also shows that for any real number w, if $w > 0$, then there is some $n \in \mathbb{N}$ such that $w \geq \frac{1}{n}$. (Just take $w = |y - x|$.)

Proofs in analysis often involve "turning approximations into exact values," which is the heart of the preceding theorem. The inequality $|y - x| < \epsilon$ tells us that y approximates x within an error of ϵ. If y is closer to x than all possible errors, then the theorem says that there is no error at all and $y = x$. With the proliferation of calculators, we are surrounded by approximations, usually with more decimal places than anyone would need. Indeed, it is too easy to mistake the number on the calculator for the exact value. In analysis, we need to know when a value is an exact value and when it is an approximation.

Historical Remarks

Bishop George Berkeley (1685–1753) accurately derided the vague concept of infinitely small quantities (infinitesimals for Leibniz and fluxions for Newton), calling them "ghosts of departed quantities." Nevertheless, infinitesimals conveyed the spirit of calculus ideas so well that they were used into the nineteenth century, when, in 1821, Cauchy's definition of a limit finally offered a workable alternative. In 1817 Bernard Bolzano (1781–1848) had enunciated the least upper bound property. However, it took until 1872 before Richard Dedekind (1831–1916) published a satisfactory presentation of the real number system.

PROBLEMS

*1. For each statement that follows, state whether it is true or (at least sometimes) false. Explain your answer or cite the part of the text supporting it.

a) Every field is ordered.

b) $\mathbb{R}$ is dense.

c) $\mathbb{Q}$ is dense in $\mathbb{R}$.

d) Infinitesimals are logically impossible.

e) Infinitesimals contradict the completeness property of $\mathbb{R}$.

f) The least upper bound of a set is the largest element of the set.

g) A set with a least upper bound has a greatest lower bound.

h) $\emptyset$ is a bounded set.

2. a) Complete the argument to show that in any ordered field, $0 < 1$.

   Why can't $1 = 0$? Now, for a contradiction, suppose that $1 < 0$. Now add -1 to each side to get a new inequality. What can you now say about $(-1) \cdot (-1)$?

b) Use part a to show that for all $n \in \mathbb{N}, 0 < n$, where $\mathbb{N}$ is the usual set of natural numbers in $\mathbb{R}$.

3. Assume that F is an ordered field. You may use earlier parts in later parts.

   a) For all $x \in F$, prove that $0 < x$ iff $-x < 0$.

   b) Prove that a negative times a positive is a negative.

   *c) For all $x, y, z \in F$, prove that if $y < z$ and $x < 0$, then $xz < xy$.

   d) For all $x, y \in F$, prove that if $0 < x$ and $0 < xy$, then $0 < y$.

   e) For all $x \in F$, prove that if $0 < x$, then $0 < 1/x$. (Hint: In Problem 2, we showed that $0 < 1$.)

f) For all $x \in F$, prove that if $x < 0$, then $1/x < 0$.

g) For all $x, y \in F$, prove that if $0 < xy$, then both x and y are positive or both are negative.

h) For all $x, y \in F$, prove that if $x < y$, then $x < \frac{x+y}{2} < y$.

i) For all $x, y \in F$, prove that if $0 < x < y$, then $\frac{1}{y} < \frac{1}{x}$ (referred to in Section 8.4).

4. a) Prove that $\mathbb{Q}$ is dense. That is, for all $a, b \in \mathbb{Q}$, if $a < b$, then there is $c \in \mathbb{Q}$ such that $a < c < b$.

b) Prove that $\mathbb{R}$ is dense.

c) Prove that $\mathbb{Q}$ is dense in $\mathbb{R}$.

5. *a) For all $x \in \mathbb{R}$, prove that $|x| = |-x|$ and $0 \le |x|$.

b) For all $x \in \mathbb{R}$, prove that $-|x| \le x \le |x|$.

c) For all $x, y \in \mathbb{R}$, prove that $|x \cdot y| = |x| \cdot |y|$.

d) For all $y, z \in \mathbb{R}$, prove that $|y + z| \le |y| + |z|$.   (Hint: Use cases.)

e) For all $a, b, c \in \mathbb{R}$, prove that $|a - c| \le |a - b| + |b - c|$ (referred to in Section 8.2). Analysts call part d the *triangle inequality*, while geometers prefer to reserve that name for part e. Part e can be interpreted as follows: The distance from a to c is less than or equal to the distance from a to b plus the distance from b to c.

f) For all $x, \epsilon \in \mathbb{R}$ with $\epsilon > 0$, prove that $|x| < \epsilon$ iff $-\epsilon < x < \epsilon$.

6. Give a proof by contradiction from the following hints to show that no partial order $\le$ can make $\mathbb{C}$ an ordered field: Either $0 < i$ or $i < 0$. In the first case, consider i^2 and use Problems 2 and 3. In the second case, consider $(-i)^2$.

7. (Referred to in Sections 8.3 and 8.5.)

a) Use your understanding of decimals to explain why $0.834343434\ldots$ is rational and $0.834334333433334\ldots$ is irrational.

b) Find a rational number strictly between the numbers of part a.

c) Find an irrational number strictly between the numbers of part a.

d) Use your understanding of decimal representations to explain why, between any two distinct real numbers, there is a rational number. (Hint: You may assume that no representation uses infinitely repeating 9s.)

e) Use your understanding of decimal representations to explain why, between any two distinct real numbers, there is an irrational number. (Hint: An irrational number can't have a finite or infinite repeating decimal, but its decimal can have a pattern.)

*8. Prove Theorem 8.1.2. (Hint: For a subset B of $\mathbb{R}$, let $-B = \{-x : x \in B\}$.)

9. Determine the l.u.b. for each set, provided that it exists. Repeat for the g.l.b. If one or both do not exist, explain why not.

a) $\{z \in \mathbb{Z} : z! = z\}$.

*b) $\{x \in \mathbb{Q} : x^2 - 3x < 2\}$.

c) $\{z \in \mathbb{Z} : z^2 - 3z < 2\}$.

d) $\{x \in \mathbb{R} : x^3 - x \le 0\}$.

e) $\{x \in \mathbb{R} : x^2 + 3x + 3 \le 0\}$.

f) $\{x \in [-\frac{\pi}{2}, \frac{\pi}{2}] : -1 < \tan(x) < 1\}$.

g) $\{x \in \mathbb{Q} : -1 < \sin(x) < 1\}$.

10. a) Determine the l.u.b. and the g.l.b. of the interval $[a, b)$.

    b) Prove that your values in part a are correct.

11. In Problem 7 of Section 2.2, we showed that $\sqrt{2}$ is not a rational number. However, we didn't prove that there was such a number. We do so in this problem, as well as generalize that proof and prove some elementary properties of square roots of positive numbers (referred to in Problem 11 of Section 8.4).

    a) Prove that for all $x, y \in \mathbb{R}$, if $0 < x < y$, then $0 < x^2 < y^2$.

    b) Prove that for all $x, y \in \mathbb{R}$, if $0 < x < y$ and $\sqrt{x}$ and $\sqrt{y}$ exist, then $0 < \sqrt{x} < \sqrt{y}$.

    c) Prove that $T = \{x \in \mathbb{Q} : x^2 < 2\}$ is bounded and nonempty.

    d) Define t to be the least upper bound of T. Prove that $t^2 \leq 2$. (Hint: Assume for a contradiction that $t^2 - 2 = \epsilon > 0$, and consider $a = t - \frac{\epsilon}{2t}$.)

    e) In part d, show that $t^2 \geq 2$, and so t is, indeed, the positive square root of 2. (Hints: Assume for a contradiction that $2 - t^2 = \delta > 0$, and consider $b = t + \frac{\delta}{4t}$. Also, use the facts that $t \geq 1$ and $\delta < 1$ after you explain why they are true.)

    f) Let s be a positive real number and consider the set $S = \{x \in \mathbb{R} : x^2 < s\}$. If u is the least upper bound of S, generalize part c to prove that $u^2 \leq s$.

    g) Generalize part e for a real number s satisfying $1 < s < 2$. (You are welcome to generalize this reasoning to $s > 2$.)

    h) Prove for all $x > 0$ that $\sqrt{\frac{1}{x}} = \frac{1}{\sqrt{x}}$. Use this equality to extend part g to square roots for s between 0 and 1.

12. Use Theorem 8.1.3 to show the following:

    a) $3.14159 \neq 3.1416$.

    b) $\frac{1}{6} \neq \frac{1}{7}$.

    c) If x and y have the same decimal representation, they are equal.

    d) Explain why $0.999\ldots = 1$ should follow from Theorem 8.1.3.

13. In Example 6, β was the least upper bound of $I = \{x : 0 < x$ and for all $n \in \mathbb{N}, x < \frac{1}{n}\}$, the set of positive infinitesimals.

    a) Suppose that $\beta \in I$. Show that $2\beta \in I$ by filling out the following reasoning: Let $k \in \mathbb{N}$. Then $2k \in \mathbb{N}$ and $\beta < \frac{1}{2k}$. What does this inequality say about 2β? Does this show that $2\beta \in I$? Why is that a contradiction?

    b) Suppose that $\beta \notin I$ and $0 < \beta$. Write the working negation of the condition "For all $n \in \mathbb{N}, x < \frac{1}{n}$." Use it to show that in this case $\beta/2 \notin I$. Why is this a contradiction?

14. In parts a through d show that each of the following sets is dense:

    a) terminating decimals

    b) rationals with odd denominators

    *c) rationals with denominators a power of 2

    d) irrationals

    e) Explain why decimals with a 2 in the fourth decimal place do not form a dense subset of $\mathbb{R}$.

    f) Explain why the real numbers with no 5 in their decimal representation do not form a dense subset of $\mathbb{R}$.

15. a) Suppose that B is a bounded set of reals, $A \subseteq B$, and A is nonempty. Prove that A is bounded and the l.u.b. of A is less than or equal to the l.u.b. of B.

b) State and prove a similar proposition for the g.l.b. of these sets.

16. Suppose that A and B are bounded, nonempty sets of reals.

*a) Prove that the l.u.b. of $A \cup B$ is the maximum of the l.u.b. of A and the l.u.b. of B.

b) Prove a similar proposition about the g.l.b. of $A \cup B$.

c) Suppose that $A \cap B \neq \emptyset$. What can be said about the l.u.b. and g.l.b. of $A \cap B$ in terms of the corresponding bounds of A and of B? Explain your answers.

17. Show that for all $w \in \mathbb{R}$, there is $n \in \mathbb{N}$ such that $w < n$. (Hint: For $w > 0$, consider $1/w$ and the proof of Theorem 8.1.3 (referred to in Section 8.3).)

18. Prove the Archimedean property of the reals: For all $x, y \in \mathbb{R}$, if $x > 0$ and $y > 0$, then there is $n \in \mathbb{N}$ such that $nx > y$ (referred to in Problem 9 of Section 8.4).

The Archimedean property tells us that any positive value x, no matter how small, can be used to "measure" any other value y, however large, by the use of enough copies of x. In effect, it prohibits infinite elements, which would be the multiplicative inverses of infinitesimals.

19. The integers $\mathbb{Z}$ satisfy all but one property of a field (property ix). Does $\mathbb{Z}$ satisfy the order properties and the completeness property?

20. Critique the questionable "proofs" that follow. Point out reasoning errors and unclear presentation. If the argument is a proof, say so. (If the claim is false, there must be an error in the argument. However, it is not enough to say that the claim is false. You need to find an error in the reasoning.)

a) Claim: $0.999\ldots = 1$. "Proof": When we divide 1 by 3, we get $0.333\ldots$, implying that $0.333\ldots = \frac{1}{3}$. Now multiply both sides by 3 to get $0.999\ldots = \frac{3}{3} = 1$. ▲

*b) Claim: For all $x, y \in \mathbb{R}$, if $x \neq 0$, $y \neq 0$ and $x < y$, then $\frac{1}{y} < \frac{1}{x}$. "Proof": Case 1: $x > 0$. From Problem 3e, we know that $0 < \frac{1}{x}$. Multiplying each side of $x < y$ by $\frac{1}{x}$ gives us $1 < \frac{y}{x}$. Similarly, multiplying both sides of this inequality by the positive number $\frac{1}{y}$ gives us $\frac{1}{y} < \frac{1}{x}$. Case 2: $x < 0$. From Problem 3f, we know that $\frac{1}{x} < 0$. Therefore, part c of Problem 3 tells us to switch the inequality when multiplying by $\frac{1}{x}$. Thus, $x < y$ becomes $1 > \frac{y}{x}$. Again, multiplying both sides by $\frac{1}{y}$ switches the inequality back, giving $\frac{1}{y} < \frac{1}{x}$. ▲

c) Claim: Every decimal representation between 0 and 1 is rational. "Proof": We use induction on the number of decimals. For the initial case, a decimal of the form $0.a$, where $a \in \{0, 1, \ldots 9\}$, equals $\frac{a}{10}$, which is a rational. For the induction step, assume that every decimal $0.a_1a_2\ldots a_n$ with n places equals the fraction $\frac{a_1a_2\ldots a_n}{10^n}$, where $a_1a_2\ldots a_n$ is written in base 10. Now consider a decimal with $n + 1$ places, say, $0.a_1a_2\ldots a_na_{n+1}$. This decimal equals $0.a_1a_2\ldots a_n + \frac{a_{n+1}}{10^{n+1}} = \frac{a_1a_2\ldots a_na_{n+1}}{10^{n+1}}$, which is a rational. By PMI, this property holds for all $n \in \mathbb{N}$. Hence, every decimal representation is a rational number. ▲

d) Claim: For all $a, b > 0$, the Arithmetic Mean—Geometry Mean Inequality holds: $\sqrt{ab} \leq \frac{a+b}{2}$. "Proof": Let $a, b \in \mathbb{R}$, and assume that $0 < a$ and $0 < b$. We convert the inequality to a known true statement. Square $\sqrt{ab} \leq \frac{a+b}{2}$ to get $ab \leq \frac{a^2+2ab+b^2}{4}$. Clear the denominator to get $4ab \leq a^2 + 2ab + b^2$. Shift the $4ab$ to get $0 \leq a^2 - 2ab + b^2 = (a-b)^2$, which is always true. ▲

e) Claim: No other decimal representation can equal 0.333. ...."Proof": Suppose that some different representation $0.a_1a_2a_3 \ldots$ equaled 0.333. ... Since they are different, they must differ in some decimal place. Suppose that the first such place is the nth one, so that $a_n \neq 3$. Then I claim that $|0.a_1a_2a_3 \ldots -0.333 \ldots| \geq 10^{-n-1}$. There are two cases to consider: $a_n < 3$ and $a_n > 3$.

Case 1. If $a_n < 3$, then $0.a_1a_2a_3 \ldots < 0.333 \ldots 3$ (n 3s) $< 0.333 \ldots$. Then $|0.a_1a_2a_3 \ldots - 0.333 \ldots| > |0.333 \ldots 3 - 0.333 \ldots| > 0.000 \ldots 0333$ (n 0s) $> 3 \times 10^{-n-1} > 10^{-n-1}$.

Case 2. If $3 < a_n$, then $0.333 \ldots < 0.333 \ldots 4$ (n 3s) $< 0.a_1a_2a_3 \ldots$. Thus, $|0.a_1a_2a_3 \ldots - 0.333 \ldots| > |0.333 \ldots 4 - 0.333 \ldots| > 0.000 \ldots 0666 \ldots$ (n 0s) $> 6 \times 10^{-n-1} > 10^{-n-1}$.

In either case, $|0.a_1a_2a_3 \ldots - 0.333 \ldots| \geq 10^{-n-1}$, which from Theorem 8.1.3 implies that $0.a_1a_2a_3 \ldots$ and 0.333... differ. ▲

Zeno's Paradoxes

Since at least the time of the Greek philosopher Zeno, who lived 2450 years ago, people have pondered the nature of space, time, and motion. Zeno crafted several paradoxes in response to different conceptions of how a line is made from individual points and so of the nature of space and time. One group of Greek thinkers taught that space and time are discrete, much the way a computer screen is divided into pixels and computer time is segmented. In other words, at any given point there is a nearest neighbor in any direction. In modern terms, the model for time could be $\mathbb{Z}$ and for space we could use $\mathbb{Z} \times \mathbb{Z} \times \mathbb{Z}$. Others held that space and time are "infinitely divisible"; that is, between any two points there are other points. In modern terms, the model for time could be $\mathbb{Q}$ or $\mathbb{R}$ and the model for space could be $\mathbb{Q} \times \mathbb{Q} \times \mathbb{Q}$ or $\mathbb{R} \times \mathbb{R} \times \mathbb{R}$.

Zeno gave four arguments purporting to show that motion was impossible. We'll paraphrase three of them, starting with one of the discrete paradoxes, the Arrow. We'll concentrate on the two most famous paradoxes, Achilles and the Tortoise, and the Dichotomy, which consider infinitely divisible space and time.

The Arrow Suppose that space and time have indivisible smallest units (like atoms or pixels). Shoot an arrow. In any given unit of time, the arrow can be in only one fixed unit of space. Hence, at any time it isn't moving.

Since we don't think that space and time are discrete, it is easy to agree with Zeno that there is a problem here. After all, in a discrete world, it is hard to see how something gets "between" places if there is no "between." In a more modern setting, do the animated figures in a computer game really move? No, the computer redraws each figure in a new spot many times a second to give the illusion of motion. In some sense, each redrawing is a new figure, so there is no motion. The illusion of motion depends on our brains filling in the sequence of images that our eyes see.

Achilles and the Tortoise Suppose now that space and time are infinitely divisible. Achilles was the fastest runner in ancient Greece and agreed to give the tortoise a head start in a race. Zeno argued as follows that Achilles could never catch up with the tortoise: For Achilles to pass the tortoise, he must first get to where the tortoise started, say, to T_0; but by then, the tortoise will have crawled a bit further, say, to T_1. Again, to pass the tortoise Achilles needs to get to T_1, but when he is there the tortoise is a wee bit further, say, at T_2. This discussion can be repeated as often as you wish, but Achilles is always behind the tortoise each time. Therefore, Achilles can never pass the tortoise.

Clearly, the conclusion is false, since faster things regularly pass slower things in the real world. So there must be some error of reasoning in Zeno's argument. However, it is not enough to say that Zeno is wrong or even to point out that the distances from T_i to T_{i+1} are decreasing, as is the time involved in going from one to the next. Let me emphasize that point by quantifying the situation. Suppose that Achilles runs 100 inches per second and starts at 0, while the tortoise crawls 1 inch per second and starts at $T_0 = 100$. After 1 second, Achilles is at T_0 and the tortoise is at $T_1 = 101$. After 1.01 seconds, Achilles is at T_1 and the tortoise is at $T_2 = 1.0101$. And so on. We can use mathematics to say that Achilles will catch up with the tortoise after $1.010101\ldots$ seconds and be ahead from then on. But that doesn't say where the error is in Zeno's argument.

The Dichotomy Again assuming infinite divisibility, Zeno claims that Achilles can't even move. In order for Achilles to get anywhere, he must first get halfway, and to get halfway, he must first get one quarter of the way, and so on. Thus, Achilles would have to do an infinite number of acts in order to move at all.

In ancient times and the Middle Ages, philosophers wrestled with these paradoxes without achieving a satisfactory resolution. Many people thought that the language of calculus solved the last two paradoxes, but as it was pointed out earlier, knowing when and where Achilles passes the tortoise doesn't show where the error in Zeno's reasoning is. Some philosophers still argue about these paradoxes. (See Salmon.)

I submit one idea for resolving these last two paradoxes. I think we need to make the distinction between a potential infinity and an actual infinity. If, like Aristotle, you believe that there can be no actual infinity, then you can't get past the potentially infinite sequence of positions in the race between Achilles and the tortoise. However, allowing for an actual infinity enables us to say that Zeno has only cleverly restricted the setting. The "always" in his conclusion "Achilles is always behind the tortoise" simply doesn't follow from his argument, since there is more beyond the infinite sequence of positions that he has considered. Similarly, in the dichotomy, we can acknowledge the infinity of points between the start and the end without qualifying the motion between any two as a separate "act."

REFERENCES

ABBOTT, S. 2001. *Understanding analysis.* New York: Springer.
HEWITT, E., and K. STROMBERG. 1975. *Real and abstract analysis.* New York: Springer-Verlag.
KEISLER, J. 1976. *Foundations of infinitesimal calculus.* Boston: Prindle, Weber and Schmidt.
SALMON, W. 2001. *Zeno's paradoxes.* Indianapolis: Hackett.

8.2 LIMITS OF FUNCTIONS

We shift our focus from the real numbers to functions and limits, primary objects of study in calculus.

DEFINITION. A *real function* is a function from a subset of $\mathbb{R}$ into $\mathbb{R}$.

Calculus courses traditionally spend only enough time on limits of real functions to be able to jump into derivatives. We'll reverse the order and emphasis in this section, spending just enough time with derivatives to explain why we need the sophisticated definition of a limit.

The derivative, intuitively, is the slope of the line tangent to the graph of a function at a point. However, defining this concept for a general real function taxed mathematicians for quite some time. The ancient Greeks developed the idea of a tangent and worked with specific curves—notably, circles, ellipses, parabolas, and hyperbolas. In the seventeenth century, mathematicians found tangents for an increasing variety of curves, and their work culminated in the general idea in Newton's and Leibniz's work. The limit definition took yet another 130 years to formulate. We can immediately visualize tangents of straightforward curves, like the graph in Figure 1. However, a general definition that can handle the infinitely turning function in Figure 2

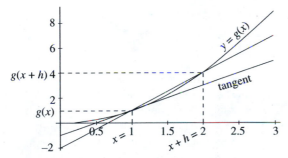

FIGURE 1. Approximating the tangent of $y = x^2$ at $x = 1$.

FIGURE 2. $y = \begin{cases} x^2 \sin(1/x) & \text{if } x \neq 0 \\ 0 & \text{if } x = 0 \end{cases}$.

requires much more care. The intuition of the general definition uses increasingly good approximations of the tangent to find the exact tangent, illustrated in Figure 1.

EXAMPLE 1

Use approximations to find the slope of the tangent to $y = g(x)$ at $x = 1$, where $g(x) = x^2$.

Solution   Recall that the slope of the line through two distinct points (x_1, y_1) and (x_2, y_2) is $\frac{\Delta y}{\Delta x} = \frac{y_2 - y_1}{x_2 - x_1}$. We approximate the tangent to $y = g(x)$ at $x = 1$ by using a line through two points on the curve, one with $x = 1$ and one nearby. The points on $y = g(x)$ are of the form $(a, g(a))$. For $x = 1$ and a nearby x-value $x = 1 + h$, we get $\frac{\Delta y}{\Delta x} = \frac{g(1+h) - g(1)}{h} = \frac{(1+h)^2 - 1^2}{h} = \frac{h^2 + 2h}{h} = h + 2$. The closer h is to 0, the closer the approximating line is to the tangent line, as seen in Figure 1. Many calculus students are only too happy to plug $h = 0$ into the formula for the approximate slope to get the exact slope of 2. But $h = 0$ is the only value of h we can't use in the formula $\frac{g(1+h) - g(1)}{h}$, since h appears in the denominator. The heart of the familiar definition of the derivative is the equation $g'(x) = \lim_{h \to 0} \frac{g(x+h) - g(x)}{h}$. Whatever the definition of a limit is, it must exclude from consideration the value of the variable "at the limit"—here, $h = 0$. Yet at the same time, it must ensure that the closer the variable comes to this value, the closer the function value gets to the supposed value. Here, the closer h gets to 0, the closer the difference quotient $\frac{g(1+h) - g(1)}{h} = h + 2$ gets to 2.   ◊

DEFINITION.   Suppose that $c \in (a, b)$ and f is a real function defined on the interval (a, b), except possibly at $x = c$. For $L \in \mathbb{R}$, $\lim_{x \to c} f(x) = L$ iff for all $\epsilon > 0$, there is $\delta > 0$ such that for all $x \in (a, b)$, if $0 < |x - c| < \delta$, then $|f(x) - L| < \epsilon$. We read $\lim_{x \to c} f(x) = L$ as "the limit of $f(x)$ as x approaches c is L."

Let's take our time to digest this very complicated definition. First an easy part: The condition $0 < |x - c|$ simply eliminates considering the value of the variable "at the limit." With derivatives, this eliminates the risk of setting $h = 0$ and so dividing by 0. Even when $f(c)$ is defined, its value is logically irrelevant to the limit, as Example 3 illustrates. Next, the Greek letters ϵ ("epsilon") and δ ("delta") measure closeness. Thus, $|f(x) - L| < \epsilon$ measures how close $f(x)$ is to the claimed value L. In Example 1, our function is the difference quotient $\frac{(1+h)^2 - 1^2}{h}$ and $L = 2$. Then $\left| \frac{(1+h)^2 - 1^2}{h} - 2 \right|$ measures how close the approximating slope is to the claimed slope of the tangent. Similarly, $|x - c| < \delta$ measures how close the variable x is to the limit value c. For Example 1, $x = h$ and $c = 0$. The implication "if $0 < |x - c| < \delta$, then $|f(x) - L| < \epsilon$" is a precise version of "when x is close to c, then $f(x)$ is close to L." Example 2 and Figure 3 illustrate this approximation idea.

Finally and most importantly, the quantified phrase "for all $\epsilon > 0$" in the definition pushes this idea beyond the intuition of Example 1. In particular, Theorem 8.1.3 allows us to say more than just "$f(x)$ is close to L." Recall that Theorem 8.1.3 states that $x = y$ iff for all $\epsilon > 0$, $|x - y| < \epsilon$. It is no accident that our definition of a limit echoes this Theorem by starting with "For all $\epsilon > 0$" and ending with "$|f(x) - L| < \epsilon$". This phrase justifies saying that $\lim_{x \to c} f(x)$ actually equals L, not just is really close to L.

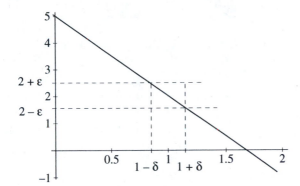

FIGURE 3. Relation of δ and ε.

EXAMPLE 2

Define $f : (0, 3) \to \mathbb{R}$ by $f(x) = 5 - 3x$. From the graph in Figure 3, we expect that $\lim_{x \to 1} f(x) = 2$. We can think of the definition of the limit in terms of a challenge and response. For any challenge ϵ, we need to give a response δ, making the implication "if $0 < |x - 1| < \delta$, then $|f(x) - 2| < \epsilon$" true. Consider, as in Figure 3, $\epsilon = 0.5$. The possible attempt of $\delta = 0.3$ fails, since $x = 1.2$ satisfies $0 < |1.2 - 1| = 0.2 < \delta$, but $|f(1.2) - 2| = |1.4 - 2| = 0.6 > \epsilon$. Let's shrink δ to 0.1. Then $|x - 1| < 0.1$ forces $0.9 < x < 1.1$, and so $1.7 = f(1.1) < f(x) < f(0.9) = 2.3$. Thus, $|f(x) - 2| < 0.3 < \epsilon$.

  A smaller challenge, such as $\epsilon = 0.001$, requires a suitably smaller response, say, $\delta = 0.0003$. To see this, suppose that $|x - 1| < 0.0003$. Then $|f(x) - 2| = |5 - 3x - 2| = |3 - 3x| = 3|1 - x| < 3\delta = 3(0.0003) = 0.0009 = \epsilon$. A microscopic challenge, such as $\epsilon = 0.00000000006$, demands a similarly tiny δ, such as 0.00000000002. Although we do not need to find δ systematically in this example, you may have already guessed how to do so. Here we seek to illustrate only that for every challenge ϵ, we can find a response δ. $\Diamond$

  As with all other mathematical definitions, the definition of a limit is designed for proofs, not for building intuition. We will restrict our examples and problems to limits that use fairly easy algebra.

EXAMPLE 3

For $j : (0, 4) \to \mathbb{R}$ given by $j(x) = \frac{2x^2 - 8x + 6}{x - 3}$ if $x \neq 3$ and $j(3) = 2$, find $\lim_{x \to 3} \frac{2x^2 - 8x + 6}{x - 3}$ and prove that your answer is correct.

Solution The graph of $j(x) = \frac{2x^2 - 8x + 6}{x - 3}$ in Figure 4 suggests that $\lim_{x \to 3} \frac{2x^2 - 8x + 6}{x - 3} = 4$. Indeed, the graph of $y = j(x)$ looks like the graph of $y = 2x - 2$ everywhere except at $x = 3$. I arbitrarily defined $j(3)$ to be 2, but that value, by definition, doesn't enter into the limit. If we pick x close to 3, say, $x = 2.99$, we find that $j(2.99) = 3.98$, close to 4. Again, for an x even closer to 3, say, $x = 3.0001$, we have $j(3.0001) = 4.0002$, which is even closer to 4. These computations make the limit plausible, but they don't prove it.

  The definition of a limit tells us the format of the proof, which will reveal where we need to do some preparatory work for the proof.

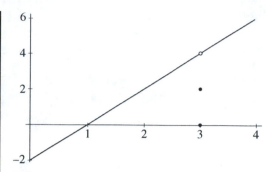

FIGURE 4. $y = j(x)$.

Format. Let $\epsilon > 0$. Pick $\delta = \underline{\quad} > 0$. Let $x \in (0, 4)$ and suppose that $0 < |x - 3| < \delta$. Then $\left| \frac{2x^2 - 8x + 6}{x - 3} - 4 \right| \ldots < \epsilon$.

Preparatory Work. We need to "guess" δ, which in general is in terms of ϵ, to ensure that the last inequality holds. For thousands of years, mathematicians have approached such challenges the same way: Start at the answer and work backwards. This method works, but it is not part of the proof. After all, by starting at the answer, we are assuming that there is such a δ, which is what we are supposed to prove.

We boldly start with $\left| \frac{2x^2 - 8x + 6}{x - 3} - 4 \right| < \epsilon$ and do some algebra on the left side: $\left| \frac{2x^2 - 8x + 6}{x - 3} - 4 \right| = \left| \frac{2x^2 - 8x + 6}{x - 3} - 4 \left(\frac{x - 3}{x - 3} \right) \right| = \left| \frac{2x^2 - 8x + 6 - 4x + 12}{x - 3} \right| = \left| \frac{2x^2 - 12x + 18}{x - 3} \right| = \left| \frac{(x - 3)(2x - 6)}{x - 3} \right| = |2x - 6| = 2|x - 3|$. So we need to show only the much simpler inequality $2|x - 3| < \epsilon$. Now remember, for us to show this inequality, the format asks us to pick δ with $0 < |x - 3| < \delta$. Conveniently, these inequalities are closely related. In particular, $2|x - 3| < \epsilon$ becomes $|x - 3| < \epsilon/2$, so we can safely pick $\delta = \epsilon/2$. We're ready for the proof.

Proof. Let $\epsilon > 0$. Pick $\delta = \epsilon/2 > 0$. Let $x \in (0, 4)$ and suppose that $0 < |x - 3| < \delta$. Then $\left| \frac{2x^2 - 8x + 6}{x - 3} - 4 \right| = \left| \frac{2x^2 - 8x + 6}{x - 3} - 4 \left(\frac{x - 3}{x - 3} \right) \right| = \left| \frac{2x^2 - 12x + 18}{x - 3} \right| = \left| \frac{(x - 3)(2x - 6)}{x - 3} \right| = |2x - 6| = 2|x - 3| < 2\delta \leq \epsilon$. ∎

REMARKS. First of all, we could have picked a smaller δ, such as $\delta = \epsilon/17$, and the proof would still hold. In general, if one choice for δ satisfies the inequalities, a smaller one will as well, since a smaller δ will simply restrict x further. Secondly, the domain $(0, 4)$ has little bearing on the problem except to guarantee that j is defined in a region around 3, where we are taking the limit. We could just as easily use $j : \mathbb{R} \to \mathbb{R}$ or $j : (2.9, 3.1) \to \mathbb{R}$. ◇

The proof in Example 3 should start to convince you that the definition of a limit can confirm correct values of limits. It can expose incorrect values as well, even if they are quite close to the correct value, as Example 4 illustrates.

EXAMPLE 4

For $k : \mathbb{R} \to \mathbb{R}$ given by $k(x) = 5 - 2x$, show that $\lim_{x \to 2}(5 - 2x) \neq 1.01$.

Solution If we plug 2 in for x, we see that the limit "ought" to be 1. However, this approach isn't a proof, even if it might convince us that the value 1.01 had better not work. To prove the negation of a limit, let's start by writing the working negation of the definition of a limit.

Working Negation of Limit $\lim_{x \to c} f(x) \neq L$ iff there is $\epsilon > 0$ such that for all $\delta > 0$, there is $x \in (a, b)$ such that $0 < |x - c| < \delta$, but $|f(x) - L| \geq \epsilon$.

Thus, for our proof we need to start with a choice of ϵ. This will measure how far $k(x) = 5 - 2x$ is away from the mistaken choice of 1.01. Now we believe that by picking x-values close to 2, we can get $k(x)$ close to 1, which is 0.01 away from 1.01, the given, incorrect value. Let's pick ϵ smaller than that, say, $\epsilon = 0.005$, just to be safe. The format forces us to let δ be any positive value and then choose x. If we pick x bigger than 2, but close to 2, $5 - 2x$ will be less than 1, so, surely, far from 1.01, at least in comparison with our ϵ. We're required also to have x within δ of 2. This discussion should suffice to suggest a proof.

Proof. Pick $\epsilon = 0.005 > 0$. Let $\delta > 0$. Pick $x = 2 + \delta/2$. Then $0 < |x - 2| = \delta/2 < \delta$. Further, $|(5 - 2x) - 1.01| = |3.99 - 2(2 + \delta/2)| = |-0.01 - \delta| > 0.01 > \epsilon$. $\Diamond$

Our definition of a limit does better than Example 4 indicates. Theorem 8.2.1 shows that a function can't have two different limits at the same x-value. Thus, once we have found a limit, we know that all other values will fail.

THEOREM 8.2.1. Suppose that $c \in (a, b)$ and f is a real function defined on the interval (a, b), except possibly at $x = c$, and that both $\lim_{x \to c}(f) = L$ and $\lim_{x \to c}(f) = M$. Then $L = M$.

Proof. For a contradiction, suppose that $L \neq M$. What can go wrong? From the definition of a limit, we can get $f(x)$ as close as we'd like to both L and M. But L and M are a certain distance apart, which ought to be incompatible with that idea. Let's use the positive value $|M - L|$ to consider a specific ϵ. In the worst case, the actual limit might be halfway between L and M, so we'll pick $\epsilon = |M - L|/2 > 0$. Then we expect that $f(x)$ can't be within ϵ of both L and M.

From our two limits, we have two deltas, say, δ_L and δ_M, such that for all $x \in (a, b)$, if $0 < |x - c| < \delta_L$, then $|f(x) - L| < \epsilon$; and similarly, if $0 < |x - c| < \delta_M$, then $|f(x) - M| < \epsilon$. We want both of these at the same time, so consider $\delta = \min(\delta_L, \delta_M)$ and suppose that $0 < |x - c| < \delta$. Then we have both $|f(x) - L| < \epsilon$ and $|f(x) - M| < \epsilon$. By the triangle inequality (Problem 5e of Section 8.1), we have $|M - L| \leq |f(x) - L| + |f(x) - M| < \epsilon + \epsilon = |M - L|$, a direct contradiction. Hence, $L = M$, as claimed. ∎

Analysis texts at this point traditionally prove a collection of theorems about limits of combinations of functions. We prove one of these theorems to illustrate the process of deriving a limit from other limits. Some additional results appear in the problems.

THEOREM 8.2.2. Suppose that $c \in (a, b)$ and f and g are real functions defined on the interval (a, b), except possibly at $x = c$. Further, suppose that $\lim_{x \to c} f(x) = L$ and $\lim_{x \to c} g(x) = M$. Then $\lim_{x \to c} (f(x) + g(x)) = L + M$.

DISCUSSION. The key challenge is to choose a successful δ for the limit $\lim_{x \to c} (f(x) + g(x))$ based on the corresponding δ_f and δ_g for the given limits $\lim_{x \to c} f(x)$ and $\lim_{x \to c} g(x)$. To be successful, we need to show, for any $\epsilon > 0$, that $|f(x) + g(x) - (L + M)| < \epsilon$. Given the separate limits of f and g, it is reasonable to try to split the left side of this inequality. From the triangle inequality (Problem 5e of Section 8.1), $|f(x) + g(x) - (L + M)| = |(f(x) - L) + (g(x) - M)| \leq |f(x) - L| + |g(x) - M|$. However, we need this whole expression to be less than ϵ, meaning that each of the parts must be considerably less than ϵ. Thus, we'll require each of $|f(x) - L|$ and $|g(x) - M|$ to be less than $\frac{\epsilon}{2}$.

Proof. Suppose all the hypotheses. Let $\epsilon > 0$. From $\lim_{x \to c} f(x)$, there is $\delta_f > 0$ such that for all $x \in (a, b)$, if $0 < |x - c| < \delta_f$, then $|f(x) - L| < \frac{\epsilon}{2}$. Similarly, there is $\delta_g > 0$ such that for all $x \in (a, b)$, if $0 < |x - c| < \delta_g$, then $|g(x) - L| < \frac{\epsilon}{2}$. Pick $\delta = \min(\delta_f, \delta_g)$. Let $x \in (a, b)$ and suppose that $0 < |x - c| < \delta$. Then $|f(x) + g(x) - (L + M)| \leq |f(x) - L| + |g(x) - M| < \frac{\epsilon}{2} + \frac{\epsilon}{2} = \epsilon$. ∎

Since the derivative is a limit, we include its definition in this section.

DEFINITION. For a function $f : (a, b) \to \mathbb{R}$ and $x \in (a, b)$, the *derivative* $f'(x)$, if it exists, is $\lim_{h \to 0} \frac{f(x+h) - f(x)}{h}$.

Problem 11 asks you to show that the limit in this definition is equivalent to the limit $\lim_{w \to x} \frac{f(w) - f(x)}{w - x}$.

A limit can fail to exist for a number of reasons. Examples 5 and 6 illustrate two situations where the definition of a limit fails, although we can describe what is happening to the function as x approaches the value in question. The examples develop modifications of the definition of a limit for these cases. The problems explore these and other situations.

EXAMPLE 5

Figure 5 gives the graph of $j : \mathbb{R} \to \mathbb{R}$ given by $j(x) = x + \lfloor x \rfloor$, where $\lfloor x \rfloor$ is the floor function (greatest integer less than or equal to x). The jump at $x = 1$ should convince you that $\lim_{x \to 1} j(x)$ does not exist, even though from each side there is a clear candidate for a limit. We'll show that the limit doesn't exist and develop the concept of a one-sided limit.

CLAIM. $\lim_{x \to 1} j(x)$ does not exist.

Proof. For a contradiction, suppose that $\lim_{x \to 1} j(x) = L$. We'll employ the working negation of a limit from Example 4 to eliminate L. Case 1. $L > 1.5$. Consider $\epsilon = 0.4 > 0$, which is less than the jump from the left side of the graph to wherever L is. Let $\delta > 0$. Pick $x = 1 - \delta/2$. Then $0 < |x - 1| = \delta/2 < \delta$. Further, $|j(x) - L| = |1 - \delta/2 + \lfloor 1 - \delta/2 \rfloor - L| = |1 - \delta/2 - L| \geq |1 - L| > 0.5 > \epsilon$.

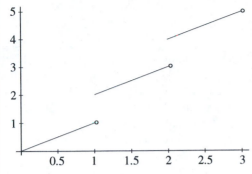

FIGURE 5. $y = x + \lfloor x \rfloor$.

We leave Case 2 to Problem 16. ∎

Even though there is no limit, this function clearly approaches a height of 1 as x approaches 1 from the left. Similarly, the function approaches a height of 2 as x approaches 1 from the right. A slight modification of the definition of a limit gives us the one-sided limit from the left. We leave the definition of a limit from the other side to Problem 16.

DEFINITION. Suppose that f is a real function defined on the interval (a, c). For $L \in \mathbb{R}$, $\lim_{x \to c^-} f(x) = L$ iff for all $\epsilon > 0$, there is $\delta > 0$ such that for all $x \in (a, c)$, if $0 < c - x < \delta$, then $|f(x) - L| < \epsilon$. We read $\lim_{x \to c^-} f(x) = L$ as "The limit of $f(x)$ as x approaches c from the left is L." ◇

EXAMPLE 6

As the graph in Figure 6 indicates, the function $k : \mathbb{R}^* \to \mathbb{R}$ given by $k(x) = 1/x^2$ has no limit as x approaches 0. (Recall that $\mathbb{R}^* = \{x \in \mathbb{R} : x \neq 0\}$.) The values of $k(x)$ surpass any bound the closer x gets to 0. The next definition makes precise the unbounded nature of k near 0.

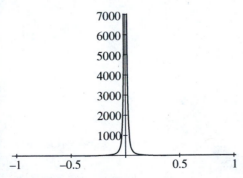

FIGURE 6. $y = 1/x^2$.

DEFINITION. Suppose that $c \in (a, b)$ and f is a real function defined on the interval (a, b), except possibly at $x = c$. Then $\lim_{x \to c} f(x) = \infty$ iff for all $B > 0$, there is $\delta > 0$ such that for all $x \in (a, b)$, if $0 < |x - c| < \delta$, then $f(x) > B$. We read $\lim_{x \to c} f(x) = \infty$ as "The limit of $f(x)$ as x approaches c is infinity."   ◇

EXERCISE. Modify the previous definition to define $\lim_{x \to c} f(x) = -\infty$.

Limits provide local understanding about a function—what is happening "near" the c in $\lim_{x \to c}$. For instance, a derivative is local information, the instantaneous rate of change at a point, not the overall change. Of course, the many vital applications of derivatives teach us how important such local information is.

Historical Remarks

In 1821, Augustin Cauchy (1789–1857) published his influential calculus text in Paris. In this book, he defined limits and many other terms and raised the standards of reasoning in analysis. (Although others in more obscure places had published some similar ideas earlier, mathematicians paid attention to Cauchy, who was a leading mathematician in a major mathematical center.)

PROBLEMS

*1. For each statement that follows, state whether it is true or (at least sometimes) false. Explain your answer or cite the part of the text supporting it.

   a) In the notation $\lim_{x \to c} f(x) = L$, the $=$ sign is not really correct, since $f(x)$ never has to equal L.

   b) If $\lim_{x \to c} f(x) = L$, then $f(c) = L$.

   c) In a proof of a limit, we can pick δ in terms of ϵ.

   d) In the definition of a limit, $|f(x) - L|$ tells us how close the function is to its limit at x.

   e) To disprove a conjectured limit, we can pick ϵ.

   f) To disprove a conjectured limit, we can pick ϵ in terms of δ.

   g) To disprove a conjectured limit, we can pick x in terms of δ.

*2. For $g : \mathbb{R} \to \mathbb{R}$ given by $g(x) = 7x - 17$, and for each ϵ that follows, find a $\delta > 0$ satisfying the conditions in the definition for $\lim_{x \to 3} g(x) = 4$.

   a) $\epsilon = 0.1$.

   b) $\epsilon = 0.0002$.

   c) $\epsilon = 10^{-15}$. (10^{-15} meters is approximately the diameter of a proton.)

   d) $\epsilon = 53$.

3. Assume that the functions in the limits that follow are defined on all of $\mathbb{R}$. Use the definition of a limit to prove these limits.

   *a) $\lim_{x \to -1} (5x + 3) = -2$.

   b) $\lim_{y \to 3} (4 - 3y) = -5$.

   c) $\lim_{z \to \sqrt{2}} (2 + \frac{1}{2}z) = 2 + \frac{\sqrt{2}}{2}$.

   d) $\lim_{w \to 1/\pi} (\pi w - 7) = -6$.

   e) For $m > 0$ and $b \in \mathbb{R}$, $\lim_{v \to 3} (mv + b) = 3m + b$.

4. **a)** Find $\lim_{x\to-1}(\frac{1}{2} - 4x)$, and use the definition of a limit to prove that your value is correct.

   b) Repeat part a for $\lim_{x\to-2} 2 - \frac{1}{4}x$.

5. Modify Example 4 to show that $\lim_{x\to2}(5 - 2x) \neq 1.0000000000000001$. (Many calculators can't distinguish between this value and 1.)

6. ***a)** Define $k : \mathbb{R}^* \to \mathbb{R}$ by $k(x) = \frac{|x|}{x}$. Prove that $\lim_{x\to0} k(x)$ does not exist. (Hint: What is $k(x)$ if $x > 0$?)

   b) Define $h : \mathbb{R}^* \to \mathbb{R}$ by $h(x) = \frac{1}{x}$. Prove that $\lim_{x\to0} h(x)$ does not exist.

7. Use the definition of a limit to prove that for every $c, m, b \in \mathbb{R}$, $\lim_{x\to c}(mx + b) = mc + b$. (Hint: Consider Problem 3e and do the case $m = 0$ separately (referred to in Section 8.3).)

8. Use the definitions of a derivative and a limit to prove that at every c, the derivative of the function f given by $f(x) = (mx + b)$ is m.

9. Use the definitions of a derivative and a limit to prove that at every c, the derivative of the function g given by $g(x) = x^2$ is $2c$.

10. Let $p, q, r \in \mathbb{R}$. Use the definitions of a derivative and a limit to prove that at every c the derivative of the function h given by $h(x) = px^2 + qx + r$ is $2pc + q$. (Hint: Do the case $p = 0$ separately. How is this problem similar to Problem 7?)

11. Show that the limit in the definition of a derivative is equivalent to $\lim_{w\to x} \frac{f(w)-f(x)}{w-x}$.

12. Suppose that f and g are functions defined on the interval (a, b), except possibly at c, and suppose also that the limits $\lim_{x\to c} f(x) = L$ and $\lim_{x\to c} g(x) = M$ exist. Show that $\lim_{x\to c}(f(x) - g(x)) = L - M$ (referred to in Problem 6 of Section 8.3).

13. Suppose that f is a function defined on the interval (a, b), except possibly at c, and suppose also that the limit $\lim_{x\to c} f(x) = L$ exists and $m, b \in \mathbb{R}$. Prove that $\lim_{x\to c}(mf(x) + b) = mL + b$.

14. Use Theorem 8.2.2 to prove that the derivative of the sum of two functions is the sum of their derivatives. More succinctly, $(f + g)'(x) = f'(x) + g'(x)$.

15. (Squeeze Theorem.) Suppose that $c \in (a, b)$ and f, g, and h are real functions defined on the interval (a, b), except possibly at $x = c$. Further, suppose that for all $x \in (a, b)$, with $x \neq c$, we have $f(x) \le g(x) \le h(x)$ and $\lim_{x\to c} f(x) = L = \lim_{x\to c} h(x)$. Prove that $\lim_{x\to c} g(x) = L$ (referred to in Section 8.3).

    (Hints: Since $g(x)$ is "squeezed" between $f(x)$ and $h(x)$, it seems reasonable that $|g(x) - L|$ can't be any bigger than the larger of $|f(x) - L|$ or $|h(x) - L|$. Consider two cases: $L \le g(x)$ and $g(x) < L$.)

16. **a)** In Example 5, prove that $\lim_{x\to1^-}(x + \lfloor x \rfloor) = 1$.

    b) Define $\lim_{x\to c^+} f(x) = L$.

    c) Use your definition in part b to prove that $\lim_{x\to1^+}(x + \lfloor x \rfloor) = 2$.

    d) Prove that $\lim_{x\to c} f(x) = L$ iff both $\lim_{x\to c^+} f(x) = L$ and $\lim_{x\to c^-} f(x) = L$.

17. ***a)** In Example 6, prove that $\lim_{x\to0} k(x) = \infty$.

    *b) Why is $\lim_{x\to0} 1/x = \infty$ false?

c) Define $\lim_{x \to c^+} f(x) = \infty$.

d) Use your definition in part c to prove that $\lim_{x \to c^+} 1/x = \infty$.

18. Prove that $\lim_{x \to c} f(x) = L$ iff $\lim_{x \to c} |f(x) - L| = 0$.

19. a) For $L \in \mathbb{R}$ and a suitable function, f, define $\lim_{x \to \infty} f(x) = L$. What conditions must the domain of f satisfy?

b) Use your definition to prove that $\lim_{x \to \infty} \frac{1}{x} = 0$.

c) Repeat part b for $\lim_{x \to \infty} \frac{2x}{x-1} = 2$.

20. Critique the questionable "proofs" that follow. Point out reasoning errors and unclear presentation. If the argument is a proof, say so. (If the claim is false, there must be an error in the argument. However, it is not enough to say that the claim is false. You need to find an error in the reasoning.)

a) Define $k : \mathbb{R} \to \mathbb{R}$ by $k(x) = 17$. Claim: For all $c \in \mathbb{R}$, $\lim_{x \to c} k(x) = 17$. "Proof": Regardless of c and ϵ, pick $\delta = 17$. Suppose that $0 < |x - c| < 17$. Then $|k(x) - k(c)| = |17 - 17| = 0 < \epsilon$. ▲

*b) Claim: $\lim_{x \to 2} x^2 = 4$. "Proof": Let $\epsilon > 0$. Pick $\delta = \frac{\epsilon}{4} > 0$. Let x satisfy $0 < |x - 2| < \delta$. Then $|x^2 - 4| = |(x + 2)(x - 2)| = |x + 2| \cdot |x - 2| \le |2 + 2| \cdot |x - 2| = 4|x - 2| < 4\delta = \epsilon$. ▲

c) Claim: $\lim_{x \to 2} x^2 \ne 3$. "Proof": Pick $\epsilon = 1 > 0$. Let $\delta > 0$. Pick $x = 2 + \delta/3$. Then $0 < |x - 2| = \delta/3 < \delta$, but $|x^2 - 3| \ge 1 + \delta > \epsilon$. ▲

d) Claim: If $\lim_{x \to c} f(x) = L$ and $\lim_{x \to c} g(x) = M$, then $\lim_{x \to c} f(x) \cdot g(x) = L \cdot M$. "Proof": Suppose that $c \in (a, b)$ and f and g are real functions defined on the interval (a, b), except possibly at $x = c$, and suppose also that $\lim_{x \to c} f(x) = L$ and $\lim_{x \to c} g(x) = M$. Let $\epsilon > 0$. Then, for $\epsilon_f = \sqrt{\epsilon} > 0$, there is $\delta_f > 0$ such that for all $x \in (a, b)$, if $0 < |x - c| < \delta_f$, then $|f(x) - L| < \epsilon_f$. Similarly, for $\epsilon_g = \sqrt{\epsilon}$, there is $\delta_g > 0$, making $|g(x) - M| < \epsilon_g$. Pick $\delta = \min(\delta_f, \delta_g)$. Let $x \in (a, b)$ and suppose that $0 < |x - c| < \delta$. Then $|f(x) \cdot g(x) - L \cdot M| = |f(x) - L| \cdot |g(x) - M| < \epsilon_f \cdot \epsilon_g = \sqrt{\epsilon} \cdot \sqrt{\epsilon} = \epsilon$. ▲

8.3 CONTINUOUS FUNCTIONS AND COUNTEREXAMPLES

"Everyone knows what a curve is, until he has studied enough mathematics to become confused through the countless number of possible exceptions."

—Felix Klein (1849–1925)

In the previous section, we explored the definition of a limit, proving that a particular value was or wasn't the limit of a function. Unfortunately, our definition of a limit doesn't give a hint about how to find what value to try for the limit L. However, for many familiar functions, we can easily "guess" limit values. For instance, $\lim_{x \to 7} 4 + 3x$ ought to be 25, because $4 + 3 \cdot 7 = 25$. That is, we simply "plug" 7 into the function, even though the definition of a limit carefully avoids that option. Colloquially, this function "gets where it is going." That is, not only does the limit exist (the function is going somewhere), but that limit is the value of the function at the point in question. Such functions are called *continuous*, as defined shortly in a modern version of Cauchy's 1821 definition. Intuitively, the graph of a continuous function can be drawn without having to lift the pen or pencil at any point. That is, it has no "jumps" or "gaps," unlike Examples 3 and 5 of Section 8.2. However, as we will see later in the subsection on counterexamples, this intuitive notion doesn't always suffice.

DEFINITION. Suppose that f is a real function defined on the interval (a, b) and $c \in (a, b)$. Then f is *continuous* at c iff $\lim_{x \to c} f(x) = f(c)$. We say that f is *continuous on a set A* iff f is continuous at each point of A.

REMARKS. The equation $\lim_{x \to c} f(x) = f(c)$ requires $f(c)$ to be defined, unlike the definition of a limit. Further, it requires the limit to exist. Only then does it make sense to say that these quantities are equal. Analysis texts use one-sided limits, as in Example 5 of Section 8.2, to define continuity at the endpoints of intervals.

EXAMPLE 1

Show that all linear functions $g : \mathbb{R} \to \mathbb{R}$ given by $g(x) = mx + b$ are continuous on all of $\mathbb{R}$.

Solution   Problem 7 of Section 8.2 showed that $\lim_{x \to c} g(x) = \lim_{x \to c} mx + b = mc + b = g(c)$, regardless of the choices for c, m, and b. This limit matches the requirement for g to be continuous on all of $\mathbb{R}$. $\Diamond$

Counterexamples

The problems extend the family of continuous functions beyond Example 1 to other familiar functions, although not nearly as many as calculus (and analysis) texts consider. Instead, like analysis texts, this book investigates less-well-behaved functions, which provide counterexamples to some flawed intuitions about functions. Counterexamples lie at the heart of analysis more than in many other areas of mathematics. This important role of counterexamples deserves explanation and exploration. Real analysis explores deeply an area students feel at home in—after all, calculus texts present statements of many theorems, and the familiar functions clearly fit the hypotheses and conclusions of these theorems. However, it may not be clear why all the hypotheses of these theorems appear. Further, these well known examples can also suggest many more properties, and many of these turn out to be false in general. Analysts, as well as beginners in analysis, need to study counterexamples to build intuition about the structure of real numbers and real functions. In fact, there is an important analysis resource (Gelbaum and Olmstead) devoted just to counterexamples.

EXAMPLE 2

An attempt at the graph of $p : \mathbb{R} \to \mathbb{R}$ given by $p(x) = \begin{cases} \sin\left(\frac{1}{x}\right) & \text{if } x \neq 0 \\ 0 & \text{if } x = 0 \end{cases}$ appears in Figure 1. This graph wiggles infinitely many times between $y = 1$ to $y = -1$ as x approaches 0 from either side, making it impossible to "be drawn without having to lift the pencil or pen at any

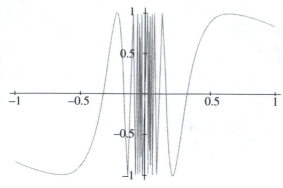

FIGURE 1. $y = \begin{cases} \sin(1/x) & \text{if } x \neq 0 \\ 0 & \text{if } x = 0 \end{cases}$.

point." However, as x approaches 0 there is no "jump" up or down from the left to the right. It also doesn't have a "gap" we can fill in to make it continuous. Instead, the graph seems to approach every single y-value between -1 and 1. Thus, discontinuities can be more subtle than our initial intuition may suggest.

CLAIM. p has no limit at $x = 0$, and so p is not continuous at 0.

DISCUSSION.   We need to show that no possible value of L can be the limit. The key is the existence of x-values arbitrarily close to 0, with images at 1 and others with images at -1. The sine function periodically has height 1: $\sin(\frac{\pi}{2}) = 1 = \sin(2\pi + \frac{\pi}{2}) = \sin(4\pi + \frac{\pi}{2})$, etc. In general, $\sin(2n\pi + \frac{\pi}{2}) = 1$ for any integer n. Similarly, for any integer n, $\sin(2n\pi + \frac{3\pi}{2}) = -1$. The $1/x$ in the definition of p means that we use $\frac{1}{2n\pi + \frac{\pi}{2}}$ for x to get $\sin(\frac{1}{x}) = 1$, and similarly, $x = \frac{1}{2n\pi + \frac{3\pi}{2}}$ to get an image of -1.

Proof.   Let L be any real number. Case 1: $L \leq 0$. To show that L is not the limit, we'll use the working negation from Example 4 of Section 8.2. Pick $\epsilon = 0.5 > 0$. Let $\delta > 0$. By Problem 17 of Section 8.1, there is $n \in \mathbb{N}$ such that $0 < \frac{1}{n} < \delta$. Consider $x = \frac{1}{2n\pi + \pi/2}$. Then $|x - 0| < \frac{1}{2\pi n} < \frac{1}{n} < \delta$. Further, $\sin(\frac{1}{x}) = \sin(2n\pi + \pi/2) = 1$. Then $|p(x) - L| = |1 - L| \geq 1 > \epsilon$. Thus, L is not the limit.
   Case 2: $0 < L$. See Problem 3a.   ∎

REMARK.   This function is continuous at every x-value except $x = 0$, although we won't develop the theorems to prove this fact.   ◊

EXAMPLE 3

Let's modify Example 2 slightly by defining $q : \mathbb{R} \to \mathbb{R}$ as $q(x) = \begin{cases} x \sin\left(\frac{1}{x}\right) & \text{if } x \neq 0 \\ 0 & \text{if } x = 0 \end{cases}$, whose graph is approximated in Figure 2. This graph also wiggles infinitely often as it approaches 0 from either side, something impossible to draw accurately. Nevertheless, it is continuous at $x = 0$, as Problem 3b asks you to show. Since the sine function is stuck between -1

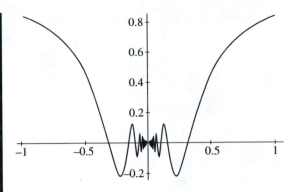

FIGURE 2. $r(x) = \begin{cases} x \sin(1/x) & \text{if } x \neq 0 \\ 0 & \text{if } x = 0 \end{cases}$.

and 1, the key is to squeeze the function q between the functions $f(x) = |x|$ and $h(x) = -|x|$ and apply the "Squeeze Theorem," from Problem 15 of Section 8.2. Problem 3c asks you to show that q does not have a derivative at $x = 0$.

The function approximated in Figure 2 of Section 8.2 goes one step farther. It is defined by $r(x) = \begin{cases} x^2 \sin\left(\frac{1}{x}\right) & \text{if } x \neq 0 \\ 0 & \text{if } x = 0 \end{cases}$ and, as Problem 3d asks you to show, has a derivative at $x = 0$. Incidentally, r has no second derivative there. $\diamondsuit$

The previous examples concerned functions that were continuous everywhere, or everywhere but one point. How about the other extreme? The next example illustrates how "bad," or at least discontinuous, functions can get. This example is due to the mathematician Peter Dirichlet (1805–1859).

EXAMPLE 4

Define the function $s : \mathbb{R} \rightarrow \mathbb{R}$ by $s(x) = \begin{cases} 1 & \text{if } x \in \mathbb{Q} \\ 0 & \text{if } x \notin \mathbb{Q} \end{cases}$. No graph can do justice to this function. Near every x-value there are both rationals and irrationals; so the graph appears in Figure 3 to be two horizontal lines at heights 0 and 1.

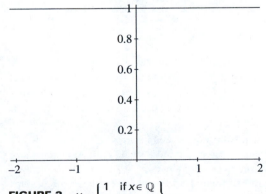

FIGURE 3. $y = \begin{cases} 1 & \text{if } x \in \mathbb{Q} \\ 0 & \text{if } x \notin \mathbb{Q} \end{cases}$.

CLAIM. For all $c \in \mathbb{R}$, the function s has no limit as x approaches c and so s is discontinuous at every x-value.

Proof. See Problem 4. ■

The next example might at first glance seem just as discontinuous as the function in Example 4.

EXAMPLE 5

Define the function $t : \mathbb{R} \to \mathbb{R}$ by $t(x) = \begin{cases} 0 & \text{if } x \notin \mathbb{Q} \\ \frac{1}{q} & \text{if } x = \frac{p}{q} \in \mathbb{Q} \text{ and} \\ & x \text{ is in lowest terms} \end{cases}$. Figure 4 attempts to graph t, which appears to have discontinuities everywhere. We'll show that it, surprisingly, has a limit of 0 at every single x-value and so is continuous at every irrational number. From Chapter 5, there are uncountably many irrational numbers and only countably many rationals. Hence, this strange function is continuous "almost all" of the time. K. J. Thomae first published properties of this function in 1875.

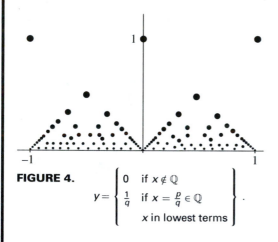

FIGURE 4. $y = \begin{cases} 0 & \text{if } x \notin \mathbb{Q} \\ \frac{1}{q} & \text{if } x = \frac{p}{q} \in \mathbb{Q} \\ & x \text{ in lowest terms} \end{cases}$.

CLAIM. For every $c \in \mathbb{R}$, $\lim_{x \to c} t(x) = 0$.

Proof. Let $c \in \mathbb{R}$ and $\epsilon > 0$. From Example 6 of Section 8.1, we know there is $n \in \mathbb{N}$ such that $\frac{1}{n} < \epsilon$. Any fraction $\frac{p}{q}$ with a reduced denominator at least as large as n will satisfy $|t(\frac{p}{q}) - 0| = \frac{1}{q} < \frac{1}{n} < \epsilon$. So we "only" need to choose δ small enough that there are no fractions in $(c - \delta, c + \delta)$ with a denominator smaller than n, with the possible exception of c. ■

As a first restriction at δ, let's say that $\delta_1 = 0.49$. Then in the interval $(c - \delta_1, c + \delta_1)$, there is at most 1 fraction with reduced denominator 1, at most 1 with reduced denominator 2, at most 2 with reduced denominator 3, and in general at most $k - 1$ with reduced denominator k. Then the number of potentially troublesome fractions in $(c - \delta_1, c + \delta_1)$ is at most $1 + 1 + 2 + 3 + \ldots + (n - 1)$, a finite number.

Let's list the finitely many fractions $\frac{p}{q}$ in this interval, with $\frac{p}{q} \neq c$ and $\frac{1}{q} \geq \epsilon$, say, a_1, $a_2, \ldots, a_k$. Consider the distances $|a_1 - c|, \ldots, |a_k - c|$ of c from each of these fractions. Since there are finitely many of these positive numbers, one of them is the smallest. Call this minimum $m = \min\{|a_i - c| : 1 \leq i \leq k\}$.

Pick $\delta = \min(m, \delta_1)$. Let x be any number satisfying $0 < |x - c| < \delta$. Either x is rational or x is irrational. If x is rational, our choice of δ guarantees that $|t(x) - 0| = t(x) < \frac{1}{n} < \epsilon$. If x is irrational, then $|t(x) - 0| = 0 < \epsilon$. Either way, t has a limit of 0 at c. $\Diamond$

Analysts have found many more curious functions that can severely stretch our intuition. In 1872, Karl Weierstrass described a family of functions continuous every-where, but differentiable nowhere. (See Abbott, 144–149.) In effect, these functions are made up entirely of corners. It is hard to imagine such a situation, let alone try, according to the naïve intuition, to draw it "without having to lift the pen or pencil at any point." The optional material following the main problem set considers a different type of counterexample, are aimed at showing why analysis needs the full power of a complete ordered field. Initial intuitions about continuity serve students well enough in calculus, but analysis requires deeper insight. A study of counterexamples leads to a more mature intuition.

PROBLEMS

*1. For each statement that follows, state whether it is true or (at least sometimes) false. Explain your answer or cite the part of the text supporting it.

  a) A function that has a limit at a point is continuous at that point.

  b) A function that is continuous at a point is defined there.

  c) A function that is continuous at a point has a derivative at that point.

  d) If a function f is not continuous at a point c, then either there is a jump at c or we can redefine $f(c)$ to make f continuous.

  e) Counterexamples provide insight in analysis.

2. *a) Show that the function v given by $v(x) = |x|$ is continuous at every real number.

  b) Show that v does not have a derivative at $x = 0$.

3. a) Prove Case 2 in Example 2.

  b) In Example 3, prove that q is continuous at 0.

  c) In Example 3, prove that q has no derivative at 0.

  d) In Example 3, prove that r has a derivative at 0.

4. Prove the claim in Example 4. (Hint: Use Problem 7 of Section 8.1.)

5. *a) If a function $f : (a, b) \to \mathbb{R}$ is continuous at $c \in (a, b)$, prove that the function $g : (a, b) \to \mathbb{R}$ given by $g(x) = |f(x)|$ is also continuous at c.

  b) Suppose that the function $f : (a, b) \to \mathbb{R}$ is continuous at c. Must the function $h : (a, b) \to \mathbb{R}$ given by $h(x) = f(|x|)$ also be continuous at c? If so, prove it; if not, give a counterexample.

6. Suppose that the functions f and g are defined on (a, b), $c \in (a, b)$, and f and g are continuous at c. Prove that their sum $f + g$ and their difference $f - g$ are continuous at c.

(Define $(f + g)(x) = f(x) + g(x)$ and $(f - g)(x) = f(x) - g(x)$.) (Hint: See Theorem 8.2.2 and Problem 12 of Section 8.2 (referred to in Section 8.5).)

7. The following parts will help you show that the squaring function $g(x) = x^2$ is continuous on all of $\mathbb{R}$:

   *a) Use the Squeeze Theorem (Problem 15 of Section 8.2) and the function $h(x) = |x|$ to show that g is continuous at 0. (Hint: Restrict the domain.)
   Note that $|x^2 - c^2| = |x + c| \cdot |x - c|$.

   *b) Use the preceding factoring to find δ in terms of ϵ, c and x so that if $|x - c| < \delta$, then $|x^2 - c^2| < \epsilon$.

   c) Explain why $|x + c| \leq 3|c|$, provided that $|x - c| < |c|$.

   d) Use part c to replace any occurrence of x in your answer to part b.

   e) For $c \neq 0$, pick δ to be the minimum of the values in parts c and d to prove that $\lim_{x \to c} x^2 = c^2$.

8. Complete the given proof outline to prove that the composition of continuous functions is continuous. More precisely, suppose that $f : (a, b) \to (d, e)$ is continuous at $c \in (a, b)$ and $g : (d, e) \to \mathbb{R}$ is continuous at $f(c)$. Then prove that $g \circ f$ is continuous at c.

   Proof. Let $\epsilon > 0$. From the continuity of g, there is $\delta_g > 0$ such that ___. For $\epsilon_f = \delta_g > 0$, the continuity of f gives us $\delta_f > 0$ such that ___. Pick $\delta = \delta_f > 0$. Let $x \in (a, b)$ and suppose that ___. Then $|f(x) - f(c)|$ ___.
   Explain why we can put $f(x)$ into the function g. Then ___. Thus, $g \circ f$ is continuous at c. ∎

9. In this problem use previous problems, theorems, and parts of this problem to show that the specified polynomials are continuous on all of $\mathbb{R}$.
   a) x^4 b) $(x^2 + x)^2$ c) x^3 d) x^6 e) x^5
   f) $px^6 + qx^5 + rx^4 + sx^3 + tx^2 + ux + v$, for any reals p, q, r, s, t, u, v.

10. The parts that follow will help you show that the function $g(x) = \frac{1}{x}$ is continuous for $c \neq 0$.
    Note that $\left| \frac{1}{x} - \frac{1}{c} \right| = \frac{|x-c|}{|x| \cdot |c|}$.
    a) Use the preceding equality to find δ in terms of ϵ, c, and x so that if $|x - c| < \delta$, then $\left| \frac{1}{x} - \frac{1}{c} \right| < \epsilon$.

    b) Explain why $c^2/2 \leq |x| \cdot |c|$, provided that $|x - c| < |c|/2$.

    c) Use part b to replace any occurrence of x in your answer to part a.

    d) For $c \neq 0$, pick δ to be the minimum of the values in parts b and c to prove that $\lim_{x \to c} \frac{1}{x} = \frac{1}{c}$.

11. a) Define $u : \mathbb{R} \to \mathbb{R}$ by $u(x) = \begin{cases} x & \text{if } x \in \mathbb{Q} \\ -x & \text{if } x \notin \mathbb{Q} \end{cases}$. Determine where u is continuous.
    b) Prove that u is continuous at each point you claimed in part a.
    c) Prove that u is not continuous at all other points.

12. Provide counterexamples to show that the hypothesis in each part that follows is needed in the statement of the Intermediate Value Theorem (IVT). That is, give a function that fulfills all of the hypotheses of the IVT, except the indicated one, and does not satisfy the conclusion (referred to in Section 3.3).

THE INTERMEDIATE VALUE THEOREM. If $f : [a, b] \to \mathbb{R}$ is continuous and Y is between $f(a)$ and $f(b)$, then there is some $X \in [a, b]$ such that $f(X) = Y$. (Informally, such a function "hits" every height Y between its starting and ending heights. We prove this as Theorem 8.5.3 at the end of Section 8.5.)

*a) f is continuous.

b) The domain is a closed interval.

c) Y is between $f(a)$ and $f(b)$.

13. Suppose that $f : [a, b] \to \mathbb{R}$ is one-to-one and continuous, and $f(a) < f(b)$. Use a proof by contradiction and the Intermediate Value Theorem (stated in Problem 12) to prove that f is increasing.

14. Provide counterexamples to show that the hypothesis in each part that follows is needed in the statement of the Extreme Value Theorem (EVT). That is, give a function that fulfills all of the hypotheses of the EVT except the indicated one and does not satisfy the conclusion (referred to in Sections 3.3 and 9.2).

THE EXTREME VALUE THEOREM. If $f : [a, b] \to \mathbb{R}$ is continuous, then there are $c, d \in [a, b]$ such that for all $x \in [a, b]$, $f(c) \le f(x) \le f(d)$. (Informally, such a function attains its maximum and minimum heights, $f(d)$ and $f(c)$. The proof of this theorem goes beyond the level of this text. See Abbott, 115.)

a) f is continuous.

b) The domain is a closed interval.

15. Provide counterexamples to show that the hypothesis in each part that follows is needed in the statement of the Mean Value Theorem (MVT). That is, give a function that fulfills all of the hypotheses of the MVT except the indicated one and does not satisfy the conclusion.

THE MEAN VALUE THEOREM. If $f : [a, b] \to \mathbb{R}$ is continuous and f has a derivative at each $c \in (a, b)$, then there is some $d \in (a, b)$ such that $f'(d) = \frac{f(b) - f(a)}{b - a}$.

(Informally, for such a function there is a place where the instantaneous slope equals the overall average slope. The proof of this theorem goes beyond the level of this text. See Abbott, 139.)

a) f is continuous.

b) The domain is a closed interval.

c) f has a derivative at every c in (a, b).

16. Critique the questionable "proofs" that follow. Point out reasoning errors and unclear presentation. If the argument is a proof, say so. (If the claim is false, there must be an error in the argument. However, it is not enough to say that the claim is false. You need to find an error in the reasoning.)

a) Claim: If $f : (a, b) \to \mathbb{R}$ is continuous at $c \in (a, b)$, then g given by $g(x) = f(|x|)$ is continuous at c. "Proof": By Problem 2, the absolute value function is continuous everywhere. We are given that f is continuous at c. Since g is the composition of continuous functions, Problem 8 shows that g is also continuous at c. ▲

b) Claim: The real function f given by $f(x) = x^n$ is continuous at 0. "Proof": Let $\epsilon > 0$. Pick $\delta = \min(1, \epsilon) > 0$. Suppose that $0 < |x - 0| < \delta$. Since $|x| < \delta \leq 1$, we have $|x^n - 0| = |x^n| \leq |x| < \delta \leq \epsilon$. ▲

c) Claim: $s : (0, \infty) \to \mathbb{R}$ given by $s(x) = \sqrt{x}$ is continuous at every $c > 0$. "Proof": Let $c, \epsilon > 0$. Pick $\delta = \min(c/2, \frac{2\epsilon}{5\sqrt{c}})$. Let $x > 0$ and suppose that $0 < |x - c| < \delta$. Note that $c/2 < x < 3c/2$ because $\delta \leq c/2$. Further, $\frac{3}{2}c < \frac{9}{4}c$, so $\sqrt{x} < \sqrt{\frac{9}{4}c} = \frac{3}{2}\sqrt{c}$ and $\sqrt{x} + \sqrt{c} < \frac{5}{2}\sqrt{c}$. Then $|x - c| = |(\sqrt{x} + \sqrt{c})(\sqrt{x} - \sqrt{c})| = (\sqrt{x} + \sqrt{c})|\sqrt{x} - \sqrt{c}| < \frac{5}{2}\sqrt{c}\delta \leq \epsilon$. ▲

*d) Claim: If a function $f : (a, b) \to \mathbb{R}$ has a derivative at $c \in (a, b)$, then f is continuous at c. "Proof": We suppose that $f'(c) = \lim_{x \to c} \frac{f(x) - f(c)}{x - c}$ exists. We need to show that $\lim_{x \to c} f(x) = f(c)$ or, equivalently, $\lim_{x \to c}(f(x) - f(c)) = 0$. Let $\epsilon > 0$. Pick $\delta = \frac{\epsilon}{|f'(c)|} > 0$. Suppose that $0 < |x - c| < \delta$. Then $|f(x) - f(c)| = |\frac{f(x) - f(c)}{x - c}(x - c)| = |\frac{f(x) - f(c)}{x - c}| \cdot |x - c| < f'(c)\delta = \epsilon$. ▲

Counterexamples in Rational Analysis

We explore some of the many properties of calculus and analysis that fail if we restrict our numbers to the rational numbers. In effect, these counterexamples argue for the essential role of the completeness property of the reals. In forming our counterexamples, we repeatedly use $\sqrt{2}$, which we showed was irrational in Problem 7 of Section 2.2, although any irrational will work. For this optional material we adjust all previous analysis definitions to restrict the domain and range to rational numbers.

DEFINITION. A *rational function* is a function from a subset of $\mathbb{Q}$ into $\mathbb{Q}$. Denote the *rational interval* $[a, b] \cap \mathbb{Q}$ by $[a, b]_\mathbb{Q}$.

EXAMPLE 6

Let $f : [0, 2]_\mathbb{Q} \to \mathbb{Q}$ be given by $f(x) = x^2$. Then f is "rationally" continuous on $\mathbb{Q}$, but it provides a counterexample to the "Rational Intermediate Value Theorem." That is, even though f is continuous on the closed bounded interval $[0, 2]_\mathbb{Q}$, f does not take on every value between $f(0) = 0$ and $f(2) = 4$. In particular, we can't solve $f(x) = 2$ with $x \in \mathbb{Q}$. (See Problem 12 for the statement of the usual Intermediate Value Theorem.) ◇

EXAMPLE 7

Show that the "Rational Extreme Value Theorem" fails. We need to find a continuous rational function from a closed bounded interval that doesn't achieve its maximum or its minimum. (See Problem 14 for the usual statement of the Extreme Value Theorem.)

Solution Let's start from a function with no maximum and no minimum and adjust it. Consider first f given by $f(x) = \frac{1}{x}$, which goes to $-\infty(+\infty)$ as x goes to 0 from the left (right). Of course, f is not continuous at 0 or even defined there. Our first adjustment is to shift the discontinuity to an irrational spot. Let $g(x) = \frac{1}{x - \sqrt{2}}$. Then g will be continuous at any rational. Unfortunately, it doesn't map rationals to rationals. We can avoid that problem by replacing the denominator with $x^2 - 2 = (x - \sqrt{2})(x + \sqrt{2})$. Define $h : [0, 2]_\mathbb{Q} \to \mathbb{Q}$ by

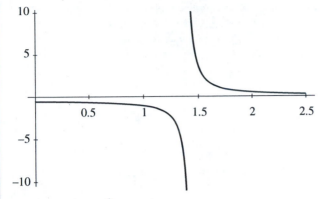

FIGURE 5. $y = \frac{1}{x^2-2}$.

$h(x) = \frac{1}{x^2-2}$. Then h has no maximum value and no minimum value, as Figure 5 illustrates. $\diamondsuit$

PROBLEMS (CONTINUED)

17. Find a counterexample for the "Rational Extreme Value Theorem" in Example 7 that has no maximum, but has a minimum.

18. Show that the "Rational Mean Value Theorem" fails. (See Problem 15 for the usual statement of the Mean Value Theorem.)

19. a) Find a nonconstant rational function with a derivative of 0 at every rational.

b) Given any rational function f with a derivative at every rational, use part a to find another rational function g with the same derivative as f, but which does not differ from f by a constant.

REFERENCES

ABBOTT, S. 2001. *Understanding analysis.* New York: Springer.

GELBAUM, B., and J. OLMSTEAD. 1964. *Counterexamples in analysis.* San Francisco: Holden Day.

8.4 SEQUENCES AND SERIES

Patterns of strings of numbers (sequences) have fascinated mathematicians for thousands of years, ranging from primes to successive approximations of irrational numbers. In this section we'll focus on the "long-term behavior" of sequences, that is, their limits, if these limits exist. As a special case of sequences, we will consider series—the accumulating sums of the terms of a sequence. Our short study of sequences and series will equip us to consider our use of the decimal system to represent real numbers.

DEFINITION. A sequence $\{a_n\}_{n\in\mathbb{N}}$ is a function from $\mathbb{N}$ into $\mathbb{R}$. A sequence $\{a_n\}_{n\in\mathbb{N}}$ is *bounded* iff the set $\{a_n : n \in \mathbb{N}\}$ is bounded; that is, there are $l, u \in \mathbb{R}$, such that for all $n \in \mathbb{N}$, $l \le a_n \le u$. A sequence $\{a_n\}_{n\in\mathbb{N}}$ is *nondecreasing* iff for all $k, n \in \mathbb{N}$, if $k < n$, then $a_k \le a_n$.

EXAMPLE 1

Let $a_n = n!$. Thus, $a_1 = 1$, $a_2 = 2$, $a_3 = 6$, $a_4 = 24$, and so on. The sequence $\{a_n\}_{n\in\mathbb{N}}$ is non-decreasing and unbounded. $\Diamond$

REMARK. The sequence with each term equal to 1 is nondecreasing by our defini-tion. Sequences such as Example 1 that satisfy the stronger condition "if $k < n$, then $a_k < a_n$" are called strictly increasing.

EXAMPLE 2

Let $b_n = -\frac{1}{n}$. The sequence $\{b_n\}_{n\in\mathbb{N}}$ is bounded and nondecreasing: $b_1 = -1$, $b_2 = -\frac{1}{2}$, $b_3 = -\frac{1}{3}$, $b_4 = -\frac{1}{4}$, and so on. Although no term ever equals 0, they are approaching 0, as Figure 1 illustrates. We formalize this idea in the definition of a limit.

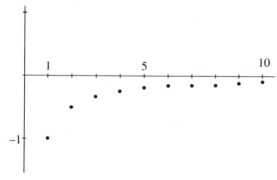

FIGURE 1. The sequence $\left\{-\frac{1}{n}\right\}_{n\in\mathbb{N}}$. $\Diamond$

DEFINITION. Given a sequence $\{a_n\}_{n\in\mathbb{N}}$ and a real number L, we say that $\lim_{n\to\infty} a_n = L$ iff for all $\epsilon > 0$, there is some $N \in \mathbb{N}$ such that for all $n \in \mathbb{N}$, if $n > N$, then $|a_n - L| < \epsilon$. If a sequence has a limit, we say that the sequence *con-verges*. Otherwise, we say that the sequence *diverges*.

Compare the logical format for the limit of a sequence, which follows, with the very similar format of a limit in Section 8.2.

FORMAT. Let $\epsilon > 0$. ... Pick $N =$ ___ . Let $n \in \mathbb{N}$ and suppose that $n > N$. Then $|a_n - L| \ldots < \ldots \epsilon$.

EXAMPLE 3

Claim: $\lim_{n \to \infty} -\frac{1}{n} = 0$.

DISCUSSION. For a given challenge, say, $\epsilon = 0.01$, we need to respond with an appropriate N. Here, $N = 100$ will work, since if $n > 100$, then $|-\frac{1}{n} - 0| = \frac{1}{n} < \frac{1}{100}$. As with limits of functions, different challenges of ϵ can require different responses N. However, in general, as ϵ gets smaller, N needs to get larger. For instance, $\epsilon = 0.0001 = \frac{1}{10,000}$ needs an N at least as large as 10,000. That is, if $n > 10,000$, then $|-\frac{1}{n} - 0| < \frac{1}{10,000} = \epsilon$. A proof provides a general way to respond to any challenge.

Proof. Let $\epsilon > 0$. From the proof of Theorem 8.1.3, we can pick $N \in \mathbb{N}$ such that $\frac{1}{N} < \epsilon$. Let $n \in \mathbb{N}$ and suppose that $n > N$. Now $N > 0$, so by Problem 3i of Section 8.1, $|-\frac{1}{n} - 0| = \frac{1}{n} < \frac{1}{N} < \epsilon$. ■ $\diamond$

EXAMPLE 4

Let $c_k = 1 + \left(\frac{-1}{2}\right)^k$. Then $c_1 = 1 + \frac{-1}{2} = 0.5$, $c_2 = 1 + \frac{1}{4} = 1.25$, $c_3 = 0.875$, $c_4 = 1.0625$, and so on. The sequence $\{c_k\}_{k \in \mathbb{N}}$ is bounded by its first two terms. Consecutive terms alternate between increasing and decreasing, but as n increases, the terms approach each other and 1.

CLAIM. $\lim_{k \to \infty} c_k = 1$.

Proof. Let $\epsilon > 0$. We know that there is $N \in \mathbb{N}$ such that $\frac{1}{N} < \epsilon$. Looking ahead, we will need $|1 + \left(-\frac{1}{2}\right)^k - 1| = \left(\frac{1}{2}\right)^k$ to be less than $\frac{1}{N}$. From Problem 2b in Section 2.4, we know that $k < 2^k$. We can invert this inequality by Problem 3i of Section 8.1 to get $\left(\frac{1}{2}\right)^k < \frac{1}{k}$. Let $k \in \mathbb{N}$ and suppose that $k > N$. Then $|1 + \left(-\frac{1}{2}\right)^k - 1| = \left(\frac{1}{2}\right)^k < \frac{1}{k} < \frac{1}{N} < \epsilon$. ■

DISCUSSION. In general, $\left(\frac{1}{2}\right)^k$ is much smaller than $\frac{1}{k}$, so we could have found a smaller value for N. However, a smaller value would not improve the proof and would require more effort to find. $\diamond$

Counterexamples have an important role to play in this section, just as they did in Section 8.3. We start with an example of a bounded sequence without a limit.

EXAMPLE 5

Let $d_i = (-1)^i + \left(\frac{-1}{2}\right)^i$. Then $d_1 = -1 - \frac{1}{2} = -1.5$, $d_2 = 1 + \frac{1}{4} = 1.25$, $d_3 = -1.125$, $d_4 = 1.0625$, and so on. The sequence $\{d_i\}_{i \in \mathbb{N}}$ is bounded, for example, by -2 and $+2$. Its terms alternate between negative numbers and positive numbers. Even though the terms get closer to each other, I claim that they don't have a limit, because the negative numbers approach -1 and the positive numbers approach 1. To prove this claim, we first write out the working negation.

WORKING NEGATION OF LIMIT OF A SEQUENCE. For a given L, there is $\epsilon > 0$ such that for all $N \in \mathbb{N}$, there is $n \in \mathbb{N}$ such that $n > N$ and $|d_i - L| \geq \epsilon$.

Proof. Case 1. $L \geq 0$. Pick $\epsilon = 1 > 0$. Let $N \in \mathbb{N}$. Consider $n = 2N + 1$, an odd integer. Then $|d_n - L| = |-1 - \left(\frac{1}{2}\right)^n - L| > 1 = \epsilon$.

Case 2. See Problem 4. ∎   ◇

The lack of convergence in Example 5 suggests that a sequence, like a function, can have only one limit. Our first theorem confirms this analogy, mimicking Theorem 8.2.1.

THEOREM 8.4.1. If a sequence converges, it has a unique limit.

Proof. See Problem 5.   ∎

Limits of sequences differ from limits of functions in significant ways, even if the formats of their definitions and Theorems 8.2.1 and 8.4.1 match well. The limit $\lim_{x \to c} f(x)$ gives us only local information about f, what is happening close to c. However, limits of sequences can tell us something about the entire sequence.

THEOREM 8.4.2. If a sequence convergen, it is bounded.

Proof. Suppose that $\{a_n\}_{n \in \mathbb{N}}$ converges, say, $\lim_{n \to \infty} a_n = L$. Then for $\epsilon = 1$, there is $N \in \mathbb{N}$ such that if $n > N$, then $|a_n - L| < 1$ or, equivalently, $L - 1 < a_n < L + 1$. Thus, we have bounded all the terms after a_N. Since the first N terms form a finite set, they have a maximum, say, M and a minimum, say, m. Then the entire sequence is bound below by the minimum of m and $L - 1$. Similarly, it is bounded above by the maximum of M and $L + 1$.   ∎

Example 5 gave a counterexample to the converse of Theorem 8.4.2. Perhaps surprisingly, putting the conditions increasing and bounded together suffices to imply convergence.

THEOREM 8.4.3. A bounded nondecreasing sequence converges.

Proof. Let $\{a_n\}_{n \in \mathbb{N}}$ be a bounded, nondecreasing sequence with upper bound u. As a set, $A = \{a_n : n \in \mathbb{N}\}$ is bounded by a_1 and u. From the completeness property of the reals, A has a l.u.b., say, b. We will show that $\lim_{n \to \infty} a_n = b$. Let $\epsilon > 0$. Since b is the least upper bound, $b - \epsilon$ is not an upper bound of A. That is, there is $k \in \mathbb{N}$ such that $b - \epsilon < a_k$. Pick $N = k$. Let $n \in \mathbb{N}$ and suppose that $k < n$. Because $\{a_n\}_{n \in \mathbb{N}}$ is nondecreasing, $a_k \leq a_n$. Hence, $b - \epsilon < a_n \leq b$, giving $|a_n - b| < \epsilon$, as required for the limit.   ∎

Series

Infinite series, an important topic in calculus since its development by Newton and Leibniz, are defined by the use of sequences. We'll start with a specific example familiar to many students.

EXAMPLE 6

What do we mean by the "infinite sum" $1 + \frac{1}{2} + \frac{1}{4} + \frac{1}{8} + \ldots = (\frac{1}{2})^0 + (\frac{1}{2})^1 + (\frac{1}{2})^2 + (\frac{1}{2})^3 + \ldots$? We can certainly add two or three terms or even finitely many terms, which motivates defining this infinite series as the sequence of its *partial sums*. Here, the initial partial sum is just 1, the next is $1 + \frac{1}{2} = 1.5$, then $1 + \frac{1}{2} + \frac{1}{4} = 1.75$, and so on. While we can't literally perform infinitely many additions, we can tell that these finite sums are increasing towards 2. In fact, each additional term adds on half the remaining distance to 2. ◇

DEFINITION. For a sequence $\{a_k\}_{k \in \mathbb{N}}$, we define the *sequence of partial sums* $\{s_n\}_{n \in \mathbb{N}}$, where $s_n = \sum_{k=1}^{n} a_k$. If $\lim_{n \to \infty} s_n$ exists, we write $\sum_{k=1}^{\infty} a_k = \lim_{n \to \infty} s_n$. We extend this notation to allow the sequence to start with a_0. Then the sums are given by $s_n = \sum_{k=0}^{n} a_k$ and the limit by $\sum_{k=0}^{\infty} a_k$. If the limit exists, we call $\sum_{k=1}^{\infty} a_k$ or $\sum_{k=0}^{\infty} a_k$ a *convergent series*. Otherwise, we call the series *divergent*.

THEOREM 8.4.4. If a series $\sum_{k=0}^{\infty} a_k$ converges, then the terms a_k go to 0. That is, $\lim_{k \to \infty} a_k = 0$.

DISCUSSION. Suppose that $\sum_{k=0}^{\infty} a_k$ converges to L. Thus, the partial sums $s_n = \sum_{k=0}^{n} a_k$ must be close to L once n is big. We need to relate these partial sums to the individual terms a_k. Now $s_n - s_{n-1} = \sum_{k=0}^{n} a_k - \sum_{k=0}^{n-1} a_k = a_n$. Since s_n and s_{n-1} are both close to L, they are close to each other. The following equalities and inequality, based on the preceding discussion, provide the core of the proof: $|a_n| = |s_n - s_{n-1}| = |(s_n - L) + (L - s_{n-1})| \leq |s_n - L| + |L - s_{n-1}|$.

Proof. See Problem 8. ∎

The converse of Theorem 8.4.4 is false. The most famous counterexample is the *harmonic series*, shown to diverge by Nicole d'Oresme (ca. 1323–1382), 300 years before Newton and Leibniz developed calculus.

EXAMPLE 7

The *harmonic series* $\sum_{k=1}^{\infty} \frac{1}{k} = 1 + \frac{1}{2} + \frac{1}{3} + \frac{1}{4} + \ldots$ diverges, even though the individual terms go to 0. The following sum arranges the terms so that each clump sums to at least $\frac{1}{2}$:

$$(1) + \left(\frac{1}{2}\right) + \left(\frac{1}{3} + \frac{1}{4}\right) + \left(\frac{1}{5} + \frac{1}{6} + \frac{1}{7} + \frac{1}{8}\right) + \left(\frac{1}{9} + \frac{1}{10} + \cdots + \frac{1}{16}\right) + \cdots$$

Clearly, the first two terms are individually at least $\frac{1}{2}$. Now $\frac{1}{3} + \frac{1}{4} > \frac{1}{4} + \frac{1}{4} = \frac{1}{2}$, so we put those together. Similarly, $\frac{1}{5} + \frac{1}{6} + \frac{1}{7} + \frac{1}{8} > \frac{1}{8} + \frac{1}{8} + \frac{1}{8} + \frac{1}{8} = \frac{1}{2}$. We can continue by putting together the fractions from $\frac{1}{2^{n-1}+1}$ to $\frac{1}{2^n}$. There will be 2^{n-1} fractions in this grouping, all at least as large as $\frac{1}{2^n}$. Further, $2^{n-1} \cdot \frac{1}{2^n} = \frac{1}{2}$, so each such grouping sums to at least $\frac{1}{2}$. Since there are infinitely many such groupings, one for each power of $\frac{1}{2}$, the sum "goes to infinity." That is, the partial sum can surpass any possible proposed limit and so the harmonic series diverges. ◇

The terms of the series in Examples 6 and 7 both go to zero, but the terms in Example 6 decrease much more quickly. This difference helps explain why $\sum_{n=0}^{\infty} \frac{1}{2^n}$ converges, but $\sum_{k=1}^{\infty} \frac{1}{k}$ diverges. In fact, each term in $\sum_{n=0}^{\infty} \frac{1}{2^n}$ is half as big as the previous one. In modern terms, we'd say that the terms decay exponentially. However, series like this one have been studied since the time of the ancient Greeks, when they called them geometric series, the name we still use.

DEFINITION.   For $a, r \in \mathbb{R}$, with $a \neq 0$ and $r \neq 0$, the series $\sum_{k=0}^{\infty} ar^k$ is a *geometric series*.

EXAMPLE 8

Consider the geometric series with $a = 3$ and $r = -0.8$. The first few terms to sum are $a_0 = 3$, $a_1 = 3(-0.8) = -2.4, a_2 = 3(-0.8)^2 = 1.92, a_3 = 3(-0.8)^3 = -1.536$. The corresponding partial sums are $s_0 = 3$, $s_1 = 0.6$, $s_2 = 2.52$, and $s_3 = 0.984$. Although the partial sums appear to approach each other, it may be unclear whether they will converge to a single value. The key is how quickly the terms $ar^k = 3(-0.8)^k$ go to 0. It is easier to show what happens in general than to work more with this particular example. Theorem 8.4.5 tells us that this series converges to $\frac{3}{1-(-0.8)} = 1.666\ldots$ $\Diamond$

THEOREM 8.4.5.

i) If $r \neq 1$, the partial sum $\sum_{k=0}^{n} ar^k$ equals $a\frac{1-r^{n+1}}{1-r}$.

ii) The geometric series $\sum_{k=0}^{\infty} ar^k$ converges iff $-1 < r < 1$.

iii) If the geometric series converges, $\sum_{k=0}^{\infty} ar^k = \frac{a}{1-r}$.

Proof.

i) See Problem 3i of Section 2.4.

ii) First consider r with $|r| > 1$ and $r \neq 0$. Then $|ar^{k+1}| = |ar^k| \cdot |r| > |ar^k| > |a| > 0$. (Recall that $a \neq 0$.) Since the terms are growing in size, their limit can't be 0. By the contrapositive of Theorem 8.4.4, we see that a geometric series with $|r| > 1$ diverges. The case $r = 1$ simply becomes the repeated sum of a, which goes off to $\pm\infty$, depending on the sign of a. Next consider the case $r = -1$. Powers of -1 simply alternate between $1 = (-1)^{2n}$ for even powers and $-1 = (-1)^{2n+1}$ for odd powers. Then the sum becomes $a - a + a - a + a - \ldots$. The partial sums alternate between a and 0. Since $a \neq 0$, again this series diverges.

So the only possible values of r giving convergence are, as claimed, $-1 < r < 1$, with $r \neq 0$. Problem 10 shows that the partial sums $a\frac{1-r^{n+1}}{1-r}$ converge to the value of part iii, namely, $\frac{a}{1-r}$, which finishes part iii as well as part ii.   ∎

Decimal Representations

In Section 8.1, we used the subtleties of decimal representations to help motivate our investigation of analysis. We are now in a position to use analysis to examine decimal representations. To spare a long preparatory excursion, we will assume that our base 10 representation of integers works as we have learned it. Thus, we can focus on what happens to the right of the decimal point. Decimal representations might well remind us of geometric series. For instance, by the right side of $\pi = 3.14159\ldots$, we mean the infinite sum $3 + \frac{1}{10} + \frac{4}{10^2} + \frac{1}{10^3} + \frac{5}{10^4} + \frac{9}{10^5} + \ldots$. The powers of 10 in the denominator act like the r of a geometric series. However, the numerators keep changing, so the series for pi isn't a geometric series. Even so, Theorem 8.4.5 can help guarantee that decimal representations match with real numbers.

DEFINITION OF DECIMAL REPRESENTATION. For any sequence $\{a_n\}_{n\in\mathbb{N}}$ with $a_n \in \{0, 1, 2, 3, 4, 5, 6, 7, 8, 9\}$, we write $0.a_1a_2a_3\ldots$ for the series $\sum_{n=1}^{\infty} \frac{a_n}{10^n}$. If $k \in \mathbb{N}$, we write $k.a_1a_2a_3\ldots$ for $k + \sum_{n=1}^{\infty} \frac{a_n}{10^n}$. Similarly, $-k.a_1a_2a_3\ldots$ is defined to be $-(k + \sum_{n=1}^{\infty} \frac{a_n}{10^n})$.

As a direct consequence of this definition, $0.999\ldots = \sum_{k=0}^{\infty} (0.9)(0.1)^k$. Since this is a geometric series, Theorem 8.4.5 tells us that $0.999\ldots = 1$. The next theorem enables us to connect decimal representations and real numbers more generally.

THEOREM 8.4.6.

i) If $\{a_n\}_{n\in\mathbb{N}}$ is any sequence with $a_n \in \{0, 1, 2, 3, 4, 5, 6, 7, 8, 9\}$, then $\sum_{n=1}^{\infty} \frac{a_n}{10^n}$ converges.

ii) Any decimal representation determines a unique real number, and every real number has at least one decimal representation.

Proof.

i) The sequence of partial sums $s_k = \sum_{n=1}^{k} \frac{a_n}{10^n}$ is nondecreasing, since we are adding a positive or zero term at each step. Further, for all $n \in \mathbb{N}$, $\frac{a_n}{10^n} \leq \frac{9}{10^n} = 9(\frac{1}{10})^n$. So the sequence $\{s_k\}_{k\in\mathbb{N}}$ is bounded by the limit of the geometric series $\sum_{n=1}^{\infty} 9(\frac{1}{10})^n$, which, as we just noted, is 1. Thus, we have a bounded nondecreasing sequence, which converges by Theorem 8.4.3.

ii) See Problem 15. ∎

Part ii of Theorem 8.4.6 summarizes the theoretical essentials of decimal representations. However, it leaves the loose end of multiple representations of a real number. Problem 16 provides an outline to show that the only time we get two representations of the same real number, such as $2.43 = 2.42999\ldots$, is when one representation has repeating nines and the other terminates (or equivalently, has repeating zeros).

Series of Functions

Series reveal their full value in advanced mathematics and its applications, rather than with decimal representations. Since the time of Newton, mathematicians and physicists have used different kinds of series to define, evaluate, and examine

functions. Before calculators and computers, the hand computation of approximations of function values, such as $\sin(1.2)$ or $e^{0.5}$, generally depended on using partial
sums of an appropriate series. In a sense, while limits "turn approximations into exact values," we can use series to turn exact values into computable approximations.
Indeed, the built-in computing programs on today's calculators use partial sums of
series to find the approximations they display so quickly. The hidden nature of this
use of series now makes this topic harder to motivate for calculus students without
reducing its continued importance. In our short excursion into sequences and series,
we will restrict the exploration of series of functions to a few problems.

Historical Remarks

Finite series and sequences can trace their ancestry to the ancient Greeks. Until
the time of Newton, infinite versions of sequences and series were often connected
with paradoxical situations, such as Zeno's Paradoxes. (See the discussion at the
end of Section 8.1.) The advent of calculus made mathematicians comfortable with
infinite processes, and they freely worked with them, often ignoring questions of
convergence. Power series, such as Newton's $\sum_{n=0}^{\infty} \frac{(-x)^{2n+1}}{(2n+1)!}$ for $\sin(x)$, were readily
accepted. Joseph Fourier (1768–1830) made amazing advances, using series involving
sines and cosines, now called Fourier series in his honor. Physicists continue to use
these series extensively. The functions he could define with these series went far
beyond what had been previously considered as functions, leading to the modern
understanding of a function. They also posed new problems for the emerging effort
to provide formal definitions and proofs in analysis. For instance, a result about
continuous functions that Cauchy had announced turned out to need an extra concept,
now called uniformly continuous.

PROBLEMS

*1. For each statement that follows, state whether it is true or (at least sometimes) false.
Explain your answer or cite the part of the text supporting it.

a) Every bounded sequence converges.

b) Every convergent sequence is bounded.

c) If a sequence diverges, its terms are unbounded.

d) An unbounded sequence diverges.

e) If a series converges, the sequence of its terms converges.

f) If the terms of a series converge to zero, then the series converges.

g) A geometric series converges.

2. Evaluate several terms of each sequence and decide whether the sequence converges or
diverges. If it converges, determine its limit as best as you can.

*a) $\{\frac{n}{2n+1}\}_{n \in \mathbb{N}}$.

b) $\{\frac{(-2)^n}{3^n}\}_{n \in \mathbb{N}}$.

*c) $\{\frac{k\sqrt{k}}{500k}\}_{k \in \mathbb{N}}$

d) $\{(\frac{-1}{\sqrt{k}} - 1)^k\}_{k \in \mathbb{N}}$

e) $\{(1 + \frac{1}{n})^n\}_{n \in \mathbb{N}}$

f) $a_1 = 1, a_2 = 2, a_{n+2} = \frac{a_n + a_{n+1}}{2}$ (the average of the two preceding terms)

3. Prove the following limits of sequences:

   *a) $\lim_{k\to\infty} \frac{k}{k+1} = 1$.

   b) $\lim_{m\to\infty} \frac{3m-1}{m+1} = 3$.

   c) $\lim_{n\to\infty}(\pi - \frac{1}{n^2}) = \pi$.

   d) $\lim_{k\to\infty} \frac{1}{\sqrt{k}} = 0$.

   e) Prove your answer in part f of Problem 2. (Hint: Use Theorem 8.4.5.)

4. Prove Case 2 of Example 5.

5. Prove Theorem 8.4.1.

6. **a)** Define a nonincreasing sequence.

   b) Prove that every bounded nonincreasing sequence converges (referred to in Section 8.5).

7. ***a)** Suppose that $\lim_{n\to\infty} x_n = L$. Show that $\lim_{n\to\infty}(2x_n + 3) = 2L + 3$. (Hint: Adjust ϵ.)

   b) Suppose that $\lim_{n\to\infty} x_n = L$ and $m, b \in \mathbb{R}$. Show that $\lim_{n\to\infty}(mx_n + b) = mL + b$. (Hint: Do the case $m = 0$ separately (referred to in Section 8.5).)

8. Prove Theorem 8.4.4.

9. Suppose that $-1 < r < 1$ and $r \neq 0$. Fill out the following steps to prove that $\lim_{n\to\infty} r^n = 0$ (referred to in Section 8.5):

   a) Prove that if $n > N$, then $|r^n| < |r^N|$.

   b) Why does it suffice to show that for each $k \in \mathbb{N}$, we can find $N \in \mathbb{N}$ such that $|r^N| < \frac{1}{k}$?

   c) Suppose that $k > 1$. Why are both $\ln(|r|)$ and $\ln(\frac{1}{k})$ negative?

   d) Recall that, $\ln(r^N) = N \ln(r)$. Modify Problem 18 of Section 8.1 to show that there is some $N \in \mathbb{N}$ such that $\ln(r^N) < \ln(\frac{1}{k})$.

   e) Prove that $\lim_{n\to\infty} r^n = 0$.

10. Use Problems 7 and 9 to complete the proof of Theorem 8.4.5. (Hint: In Problem 7b, use $x_n = r^{n+1}$.)

11. Define $\{b_n\}_{n\in\mathbb{N}}$ recursively by $b_1 = 1$ and $b_{n+1} = \sqrt{2b_n}$.

    a) Calculate the first four terms of this sequence and make a conjecture about its limit.

    b) Use induction to prove that for all $n \in \mathbb{N}$, $1 \le b_n < 2$.

    c) Prove that for all $n \in \mathbb{N}$, $b_n < b_{n+1}$. (Hint: Show that $b_n^2 < 2b_n$ and use Problem 11 in Section 8.1.)

    d) Prove that $\lim_{n\to\infty} b_n$ exists. Call this limit L.

    e) Explain why it is reasonable to say that $\lim_{n\to\infty}\sqrt{2b_n} = \sqrt{2L}$. (See Problem 19 for the general setting.)

    f) Assuming that part e is true (which it is), find L.

*12. Suppose that for all $n \in \mathbb{N}$, $a_n \le c$ and $\lim_{n\to\infty} a_n = L$. Prove that $L \le c$. (Hint: In a proof by contradiction, consider that $L - c > 0$ (referred to in Section 8.5).)

13. **a)** Suppose that $\lim_{n\to\infty} a_n = L$ and $\lim_{n\to\infty} b_n = M$. Prove that $\lim_{n\to\infty} a_n \pm b_n = L \pm M$.

    b) Suppose that for all $n \in \mathbb{N}$, $a_n \le b_n$, $\lim_{n\to\infty} a_n = L$, and $\lim_{n\to\infty} b_n = M$. Prove that $L \le M$. (Hint: Use part a and Problem 12 (referred to in Section 8.5).)

14. Suppose that for all $n \in \mathbb{N}$, $0 \le a_n \le b_n$ and $\sum_{n=1}^{\infty} b_n$ converges to M. Prove that $\sum_{n=1}^{\infty} a_n$ converges and its limit is less than or equal to M.

15. Fill out the steps that follow to prove Theorem 8.4.6 part ii. Throughout, assume that $a_i \in \{0, 1, \ldots, 9\}$. Parts a and b show the theorem for numbers in $[0, 1]$.

   a) The decimal $0.a_1a_2a_3 \ldots$ represents the real number $\sum_{n=1}^{\infty} \frac{a_n}{10^n}$. Show that this number is in $[0, 1]$.

   b) For $s \in [0, 1]$, we find a decimal representation of s. Let $S_k = \{\sum_{n=1}^{k} \frac{a_n}{10^n} : \sum_{n=1}^{k} \frac{a_n}{10^n} \leq s\}$, the k-place decimals less than or equal to s. (For instance, with the aid of my calculator I found that $s = e^{-2}$ has $S_1 = \{\frac{0}{10}, \frac{1}{10}\} = \{0.0, 0.1\}$ and $S_2 = \{0.00, 0.01, \ldots, 0.09, 0.10, 0.11, 0.12, 0.13\}$.) Why is each S_k a finite set? Why does each S_k have a least upper bound? Denote the l.u.b. of S_k by b_k. Why do b_k and b_{k+1} have the same first k decimal places $\frac{a_n}{10^n}$, for $1 \leq n \leq k$? Why does this show that s has at least one decimal representation?

   c) Prove this claim: For $r \in \mathbb{R}$ with $1 \leq r$, there are unique $n \in \mathbb{N}$ and $s \in [0, 1)$ such that $r = n + s$. (Hint: Why does $\{k \in \mathbb{N} : k \leq r\}$ have a least upper bound?)

   d) Why do parts a, b, and c guarantee that every positive real has a decimal representation?

   e) How do we handle negative real numbers?

16. a) If $x = 0.x_1x_2x_3 \ldots$ and $y = 0.y_1y_2y_3 \ldots$ have different decimal representations and neither has infinitely repeating 9s, show that $x \neq y$. (Hints: We may suppose that they first differ in, say, the kth decimal place; that is, $x_k \neq y_k$. Why may we assume that $x \leq y$? Consider cases, starting with $y_k - x_k > 1$. Next, suppose that $y_k - x_k = 1$ and look at the next decimal place, x_{k+1} and y_{k+1}. First consider the case where $x_{k+1} \neq 9$. Then consider the case where $y_{k+1} \neq 0$.)

      Consider now the hard case: $x_{k+1} = 9$ and $y_{k+1} = 0$. Since x does not have infinitely repeating 9s, there is some $j > k + 1$ where either $x_j < 9$ or x stops at the $(j-1)^{st}$ place. What happens in these cases?

   b) Suppose that x has a finite decimal representation ending with the digit d in the 10^{-k} place, $d \neq 0$. For instance, for $x = 0.515253$, we have $d = 3$ and $k = 6$. Let y be the number $x - d10^{-k} + w$, where $w = (d-1)10^{-k} + 0.999 \ldots \cdot 10^{-k}$. For instance, the previous x gives $y = 0.515253 - 0.000003 + w$, where $w = 0.000002 + 0.000000999 \ldots$. In other words, $y = 0.515252999 \ldots$. Show that $x = y$.

17. Use Problem 14 to compare the series $\sum_{n=0}^{\infty} \frac{1}{n!}$ with the geometric series $\sum_{n=0}^{\infty} 2(\frac{1}{2})^n$ to show that the first series converges. (Hint: An induction proof may help you prove the condition corresponding to $0 \leq a_n \leq b_n$. [Recall from calculus that $e = \frac{1}{0!} + \frac{1}{1!} + \frac{1}{2!} + \ldots = \sum_{n=0}^{\infty} \frac{1}{n!}$].)

18. Use Problem 14 to compare $\sum_{n=1}^{\infty} \frac{r^n}{n}$ with a geometric series to show that this series converges for r satisfying $0 < r < 1$. Explain why we can define a function $f : (0, 1) \rightarrow \mathbb{R}$ by $f(x) = \sum_{n=1}^{\infty} \frac{x^n}{n}$. (You may recall from calculus that $f(x) = -\ln(1 - x)$.)

19. Suppose that $f : (a, b) \rightarrow \mathbb{R}$ is continuous at $c \in (a, b)$ and $\{c_n\}_{n \in \mathbb{N}}$ is a sequence of elements of (a, b). Prove that if $\lim_{n \to \infty} c_n = c$, then $\lim_{n \to \infty} f(c_n) = f(c)$. (Hint: From $\epsilon > 0$, use the continuity of f to find a related δ. For this δ, use $\lim_{n \to \infty} c_n = c$ to find a related N (referred to in Section 8.5).)

20. Critique these paraphrases of arguments about the series $1 - 1 + 1 - 1 + 1 - 1 + \ldots$ from the early days of calculus, around 1700. (Leibniz favored the third argument.)

      Argument 1. The sum is clearly 0. Just group the terms as follows: $(1 - 1) + (1 - 1) + (1 - 1) + \ldots$. Since each pair adds to 0, the entire sum is 0.

Argument 2. Then the sum is just as clearly 1. Group the terms as follows: $1 + (-1 + 1) + (-1 + 1) + \dots$. Again, each pair after the initial term sums to 0, so only the first term matters, giving 1.

Argument 3. No, the sum is the average, $\frac{1}{2}$. Half of the time the partial sums are 1, and 0 the other half. Besides, this series is a geometric series with $a = 1$ and $r = -1$. From the formula for a geometric series, we see that $\frac{a}{1-r} = \frac{1}{1-(-1)} = \frac{1}{2}$ is the sum.

21. Critique the questionable "proofs" that follow. Point out reasoning errors and unclear presentation. If the argument is a proof, say so. (If the claim is false, there must be an error in the argument. However, it is not enough to say that the claim is false. You need to find an error in the reasoning.)

  a) Claim: If $\lim_{n \to \infty} a_n = L$, then $\lim_{n \to \infty} |a_n| = |L|$. "Proof": Let $\epsilon > 0$. From $\lim_{n \to \infty} a_n = L$, there is $N \in \mathbb{N}$ such that if $n > N$, then $|a_n - L| < \epsilon$. Then $||a_n| - |L|| = |a_n - L| < \epsilon$. ▲

  b) Claim: $\sum_{n=1}^{\infty} \frac{1}{n(n+1)} = 1$. "Proof": Since Problem 3a shows that $\lim_{k \to \infty} \frac{k}{k+1} = 1$, we need only show (by induction) that for all $k \in \mathbb{N}$, $\sum_{n=1}^{k} \frac{1}{n(n+1)} = \frac{k}{k+1}$. For the initial step, note that $\sum_{n=1}^{1} \frac{1}{n(n+1)} = \frac{1}{1 \cdot 2} = \frac{1}{1+1}$. Assume that $\sum_{n=1}^{k} \frac{1}{n(n+1)} = \frac{k}{k+1}$, and consider that $\sum_{n=1}^{k+1} \frac{1}{n(n+1)}$. We have $\sum_{n=1}^{k+1} \frac{1}{n(n+1)} = \sum_{n=1}^{k} \frac{1}{n(n+1)} + \frac{1}{(k+1)(k+2)} = \frac{k}{k+1} + \frac{1}{(k+1)(k+2)} = \frac{k(k+2)+1}{(k+1)(k+2)} = \frac{(k+1)^2}{(k+1)(k+2)} = \frac{k+1}{k+2}$. By PMI we have shown the needed equality. ▲

  c) Claim: $\lim_{n \to \infty} ((-1)^n + (-\frac{1}{2})^n) = 1$. "Proof": Let $\epsilon > 0$. We know from Theorem 8.1.3 that there is $N \in \mathbb{N}$ such that $\frac{1}{N} < \epsilon$. From Example 4, we know that $\frac{1}{2^N} < \frac{1}{N}$. Pick $K = 2N$ and let $n > K$. Since K is even, $(-1)^K = 1$. Then $|(-1)^n + (-\frac{1}{2})^n - 1| < |(-1)^K + (-\frac{1}{2})^K - 1| = |(-\frac{1}{2})^K| < \frac{1}{2^N} < \frac{1}{N} < \epsilon$. ▲

  *d) Claim: $1 + 2 + 4 + 8 + \dots = 1$. "Proof": Let $x = \sum_{i=0}^{\infty} 2^i$. Then $2x = 2(\sum_{i=0}^{\infty} 2^i) = 2(1 + 2 + 4 + \dots) = 2 + 4 + 8 + \dots = 1 + 2 + 4 + 8 + \dots - 1 = (\sum_{i=0}^{\infty} 2^i) - 1 = x - 1$. We can solve $2x = x - 1$ to find $x = -1$. Additionally, $\sum_{i=0}^{\infty} 2^i$ is a geometric series with $a = 1$ and $r = 2$. So its sum by Theorem 8.4.5 is $\frac{a}{1-r} = \frac{1}{1-2} = -1$. ▲

8.5 DISCRETE DYNAMICAL SYSTEMS

"... [I]t may happen that small differences in the initial conditions produce very great ones in the final phenomena."

—Henri Poincaré (1854–1912)

The field of dynamical systems studies changes over time; in particular, it investigates the long-term behavior of systems. In the long run, for instance, a swinging pendulum slows and stops. The periodic orbits of the planets illustrates another type of long-term behavior well studied since ancient times. Much more recently, mathematicians and scientists have intensely studied a radically different sort of behavior called chaos. Popular descriptions of chaos tend to emphasize the unpredictable nature of chaos and avoid the technicalities, which depend on analysis. (See Gleick for a popular introduction.)

Our short account will focus on discrete dynamical systems, building on our brief look at analysis. ("Discrete" means "separate," as opposed to continuous dynamical systems, which involve differential equations and so are beyond the level of this text.) After introducing some essential terms, we pause the development for a project. Students should, in my view, develop some intuition and conjectures about dynamical systems before continuing. Otherwise, the ensuing definitions and results may seem unmotivated, when in actuality they are a response to the patterns suggested by examples. A system will be for us a real function and we will apply the function repeatedly. The resulting sequence of images $(x, f(x), f(f(x)), \ldots,$ etc.) is what makes the system discrete. We will find functions exhibiting chaos, which we will define, as well as simpler dynamics. The use of an exponent for repeated composition in the definition that follows is now standard, even though it, unfortunately, conflicts with the usual use of exponents.

DEFINITION. Given a function $f : A \rightarrow A$, define f^n recursively by $f^1 = f$ and, for $n \geq 1$, $f^{n+1} = f \circ f^n$. For $x \in A$, the *iterates* of x under f are the images of x under the functions f^n. We often abbreviate the nth iterate $f^n(x)$ by x_n, and we write $x_0 = x$.

EXAMPLE 1

Define $g : \mathbb{R} \rightarrow \mathbb{R}$ by $g(x) = x^2$. Note that $g^2(x) = g(g(x)) = (x^2)^2 = x^4$ and $g^3(x) = g(g^2(x)) = (x^4)^2 = x^8$. Thus, the "exponents" on the function in the definition do not match ordinary exponents. (Iterations of other functions can give very different outcomes.) Table 1 gives the rounded values of the first 10 iterates of 1, 1.01, and 0.99, suggesting how different the long-term behaviors of these nearby points are. The iterates of 1 are all 1, whereas the iterates of 1.01 grow without bound and the iterates of 0.99 head towards 0. We'll return to this example after defining some terms.

TABLE 1. Iterates of Several x-Values

x_0	x_1	x_2	x_3	x_4	x_5	x_6	x_7	x_8	x_9	x_{10}
1	1	1	1	1	1	1	1	1	1	1
1.01	1.02	1.04	1.08	1.17	1.37	1.89	3.57	12.77	163.1	26.612
0.99	0.98	0.96	0.92	0.85	0.72	0.53	0.28	0.07	0.006	0.00003

$\Diamond$

DEFINITION. For $f : A \rightarrow A$ and $A \subseteq \mathbb{R}$, we say that $x \in A$ is a *fixed point* of f iff $f(x) = x$. A fixed point p is *attracting* iff there is some $\delta > 0$ such that for all $x \in A$, if $|x - p| < \delta$, then $\lim_{n \to \infty} f^n(x) = p$.

EXAMPLE 2

For g given by $g(x) = x^2$, the fixed points are 1 and 0. Our understanding of squares tells us that the dynamics of squaring mimic what happens in Table 1: Numbers bigger than 1 have increasingly larger squares, and numbers between 0 and 1 have increasingly smaller squares. (The square of a negative are positive, so the dynamics for negative numbers are reflected in the dynamics of positive numbers.) Of the two fixed points, only 0 is attracting. We can use $\delta = 1$ for the interval of attraction around 0, since for any x strictly between -1 and 1, the squaring process will send the iterates of x toward 0. Although $f(-1) = 1$ shows that -1 is "attracted" to 1, 1 is not attracting, since no interval around 1 is attracted to 1. A more detailed treatment of dynamical systems would investigate *repelling fixed points*, such as 1 here, points from which nearby points head away.

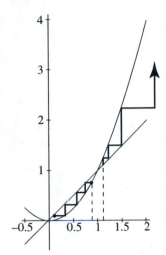

FIGURE 1. Graphical analysis for $y = x^2$.

Graphical analysis gives us a visual way to approximate the iterates x_n. Graph both $y = g(x)$ and $y = x$ on the same axes, as in Figure 1. From a given $x = x_0$, we move vertically to the point $(x, g(x))$ on $y = g(x)$. Now slide horizontally to $y = x$ to find the point $(g(x), g(x))$. Directly below this on the x-axis is $x_1 = g(x)$. Go vertically again from $(g(x), g(x))$ to $y = g(x)$, where the point is $(g(x), g^2(x))$, and horizontally to $(g^2(x), g^2(x))$ on $y = x$, which gives us $x_2 = g^2(x)$. Continue jumping back and forth between $y = g(x)$ and $y = x$ to find additional iterates. Figure 1 indicates that an initial value of x a bit less than 1 heads towards 0 as a limit, whereas an initial value of x greater than 1 heads off to infinity, matching the numerical information in Table 1. We can estimate fixed points easily on a graph, such as Figure 1, by looking for the intersection of $y = g(x)$ and $y = x$. ◇

How common are fixed points? How can we tell if a fixed point is attracting? The next theorem answers the first question. The project investigates a family of functions, which will help us approach the more interesting question of attracting fixed points.

THEOREM 8.5.1. If $f : [a, b] \to [a, b]$ is continuous, then f has a fixed point.

Proof. We apply Theorem 8.5.3, the Intermediate Value Theorem (IVT) from calculus. (Its statement and proof appear at the end of this section.) We need to find a fixed point; that is, an x-value such that $f(x) = x$. Let's shift the x to the other side to get a related equation $f(x) - x = 0$. The constant 0 now acts like the intermediate value of the IVT. For all x in $[a, b]$, we know that $a \le f(x) \le b$ so $a \le f(a)$ and

$f(b) \leq b$. In other words, $0 \leq f(a) - a$ and $f(b) - b \leq 0$. So a new function g given by $g(x) = f(x) - x$ goes from a nonnegative value at $x = a$ to a nonpositive value at $x = b$.

The IVT assures us that if g is continuous, then it has to cross $y = 0$ somewhere in between a and b. But at that spot we'd have $f(x) - x = 0$, or $f(x) = x$, our desired fixed point. Hence, we need only show that g is continuous. We are given that f is continuous. Further, from Example 1 of Section 8.3, the function given by $h(x) = x$ is continuous. From Problem 6 of Section 8.3, their difference, $g(x) = f(x) - x$, is continuous. ∎

Project

To develop some intuition about dynamical systems, we investigate the family of functions $h_k : [0, 1] \to [0, 1]$ given by $h_k(x) = kx(1 - x)$ for $0 \leq k \leq 4$. (The restriction $0 \leq k \leq 4$ ensures that $h_k : [0, 1] \to [0, 1]$.)

REMARK. This family is often called *logistic* because of a formal similarity with the logistic differential equation, $y' = ky(1 - y)$. Unfortunately, this similarity is rather misleading because the left side of the differential equation is the derivative of the function, not the original function (at a different value). Even though the discrete system has applications, it doesn't model the same sort of behavior that the differential equation models.

EXERCISE 1.

a) Graph $y = h_k(x)$ and $y = x$ on the same axes for various choices of k.

b) Use your graphs in part a to estimate the fixed points in each case.

c) For a general value of k, find a formula for the fixed point(s) of h_k in $[0, 1]$. Do your estimates in part b come close to the exact values from this formula?

To avoid the boredom of long computations to find iterations, we'll make use of a shortcut on many graphing calculators. (You are welcome to program a computer instead.) I'll illustrate, giving instructions for a TI calculator with an initial value of $x = 0.1$ for $h_{2.5}$, where $h_{2.5}(x) = 2.5x(1 - x)$. (Other calculators have similar capabilities, but may use different keystrokes.) First type in the initial value, 0.1, and press the ENTER key. Next, type in the formula for the function, using the ANSWER key as the variable. In this example, we would type $2.5(2^{nd}$ ANS$)(1 - 2^{nd}$ ANS$)$ and press ENTER. The calculator should read 0.225, the first iterate. Pressing ENTER each time from now on should give succeeding iterates 0.4359375, 0.6147 . . ., and so on.

Figure 2 gives the graphs of $y = h_{2.5}(x)$ and $y = x$. The graphical analysis in Figure 2 suggests that the fixed point 0.6 is attracting, but 0, the other fixed point, is not attracting.

Figure 3 gives the corresponding graphs of $y = h_{10/3}(x)$ and $y = x$. At first glance, the graph of $y = h_{10/3}(x)$ looks quite similar to the graph of $y = h_{2.5}(x)$ in Figure 2. However, the graphical analysis here suggests that neither fixed point

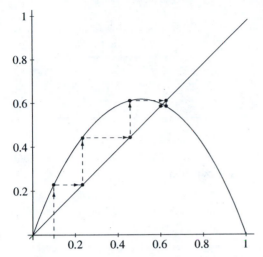

FIGURE 2. Graphical analysis for $y = 2.5x(1 - x)$.

(0 or 0.7) is attracting. Table 2 suggests a different type of dynamics, a pair of numbers that seem to attract other numbers, at least within rounding errors.

EXERCISE 2.

a) For various values of k in $h_k(x) = kx(1 - x)$ and various initial values of x_0 in [0, 1], find the iterates of x_0. (Pick x_0 different from fixed points.)

b) Based on your computations in part a, decide which, if any, of the fixed points are attracting for different values of k. Make a table with increasing values of k, their fixed points, and which, if any, are attracting.

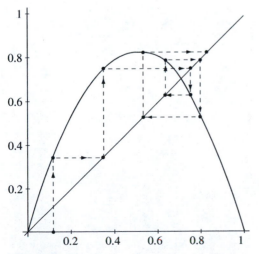

FIGURE 3. Graphical analysis for $y = \frac{10}{3}x(1 - x)$.

TABLE 2.   Selected Iterations for $h_{10/3}(x)$

x_0	x_1	x_2	x_3	x_4	x_5	x_6	x_{20}	x_{21}	x_{30}	x_{31}
0.01	0.03	0.11	0.32	0.72	0.67	0.74	0.83	0.47	0.83	0.47
0.11	0.33	0.73	0.65	0.76	0.62	0.79	0.83	0.47	0.83	0.47
0.73	0.65	0.75	0.62	0.78	0.57	0.82	0.83	0.47	0.83	0.47

c) Make a conjecture about the long-term behavior of the system h_k for k in different intervals between 0 and 4. (For a range of values of k, it is acceptable to say that it is too complicated to discern from a calculator.)

d) Do graphical analyses for values of k in the different intervals you conjectured in part c. How and why do the graphical analyses seem to differ?

e) For a general k, find the value of $h'_k(p)$ and $h'_k(q)$, where p and q are the fixed points from your formula in part c of Exercise 1. Add to your table in part c of this exercise a column for these derivatives. Make a conjecture about how the derivative is related to whether a fixed point is attracting.

The Dynamics of Attraction

What is the key difference between functions like $h_{2.5}$, with an attracting fixed point, and functions like $h_{10/3}$, with no attracting fixed point? Theorem 8.4.2 shows that the value of the derivative at a fixed point determines whether the point attracts. The graphical analyses in Figures 4 and 5 for two straight lines with negative slopes illustrate the essential idea of the theorem. In Figure 4, where the slope is -0.7, the pattern of iterates spirals in toward the fixed point. Because the slope is between -1 and 1, each horizontal shift from that line is followed by a shorter vertical shift back

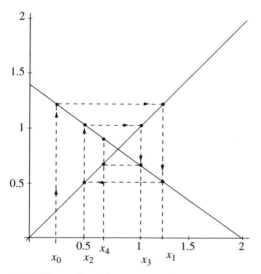

FIGURE 4.   Graphical analysis for $y = 1.4 - 0.7x$.

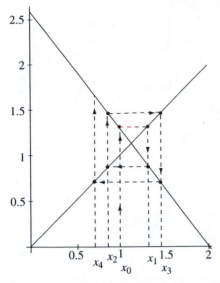

FIGURE 5. Graphical analysis for $y = 2.6 - 1.3x$.

to the line. In turn, each vertical shift from $y = x$ is followed by an equally long horizontal shift to get to another point on $y = x$. In Figure 5, where the slope is -1.3, the spiral heads outward. Here, each horizontal shift is followed by a larger vertical shift, not a smaller one. From elementary calculus, we see that $h'_{2.5}(0.6) = -0.5$, so the derivative at the fixed point is between -1 and 1 and we spiral in. However, $h'_{10/3}(0.7) = -1.333$, which forces values close to this fixed point to be repelled.

The case for positive slopes is effectively illustrated in Figure 1, where one sequence of iterates zigzags in to a fixed point and the other zigzags out to infinity. For both functions $h_{2.5}$ and $h_{10/3}$, the derivative at 0 is greater than 1, which is outside the range of attraction, according to the theorem.

THEOREM 8.5.2. Suppose that $f : [a, b] \to [a, b]$ has a fixed point at c and $|f'(c)| < 1$. Then c is an attracting fixed point. If $|f'(c)| > 1$, then c is not attracting.

DISCUSSION. The idea is to shift from the derivative $f'(c)$ to the difference quotient $\frac{f(x) - f(c)}{x - c}$. From $|f'(c)| < 1$, we'll get $\left| \frac{f(x) - f(c)}{x - c} \right| < 1$ or $|f(x) - f(c)| < |x - c|$. In effect, this will mean that the images are closer together than the original values. Since $f(c) = c$, this means that $f(x)$ is closer to c than x was. That isn't quite enough to show that $\lim_{n \to \infty} f^n(x) = c$, but the proof will provide more precision.

Proof. Suppose that $|f'(c)| < 1$. Then there is some $t > 0$ so that $-1 < -t < f'(c) < t < 1$. Since $f'(c)$ is a limit, we need to spread out a bit from the interval $(-t, t)$ to get a corresponding interval for the difference quotient. Let $r = \frac{t+1}{2}$, halfway between t and 1. For $\epsilon = r - |f'(c)|$, the definition of the limit for $f'(c)$ tells us there is $\delta > 0$ such that if $0 < |x - c| < \delta$, then $\left| \frac{f(x) - f(c)}{x - c} - f'(c) \right| < \epsilon$. Thus, $\left| \frac{f(x) - f(c)}{x - c} \right| < |f'(c)| + \epsilon = r < 1$. Then

$|f(x) - f(c)| < r|x - c|$ and $|f^n(x) - f^n(c)| < r^n|x - c|$. Since c is fixed, we can substitute c for $f^n(c)$ to get $|f^n(x) - c| < r^n|x - c|$, showing that the distance shrinks by a factor of at least r with each iteration. By Problems 7 and 9 of Section 8.4, we know that $\lim_{n \to \infty} r^n|x - c| = 0$, so $\lim_{n \to \infty} |f^n(x) - c| = 0$. Equivalently, $\lim_{n \to \infty} f^n(x) = c$, showing that c is an attracting fixed point.

If a fixed point c satisfies $|f'(c)| > 1$, the inequalities in the previous argument reverse to show that c can't be attracting. ∎

REMARKS. If the derivative equals 1 at a fixed point, the theorem doesn't determine whether the fixed point is attracting or not. Further, some such fixed points are attracting and some are not. The value of the derivative at an attracting fixed point governs how quickly the iterates converge toward that point. In general, the closer the derivative is to 0, the more quickly the values are attracted to the fixed point.

While the function $h_{10/3}$ in Example 2 has no attracting fixed point, it appears to have a pair of attracting points, a more complicated dynamic.

DEFINITION. For $f : A \to A$ where $A \subseteq \mathbb{R}$, we say that x is *periodic* iff there is some $n \in \mathbb{N}$ such that $f^n(x) = x$. The smallest such n is the *period* of x. A point p of period n is *attracting* iff there is some $\delta > 0$ such that for all $x \in A$, if $|x - p| < \delta$, then $\lim_{k \to \infty} f^{kn}(x) = p$.

EXAMPLE 3

From Table 2, the function $h_{10/3}$ given by $h_{10/3}(x) = \frac{10}{3}x(1 - x)$ appears to have 0.83 and 0.47 as approximations of attracting points of period 2. We know how to find the fixed points of $h_{10/3}$: Solve $x = \frac{10}{3}x(1 - x)$ to find 0 and 0.7. Theorem 8.5.2 assures us that these are not attracting. A similar procedure gives the exact value of the period two points and determines that they are attracting. However, instead of using the function $h_{10/3}$, we compose it with itself.

Consider $j(x) = h_{10/3}(h_{10/3}(x)) = \frac{10}{3}[\frac{10}{3}x(1 - x)][1 - \frac{10}{3}x(1 - x)]$. Figure 6 illustrates the four fixed points of j. Unfortunately, for hand calculation, $j(x) - x = 0$ is a fourth degree

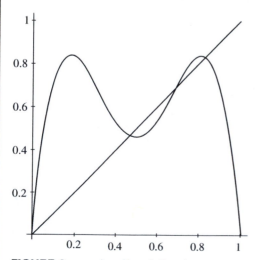

FIGURE 6. $y = h_{10/3}(h_{10/3}(x))$ and $y = x$.

equation. However, we know two of the possible roots, namely, the fixed points of $h_{10/3}$, which j also fixes. We can use a computer algebra system (or careful factoring and the quadratic formula) to find the other two roots of $j(x) - x = 0$, which are $\frac{13 \pm \sqrt{13}}{20}$, or approximately 0.8303 and 0.4697. These numbers are the two period 2 points of $h_{10/3}$. Further, we can apply Theorem 8.5.2 to j at these points to determine whether they are attracting. A calculator or computer gives the derivative of j at both of these points to be approximately -0.444, which means that they are both attracting fixed points for j. Thus, they are indeed attracting period 2 points for $h_{10/3}$. $\diamond$

We can apply a similar procedure to other functions in the family h_k to find in principle any points with periods and whether these points are attracting or not. However, even for period 2 points, the algebra is already more difficult than enlightening. Computers are essential aids for investigating period points and other aspects of dynamics, except for carefully chosen functions. We will turn instead to the idea of chaos. While the function h_4 exhibits chaos, we will use a function easier to fit with the technical definition of chaos. (Problem 16 investigates h_4.)

Chaos

The everyday meaning of the word "chaos" suggests the opposite of order and structure; it even suggests randomness. However, the formal term is much more complicated. While chaotic systems in the real world defy prediction, they are strictly deterministic, not random. A chaotic system requires infinite precision to make long-term predictions, but real-world measurements can have only so much precision. This limit of precision makes long-term prediction impossible in applied chaotic systems. Let's consider a particular function to work through the mathematical definition of chaos.

EXAMPLE 4

Define $s : [0, 1) \to [0, 1)$ by $s(x) = 10x - \lfloor 10x \rfloor$. (Recall that $\lfloor w \rfloor$ is the greatest integer less than or equal to w, as defined in Problem 13 of Section 3.1.) The letter s stands for "shift," because this function shifts the decimal representation by one, dropping the leftmost decimal place. For instance, for $x = 0.2357$, we have $10x = 2.357$ and $\lfloor 10x \rfloor = 2$. So, $s(x) = 2.357 - 2 = 0.357$, a shifting of the last terms of the original value after the first decimal is dropped. Similarly, $s(0.123412341234\ldots) = 0.23412341234\ldots$. Figure 7 gives part of the graph of $y = s(x)$, together with $y = x$. The fixed points have repeated digits, which are the ninths, such as $0.111\ldots = \frac{1}{9}$. Since the derivative at each of these points (except 0) is 10, by Theorem 8.5.2 none of them are attracting fixed points. (There is a one-sided derivative at 0, which is not attracting, for essentially the same reason.)

The many period 2 points must be of the form $0.ababab\ldots$ so that two applications of the shift function give us back the same number. Further, we see that the slope of s^2, where it exists, is 100, since $s(s(x)) = 10(10x - \lfloor 10x \rfloor) - \lfloor 10x - \lfloor 10x \rfloor \rfloor = 100x - 10\lfloor 10x \rfloor$. Hence, none of the period 2 points are attracting.

As we take more iterates, s^n becomes increasingly steep. In short, there are numbers of every period, and none of them can be attracting. The decimal representations of rational numbers are either terminating or, eventually, infinitely repeating. Thus, the iterates of rationals go either to a fixed point or to a cycle of periodic points.

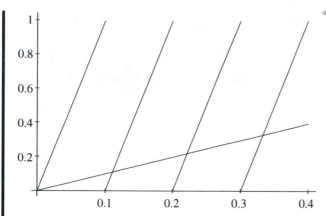

FIGURE 7. $y = s(x)$ and $y = x$.

In contrast, the irrationals have more interesting iterates, since their decimal representations never repeat. For $x_0 = \pi - 3 = 0.14159\ldots, x_1 = s(x) = 0.41592\ldots, x_2 = s^2(x) = 0.15926\ldots, x_{10} = s^{10}(x) = 0.89793\ldots, x_{100} = s^{100}(x) = 0.82148\ldots, x_{1000} = s^{1000}(x) = 0.38095\ldots$, and so on. The function s is completely definite, but the long-term behavior of it is quite complicated. Indeed, since we don't know all the infinitely many decimals of π, we don't really know its long-term behavior. We'll illustrate how s is an example of chaos after we define this technical term and give some explanation. ◇

DEFINITION. A function $f : [a, b] \to [a, b]$ is *chaotic* iff the following three conditions hold:

  i) The periodic points of f are *dense* in $[a, b]$. That is, for all x and y with $a < x < y < b$, there is a periodic point p such that $x < p < y$.

  ii) f is *transitive*. That is, for all $\epsilon > 0$ and all $x, y \in [a, b]$, there are $z \in [a, b]$ and $k \in \mathbb{N}$ such that $|z - x| < \epsilon$ and $|f^k(z) - y| < \epsilon$.

  iii) f has *sensitive dependence on initial conditions*. That is, there is $\epsilon > 0$ such that for all $x \in [a, b]$ and all $\delta > 0$, there are $y \in [a, b]$ and $k \in \mathbb{N}$ such that $|y - x| < \delta$ and $|f^k(y) - f^k(x)| > \epsilon$.

We encountered the concept of density in Sections 4.4 and 8.1. The density of the periodic points in $[a, b]$ is like the density of the rationals in the real numbers. The other two criteria for chaos deserve some explanation.

Transitivity in a dynamical system essentially means that the system has extremely good mixing. Imagine a small bit of red paint in a bucket of white paint. (See Figure 8.) Now imagine that we stir the paint once with a stick. The red has streaked in a rough circle through the white paint, although it is still fairly distinct from the white. With each additional stir, the red paint particles are more dispersed among the white ones. After some time, the entire bucket is a light pink because there are red particles of paint in virtually every location. The points z with $|z - x| < \epsilon$ in the definition of transitivity are like the bit of red paint, all close to x at the start. The stirring is the application of the function f multiple times, and the k counts the

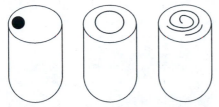

FIGURE 8. Transitivity condition for paint.

number of times we stir the paint. Since the y can be anywhere, the final inequality $|f^k(z) - y| < \epsilon$ requires that there is red arbitrarily close to any place y. That is, the paint is essentially pink, no longer swirls of red and white.

The sensitive dependence on initial conditions sets serious limitations on predictions in actual chaotic systems. It says that no matter how precisely you measure your variables at the start, you can't be sure of the status of the system substantially later. That is, even though x and y are very close together at the start (written as $|y - x| < \delta$ in the definition), their iterates can be quite far apart ($|f^k(y) - f^k(x)| > \epsilon$ in the definition). The logical format of sensitive dependence on initial conditions is quite close to the working negation of a limit, given in Example 4 of Section 8.2.

REMARKS. One can deduce sensitive dependence on initial conditions from the other two properties of chaos. However, this dependence is probably the most characteristic aspect of chaos, and so we will retain it as part of the definition.

In popular explanations of chaos, the sensitive dependence on initial conditions is often called the "butterfly effect." It suggests that differences as small as the flap of a butterfly's wing in one place can compound into dramatically different outcomes in the weather somewhere else. In spite of simplistic portrayals, it does **not** mean that a butterfly's wing beat causes a storm many days later. Rather, it means that a little bit of error in our initial measurements can lead to huge uncertainty in a matter of weeks in weather prediction. As a result, weather predictions beyond two weeks have little value.

EXAMPLE 4 (CONTINUED)

We saw that the rationals were dense in the reals in Problem 7 of Section 8.1. For s, we know that the rational points are "eventually" periodic points. Problem 12 shows that there are dense periodic points. We'll simply illustrate the idea here. Suppose that we are given the two numbers 0.81735 and 0.81736. Then 0.817358173581735 . . . is between them and is periodic.

To illustrate transitivity, consider any two numbers in $[0, 1)$, say, $x = 0.103410341034 \ldots$ and $y = 0.86092771$, and the allowable error of $\epsilon = 0.0000001$. Now we need to find z close to x that will eventually be close to y. To be close here, z and x need to agree for more than the first six decimal places. Our choice of z must start out as 0.1034103, but after that we are free to pick the decimals. Looking at our target of y, let's pick $z = 0.10341038609277$. Then we certainly have $|z - x| < \epsilon$, and letting $k = 7$, we have $|f^7(z) - y| = |0.8609277 - 0.86092771| = 0.00000001 < \epsilon$.

In effect, the multiplication by 10 in the formula for s ensures that after one iteration each tenth of the original domain is spread out over the entire range. Similarly, after two iterations, each one-hundredth of the domain is spread over the range, and so on. In other words, the function s achieves the thorough mixing required by the definition of transitivity. Even better, the structure of s enables us to predict the iterations needed to accomplish any desired mixing. Problem 13 generalizes our illustration.

We can see the essentials of confirming the sensitive dependence on initial conditions for s through another numerical example. The choice of ϵ is ours to make. It says how far apart two images are. Since the whole range is just $[0, 1)$, we need ϵ noticeably smaller than 1, the width of the entire interval. Let's use $\epsilon = 0.4$. Next we are challenged with some random x, say, $x = 0.36336333633336\ldots$, and some small δ, say, $\delta = 0.0000000001$. We need to pick some y close to x, one of whose iterates will eventually be far from the corresponding iterate of x. In order for $|y - x| < \delta$, our choice of y has to agree with x for more than the first 10 places. Let's try $y = 0.3633633363339$. After $k = 12$ shifts, we'll have $s^{12}(x) = 0.36333336\ldots$, and $s^{12}(y) = 0.9$. Therefore, $|s^{12}(y) - s^{12}(x)| = 0.53666663\ldots > \epsilon$. Problem 14 asks you to generalize this situation.

The definition of s allows us easily to determine any particular iterate of any x we wish, assuming that we know x exactly. Thus, we can readily fulfill all the conditions of chaos, which are given in the language of approximations, with δ and ϵ. $\diamondsuit$

Dynamical systems and chaos draw deeply on analysis. But even this short excursion has revealed the value of analytical thinking to understanding the definitions we have considered. For more on dynamical systems, see Devaney.

Historical Remarks

Henri Poincaré's prize-winning essay in 1890 on the three-body problem gave the first instance of what we now call chaos. Prior to that time, mathematicians had successfully used Newtonian mechanics to predict the (periodic) paths of two bodies, such as a star and a planet, given their initial positions, velocities, and mutual gravitational attraction. They also knew that small errors in observation of that data had only limited influence on the predicted paths. However, Poincaré realized (after he won the prize, but before the essay was published) that the corresponding situation for three such bodies as the sun, the earth, and the moon could be much more complicated. In effect, he found that the general system could be chaotic. Instead of looking for an exact solution (which remains elusive to this day), he suggested classifying the types of solutions systems could have. While a number of mathematicians made significant theoretical contributions to dynamical systems over the next 70 years, the subject didn't really blossom until the 1960s. Theoretical advances, such as those by Stephen Smale, facilitated the classification that Poincaré could only hope for. Also, the huge increase in computing speed and power made computer experimentation and visualization of dynamical systems possible. Once mathematicians could see and test examples, they were able to make insightful definitions and conjectures.

PROBLEMS

*1. For each statement that follows, state whether it is true or (at least sometimes) false. Explain your answer or cite the part of the text supporting it.

a) f^2 is the square of the function f.

b) If p is a fixed point of f, then $f(f(p)) = p$.

c) Continuous functions always have a fixed point.

d) If the derivative at a point is between -1 and 1, that point is attracting.

e) A chaotic dynamical system is random.

f) A chaotic dynamical system is completely determined.

g) Sensitive dependence on initial conditions means that small changes in the initial values can have large long-term consequences.

h) According to the butterfly effect, a butterfly flapping its wings in one country can cause a hurricane in another country.

2. Find the fixed points of the following functions defined on all of $\mathbb{R}$:

a) $g(x) = 0.6 - 0.5x$

b) $h(x) = |1.5(x - 0.6)|$

*c) $j(x) = 1 + x - x^2/2$

d) $k(x) = 3 + x - x^2/2$

e) $m(x) = x^3 + 3x + 3$

f) $n(x) = \frac{1-x}{x+2}$ for $x \neq -2$ and $n(-2) = -1$

3. For each part of Problem 2, determine which fixed points are attracting. For those that are attracting, find some interval of attraction.

4. For each given function, find its fixed point and the derivative at this fixed point. Then, from numerical or graphical evidence, decide whether the fixed point is attracting or not. Explain your decision.

*a) $-x^2 + 5x - 4$

b) $\sin(x)$

c) $x^3 - 3x^2 + 4x - 1$

5. In Example 4, describe all points of period 3. What is the slope of s^3 at such points?

6. Let $p(x) = (x - 1)^2$ on $\mathbb{R}$.

a) Find the fixed points of p and show that they are not attracting.

*b) Find the period 2 points of p. (Hint: Graph $p \circ p = p^2$ and $y = x$.)

c) Determine whether the points you found in part b are attracting period points.

7. Repeat Problem 6, using $q(x) = (x - 2)^2$. (Hints: In part b, find a formula for $q(q(x))$. The fixed points of q must be roots of $q(q(x)) - x = 0$. Thus, you can find two factors of this fourth degree polynomial, corresponding to the fixed points. What remains can be solved by the quadratic formula.)

8. Suppose that the point p has period n for the function f. Show that for all $k \in \mathbb{N}$, $f^{kn}(x) = x$.

9. a) On $\mathbb{R}$ except 1, define $r(x) = \frac{1}{1-x}$. Pick a particular value for x other than 1 or 0 and find $r^3(x)$.

b) For x not equal to 1 or 0, find a formula for $r^3(x)$. Describe the dynamics of r. Why must we omit both 1 and 0?

10. Let $v : \mathbb{R} \rightarrow \mathbb{R}$ be given by $v(x) = x + \sin(x)$.
    a) Graph $y = v(x)$ and $y = x$.
    b) Describe all the fixed points of v.
    c) Describe all of the attracting fixed points of v.
    d) For each attracting fixed point from part c, describe all the numbers attracted to that fixed point. Explain your answer.

11. Newton's method is an important dynamical system approximating the roots of many functions quite quickly. We will investigate why this method works so well. (Any standard calculus text derives the formula for Newton's method, also called the New-ton Raphson method.) Given a differentiable function f, recall that we can approximate a root of f by using the iterative formula $x_{n+1} = x_n - \frac{f(x_n)}{f'(x_n)}$. For instance, let $f(x) = x^3 - x + 1$, whose graph in Figure 9 suggests that its root is somewhat less than -1. From $f(x) = x^3 - x + 1$ we obtain the derivative $f'(x) = 3x^2 - 1$. Using the initial value $x_0 = -1$, we find that $x_1 = -1 - (\frac{-1+1+1}{3-1}) = -1.5$. In turn, $x_2 = -1.3478\ldots$, $x_3 = -1.3252\ldots$, $x_4 = -1.32471817412\ldots$. As far as my calculator is concerned, every further iteration equals x_4. Given the function f, we define the related "Newton function" $N(x) = x - \frac{f(x)}{f'(x)}$, which simply embodies Newton's method. For parts a through d, assume that $f'(p) \neq 0$.

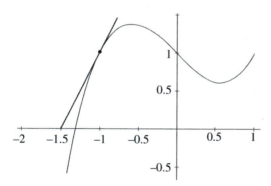

FIGURE 9. Newton's method for $y = x^3 - x + 1$.

    a) Show that a value p is a root of $f(x) = 0$ iff p is a fixed point of N.
    *b) Find the derivative N'.
    c) For p a root of f, find $N'(p)$. Simplify your answer, using the fact that $f(p) = 0$.
    d) What do Theorem 8.5.2 and part c tell us about Newton's method when $f'(p) \neq 0$?
        The remark after Theorem 8.5.2 and the value of $N'(p)$ you found explain why, in general, Newton's method converges quite quickly after starting from a reasonable initial value. The exceptions often work, but not necessarily quickly, as illustrated in the following parts:
    e) Find N for $f(x) = x^2$. Does this N have an attracting fixed point p? What is $f'(p)$? How quickly does the sequence $\{x_n\}_{n \in \mathbb{N}}$ converge to the root?
    f) Repeat part e for $f(x) = x^k, k > 2$.

12. Show that the periodic points of s in Example 4 are dense in $[0, 1]$. (Hint: Use your understanding of decimal representation to argue that between any two real numbers

there is a rational number whose representation is strictly repeating, like $0.234234234\ldots$, rather than eventually repeating, like $0.56234234234\ldots$.)

*13. Generalize the illustration of transitivity in Example 4. Given any $x, y \in [0, 1]$ and $\epsilon > 0$, describe how to determine z and k.

14. Generalize the illustration in Example 4 of sensitive dependence on initial conditions. Keep ϵ as 0.4. Given any $x \in [0, 1]$ and $\delta > 0$, describe how to determine y and k.

15. Figure 10 gives the graph of a continuous function t exhibiting chaotic dynamics, which we explore in this problem. To simplify the analysis, use repeating 9s rather than terminating decimals. For instance, use $0.36999\ldots$ instead of 0.37.

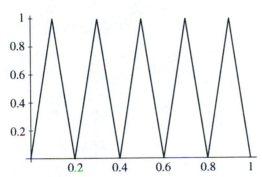

FIGURE 10. $y = t(x)$.

a) Complete the following definition of $t : [0, 1] \to [0, 1]$:

$$t(x) = \begin{cases} 10x & \text{if } 0 \le x \le 0.1 \\ 2 - 10x & \text{if } 0.1 < x \le 0.2 \\ 10x - 2 & \text{if } 0.2 < x \le 0.3 \\ 4 - 10x & \text{if } 0.3 < x \le 0.4 \\ \cdots & \cdots \end{cases}.$$

b) If the first digit of x is even, explain why $t(x) = s(x)$, where s is the shift function of Example 3.

c) If the first digit of x is odd, explain why $t(x)$ is the decimal obtained by deleting the first digit of x and changing all the remaining digits from d to $9 - d$.

d) List several of the fixed points of t, describe their general form(s), and explain why they are not attracting fixed points.

e) Describe what the graph of $t^2 = t \circ t$ looks like and what possible values its derivative can be.

f) List several of the period 2 points of t, describe their general form(s), and explain why they are not attracting.

g) Explain why the periodic points of t are dense.

h) Use an example to illustrate why t is transitive.

i) Explain how to generalize part h.

j) Use an example to illustrate why t has sensitive dependence on initial conditions.

k) Explain how to generalize part j.

16. This problem investigates the function h_4 given by $h_4(x) = 4x(1 - x)$, which exhibits chaos.

 a) Find $h_4 \circ h_4 = (h_4)^2$ and graph $y = h_4 \circ h_4$, $y = h_4$, and $y = x$ on the same axes. Find $h_4'(p)$ for the fixed points p of h_4. Explain why $h_4 \circ h_4$ is even steeper at its fixed points than h_4 is at its fixed points.

 b) Describe the general shape of the graph of $(h_4)^n$. Why does it start to look like t^k in Problem 15, for appropriate k?

 c) Why should your answer in part b lead you to think that h_4 exhibits chaotic dynamics? (The techniques to show that this function is chaotic go beyond this text. See Devaney.)

The Intermediate Value Theorem

Before we state and prove the Intermediate Value Theorem, let's look at an example illustrating both the statement and the idea of the proof. The theorem turns the approximation of the example into an exact value.

EXAMPLE 5

Approximate the root of $x^3 + x + 1 = 0$.

Solution   For $f(x) = x^3 + x + 1$, we have $f(0) = 1$, which is too big and $f(-1) = -1$, which is too small. (See Figure 11.) In between, we have $f(-0.5) = 0.385$, still too big. We can go halfway between -1 and -0.5 to get $f(-0.75) = -0.171875$, too small. We can keep halving the distance between x-values whose images are too big and too small. The next step will consider $f(-0.625) > 0.13$, which is again too big. We continue honing in on what will be "just right." My calculator tells me that the root is approximately -0.682327803828. (Newton's method, discussed in Problem 11, provides a faster way to approximate this root.)

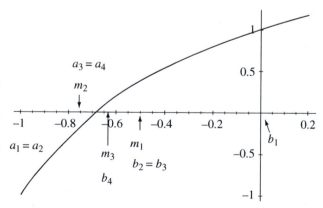

FIGURE 11.   Approximating the root of $x^3 + x + 1 = 0$.   ◇

THEOREM 8.5.3. (The Intermediate Value Theorem).   If $f : [a, b] \to \mathbb{R}$ is continuous and Y is between $f(a)$ and $f(b)$, then there is some $X \in [a, b]$ such that $f(X) = Y$.

Proof. Without loss of generality, we may assume that $f(a) \leq Y \leq f(b)$. We recursively define a lower sequence $\{a_n\}_{n \in \mathbb{N}}$ and an upper sequence $\{b_n\}_{n \in \mathbb{N}}$ of approximations of X by using averages $m_n = (a_n + b_n)/2$. We let $a_1 = a$ and $b_1 = b$. Given a_n and b_n, with $f(a_n) \leq Y \leq f(b_n)$, we define a_{n+1} and b_{n+1} as follows: If $f(m_n) \leq Y$, let $a_{n+1} = m_n$ and $b_{n+1} = b_n$. If $Y < f(m_n)$, let $a_{n+1} = a_n$ and $b_{n+1} = m_n$.

   (In terms of Example 5 and Figure 11, $a_1 = -1$ and $b_1 = 0$. Then $a_2 = a_1 = -1$ and $b_2 = m_1 = -0.5$. Similarly, $a_3 = m_2 = -0.75$ and $b_3 = b_2 = -0.5$; and in the next step, $a_4 = a_3$ and $b_4 = m_3 = -0.625$.)

   By construction, for all $n \in \mathbb{N}$, we have $a \leq a_n \leq a_{n+1} \leq b_{n+1} \leq b_n \leq b$. Hence, the sequence $\{a_n\}_{n \in \mathbb{N}}$ is increasing and bounded; so, by Theorem 8.4.3, it has a limit, say, L_a. Similarly, the sequence $\{b_n\}_{n \in \mathbb{N}}$ is decreasing and bounded, so it has a limit, say, L_b.

   Claim 1: $L_a = L_b$.

Proof. From the proof of Theorem 8.4.3, for all n, $a \leq a_n \leq L_a$; and similarly, by Problem 6 of Section 8.4, $L_b \leq b_n \leq b$. Also, Problem 13 of that section tells us that $L_a \leq L_b$. Thus, for all $n \in \mathbb{N}$, $|L_b - L_a| \leq |b_n - a_n|$. Further, by construction, $|b_{n+1} - a_{n+1}| = 0.5|b_n - a_n|$, and so $|b_n - a_n| = 0.5^{n-1}(b - a)$. By Problem 9 of Section 8.4, $\lim_{n \to \infty} |b_n - a_n| = 0$. Theorem 8.1.3 gives us $L_a = L_b$.

   By Problem 19 of Section 8.4, $\lim_{n \to \infty} f(a_n) = f(L_a) = f(L_b) = \lim_{n \to \infty} f(b_n)$.

   Claim: $f(L_a) = Y$.

Proof. For all $n \in \mathbb{N}$, $f(a_n) \leq Y$. By Problem 12 of Section 8.4, $f(L_a) \leq Y$. Similarly, $f(L_b) \geq Y$. Together these give us $f(L_a) = f(L_b)$, which is, therefore, Y. ∎

REFERENCES

DEVANEY, R. 1992. *A first course in chaotic dynamical systems: theory and experiment.* Reading, Mass.: Addison-Wesley.

GLEICK, J. 1987. *Chaos: making a new science.* New York: Viking.

DEFINITIONS FOR CHAPTER 8

Section 8.1

- A subset W of a partially ordered set X is *dense in X* iff for all $x, y \in X$, if $x < y$, then there is $w \in W$ such that $x < w$ and $w < y$. If X is dense in itself, we simply say that X is *dense.*

- A *field F* is a set with two operations, written $+$ and $\cdot$ satisfying these properties: For all $x, y, z \in F$,

i) $x + y = y + x$,	ii) $x \cdot y = y \cdot x$ (commutativity),
iii) $x + (y + z) = (x + y) + z$,	iv) $x \cdot (y \cdot z) = (x \cdot y) \cdot z$ (associativity),
v) $x \cdot (y + z) = (x \cdot y) + (x \cdot z)$,	(distributivity),

vi) there is $0 \in F$ such that for all $x \in F$,
$$x + 0 = x = 0 + x \qquad \text{(additive identity)},$$

vii) there is $1 \in F$ such that $1 \neq 0$ and for all
$$x \in F, x \cdot 1 = x = 1 \cdot x \qquad \text{(multiplicative identity)},$$

viii) for all $x \in F$, there is $y \in F$
such that $x + y = 0 = y + x$ (additive inverses),

ix) for all $x \in F$, if $x \neq 0$, there is $y \in F$
such that $x \cdot y = 1 = y \cdot x$ (multiplicative inverses).

- An *ordered field* F is a field with a partial order $\leq$ such that for all $x, y, z \in F$,
  i) $x < y$ or $x = y$ or $y < x$,
  ii) if $x \leq y$, then $x + z \leq y + z$,
  iii) if $x \leq y$ and $0 \leq z$, then $x \cdot z \leq y \cdot z$.

- Let $\leq$ be a partial order on a set X.
  - For a subset S of X, an element u of X is an *upper bound* of S iff for all $s \in S$, $s \leq u$.
  - For a subset S of X, an element l of X is a *lower bound* of S iff for all $s \in S, l \leq s$.
  - A subset S of X is *bounded* iff S has an upper bound and a lower bound.
  - For a subset S of X, an element b of X is a *least upper bound* of S iff b is an upper bound of S and for every upper bound u of S, $b \leq u$.
  - For a subset S of X, an element g of X is a *greatest lower bound* of S iff g is a lower bound of S and for every lower bound l of $S, l \leq g$.

We often abbreviate "least upper bound" and "greatest lower bound" by l.u.b. and g.l.b., respectively.

- A *complete ordered field* is an ordered field satisfying the *completeness property*: Every nonempty bounded set has a least upper bound in the field.

- For all $x \in \mathbb{R}$, $|x| = \left\{ \begin{array}{c|c} x & \text{if } 0 \leq x \\ \hline -x & \text{if } x < 0 \end{array} \right\}.$

Section 8.2

- A *real function* is a function from a subset of $\mathbb{R}$ into $\mathbb{R}$.

- Suppose that $c \in (a, b)$ and f is a real function defined on the interval (a, b), except possibly at $x = c$. For $L \in \mathbb{R}$, $\lim_{x \to c} f(x) = L$ iff for all $\epsilon > 0$, there is $\delta > 0$ such that for all $x \in (a, b)$, if $0 < |x - c| < \delta$, then $|f(x) - L| < \epsilon$. We read $\lim_{x \to c} f(x) = L$ as "The limit of $f(x)$ as x approaches c is L."

- $\lim_{x \to c} f(x) \neq L$ iff there is $\epsilon > 0$ such that for all $\delta > 0$, there is $x \in (a, b)$ such that $0 < |x - c| < \delta$, but $|f(x) - L| \geq \epsilon$.

- For a function $f : (a, b) \to \mathbb{R}$ and $x \in (a, b)$ the *derivative* $f'(x)$, if it exists, is $\lim_{h \to 0} \frac{f(x+h) - f(x)}{h}$.

■ Suppose that f is a real function defined on the interval (a, c). For $L \in \mathbb{R}$, $\lim_{x \to c^-} f(x) = L$ iff for all $\epsilon > 0$, there is $\delta > 0$ such that for all $x \in (a, c)$, if $0 < c - x < \delta$, then $|f(x) - L| < \epsilon$. We read $\lim_{x \to c^-} f(x) = L$ as "The limit of $f(x)$ as x approaches c from the left is L." ◇

■ Suppose that $c \in (a, b)$ and f is a real function defined on the interval (a, b), except possibly at $x = c$. Then $\lim_{x \to c} f(x) = \infty$ iff for all $B > 0$, there is $\delta > 0$ such that for all $x \in (a, b)$, if $0 < |x - c| < \delta$, then $f(x) > B$. We read $\lim_{x \to c} f(x) = \infty$ as "The limit of $f(x)$ as x approaches c is infinity."

Section 8.3

■ Suppose that f is a real function defined on the interval (a, b) and $c \in (a, b)$. Then f is *continuous* at c iff $\lim_{x \to c} f(x) = f(c)$. We say that f is *continuous on a set A* iff f is continuous at each point of A.

■ A *rational function* is a function from a subset of $\mathbb{Q}$ into $\mathbb{Q}$. Denote the rational interval $[a, b] \cap \mathbb{Q}$ by $[a, b]_\mathbb{Q}$.

Section 8.4

■ A *sequence* $\{a_n\}_{n \in \mathbb{N}}$ is a function from $\mathbb{N}$ into $\mathbb{R}$. A sequence $\{a_n\}_{n \in \mathbb{N}}$ is *bounded* iff the set $\{a_n : n \in \mathbb{N}\}$ is bounded; that is, there are $l, u \in \mathbb{R}$ such that for all $n \in \mathbb{N}$, $l \leq a_n \leq u$. A sequence $\{a_n\}_{n \in \mathbb{N}}$ is *nondecreasing* iff for all $k, n \in \mathbb{N}$, if $k < n$, then $a_k \leq a_n$.

■ Given a sequence $\{a_n\}_{n \in \mathbb{N}}$ and a real number L, we say that $\lim_{n \to \infty} a_n = L$ iff for all $\epsilon > 0$, there is some $N \in \mathbb{N}$ such that for all $n \in \mathbb{N}$, if $n > N$, then $|a_n - L| < \epsilon$. If a sequence has a limit, we say that the sequence *converges*. Otherwise, we say that the sequence *diverges*.

■ For a sequence $\{a_k\}_{k \in \mathbb{N}}$, we define the *sequence of partial sums* $\{s_n\}_{n \in \mathbb{N}}$, where $s_n = \sum_{k=1}^{n} a_k$. If $\lim_{n \to \infty} s_n$ exists, we write $\sum_{k=1}^{\infty} a_k = \lim_{n \to \infty} s_n$. We extend this notation to allow the sequence to start with a_0 and so the sums to be $s_n = \sum_{k=0}^{n} a_k$ and the limit to be $\sum_{k=0}^{\infty} a_k$. If the limit exists, we call $\sum_{k=1}^{\infty} a_k$ or $\sum_{k=0}^{\infty} a_k$ a *convergent series*. Otherwise, we call the series *divergent*.

■ For $a, r \in \mathbb{R}$ with $a \neq 0$ and $r \neq 0$, the series $\sum_{k=0}^{\infty} ar^k$ is a *geometric series*.

■ For any sequence $\{a_n\}_{n \in \mathbb{N}}$ with $a_n \in \{0, 1, 2, 3, 4, 5, 6, 7, 8, 9\}$, we write $0.a_1 a_2 a_3 \ldots$ for the series $\sum_{n=1}^{\infty} \frac{a_n}{10^n}$. If $k \in \mathbb{N}$, we write $k.a_1 a_2 a_3 \ldots$ for $k + \sum_{n=1}^{\infty} \frac{a_n}{10^n}$. Similarly, $-k.a_1 a_2 a_3 \ldots$ is defined to be $-(k + \sum_{n=1}^{\infty} \frac{a_n}{10^n})$.

Section 8.5

■ Given a function $f : A \to A$, define f^n recursively by $f^1 = f$ and, for $n \geq 1$, $f^{n+1} = f \circ f^n$. For $x \in A$, the *iterates* of x under f are the images of x under the functions f^n. We often abbreviate the nth iterate $f^n(x)$ by x_n, and we write $x_0 = x$.

▪ For $f : A \to A$ and $A \subseteq \mathbb{R}$, we say that $x \in A$ is a *fixed point* of f iff $f(x) = x$. A fixed point p is *attracting* iff there is some $\delta > 0$ such that for all $x \in A$, if $|x - p| < \delta$, then $\lim_{n \to \infty} f^n(x) = p$.

▪ For $f : A \to A$ and $A \subseteq \mathbb{R}$, we say that x is *periodic* iff there is some $n \in \mathbb{N}$ such that $f^n(x) = x$. The smallest such n is the *period* of x. A point p of period n is *attracting* iff there is some $\delta > 0$ such that for all $x \in A$, if $|x - p| < \delta$, then $\lim_{k \to \infty} f^{kn}(x) = p$.

A function $f : [a, b] \to [a, b]$ is *chaotic* iff the following three conditions hold:

i) The periodic points of f are *dense* in $[a, b]$. That is, for all x and y with $a < x < y < b$, there is a periodic point p such that $x < p < y$.

ii) f is *transitive*. That is, for all $\epsilon > 0$ and all $x, y \in [a, b]$, there are $z \in [a, b]$ and $k \in \mathbb{N}$ such that $|z - x| < \epsilon$ and $|f^k(z) - y| < \epsilon$.

iii) f has *sensitive dependence on initial conditions*. That is, there is $\epsilon > 0$ such that for all $x \in [a, b]$ and all $\delta > 0$, there are $y \in [a, b]$ and $k \in \mathbb{N}$ such that $|y - x| < \delta$ and $|f^k(y) - f^k(x)| > \epsilon$.

CHAPTER 9

METAMATHEMATICS AND THE PHILOSOPHY OF MATHEMATICS

"There are more things in heaven and earth, Horatio,
Than are dreamt of in your philosophy."

—Shakespeare (Hamlet) (1564–1616)

I**N THE PREVIOUS** chapters, we studied proofs and a number of concepts common to many areas of mathematics, topics essential to the foundation of doing mathematics. We take a step back from doing mathematics in this chapter in order to look at the philosophical and metamathematical foundations of mathematics. For over 2000 years, philosophy has sought fundamental insights about the nature of things. After having done mathematics in earlier chapters, we will reflect from a philosophical viewpoint in Section 9.2 on what mathematics is.

Before we look at mathematics from the "outside" in that section, we investigate the foundations of mathematics from within mathematics. During the twentieth century, mathematicians developed sophisticated means, called *metamathematics*, to investigate mathematics itself by using mathematics. In the process, they proved a number of deep results about logic and mathematics, including some inherent limitations to what mathematics can accomplish.

9.1 METAMATHEMATICS

"God exists since mathematics is consistent, and the Devil exists since we cannot prove it."

—André Weil (1906–1998)

"Either mathematics is too big for the human mind or the human mind is more than a machine."

—Kurt Gödel (1906–1978)

Metamathematics is the investigation of mathematical systems from a mathematical viewpoint outside those systems. (The Greek prefix "meta" means "beyond.") In the twentieth century, mathematicians proved many profound metamathematical results. While the proofs go well beyond the level of this text, in my view students deserve at least a brief introduction to these results. In metamathematics we place an intellectual "box" around the formal mathematical system we're studying and use general mathematical reasoning to investigate the formal system. Since we explicitly describe the formal system by means of its starting assumptions, or axioms, we call such a system an *axiomatic system*. Axiomatic systems also include grammar and other things, but we will not worry about them in this brief survey. See DeLong for a deeper introduction, including more metamathematical results than we can discuss here.

David Hilbert (1862–1943) led the effort to develop metamathematics as an attempt to establish mathematics on an unquestionable footing. Controversies, such as Russell's Paradox discussed in Section 1.4, had made the foundations of mathematics suspect, even though no one doubted the mathematical details of individual proofs. What if, for instance, the axioms for the real numbers or Euclidean geometry or set theory, developed between 1890 and 1910, contained hidden contradictions? How could we be sure that there were no contradictions? Hilbert's program sought to end such concerns forever by establishing the absolute consistency of mathematics by the use of means no one would doubt. While it ultimately failed, this program accomplished much and gave us deep insight into mathematics.

Metalogic

The first step of Hilbert's program looked at logic, which underlies all mathematics. The initial layer of logic, called the *propositional calculus*, considers propositions built from the basic logical words "and," "or," "not," "if, then," and so on, but without allowing quantifiers. (Sections 1.1 and 1.2 discuss the propositional calculus.) Emil Post (1897–1954) showed for his Ph.D. dissertation in 1920 that logic at this level is *consistent*, as defined next.

DEFINITION. A system is *consistent* iff no contradiction can be proven in the system.

It is worth asking how he could prove something about logic without risking assuming the logic he was trying to investigate. The truth tables we used in Sections 1.1 and 1.2 provide the key. Recall that the inputs (P and Q, etc.) and the output (an expression such as $P \Rightarrow (\neg P \Rightarrow Q)$) have possible truth values of 0 or 1. Post proposed to show that provable propositions are always tautologies; that is, their output truth values are always 1. Post, in effect, described a machine that could mechanically check the truth values of any propositional form. The machine is not using logic, although we would use logic to build the machine. Such a machine can check that every entry in the truth table of a logical axiom has an output of 1, as needed for tautologies. Similarly, it can check that the explicit proof formats of the system transform tautologies into tautologies. Finally, a machine can check that a contradiction has only outputs of 0. Thus, a contradiction can't be proven from the axioms of logic by the use of the proof formats. Post also gave a method to prove every tautology mechanically from the axioms. (See DeLong, 134–141.)

The next step, logic with quantifiers, is called the *predicate calculus*. (See Section 1.3.) Without truth tables for "there exists" and "for all," it might seem hard to approach a proof of consistency. However, the tautologies of the predicate calculus must remain valid regardless of what we consider the "universe" over which the quantifiers range. Mathematicians, being clever, used a universe with just one object, which allows mechanical checking to show consistency.

Kurt Gödel (1906–1978), for his Ph.D. dissertation in 1930, connected logic with the concept of a model of an axiomatic system. His dissertation connected metalogic with metamathematics proper.

Axiomatic Systems and Models

Since the time of Euclid, mathematicians have used axioms as starting points for proving theorems. Whether axioms are considered "self-evident truths" or simply assumptions, we use them to prove other results. An axiomatic system has other aspects, notably, *undefined terms*. Just as we need to have some starting assumptions to prove other propositions, we need starting terms from which to define other terms. Otherwise, we could never get started, needing to define each term on the basis of previous ones. Hilbert realized that undefined terms had to be free of interpretation inside the axiomatic system. In the system, only the axioms constrain what qualifies as an example of a term. Since axiomatic systems first arose in geometry, we'll use a geometric example, called an *affine plane*.

EXAMPLE 1

Consider the axiomatic system with undefined terms "point," "line," and "is on" and these axioms:

i) There are a line and a point not on that line.

ii) For every line there are at least two points on that line.

iii) For every two distinct points there is a unique line that is on both of them.

iv) For every line l and point P not on l, there is a unique line k such that P is on k and no point is on both k and l.

REMARKS. To avoid unnecessary formality, we use the phrases "a point is on a line" and "a line is on a point" interchangeably. Axiom iv asserts the existence of what are called *parallel* lines, as defined next.

DEFINITION. Two lines l and k are *parallel* iff $l = k$ or there is no point on both l and k.

Even though this system is much simpler than ordinary Euclidean geometry, we can still prove some theorems.

CLAIM 1: For any point P there is a line l such that P is not on l.

Proof. See Problem 3. ■

CLAIM 2: For every point there are at least three lines on that point.

Proof. Let P be any point. By Claim 1, there is a line l with P not on l. By axiom ii, l has at least two points on it, say, Q and R. By axiom iii, there is a line k on P and Q and a line m on P and R. Further, axiom iv gives us a line j on P parallel to l. We have the three desired candidates for lines. Problem 3 asks you to prove that they are distinct lines. ■ ◇

The proofs in Example 1 are inside the axiomatic system. Although we could prove other propositions, such proofs cannot begin to answer the metamathematical question, "Is the system consistent?" A model in the metamathematical sense can show consistency.

DEFINITION. A *model* of an axiomatic system is a set of objects and an interpretation of the undefined terms of the system so that all the axioms of the system are true about the objects.

We definitely need to be in the realm of metamathematics to talk about axiomatic systems and models. In effect, the axiomatic system is in one "box" and the model is in another one. (See Figure 1.) We have to be outside of both boxes to check whether the objects and interpretations in the model box satisfy the axioms in the axiomatic system box.

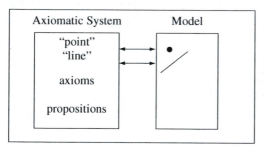

FIGURE 1

EXAMPLE 2

Figure 2 gives a picture of a model of the axiomatic system of Example 1. In particular, the four dots are the "points," the six line segments are the "lines," and when a dot and a line segment

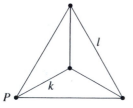

FIGURE 2

touch, one "is on" the other. As we can readily verify, all of the axioms are true in this model. Let's illustrate with one of the instances of axiom iv. The "point" labeled P is not on the "line" l, but P is on k. Furthermore, k and l have no "point" on both of them. Even though k and l don't look parallel to our eyes, they satisfy the definition. Actually, the dots are too big and the segments are too thick and short to qualify as what we usually consider points and lines. However, the terms "point" and "line" are undefined. As long as the objects satisfy the axioms, we can use whatever we like as interpretations of the undefined terms. $\Diamond$

Mathematicians have realized for some time that if an axiomatic system has a model, then the system has to be consistent. Basically, a proposition can't be both true and false in the same model. Hence, the set of all propositions true about a model can't contain a contradiction. That is, they must be consistent. In particular, the axioms for the model must be consistent. Gödel formalized this idea and, more surprisingly, showed the converse.

METATHEOREM 9.1.1. (GÖDEL COMPLETENESS THEOREM, 1930). An axiomatic system is consistent iff it has a model.

EXAMPLE 3

The model of Figure 2 shows that the axiomatic system of Example 1 is consistent. $\Diamond$

Gödel's Completeness Theorem accomplishes more than giving us a way to show systems consistent. It provides us with a metamathematical way to prove that some propositions are unprovable in an axiomatic system. At first, showing a proposition unprovable within a system would seem to require knowing all the possible proofs. Instead, we use a model to help show that a proposition, say, P, can't be proven within an axiomatic system, say, A. The idea is to find a model of A in which $\neg P$, the negation of the proposition, is true. Such a model forces A and $\neg P$ to be consistent. By definition, P and $\neg P$ together form a contradiction and so are inconsistent. If we could prove P from A, we could prove it from A and $\neg P$, getting a contradiction even though we know that A and $\neg P$ have no contradictions. So, the model shows that P can't be proven from A. The following definition takes this unprovability one step farther:

DEFINITION. A proposition is *independent* of an axiomatic system iff neither the proposition nor its negation can be proven in the axiomatic system.

EXAMPLE 4

We show the independence of the first axiom from the other three axioms in Example 1. Figure 2 gives a model in which all four are true. So the negation of axiom i can't be proven from the other axioms. Figure 3 pictures a model with three dots, all on the same line segment. If we use the same interpretations (dot as "point," etc.), axiom i is clearly false. But just as clearly, axioms ii and iii are true. Finally, axiom iv is also true because there can be no instances of a point not on a line, and so no counterexamples to axiom iv. Thus, axiom i can't be proven from

FIGURE 3

the other axioms. These two models show the independence of axiom i from the other axioms. See Problem 4 for the independence of each of the other axioms. ◊

Historically, the first important instance of proving a proposition independent of some axioms precedes Gödel's theorem by 60 years. From the time of Euclid (300 B.C.), mathematicians had tried to prove one of his axioms (the parallel postulate) from the others. In 1868, Eugenio Beltrami (1835–1900) built (inside Euclidean geometry) the first model of a "non-Euclidean" geometry in which the parallel postulate was false. (See Sibley, Chapters 1 and 3, for more on this geometry, axiomatic systems, and models.) More recently, Gödel and Paul Cohen proved the independence of the Continuum Hypothesis in set theory by building different models. (See Section 5.3 for more information.)

Relative Consistency

Finite models can be thoroughly checked, as in Example 1, giving us total confidence in the consistency of the corresponding axiomatic systems. However, most mathematical systems of interest involve infinitely many objects. Are the corresponding axiomatic systems consistent? For instance, it might seem from Gödel's Completeness Theorem that the axioms of Euclidean plane geometry are consistent, since we have the familiar model whose points are ordered pairs $(x, y) \in \mathbb{R} \times \mathbb{R}$. However, this model depends on the consistency of the real numbers $\mathbb{R}$. So in turn, we need a model of $\mathbb{R}$. What exactly are numbers in our model? We can use the points on a line to model $\mathbb{R}$, but we get a logical problem: The system of real numbers would depend on geometry, and geometry would depend on the real numbers. We can define numbers as sets, but then we need to show the consistency of set theory. Unfortunately, models of set theory are either constructed in set theory (as Gödel and Cohen did for the Continuum Hypothesis) or in a system whose consistency is at least as questionable. Consistency of any of these systems seems dependent on the consistency of another system, as in the next definition.

DEFINITION. An axiomatic system is *relatively consistent* iff it has a model built assuming the consistency of another axiomatic system.

Relative consistency of a system seems an unsatisfying endpoint. Hilbert's research program sought to do better, to prove the absolute consistency of mathematics. The accomplishments of Post and Gödel in logic, discussed earlier, were important steps toward this goal. The year after developing his Completeness Theorem, Gödel dashed Hilbert's lofty goal with his justly famous Incompleteness Theorems. While Hilbert's vision was ill-fated, it inspired some of the most profound mathematics ever written.

Gödel's Incompleteness Theorems

After showing the consistency of the predicate calculus, Gödel turned to a more challenging system, elementary arithmetic. The intended model is the set $\mathbb{N}$ of natural numbers with the familiar operations of addition and multiplication. While this system is infinite, it is far less complicated than all of set theory. Even so, Gödel proved two theorems about inherent limitations of this system and, by extension, of mathematics. We'll first paraphrase these in somewhat imprecise language to convey the central ideas, before presenting a more careful account.

The First Incompleteness Theorem tells us that no consistent axiomatization of arithmetic can prove all the true statements of arithmetic (and no false ones). In effect, Gödel constructs a proposition Q asserting that it is not provable. (If Q were provable, it would be false, since it says that it is unprovable. If it is unprovable, what it says is true. This situation might well remind you of the odd sentence, "This sentence is false," discussed in Section 1.1. See Problem 11.) Gödel's second theorem, in effect, shows that we can't prove the absolute consistency of arithmetic in arithmetic, unless arithmetic is inconsistent. (See DeLong, 160 ff, for the technical hypotheses in these theorems that are beyond the level of this text.)

Gödel achieved these profound results by finding a way to get arithmetic to talk about itself. More accurately, he encoded mathematical statements as numbers. In the first layer of this encoding, we assign each of the symbols of the language a number. For instance, we can encode $=$ as 1, $+$ as 2, and 0 as 3. Next, we encode a string of symbols, using exponents of primes in increasing order. For instance, the five symbols in the sentence $0 + 0 = 0$ have the corresponding numbers 3, 2, 3, 1, and 3, respectively. Then these numbers become exponents to encode the sentence as the rather large number $2^3 \cdot 3^2 \cdot 5^3 \cdot 7^1 \cdot 11^3 = 83{,}853{,}000$. A proof of that sentence involves a much longer string of symbols, encoded as a huge number. By the Fundamental Theorem of Arithmetic (Theorem 2.4.7), positive integers have unique factorizations. Thus, given a number, we can figure out the symbols and their order that this number represents.

Among many other things, Gödel showed, given numbers x and y, how to recognize whether x represents the symbols that provide a proof in this system for the symbols encoded as y. Then he cleverly devised a proposition Q "saying," "For all numbers x, x does not encode in this system a proof of Q." I put the word "saying" in quotes because the meaning of the proposition Q happens outside of the system, although proofs and the number of a proof are in the system. Even so, the proposition Q asserts that it can't be proven.

In a way similar to Gödel's first theorem, his second theorem involves finding a proposition describing consistency. Since "$0 + 0 = 0$" is a theorem of arithmetic, its negation "$\neg (0 + 0 = 0)$" had better not be a theorem. So, the proposition C, "For all numbers x, x does not encode in the system the proof of $\neg (0 + 0 = 0)$," effectively asserts the consistency of the system. Gödel showed that if C were provable in the system, then Q would also be provable, making the system inconsistent. Hence, if arithmetic is consistent, we can't prove its consistency in arithmetic. A similar result holds in any system sophisticated enough to include arithmetic.

Since Gödel's work, others have found proofs of the consistency of arithmetic by using systems beyond arithmetic; but the systems have to be at least as open to doubt as arithmetic. (See DeLong, 160–187, for a more thorough treatment of Gödel's work and related matters. See Franzén for good explanations of Gödel's ideas and how they have been misinterpreted.)

Gödel's results ended the hope of proving the absolute consistency of mathematics. However, the consequences are not all bad. First of all, the lack of a proof of consistency has not slowed mathematical progress, and no one seriously doubts the consistency of mathematics. Secondly, Gödel's proofs attest to the incredible ingenuity and insight of mathematicians. Indeed, such human ingenuity seems to suggest, in light of the quote from Gödel at the start of this section, that we are more than machines. And over the last 70 years, mathematicians have continued to transcend the limitations of formal systems in ways that even the most sophisticated computers are (as yet) unable to do. See Hofstadter for an imaginative exploration of the implications of Gödel's work.

PROBLEMS

*1. For each statement that follows, state whether it is true or (at least sometimes) false. Explain your answer or cite the part of the text supporting it.

a) Nothing can be proven about logic because we must use logic in any proof.

b) Truth tables establish that propositional calculus is true.

c) In metamathematics, truth is a relationship between a model and an axiomatic system.

d) An axiomatic system has a model iff it is consistent.

e) The axiomatic system of the real numbers is relatively consistent.

f) Gödel's Incompleteness Theorems imply that no mathematical system can be shown to be consistent.

2. a) Use a truth table to verify that $(P \wedge \neg P) \Rightarrow Q$ is a tautology. (See Sections 1.1 and 1.2.)

b) Interpret the symbols $(P \wedge \neg P) \Rightarrow Q$ in everyday English.

c) Use your answer in part b to explain why we need axiomatic systems to be consistent.

3. a) Prove Claim 1 in Example 1. (Hint: For Case 1, consider P not on the line of axiom i.)

b) Finish the proof of Claim 2 in Example 1.

4. a) Prove with a model that axiom ii in Example 1 is independent of the axiomatic system formed from the other three axioms.

b) Repeat part a for axiom iii.

c) Repeat part a for axiom iv.

5. For the axiomatic system in Example 1, fill out the incomplete ideas that follow to prove the following claim:

Claim 3. If a line l has exactly n distinct points on it, then every line has exactly n distinct points on it.

Let $P_1, \ldots, P_n$ be the points on l. Case 1. Let m be a line with a point, say, P_1, on both m and l. By ___, m has another point, say, Q_2, on it. For each i, consider the line k_i

on P_i parallel to the line on both P_2 and Q_2. Use axiom iv to show that each of the lines k_i has a point Q_i on m. Show that these points on m are all distinct, again using axiom iv. Use symmetry to show that m can't have more than n points on it. What is Case 2?

6. Consider the axiomatic system with the undefined terms "point" and "midpoint" and these axioms:

   i) All points x, y have a unique midpoint, written xMy, which is a point;

   ii) for all points x; $xMx = x$;

   iii) for all points x and y, $xMy = yMx$; and

   iv) for all points s and t, there is a unique point x such that $sMx = t$.

   a) Show that the set of rational numbers $\mathbb{Q}$, with $xMy = \frac{x+y}{2}$, the average of x and y, is a model of this system.

   *b) Fill out Table 1 so that the set $\{a, b, c\}$ and M as determined by the table satisfy all of the axioms. (We read this table as follows: The letters in the first column and first row are the input "points." The other letters are their "midpoints." For instance, the c in the second row and third column is the midpoint of the a from the start of that row and the b from the top of that column. In other words, $aMb = c$. The a to the left of that c represents the equation $aMa = a$.)

TABLE 1. Midpoints

M	a	b	c
a	a	c	
b			
c			

   *c) Can we define midpoints on the set $\{a, b\}$ that satisfy all of the axioms? Explain your reasoning.

   d) Explain why we can't define midpoints on the set $\{a, b, c, d\}$ to satisfy all of the axioms.

   e) Make a table similar to the one in part b to give a model of midpoints on the set $\{a, b, c, d, e\}$ satisfying all of the axioms.

   f) Make a conjecture about what size models can satisfy all of the axioms.

7. *a) Prove with a model that axiom ii in Problem 6 is independent of the axiomatic system formed from the other three axioms.

   b) Repeat part a for axiom iii.

   c) Repeat part a for axiom iv.

   (Hint: Use tables similar to the one in Problem 6b. None of the tables need more than four elements.)

8. Consider the axiomatic system with the undefined terms "vertices," "edges," and "is on" and these axioms:

   i) For each edge e, there are exactly two distinct vertices v_1 and v_2 such that v_1 is on e and v_2 is on e.

   ii) For each vertex v, there are exactly three edges e_1, e_2, and e_3 such that v is on e_1, v is on e_2, and v is on e_3.

   a) Find a model of this axiomatic system with four vertices.

   b) Find a model of this system with six vertices.

    c) Can there be a model with five vertices? Explain your reasoning.

    d) Based on parts a, b, and c, make a conjecture and try to prove it.

9. a) Prove with a model that axiom i in Problem 8 is independent of the axiomatic system formed from the other axiom.

    b) Repeat part a with axiom ii.

10. In this problem you are to develop an axiomatic system with undefined terms "spy," "plot," and "is involved in." (You may add more undefined terms, such as "double agent.")

    a) Device axioms using the given undefined terms and logical terminology. Make sure that your axioms are consistent.

    b) Give a model for the axiomatic system in part a.

    c) Device and prove theorems for the axiomatic system in part a.

    d) Use models to investigate the independence of the axioms in part a.

11. Repeat Problem 10 to develop an axiomatic system with undefined terms of your choice.

12. Let S be any axiomatic system whose axioms use only the logical symbols $\forall$, $\vee$, $\wedge$, and $\Rightarrow$. Explain why the empty set is a model of S and so S is consistent.

13. In an essay, discuss the relationships among the following: The sentence "This sentence is false" in Section 1.1, Russell's Paradox in Section 1.4, Cantor's diagonal argument in Section 5.3, and proposition Q of this section.

*14. Explain what is wrong with the following reasoning: The proposition Q is unprovable and true. Hence, we can, without contradiction, add it to the list of axioms of arithmetic. Now we can prove it. So Gödel's Incompleteness Theorems don't really show that arithmetic is incomplete. Rather, only Gödel's particular formulation of arithmetic is incomplete.

15. Critique the following reasoning: No computer can transcend the system (language) in which it is formulated. Humans can transcend systems, as evidenced by metamathematics and, in particular, the reasoning in Gödel's Incompleteness Theorems. Hence, no computer that could ever be built, no matter how sophisticated or fast, could achieve human intelligence.

16. Critique the following reasoning: By Gödel's Incompleteness Theorems, we can never be sure that the mathematical systems we use won't have contradictions. Therefore, "proofs" of the relative consistency of systems are worse than useless, since they convey a false sense of confidence in mathematical systems.

17. Critique the following reasoning: We can prove the absolute consistency of finite mathematical systems (provided, of course that, they are consistent). We can't prove the absolute consistency of infinite systems. Therefore, mathematics should restrict itself to provably consistent finite systems.

REFERENCES

DeLong, H. 1970. *A profile of mathematical logic.* Reading, Mass.: Addison-Wesley.

Franzén, Torkel. 2005. *Gödel's theorem: An incomplete guide to its use and abuse.* Wellesley, Mass.: A. K. Peters.

Hofstadter, D. 1979. *Gödel, Escher, Bach: An eternal golden braid.* New York: Basic Books.

Sibley, T. 1997. *The geometric viewpoint: A survey of geometries.* Reading, Mass.: Addison-Wesley.

9.2 THE PHILOSOPHY OF MATHEMATICS

"Every good mathematician is at least half a philosopher, and every good philosopher is at least half a mathematician."

—Gottlob Frege (1848–1925)

In the previous section, we took a step back from mathematics to a metamathematical look at mathematical systems. We now take another step back to look at mathematics as a whole, from a philosophical point of view. The widely acclaimed certainty of mathematics and its amazing applicability single mathematics out as substantially different from other human knowledge. The abstract nature of mathematical objects also differentiates mathematics from science. For over 2000 years, philosophers have pondered the special role of mathematical knowledge, looking for insight into what mathematics is. Given how much people agree about answers to mathematical questions, it may be surprising how little people can agree about answers to philosophical questions about mathematics. This section briefly describes several influential philosophies of mathematics, presenting their basic understanding of mathematics as well as a discussion of some of their weaknesses. After presenting these established philosophies, we consider a recent philosophical discussion about the nature of mathematical proof raised by some recent mathematical results.

Philosophy seeks insight into fundamental questions about the nature of the world. With regard to mathematics, people have often asked three prominent philosophical questions:

1. To what extent and why are mathematical results certain or true?
2. Why is mathematics applicable to the world around us?
3. In what sense do mathematical objects exist?

In your own reflections, it might be helpful to make these questions more specific. For instance, does the result $1 + 1 = 2$ have the same status of certainty or truth as Theorem 5.3.3, which states that the set of reals $\mathbb{R}$ and the power set $\mathcal{P}(\mathbb{N})$ have the same infinite size? Why can NASA confidently use mathematics to predict a satellite's orbit around one of the moons of Jupiter? What does it mean to say that the number 2 exists? Do more abstract mathematical objects, like the number i and a 17-dimensional vector space, exist in the same sense as 2? If mathematical entities exist, where do they exist?

As you read the descriptions of philosophies of mathematics, keep in mind that philosophy is a very different enterprise than mathematics. While we will not see any resolution of competing views, let alone satisfying proofs, philosophy of mathematics is not just a matter of opinion. The goal is a coherent understanding that provides insight into the nature of mathematics. Such a goal is extremely difficult to achieve; at the same time, it is easy to find weaknesses in other people's attempts. Each philosophical approach deserves sympathetic consideration of the insight it provides along with an acknowledgement of the difficulties it doesn't answer satisfactorily.

Platonism (Realism)

"The knowledge at which geometry aims is the knowledge of the eternal."
—Plato (429–348 B.C.)

Many professional mathematicians and students who work for an extended period on a mathematics problem have the strong feeling that mathematical objects exist outside of our limited conceptions. People often feel that they know a mathematical statement is true before they devise a proof of it. Plato (429–348 B.C.), one of the earliest and most influential philosophers, developed an understanding of mathematics by building on these insights, and so this philosophical outlook is often called Platonism. Plato strongly distinguished between appearances and reality, opinion and knowledge. Plato held that mathematics provided one of the few ways for mortals to know the ideal reality beyond appearances. For a Platonist, mathematical objects exist in an ideal, eternal realm independently of humans. Mathematical theorems, for Platonists, are eternal truths. (After Plato, some have interpreted Plato's ideal realm as the mind of God.)

Realism is a related, but more modern, philosophical stance. Realists assert that mathematical objects exist objectively and independently of humans, without requiring an ideal, eternal realm where they exist. For both groups, a proof isn't necessary for a statement to be true, although fallible humans need proofs in order to recognize true statements.

Plato's famous "cave allegory" (Plato, Book VII, 205–211) describes the physical world as an imperfect imitation of the ideal world. For a Platonist, mathematics fits the ideal world perfectly and so fits the physical world as well as the perfect can match the imperfect. For instance, because we reason about ideal circles, not humanly made ones, our mathematical calculations are more reliable than measurements. A realist modifies these ideas and speaks of the perfect theoretical fit of mathematics with the principles underlying the real world. (For more on Plato, see Baum, 15–38.)

WEAKNESSES. How can humans have any access to ideal mathematical objects that are independent of humans? Where can infinite mathematical objects, such as an infinite dimensional vector space, exist? How can we determine the truth of relationships among such ideal objects?

The Platonic confidence in the eternal truth of mathematics received a rude shock with the development of non-Euclidean geometries in the nineteenth century. Mathematicians proved theorems in these new geometries, flatly contradicting corresponding Euclidean theorems, leading people to ask whether either set of theorems was true. (See Sibley, Chapter 3.) Since that time, mathematicians have explored a variety of geometric and algebraic systems. Such competing systems make it harder for realists to claim a "perfect theoretical fit of mathematics with the principles underlying the real world."

Formalism

"The unproved postulates with which we start are purely arbitrary. They must be consistent, but they had better lead to something interesting."

—Julian Coolidge (1873–1954)

"Mathematics is a game played according to certain simple rules with meaningless marks on paper."

—David Hilbert (1862–1943)

Formalism developed along with metamathematics and the attempt to prove the absolute consistency of mathematics. The profound insights from metamathematics described in Section 9.1 required us to separate mathematics into uninterpreted axiomatic systems and their models. Formalism claims that mathematics is nothing more than these axiomatic systems and models. For a formalist, mathematical existence is quite minimal. In an axiomatic system, an object's existence asserts only that this existence does not lead to a contradiction. Models do have "objects" in them, although they are typically elements of other formal systems. For instance, John von Neumann developed a model of the natural numbers within set theory, using $\emptyset$ as 0, $\{\emptyset\}$ as 1, $\{\emptyset, \{\emptyset\}\}$ as 2, and so on. The only "things" required to exist are the empty set—nothing—and the ability to form sets out of already existing ones.

Formalists restrict truth to the connection between a model and an axiomatic system. For instance, the Pythagorean Theorem is true in a model of Euclidean geometry and false in a model of spherical geometry. It makes no sense to a formalist to ask whether the Pythagorean Theorem is true or false on its own. However, proofs are a central concern to a formalist. Because the requirements of a formalist proof are so explicit, we can be absolutely certain of its validity.

In spite of the quote from Coolidge, formalists know that mathematicians do not create totally arbitrary systems. Indeed, mathematicians often derive their inspiration from actual problems by abstracting out some particular regularity. A formalist would assert that, given the extraordinary range of possible axiomatic systems, it should not be surprising to find a system to fit any particular pattern. Formalists welcome systems describing almost anything, real or imagined, as long as the mathematics is interesting, as in Coolidge's quote. Indeed, formalists have developed an axiomatic system for intuitionistic logic, to the consternation of some intuitionists who intentionally avoided formalizing their views. (Intuitionism is a very different philosophy of mathematics; it will be discussed shortly; for more on formalism, see Körner, Chapters 4 and 5.)

WEAKNESSES. Much of what mathematicians do as mathematicians doesn't fit within formal systems and models. Few, if any, mathematicians actually do mathematics in an abstract formal system. Instead, we develop and work with intuitions about mathematical objects, sketch informal proofs, draw pictures, and work examples before writing out proofs. Even a proof published in a research journal is not written in a formal system, since such formal proofs would be essentially unreadable. Few people would devote their professional life to "a game ... with meaningless marks on paper," as in Hilbert's quote. (In fairness, Hilbert didn't treat the substantial mathematics he developed as meaningless.)

The variety of mathematical systems may make it possible to model almost anything, but someone who isn't a formalist might want to press the issue of applications further. Why are our most important and long-studied systems the most applicable ones? As just one example, ellipses were studied for more than 1800 years before Johann Kepler (1571–1630) realized that they modeled the orbits of planets. Also, how is it that systems developed for one use also often turn out to fit other applications very well?

Some people claimed formalism failed completely because Gödel's Incompleteness Theorems, discussed in Section 9.1, showed the impossibility of its key goal, proving the consistency of mathematics. Although disappointed, formalists consider Gödel's work an important accomplishment of the formalist approach and carefully separate metamathematics from philosophy.

Intuitionism (Constructivism)

"God made the integers, all else is the work of man."
—Leopold Kronecker (1823–1891)

"The primary concern of mathematics is number, and this means positive numbers. . . .
    Our program is simple: to give numerical meaning to as much as possible of [mathematics]."
—[Bishop and Bridges, 3–4]

In a multitude of languages and cultures, children readily learn to count. This ease suggests a universal intuition of the natural numbers. Compare that ease with how hard, even unnatural, abstract mathematics is for so many people. Intuitionists reacted against the formal axiomatic approach to mathematics developed in the late nineteenth century. Instead, they held that all mathematics should be explicitly "constructed" from the natural numbers. This insistence on construction led to the modern philosophical stance of mathematical constructivism, which doesn't depend on the innate intuition of the natural numbers. (Neither philosophy explicitly describes what a construction is, although they provide useful examples.)

Constructivists and intuitionists argue that we can be sure of the truth of explicitly constructed mathematics, but all other mathematics is suspect. For instance, the proof of the Extreme Value Theorem (stated in Problem 14 in Section 8.3) gives no way to find where the maximum or minimum of a function actually occurs. So, a constructivist would say that the theorem isn't proven. (A constructivist wouldn't say that it is false. Indeed, for many functions, calculus gives us an explicit way to find the maximum or minimum.) The many theorems dependent on the explicitly nonconstructive Axiom of Choice, discussed in Section 5.4, have no place in constructive mathematics.

Constructivists and intuitionists also avoid proofs by contradiction, since such proofs don't construct an object. The basis of a proof by contradiction is the law of the excluded middle. It asserts that either a proposition is true or its negation is true;

there is no "middle" option. For this reason, we call standard logic 2-valued logic. In Section 1.1, we formulated this law as $P \vee \neg P$, which is a tautology. Thus, in 2-valued logic, eliminating one of the options P or $\neg P$ forces the other one to be true, the essence of a proof by contradiction. For instance, the proof by contradiction for Theorem 2.5.1 establishes the infinitude of primes without constructing any primes. (For more, see Körner, Chapters 6 and 7, and Bishop and Bridges.)

Note: In mathematics education, a constructivist approach refers to a very different idea. In this pedagogical view, students must construct their own individual understanding of mathematical concepts, as opposed to constructing mathematical objects.

WEAKNESSES. Few mathematicians are willing to give up the many lovely results that require nonconstructive proofs. Further, the consistency of 2-valued logic, which includes the law of the excluded middle, argues against the need for any logical restrictions. (See Section 9.1.) In addition, the need for explicit constructions, while useful for computers, makes mathematical statements more cumbersome to write and harder to understand. Some people criticize intuitionists and constructivists for not explicitly stating all types of constructions and logical methods they find acceptable.

Social Constructivism, or Humanism

"The observable reality of mathematics is this: an evolving network of shared ideas with objective properties. . . .

[D]aily experience finds mathematical truth to be fallible . . . , like other kinds of truth."

—Reuben Hersh (1927–   )

Social constructivists and humanists see mathematics as a broad cultural activity. As students learn mathematics, their concepts are molded by teachers and parents, as are their understandings of other aspects of culture. Ordinary people's conceptions (and misconceptions) about mathematics are also part of mathematics, and these conceptions vary both across and within cultures. The history of mathematics reveals changing standards of what counts as mathematics even among mathematicians. For instance, the infinitesimals used in the eighteenth century to explain calculus became unacceptable in the nineteenth century with the rise of analysis. (See Section 8.1.) And today professional mathematicians use a social process to decide what mathematics is important, especially what gets funded by governmental agencies.

While almost everyone acknowledges the social and historical reality of the previous statements, social constructivists and humanists go further. They say that mathematics is nothing more than its social, historical, and cultural aspects. Mathematical propositions in this view have no more intrinsic claim to truth than culturally shared opinions. Indeed, truth for them is a social consensus. Mathematical statements do have greater reliability than everyday opinions, in their view. However, they would argue that the greater reliability comes from how mathematicians train new mathematicians about the acceptable rules for doing mathematics. For them, mathematics fits the real world because humans live in the real world: We have adjusted our mathematics to fit the world. (For more, see Hersh and Ernest.)

WEAKNESSES. The certainty of mathematics seems beyond the ability of social constructivism and humanism to explain. Scientists and lay people, as well as mathematicians, generally consider mathematics more firmly based than scientific theories, let alone opinions. Euclid's proof of the Pythagorean Theorem is as convincing today as it was 2300 years ago, even if we now realize the hidden assumptions behind Euclid's axioms. However, Newton's wonderful contributions to physics 300 years ago are now seen as very good first approximations, compared with the theories of relativity and quantum mechanics. In addition, the unrivaled power of mathematics in a huge variety of applications points to an objectivity beyond social consensus or adjustment to reality. It is not just social consensus enabling scientists to predict the trajectory of a rocket to a distant planet.

Immanuel Kant

What must exist in order for people to experience the world? Kant (1724–1804) held that we must have an innate sense of time and space in which experiences happen. Thus, he thought we needed the mathematical intuitions of number (for time) and geometry (for space) in order to experience the world. Since in his view these intuitions came prior to experience, he used the Latin term *a priori* for them. (For knowledge dependent on experience, he used another Latin term: *a posteriori*.) He also viewed logic as *a priori*, a necessary prerequisite to our experience.

He distinguished logic from mathematics by using a different criterion. He viewed all logical truths as *analytical*; that is, the conclusion is inherent in the terms. For instance, the tautology "It is snowing or it is not snowing" is an analytic truth. Simply by knowing the meanings of "or" and "not," we know that the sentence is true, even without experiencing snow. In this view, analytic *a priori* knowledge is not new knowledge.

Mathematical propositions, however, lead to new knowledge for Kant. Just by knowing the definition of a triangle as a shape with three straight sides, you would not immediately know that the sum of its angles (in Euclidean geometry) is 180°. We gain this new knowledge by reasoning and proof. Kant called such new knowledge *synthetic*. Facts, such as the fact that water flows downhill, and scientific theories are also synthetic, but they are *a posteriori*, dependent on experience. And because they depend on experience, their truth status is always open to revision in the face of new experiences and experiments. Thus, mathematics held a critical place for Kant. Only with mathematics can we gain genuinely new knowledge (synthetic) that we know is absolutely true (*a priori*). Table 1 summarizes the kinds of knowledge, including one other logical possibility, analytic *a posteriori* knowledge. However, Kant argued that no knowledge fit this remaining category. (For more, see Baum, 212–234.)

WEAKNESSES. Developments in the 200 years since Kant's death call into question Kant's characterizing mathematics as synthetic *a priori*. Kant took Euclidean geometry as the necessary intuition underlying spatial experience. The advent in 1829

TABLE 1. Kantian Categories of Knowledge

	a priori	*a posteriori*
analytic	logic, definitions	Ø
synthetic	mathematics	facts, science

of non-Euclidean geometry, which contradicts Euclidean geometry, led to questions of the role of Euclidean geometry. A century after Kant's death, Einstein's theory of relativity related the geometry of the universe to gravitational fields, arguing against the *a priori* nature of geometry. More recent psychology experiments suggest that our intuitions of space and time develop in response to our experience of the world. Kant's distinctions, however, still influence philosophical discussions and provide insight.

Logicism

"All mathematics is tautology."

—Ludwig Wittgenstein (1889–1951)

"The fact that all Mathematics is Symbolic Logic is one of the greatest discoveries of our age ... "

—Bertrand Russell (1872–1970)

Gottlob Frege (1848–1925), the founder of logicism, wanted to give mathematics a solid foundation. He thought that all concepts and theorems of mathematics could be derived from logic. Certainly, mathematical arguments are supremely logical and rest on explicit assumptions. According to logicists, nothing but logic and explicit assumptions go into mathematics; so mathematics is just a branch of logic. In particular, it is independent of human experience. To avoid Russell's Paradox, discussed in Section 1.4, Russell and Whitehead developed a hierarchy of logical "types." We can think of these types as different kinds of sets. The number 1, for instance, was a higher type (or set) containing all first level sets with a single element. Thus, $\{\pi\}$, $\{\emptyset\}$, $\{\mathbb{R}\}$, and many other sets, each with a single element, would be elements of the class of 1. While such a definition was impractical for ordinary arithmetic, it did allow logicists to formulate many mathematical concepts, strictly in terms of logical ones. (For more, see Körner, Chapters 2 and 3.)

WEAKNESSES. Logicism makes no attempt to explain the applicability of mathematics. More critically for logicists, their own goal of reducing mathematics to logic appears to fail. Mathematics needs an explicit axiom stating the existence of some infinite set, such as $\mathbb{N}$, whereas the predicate calculus doesn't need an infinite universe. Even more, Gödel's proofs of the completeness of logic (the predicate calculus) and the essential incompleteness of arithmetic make it hard to argue that we can reduce mathematics to logic. (See Section 9.1.)

Empiricism

For empiricists, mathematical ideas are abstracted from concrete facts. Thus, $2 + 3 = 5$ might come from seeing two apples together with three apples amounting to five apples. More generally, empiricists argue that humans need mathematics to describe and investigate the world around them. Thus, mathematical objects exist only to the extent that they are necessary to explain the physical world. Higher mathematics becomes for them a tool of science. Mathematical truth is dependent on its grounding in experience. Mathematics derives its certainty from how well it fits everyday life and how well it enables science to make accurate predictions. In this view, mathematical proofs have a role in bolstering empirical verification of mathematics.

WEAKNESSES. Very little mathematics of the last 400 years can be reduced to abstractions from our concrete sense experiences. Even the role of a tool for science is far too restrictive for much of mathematics. Without a central place for proofs, this view of mathematics seems restricted to elementary and secondary school mathematics and algorithms.

Embodied Mind Theories

"[M]athematics as we know it arises from the nature of our brains and our embodied experience. . . .

[A] great many of the most fundamental mathematical ideas are inherently metaphorical in nature."

—Lakoff and Núñez, xvi

Common sense tells us that mathematics takes place in our minds or, more scientifically, in our brains. Recent scientific investigations of human learning and the brain have started to give insights into how people do mathematics. There is some evidence that babies only a few days old can distinguish among one object, two objects, and three objects. This suggests that at least small numbers might be innate concepts, hardwired into our brains. After several months, babies recognize some rudimentary aspects of addition, such as one object and one object give two objects, not one or three objects. (See Lakoff and Núñez, 15–19.)

Some people have extrapolated intriguing experimental results into a preliminary description of mathematics. They aim to answer scientifically many of the questions philosophy of mathematics explores. (See Lakoff and Núñez.) In this view, we develop our (personal) concepts of numbers and space as metaphors extending our experiences. Metaphors for them represent "abstract concepts in concrete terms" and ground "modes of reasoning . . . in the sensory-motor system" (Lakoff and Núñez, 5). In Section 1.4, for instance, I used a concrete metaphor to naïvely introduce a set "as something like a bag and its elements as the objects in the bag."

As we extend these metaphors and use them to order the world around us, they acquire a certainty equal to that of our physical perceptions. From the point of view of an embodied mind theory, mathematical proofs do not confer certainty on mathematical results. The feeling of certainty in mathematics is for them psychological, based on our mental metaphors conforming with our physical experiences. Existence of mathematical objects is the same as other ideas, all in our heads. Presumably, evolutionary forces have resulted in our predisposition to develop and learn mathematical concepts. (For more, see Lakoff and Núñez.)

WEAKNESSES. Someday, we may know enough to augment or even replace philosophy of mathematics with a neuropsychology of mathematics. Currently, the experimental evidence only gives hints. The concept of a metaphor may provide a fruitful basis for psychology research, but for now it seems insufficient to explain the certainty and applicability of mathematics. While mathematical ideas, like any ideas, may well be only neurological messages, we still need to explain the qualitative difference between mathematics and other conceptions. Why do mathematical metaphors enable us to make precise calculations and predictions beyond other metaphors?

The Nature of Mathematical Proof

In Section 2.1, I said, "Mathematicians demand the standard of logical necessity" for proofs. I hope that you think the proofs in this text and those you have written for the problems meet such a high standard. However, a number of proofs published since 1970 call this understanding of proof into question. We'll briefly consider three widely discussed proofs of significant mathematical results. (See the references to learn more on these theorems.)

From 1852, when it was first conjectured, until 1976, mathematicians had offered several incorrect proofs of what we now call the Four Color Theorem. (The Four Color Theorem states that for every [ordinary] map drawn in the plane, the regions can be colored with at most four colors so that no adjacent regions have the same color. See Beck et al., Chapter 1.) In 1976, Ken Appel and Wolfgang Haken finally published a proof of this theorem, notable because for the first time in history the proof required the use of a computer in an essential way. Their proof broke the problem into nearly 2000 cases and employed a computer to list and check them. Not all mathematicians were comfortable calling something a proof if no one could possibly check it all. Among other questions they asked was, "How do we know that a computer glitch didn't cause a mistake in some case?"

In 1981, after decades of work by dozens of mathematicians, a small team of mathematicians announced the classification of all finite simple groups. (Even an informal statement of this classification goes beyond the level of this text. See Gallian, 413–417.) The original proofs of the many parts of the classification fill over 10,000 journal pages, far more than any one person could digest. Between 1981 and 2004, a number of mathematicians reviewed many of the individual proofs, correcting several mistakes and omissions along the way. These mathematicians announced the verification of the classification, although they admit that some minor errors likely persist among all the thousands of pages of proofs. In what sense is this classification proven if, practically speaking, no one person can completely check the proof?

The *Annals of Mathematics* published, in 2005, a 120-page condensation of a proof of the Kepler Conjecture by Thomas Hales and Samuel Ferguson. (Kepler conjectured that the most efficient way to pack spheres in space was in nested layers, where the spheres in each layer fit the way cells in a honeycomb fit. See Casti,

Chapter 4.) The length of the proof (increased by the 40,000 lines of computer code needed) is not nearly as remarkable as the qualification that the editors attached to the paper. According to the disclaimer, 12 referees labored over a four-year period, trying to verify the proof. While these experts in this area of mathematics found no errors, they could claim to be only 99% confident that the proof was correct. In what sense is something a proof if well-qualified experts can't be confident that it is a proof?

Mathematicians, scientists, and others count on the certainty of mathematics. Even if we resolved the particular proofs discussed here, how should we decide what counts as a mathematical proof? This question is both mathematical and philosophical. However, these recent results came after the enunciation of most of the philosophies described previously. Thus, these philosophies don't give explicit counsel on where to draw the line. I leave the resolution to you and the mathematical community in general.

PROBLEMS

*1. For each statement that follows, state whether it is true or (at least sometimes) false. Explain your answer or cite the part of the text supporting it.

 a) A philosophy of mathematics is just someone's opinion about what mathematics is.

 b) To a Platonist, the truth of a proposition does not depend on the existence of a proof.

 c) To a formalist, a proof of a proposition does not depend on the truth of the proposition.

 d) In intuitionism, our true intuitions replace the need for proof.

 e) In Kant's terms, a logicist would say that mathematics is analytic *a priori.*

 f) In Kant's terms, an empiricist would describe mathematics as synthetic *a posteriori.*

 g) Both constructivism and social constructivism hold that a mathematical proposition is true because it is constructed.

 h) Psychological and biological experiments have made philosophy of mathematics obsolete.

Essay Topics

2. Summarize how the different philosophies of mathematics described in this section view mathematical truth and certainty.

3. Discuss which insights of the various philosophies you find compelling. Give examples from your own mathematical experience that substantiate these insights.

4. Discuss which insights of the various philosophies you find unconvincing. Give examples from your own mathematical experience that run counter to these insights.

5. How would various philosophies respond to the question "Is mathematics discovered or created?"

6. What is the status of a conjecture for the various philosophies of mathematics? (A conjecture, for the purpose of this essay, is an as-yet-unproven proposition that someone thinks is correct or true.)

7. There are at least two pedagogical trends in elementary school mathematics. One stresses rote learning of mathematical facts in order to acquire mathematical understanding.

Another emphasizes the assisted development of a child's own conception and articulation of mathematics. (Both emphasize problem solving, sooner or later.) Discuss which philosophies of mathematics might support one or another of these pedagogical approaches. Explain why you think that they support the ones you say they support.

8. There is now a journal called *Experimental Mathematics*. Among other articles, it publishes "formal results inspired by experimentation, conjectures suggested by experiments, descriptions of algorithms and software for mathematical exploration." Discuss how the various philosophies would view these topics as a part of mathematics. (See the journal's web page www.expmath.org for this quote and further information.)

9. Discuss how the different philosophies of mathematics might respond to the controversial proofs discussed in this section.

10. Most philosophies of mathematics predate the advent of computers. Discuss how the prevalence of computers and their ability to do even some symbolic mathematics might lead to a modification of at least one of the philosophies presented.

11. Discuss how the various philosophies of mathematics would view this quote by Georg Cantor: "The essence of mathematics lies in its freedom."

12. Discuss how the various philosophies of mathematics might view the following quote by Humpty Dumpty, applied to mathematical terms: "When I use a word it means just what I choose it to mean—neither more nor less" (Carroll, 94).

13. Explain why the meaning of the terms "analytic" and "*a posteriori*" made it reasonable for Kant to claim that no knowledge was analytic *a posteriori*.

14. Formalism and logicism were directly affected by Gödel's Incompleteness Theorems. How do you think other philosophies of mathematics are affected by the proofs of these theorems?

15. Discuss how mathematicians' personal philosophies of mathematics might influence the mathematics they do.

16. Most of the philosophies of mathematics presented in the text ignore human differences such as gender, ethnic, economic, or religious differences. Discuss how these differences might influence one's understanding of mathematics. For instance, is the relatively small number of women and minorities among professional mathematicians related to something about mathematics itself or just to the wider society?

17. Discuss your views on the comparative truth and certainty of the following statements: "Daimler Chrysler produced 211,365 vehicles worldwide in April 2006."
    (Source: http://www.marketwatch.com/News/Story/Story.aspx?guid=%7BAD FFBCFA-D77D-42C2-A563-3BE720AAB606%7D&siteid=mktw&dist=)
    "There are infinitely many primes." (Source: Theorem 2.5.1 and its proof.)

18. Discuss your view of how broadly the concept of proof can apply. In particular, discuss proofs involving computers, and proofs too long and/or too complicated to be understood by any one human being.

19. Develop and describe your own philosophy of mathematics. Use examples and reasoning to defend your choices. Make an effort to address weaknesses that your philosophy may have.

20. Write a response to Judith Grabiner's article. (See Grabiner.)

21. Write a response to the sonnet *Paradox*, written by Clarence R. Wylie, Jr., about the nature of mathematics.

> ### PARADOX
> *Not truth, nor certainty. These I forswore*
> *In my novitiate, as young men called*
> *To holy orders must abjure the world.*
> *'If . . . , then . . . ,' this only I assert;*
> *And my successes are but pretty chains*
> *Linking twin doubts, for it is vain to ask*
> *If what I postulate be justified,*
> *Or what I prove possess the stamp of fact.*
> *Yet bridges stand, and men no longer crawl*
> *In two dimensions. And such triumphs stem*
> *In no small measure from the power this game,*
> *Played with the thrice-attenuated shades*
> *Of things, has over their originals.*
> *How frail the wand, but how profound the spell!*

22. Write a dialog involving people representing at least two of the philosophies presented. Seek in your dialog some fruitful outcome on a particular topic of philosophical interest, not just disagreement.

23. Mathematicians often use the words "beautiful" and "elegant" to describe mathematics they admire. Describe what you think counts as mathematical beauty and why it should be valued.

24. Discuss the ethical or moral issues that mathematicians as mathematicians need to address.

REFERENCES

BAUM, R. 1973. *Philosophy and mathematics from Plato to the present*. San Francisco: Freeman, Cooper and Co.

BECK, A., et al. 2000. *Excursions into mathematics*. Natick, Mass.: A. K. Peters.

BISHOP, E., and D. BRIDGES. 1985. *Constructive analysis*. New York: Springer Verlag.

CARROLL, L. 1965. *Through the looking-glass*. New York: Random House.

CASTI, J. 2001. *Mathematical mountaintops : The five most famous problems of all time*. New York: Oxford University Press.

ERNEST, P. 1998. *Social constructivism as a philosophy of mathematics*. Albany: State University of New York Press.

GALLIAN, J. 2002. *Contemporary abstract algebra*, 5th ed., Boston: Houghton Mifflin.

GRABINER, J. Oct. 1988. "The centrality of mathematics in the history of western thought," *Math. Mag.* vol. 61, no. 4, 220–230.

HERSH, R. 1997. *What is mathematics, really?*, New York: Oxford University Press.

KÖRNER, S. 1968. *The philosophy of mathematics: An introductory essay*. New York: Dover.

LAKOFF, G., and R. NÚÑEZ. 2000. *Where mathematics comes from: How the embodied mind brings mathematics into being*. New York: Basic Books.

PLATO, 1989. *The Republic and other works*. Translated by Jowett. New York: Double Day.

SIBLEY, T. 1997. *The geometric viewpoint: A survey of geometries*. Reading, Mass.: Addison-Wesley.

DEFINITIONS FOR CHAPTER 9

Section 9.1

- A system is *consistent* iff no contradiction can be proven in the system.

- Two lines l and k are *parallel* iff $l = k$ or there is no point on both l and k.

- A *model* of an axiomatic system is a set of objects and an interpretation of the undefined terms of the system so that all the axioms of the system are true about the objects.

- A proposition is *independent* of an axiomatic system iff neither the proposition nor its negation can be proven in the axiomatic system.

- An axiomatic system is *relatively consistent* iff it has a model built assuming the consistency of another axiomatic system.

THE GREEK ALPHABET

Small	Capital	Name
α	A	alpha
β	B	beta
γ	Γ	gamma
δ	Δ	delta
ε	E	epsilon
ζ	Z	zeta
η	H	eta
θ	Θ	theta
ι	I	iota
κ	K	kappa
λ	Λ	lambda
μ	M	mu
ν	N	nu
ξ	Ξ	xi
o	O	omicron
π	Π	pi
ρ	P	rho
σ	Σ	sigma
τ	T	tau
υ	Υ	upsilon
ϕ	Φ	phi
χ	X	chi
ψ	Ψ	psi
ω	Ω	omega

SELECTED ANSWERS

Chapter 1

Section 1.1
1. a) False   b) True   c) False
   d) False ("Not A or not B")   e) True
   f) False   g) True
2. c) proposition   d) not a proposition
3. a)

A	B	A∨(B∧¬A)
T	T	T
T	F	T
F	T	T
F	F	F

c)

A	B	C	(A∨B)∧(A∧C)
T	T	T	T
T	T	F	F
T	F	T	T
T	F	F	F
F	T	T	F
F	T	F	F
F	F	T	F
F	F	F	F

5. a) not equivalent
   c) logically equivalent
6. a) We won't win the next game and we
   will win the tournament.
   e) $(\neg X \wedge Y) \vee (Z \wedge \neg X)$.
7. a) contradiction   c) neither
   e) tautology
8. a) There need to be two rows because
   the input has two options, T and F.
   Each row has two possible outputs,
   T or F. Since $2 \times 2 = 4$, there are four
   different last columns, given by the
   four given forms.
   e) Key: Use De Morgan's Laws.
9. a) Use truth tables, $P \vee Q = P + Q -$
   $P \cdot Q, \neg P = 1 - P.$ c) All hold, no.

Section 1.2
1. a) True   b) False   c) True   d) False
   e) False   f) True   g) False
2. a) i) If your grade suffers, you didn't
   study.   b) i) If your grade didn't
   suffer, you studied.
3. a) converse   c) original
7. b)

F	G	H	G∧(H ⇒ ¬F)
T	T	T	F
T	T	F	T
T	F	T	F
T	F	F	F
F	T	T	T
F	T	F	T
F	F	T	F
F	F	F	F

8. Use truth tables.   a) not equivalent
   c) equivalent
9. Use truth tables.   d) tautology
11. a) $X \wedge \neg Y \wedge \neg Z$ e) $(X \Rightarrow Y) \wedge \neg Z$

12. a)

1	4	2	3
2	3	4	1
4	1	3	2
3	2	1	4

14. d) $(A \wedge (J \vee S)) \Rightarrow (N \vee I)$ negates to
    $(A \wedge (J \vee S)) \wedge \neg N \wedge \neg I$, where A
    is "Your spouse is a nonresident alien,"
    J is "You file a joint return," S is "You
    file a separate return," N is "Your
    spouse must have an SSN" and I is
    "Your spouse must have an ITIN."

Section 1.3
1. a) False   b) True   c) False   d) True
   e) False (The first can be vacuously true.)
   f) True   g) True   h) False

2. a) True   d) False

3. b) True   d) False

4. a) True   f) False

5. a) Let P be the set of people, $B(x)$ mean "x is a baby," and $L(x)$ mean "x is logical." $\forall x \in P,\ B(x) \Rightarrow \neg L(x)$.

7. a) $\forall l \in \mathcal{L}\ \forall P \in \mathcal{P}\ (P \notin l \Rightarrow \exists! k \in \mathcal{L}:$ $(k \parallel l \wedge P \in k))$, for $\mathcal{L}$ the set of lines and $\mathcal{P}$ the set of points.

9. a) There is an integer that is neither even nor odd.

   c) $\exists B \in \mathbb{R}\ s.t.\ \forall \delta > 0\ \exists x \in \mathbb{R}\ s.t.\ |x| < \delta \wedge \frac{1}{x^2} \leq B$.

11. a) Only $\forall s \in \mathbb{R}\ \exists t \in \mathbb{R}: s + t = s^2$ is true because t depends on s.

14. b) #

15. b) A quadrilateral is a *trapezoid* iff it has exactly one pair of opposite sides parallel.

16. c) An integer d is a *divisor* of an integer k iff there is an integer n such that $dn = k$.

   g) An integer z is a *greatest common divisor* of the integers k and n iff z is a common divisor of k and n and for all common divisors y of k and n, $y \leq z$.

17. b) For $n \in \mathbb{N}$, a real number y is an nth *root* of the real number x iff $y^n = x$.

   c) A real number u is an *upper bound* of a set S, for $S \subseteq \mathbb{R}$, iff for all $x \in S$, $x \leq u$.

20. a) "Babies cannot manage crocodiles." (Carroll, 1263)

Section 1.4

1. a) False   b) True   c) False   d) False
   e) True   f) False   g) False

2. a) $\{1, 2, 3, 4, 5, 6, 8, 10\}$ h) $\{6, 8, 9, 10\}$

3. e) $(1, 3)$ h) $(-\infty, 1] \cup (5, \infty)$

5. a) False, $(J \cup K) \cap L \subseteq J \cup (K \cup L)$
   c) equal

8. a) 17, 10

9. a) $\emptyset, \{1\}, \{2\}, \{1, 2\}$

11. a) 3

12. b) $\{\pi\}$ is a subset of $[1, 4]$.   d) $\{\emptyset\}$ is both an element and a subset of $\{\emptyset, \{\emptyset\}\}$.

13. a) i) Must be true.   ii) Might be true.
   b) iv) Must be false.

Section 1.5

1. a) False (not meaningful)
   b) True (0 and 0)   c) True   d) False
   e) True   f) False   g) True

2. a) iv) $441 = 3^2 7^2$ b) iv) 9

4. a) i) 24

5. b) i) 144

6. b) False for $\mathbb{Z}$, with $y = 2$ and $z = -6$. True for $\mathbb{N}$.

9. a) 9, 2. If r is remainder of x divided by n, then $n - r$ is the remainder of $-x$ when divided by n.

11. b) False

12. a) True

13. Suppose that $a \equiv b\ (\mathrm{mod}\,n)$ and $c \equiv d$ $(\mathrm{mod}\,n)$.
   a) Then $(a - c) \equiv (b - d)(\mathrm{mod}\,n)$.

14. a) i) 51

16. a) 1 1 2 3 5 8 13 21 34 55 89 144

18. b) $\sum_{i=1}^{1} a_i = a_1$ and, for $n \in \mathbb{N}$, $\sum_{i=1}^{n+1} a_i = a_{n+1} + \sum_{i=1}^{n} a_i$.

Section 1.6

1. a) True   b) False   c) False
   d) True   e) True   f) False   g) False
   h) False (unless $1 \in I$)

2. a) $T \times V = \{(3, \#), (3, \$), (5, \#), (5, \$),$ $(7, \#), (7, \$)\}$

3. a) $F = \{(1, 2), (2, 1)\}$

4. a) $(D \times F) \cup (E \times G) \subseteq (D \cup E) \times$ $(F \cup G)$

5. a) $\{\emptyset, \{x\}, \{y\}, \{x, y\}\}$

6. a) $\mathcal{P}(A) \cup \mathcal{P}(B) \subseteq \mathcal{P}(A \cup B)$.

9. a) $\bigcup_{i \in \mathbb{N}} V_i = (1, 7)$ $\bigcap_{i \in \mathbb{N}} V_i = [2, 6]$

12. b) Not a partition: Each B_i misses 0, add $B_0 = \{0\}$.

13. a) Partition of concentric circles and a point.
   b) Not a partition, since all lines intersect in $(0, 0)$ and the points $(0, y)$ are missed, for $y \neq 0$.

Chapter 2

Section 2.1

1. a) False   b) False   c) True   d) True
   e) False   f) False

2. a) Proof. Let x and y be integers. Then $x \cdot y$ is an integer. Suppose that x and y are even. From the definition of even,

we can find $j, k \in \mathbb{Z}$ so that $x = 2j$ and $y = 2k$. Then $x \cdot y = 2j2k = 2(j2k)$. Since $j2k \in \mathbb{Z}$, xy is even. ∎

3. a) Proof. Let $x, y \in \mathbb{Z}$, and suppose that they are even. Then $\exists j, k \in \mathbb{Z}$ s.t. $x = 2j \land y = 2k$. So $x - y = 2j - 2k = 2(j - k)$. Since $j - k \in \mathbb{Z}$, $x - y$ is even. ∎

4. a) Proof. Let S and T be sets. Suppose that $x \in S \cap T$. Then $x \in S$ (and $x \in T$). Thus, $S \cap T \subseteq S$. ∎

5. a) Proof. Let $p, q, r \in \mathbb{Z}$, and suppose that $p|q$ and $q|r$. Then $\exists j, k \in \mathbb{Z}$ s.t. $q = pj$ and $r = kq = kjp$. Since $kj \in \mathbb{Z}$, $p|r$. ∎

6. a) Proof. Let $a, b \in \mathbb{Z}$, and suppose that $a \equiv b \pmod{12}$. Then $\exists j \in \mathbb{Z}$ s.t. $b - a = 12j = 6(2j)$. Since $2j \in \mathbb{Z}$, $a \equiv b \pmod 6$. ∎

8. a) Hint: Split 3θ into $2\theta + \theta$ and then 2θ into $\theta + \theta$.

14. a) The proof can't specify whether n is even or odd. It could be any integer, either even or odd.

15. a) An example is not a proof. The general statements aren't proven.

    f) This is a proof, assuming Problem 4a.

Section 2.2

1. a) True b) False c) True d) False
   e) True f) False

2. a) Proof. Let $r \in \mathbb{R}$. For the contrapositive, we will show that "if $(r - 14) \in \mathbb{Q}$, then $r \in \mathbb{Q}$." So, suppose that $(r - 14) \in \mathbb{Q}$. Now 14 is also a rational, and $r = (r - 14) + 14$. So $r \in \mathbb{Q}$. ∎

3. a) Proof. Let $z \in \mathbb{Z}$, and for the contrapositive, suppose that z is even, say, $z = 2k$, where $k \in \mathbb{Z}$. Then $z^2 = (2k)^2 = 4k^2$ and $k^2 \in \mathbb{Z}$, so 4 divides z^2, showing the contrapositive. ∎

4. a) Hint: Use contradiction. If $x > 0$, what can you say about x^3 and x^6?

7. b) Proof. Suppose that $\frac{p}{q} \in \mathbb{Q}$, and for a contradiction $\frac{p}{q} = \sqrt{3}$. Square both sides and multiply by q to get $p^2 = 3q^2$. Now $q > 0$, and since $\sqrt{3} > 1$, we have $p > q > 0$

as well. If $q = 1$, then $\frac{p}{q} = p \in \mathbb{Z}$, but $1 < \sqrt{3} < 2$, so this can't happen. So both p and q are at least 2. By the Fundamental Theorem of Arithmetic, we can factor them, say, $p = p_1 p_2 \ldots p_n$ and $q = q_1 q_2 \ldots q_k$. Then $p^2 = p_1^2 p_2^2 \ldots p_n^2$, showing that each prime factor of p appears an even number of times. But we also have $p^2 = 3q^2 = 3q_1^2 q_2^2 \ldots q_k^2$, in which 3 appears an odd number of times. Since the Fundamental Theorem of Arithmetic guarantees unique factorizations, we have a contradiction. ∎

8. a) The argument shows that "if $r < -1$, then $r^2 < r^4$." However, that proposition isn't the contrapositive of the original proposition, which is false.

9. b) Wrong format. One should negate the conclusion, assuming that $v + w$ is rational.

Section 2.3

1. a) False b) False c) True d) True
   e) False

2. b) Proof. Let $s \in \mathbb{R}$. For existence, pick $t = e^s$, which is in $\mathbb{R}$. Then $\ln(t) = \ln(e^s) = s$. For uniqueness, suppose that v and w both satisfy $\ln(v) = s = \ln(w)$. Then $e^{\ln(v)} = e^{\ln(w)}$. But $e^{\ln(v)} = v$ and similarly, $e^{\ln(w)} = w$. So $v = w$, showing uniqueness. ∎

3. a) Proof. Pick $r = 1.5 \in \mathbb{R}$. Then $r^2 = 2.25$ and $3r - 2 = 2.5$, showing that $r^2 < 3r - 2$ in this case. ∎

4. a) Let $j, k \in \mathbb{Z}$ and suppose that $j + k$ is even, say, $j + k = 2z$, for $z \in \mathbb{Z}$. If j and k are even, we are done. So we may suppose that at least one is odd, say, $j = 2i + 1$. Then we can solve $j + k = 2z$ for k to get $k = 2(z - i) - 1 = 2(z - i - 1) + 1$, which is odd. ∎

5. a) $\exists s \in \mathbb{R}$ s.t. $\forall t \in \mathbb{R}, st \neq \pi$. Proof. Pick $s = 0 \in \mathbb{R}$ and let $t \in \mathbb{R}$. Then $st = 0 \neq \pi$. ∎

6. a) True ($a = 0$)

7. a) Proof. Let $p \in \mathbb{Z}$. Then $1 \in \mathbb{Z}$ and $1p = p$. So $p|p$. ∎

10. a) Hint: $(x \in S) \Rightarrow (x \in S \lor x \in T)$.

12. a) $(D \times F) \cup (E \times G) \subseteq (D \cup E) \times (F \cup G)$, but the sets are not equal. Example. Let $D = \{d\}$, $E = \{e\}$, $F = \{f\}$, and $G = \{g\}$. Consider (d, g).

14. a) Proof. Let $\epsilon > 0$. Pick $\delta = \epsilon/2 > 0$. Let $x \in \mathbb{R}$ and suppose that $0 < |x-3| < \delta$. Then $|2x + 4 - 10| = |2x - 6| = 2|x - 3| < 2\delta = \epsilon$. (A smaller positive δ also works.) ∎

15. d) The argument shows the converse of the claim. (The claim needs a quantifier.)

16. c) The inference from $A \subseteq X \cup Y$ to $A \subseteq X$ or $A \subseteq Y$ is incorrect.

Section 2.4

1. a) False b) True c) False
  d) False ($\emptyset$) e) True f) False

2. b) Proof by induction. For the initial case, we use $n = 0$, in which case 0 is indeed less than $2^0 = 1$. Suppose for $n = k$, with $k \geq 0$, that $k < 2^k$. Then $k + 1 \leq k + k = 2k < 2 \cdot 2^k = 2^{k+1}$. By PMI we can conclude that $n < 2^n$ for n a nonnegative integer. ∎

3. a) Proof. For $n = 1$, $\sum_{i=1}^{1}(3i - 1) = 3 \cdot 1 - 1 = 2$ and $(3 \cdot 1 + 1)1/2 = 2$. Suppose for $n = k$, where $k \geq 1$, that $\sum_{i=1}^{k}(3i - 1) = (3k + 1)k/2$ and consider $n = k + 1$. Then $\sum_{i=1}^{k+1}(3i - 1) = \sum_{i=1}^{k}(3i - 1) + (3(k + 1) - 1) = (3k + 1)k/2 + (3(k + 1) - 1) = \frac{3}{2}k^2 + \frac{7}{2}k + 2$. Also, $(3(k + 1) + 1) \times (k + 1)/2 = \frac{3}{2}k^2 + \frac{7}{2}k + 2$. By PMI, $\sum_{i=1}^{n}(3i - 1) = (3n + 1)n/2$ for all $n \in \mathbb{N}$. ∎

l) Rewrite "7 divides $11^n - 4^n$" as "$11^n = 7j + 4^n$." Multiply by $11 = 7 + 4$ to get $11^{n+1} = (7 + 4) \times (7j + 4^n) = 7(7j + 4j + 7 \cdot 4^n) + 4^{n+1}$.

4. a) Proof. For the initial step, $n = 5$ and $2^5 = 32 > 25 = 5^2$. For the induction step, suppose, for $n = k \geq 5$, that $2^k > k^2$. Then for $n = k + 1$, $2^{k+1} = 2 \cdot 2^k > 2 \cdot k^2$. Now $(k + 1)^2 = k^2 + 2k + 1$. Since $n \geq 5$, $k^2 \geq 5k > 2k + 1$. So $2k^2 > k^2 + 2k + 1 = (k + 1)^2$. By PMI, $2^n > n^e$ for all $n \in \mathbb{N}$ with $n \geq 5$. ∎

7. a) Note that there are $(n + 1 - k)^2$ possible lower left corners for squares of size k^2. Hence, we can use induction to show that the total number of squares is $\sum_{k=1}^{n}(n + 1 - k)^2 = \sum_{j=1}^{n} j^2 = n(n + 1)(2n + 1)/6$, by Problem 3e.

9. a) For the induction step, note that $\sum_{i=1}^{k+1} f_i = \sum_{i=1}^{k} f_i + f_{k+1} = (f_{k+2} - 1) + f_{k+1} = f_{k+3} - 1$.

11. Check the cases $n = 1, 2$, and 3. For the induction step, note that $b_{n+3} = b_n + b_{n+1} + b_{n+2} \leq 2^{n-1} + 2^n + 2^{n+1} = 2^{n-1} + 2 \cdot 2^{n-1} + 4 \cdot 2^{n-1} = 7 \cdot 2^{n-1} < 8 \cdot 2^{n-1} = 2^{n+2}$.

15. 17 cents. For strong induction, do initial cases for $18 \leq n \leq 21$. For the induction step, add 4.

20. a) inductive b) not inductive

21. a) $2an - a + b$

23. c) The initial step is omitted. (The claim is false.)

24. e) The argument switches from the numbers up to $k + 1$ to the numbers up to $(k + 1)!$, which is generally much bigger.

Section 2.5

1. a) True b) False c) False d) False
  e) True f) False

2. Claim: For all sets A, B, and C, if $B \subseteq C$, then $A \cup B \subseteq A \cup C$. Proof. Let A, B, and C be sets and suppose $B \subseteq C$. Now let $x \in A \cup B$. Then, by the definition of union, $x \in A$ or $x \in B$. In the first case, $x \in A \cup C$ immediately. In the second case, our hypothesis gives us $x \in C$, and so $x \in A \cup C$. Thus, either way, if $B \subseteq C$, then $A \cup B \subseteq A \cup C$. ∎

6. a) Use one of De Morgan's Laws.

7. b) Proof. Let J, K, and L be sets and suppose that $L \subseteq K$ and $J \subseteq K$. Let $x \in J \cup L$. Then $x \in J$ or $x \in L$. In either case, $x \in K$. ∎

10. d) Proof. Suppose that $J \subseteq I$ and $J \neq \emptyset$. Let $x \in \bigcup_{j \in J} S_j$. So, $\exists j \in J$ s.t. $x \in S_j$. For this j, $j \in I$ because $J \subseteq I$. So, $\exists j \in I$ s.t. $x \in S_j$ and $x \in \bigcup_{i \in I} S_i$. ∎

14. a) Note: 2 divides 10, so 2 divides any power of 10. Now use Problem 5 from Section 2.1.

17. a) $A_1 = \{\emptyset\}$, $A_2 = \{\emptyset, \{\emptyset\}\}$, $A_3 = \{\emptyset, \{\emptyset\}, \{\emptyset, \{\emptyset\}\}\}$. $|A_i| = i$.

20. a) The negation of "Every even number greater than 2 is the sum of two primes" is "There exists some even number greater than 2 that is not the sum of two primes."

21. a) This is essentially a proof, although the argument didn't show all of the details.

Chapter 3

Section 3.1

1. a) False   b) False   c) True   d) False
   e) True   f) True   g) False   h) True

2. a) not a function, $f(3)$ undefined.
   b) function, $\{1, 2, 6\}$.

3. a) domain: $(-\infty, -2] \cup [2, \infty)$, range: $[0, \infty)$

4. a) function   b) not a function (undefined at $x = -1$)

6. b) both   c) neither

7. b) both   f) onto

8. a) $f(x) = x^3$ $g(x) = \sin(x)$

10. b) inverse, $g^{-1}(x) = x$

12. a) $[-5, 3]$ d) $[2 - \sqrt{5}, 2 - \sqrt{3}] \cup [2 + \sqrt{3}, 2 + \sqrt{5}]$ h) $[-1, 1]$

14. b) $f[A \cap B] = [0, 4]$, but $f[A] \cap f[B] = [0, 9]$

15. b) both are $[-\sqrt{2}, \sqrt{2}]$.

17. a) 8

19. b) 60

Section 3.2

1. a) False   b) False   c) True   d) True
   e) False

2. a) Proof. Let $x, y \in \mathbb{Z}$, and suppose that $a(x) = a(y)$. Then $23 - x = 23 - y$, so $x = y$, and a is 1-1. Let $z \in \mathbb{Z}$ and pick $x = 23 - z$. Then $a(x) = 23 - (23 - z) = z$, showing onto. ∎

5. a) $f|_W$ is 1-1, so we can define its inverse, called arccos, or $\cos^{-1}$.

6. a) Use contrapositive and cases.

11. c) Note that $f \circ i_X(z) = f(z) = i_X \circ f(z)$.

13. a) Wrong format. Start with $z \in Z$ and find element in X.

14. b) This argument assumes that f has an inverse.

Section 3.3

1. a) False   b) True   c) True   d) False

2. a) $[-4, 14]$ d) $[-\sqrt{2}, \sqrt{2}]$

5. a) $f : \mathbb{R} \to \mathbb{R}$ with $f(x) = x^2$, $A = [0, 1]$, $B = [-1, 0]$. $f[A \cap B] = \{0\}$, $f[A] = f[B] = [0, 1]$

6. a) Hint: $A - B \subseteq A$.

11. a) Yes

14. a) The assumption that $A \subseteq B$ is completely unjustified. Similarly, $f[B] \subseteq f[A]$ doesn't imply that $B \subseteq A$.

15. c) This argument assumes that there is an $x \in X$ with $f(x) = y$. But y need not be in the range of f.

Chapter 4

Section 4.1

1. a) True   b) False   c) False   d) False
   e) False   f) True   g) True   h) True

2. c) R, S, T, not A

3. a) 1st and 4th relations are reflexive, 1st and 2nd are symmetric, 1st and 3rd are transitive, 3rd and 4th are antisymmetric.

7. a) 20, 40

10. b) $\in$, $<$, "is sister of"

15. a) 16

Section 4.2

1. a) True   b) False   c) False   d) False
   e) True

2. a) Proof. Reflexive: Let $x \in \mathbb{R}$. Then $x - x = 0 \in \mathbb{Z}$, so $x \equiv x \pmod{\mathbb{Z}}$. Symmetric: Let $x, y \in \mathbb{R}$ and suppose that $x \equiv y \pmod{\mathbb{Z}}$. Then $y - x \in \mathbb{Z}$, so $x - y = -(y - x) \in \mathbb{Z}$, so $y \equiv x \pmod{\mathbb{Z}}$.
      Transitive: Let $x, y, z \in \mathbb{R}$ and suppose that $x \equiv y \pmod{\mathbb{Z}}$ and $y \equiv z \pmod{\mathbb{Z}}$. Then $(y - x), (z - y) \in \mathbb{Z}$, so $z - x = z - y + y - x = (z - y) + (y - x) \in \mathbb{Z}$; hence, $x \equiv z \pmod{\mathbb{Z}}$. ∎

   c) 0.2, 1.2, 2.2, −0.8, −3.8

5. a) Proof of i. Let $a, b, c, d \in \mathbb{Z}$, and $n \in \mathbb{N}$, and suppose that $a \equiv c \pmod{n}$ and $b \equiv d \pmod{n}$. Then $\exists p, q \in \mathbb{Z}$

$s.t. c = pn + a \wedge d = qn + b$. Then $c + d = (p + q)n + (a + b)$, showing that $a + b \equiv c + d \pmod{n}$. ∎

9. b) Represent nonvertical lines by $y = mx + b$ and vertical lines by $x = b$, for $m, b \in \mathbb{R}$. Let $\mathbb{R}_\infty = \mathbb{R} \cup \{\infty\}$. Map $y = mx + b$ to m and $x = b$ to ∞.

12. b) Proof. For $(p, q), (r, s), (t, u), (y, z) \in \mathbb{Z} \times \mathbb{N}$, suppose that $(p, q)F(r, s)$ and $(t, u)F(y, z)$. Then $ps = rq$ and $tz = yu$. Multiplying left sides and right sides gives $pstz = rqyu$. By commutativity, this gives $ptsz = ryqu$, showing that $(pt, qu)F(ry, sz)$, and so $(p, q) \times (t, u)F(r, s) \times (y, z)$, as desired. ∎

18. c) From bSa and the definition of S^{-1}, we'd get $aS^{-1}b$. Also only considers one direction.

19. a) It is a proof, but perhaps overly brief.

Section 4.3

1. a) $\{1, 2, 3, 4, 5, 6, 7, 8, 9, 10\}$

2. a) Proof. For all i, $i + 0.5 \in B_i$, so $B_i \neq \emptyset$. $\forall x \in \mathbb{R}, (\lfloor x \rfloor \leq x < \lfloor x \rfloor + 1) \wedge (\lfloor x \rfloor \in \mathbb{Z})$, so $\forall x \in \mathbb{R}, x \in B_{\lfloor x \rfloor}$. So $\mathbb{R} \subseteq \bigcup_{i \in \mathbb{Z}} B_i$. Clearly, each $B_i \subseteq \mathbb{R}$, so we get equality. Finally, let $i, j \in \mathbb{Z}$. If $i \neq j$, then one is bigger, say, $i < j$. Since they are integers, $i + 1 \leq j$ and so every x in $B_i = [i, i + 1)$ is less than every y in $B_j = [j, j + 1)$, showing disjoint. ∎ xRy iff $\lfloor x \rfloor = \lfloor y \rfloor$.

4. a) j, k, where $j(x) = x^2 + \sin(x) + 17$ and $k(x) = x^2 + \sin(x) + 23$.

5. a) 6, 24, 54

12. a) Transitivity need not apply to "not R." (The proposition is false.)

13. b) If X and Y are not disjoint, any element in both X and Y appears in some A_i and some B_k, disproving disjointness.

Section 4.4

1. a) True b) False c) False d) True e) False f) False g) False

3. a) Proof. Suppose that $\trianglelefteq$ is a partial order on X and $S \subseteq X$ has a and b as least upper bounds. Since a is an l.u.b. and b is an upper bound, $a \trianglelefteq b$. Similarly, $b \trianglelefteq a$. So $a = b$. ∎

5. a) Proof. Suppose that g is a greatest element of X for the partial order $\sqsubseteq$. Let $x \in X$ and suppose that $g \sqsubseteq x$. Because g is the greatest, we also have $x \sqsubseteq g$. By antisymmetry, $x = g$, showing that g is maximal. ∎

6. a) Proof of transitivity. Let $y, z, w \in Y$, and suppose that ySz and zSw. Then yRz and zRw, so yRw. Since $(y, w) \in Y \times Y$ as well, ySw. ∎

d) R^{-1} is dense. Proof. Let $a, b \in X$, and suppose that $a \neq b$. WLOG $bR^{-1}a$, and so aRb. By the density of R, $\exists c \in X$ s.t. $aRc \wedge cRb \wedge a \neq c \wedge b \neq c$. But then $bR^{-1}c \wedge cR^{-1}a \wedge b \neq c \wedge a \neq c$, showing that R^{-1} is dense. ∎ $R \cap Y \times Y$ need not be dense. Consider $X = \mathbb{R}$ and $Y = \mathbb{Z}$, using the usual order $\leq$ for R.

17. a) Proof. Let $x, y \in \mathbb{R}$. By axiom 1, there are three cases for $y - x$. Case 1. $y - x \in \mathbb{P}$, showing that $x \leq y$. Case 2. $-(y - x) \in \mathbb{P}$, giving $x - y \in \mathbb{P}$, showing that $y \leq x$. Case 3. $x = y$, showing that both $x \leq y$ and $y \leq x$. ∎

18. a) 5

23. a) The argument assumes that $\trianglelefteq$ is linear. Also, $m \trianglelefteq x$ doesn't contradict the definition of maximal, since $m \trianglelefteq x$ allows $m = x$.

24. a) Not every subset of $\mathbb{R}$ is an interval.

Chapter 5

Section 5.1

1. a) False (not on a set) b) False c) True d) True e) False f) True g) False h) True i) True

5. a) $f(x) = 3x - 1$. d) $f(x) = \frac{x}{1-x^2}$.

7. a) Each radius of C_2 crosses both circles in a unique point. Match these points.

13. a) Proof. Define $f : A \times A \to A^{\{1,2\}}$ so that $f(x, y)$ is the function h such that $h(1) = x$ and $h(2) = y$. Let $(x, y) \in A \times A$. Then $f(x, y)$ is a uniquely defined function from $\{1, 2\}$ to A: Each element of $\{1, 2\}$ has exactly one image. So f is in turn a function from $A \times A$ to $A^{\{1,2\}}$. Let

$f(x, y) = h$ and $f(z, w) = j$ and $f(x, y) = f(z, w)$. Then $x = h(1) = j(1) = z$ and $y = h(2) = j(2) = w$. Thus, $(x, y) = (z, w)$, showing f is 1–1. For onto, let $k : \{1, 2\} \to A$. Pick $(k(1), k(2)) \in A \times A$. Then $f(k(1), k(2))$ is a function mapping 1 to $k(1)$ and 2 to $k(2)$, which makes it identical to k. ∎

16. a) An example does not suffice. While a particular 1–1 function may not be onto, another 1–1 function could be.

17. b) This is a proof.

Section 5.2

3. a) For $\frac{p}{q} \in \mathbb{Q}$ and $n \in \mathbb{N}$, note that $\frac{p}{q} = \frac{np}{nq}$. Map $\mathbb{N}$ to $\{\frac{np}{nq} : n \in \mathbb{N}\}$, using the obvious bijection.

4. a) From Theorem 5.2.1 and Problem 11a of Section 5.1, we have $\mathbb{N} \sim \mathbb{N} \times \mathbb{N} \sim (\mathbb{N} \times \mathbb{N}) \times \mathbb{N}$.

6. a) Let $\mathbb{Q}^* = \{x \in \mathbb{Q} : x \neq 0\}$. Why is $\mathbb{Q}^*$ countable? Let P_n be the set of all n^{th} degree rational polynomials. Show that $f : P_n \to \mathbb{Q}^* \times \mathbb{Q}^n$ mapping the polynomial $a_n x^n + \dots a_1 x + a_0$ to $(a_n, \dots, a_1, a_0)$ is a bijection. Use a problem and a theorem to show that P_n is countable.

8. a) Hint: Consider the mapping $g(\{a, b\}) = (\min(a, b), \max(a, b))$.

10. a) Define $f : \mathbb{N} \times \mathbb{N} \to \mathbb{N}$ by, for example, $f(a, b) = 2^a 3^b$.

14. a) The negation of "A and B are infinite" is "A is finite or B is finite," instead of "A is finite and B is finite."

b) This is a reasonable argument, although it skips over details.

Section 5.3

1. b) c e) 3   k) 2^c

3. Use $B \subseteq S$ and the function $f : S \to B$ given by $f(x, y) = (x/2, y/2)$.

9. For example, for $n + \aleph_0 = \aleph_0 + \aleph_0$, use $\{1, \dots, n\} \cup \{n + 1, \dots\} \to \{$odds in $\mathbb{N}\} \cup \{$evens in $\mathbb{N}\}$ by $f(x) = x$. When possible, make one set a subset of the next one and use the identity map.

12. Hint: For $A \subseteq \mathbb{R}$, define $f_A : \mathbb{R} \to \mathbb{R}$ by $\begin{cases} 1 & \text{if } x \in A \\ 0 & \text{if } x \notin A \end{cases}$.

15. b) Hints for a proof. f_S is a function because for each $k \in \mathbb{N}$, k is in exactly one A_n and so has a unique image, either itself or the other element of A_n, depending on whether $n \in S$ or $n \notin S$. For 1–1, suppose that $f_S(j) = f_S(k)$. Then $f_S(j)$ and $f_S(k)$ are in the same A_n, forcing $j, k \in A_n$, and so $j = k$. Thus f_S is 1–1. For onto, let $k \in \mathbb{N}$. Then there is a unique $n \in \mathbb{N}$ such that $k \in A_n$. Use cases $n \in S$ or $n \notin S$ and pick j accordingly.

18. a) Listing the elements of A and B assumes that they are countable.

b) Hint: What would the image of $(17, 28534)$ be?

Section 5.4

1. a) Pick $(a + b)/2$.   c) no explicit rule.

2. a) any s with $s \geq 49$.

b) no maximal element

3. a) Proof. Let A be an infinite set. Hence, $A \neq \emptyset$, so there exists $a_1 \in A$. Let $A_1 = A - \{a_1\}$. If $|A_1| = n$ were finite, $|A|$ would be $n + 1$, a finite number. So A_1 is infinite. We use the Axiom of Choice to define a_i and A_i recursively for all $i \in \mathbb{N}$ and show that each A_i is infinite: Suppose that we have defined a_i and A_i for $1 \leq i \leq n$, and A_i is infinite for these i. Then there is $a_{n+1} \in A_n$, because A_n is infinite. Define $A_{n+1} = A_n - \{a_{n+1}\}$. As before, A_{n+1} is infinite. Let $B = \{a_i : i \in \mathbb{N}\}$. If $i \neq j$, then $a_i \neq a_j$ because if $i < j$, then $a_i \in A_{i+1}$, but $a_j \notin A_{i+1}$, and similarly if $j < i$. So $B \sim \mathbb{N}$ is a countably infinite subset of A. ∎

Chapter 6

Section 6.1

1. a) False   b) True   c) True   d) False
e) False

3. a) hexagon with 2 diagonals
c) impossible

4. b) hexagon versus two triangles

5. a) 15

11. b) $d = n - 1$, $n(n - 1)/2$ edges. Proof. Suppose that a complete graph has n vertices. For each vertex, there are $n - 1$ other vertices, which is the maximum number of edges that can connect to that vertex. Hence, each vertex is of degree $n - 1$. The reasoning of Theorem 6.1.1 shows that the number of edges is $nd/2 = n(n - 1)/2$.

18. Proof of transitivity. On V, the vertices of a graph G, define $v \sim w$ iff there is a path from v to w. For transitivity, suppose that $v \sim w$ and $w \sim x$. That is, there are paths from v to w and from w to x. By Problem 17, there is a path from v to x; so $v \sim x$, showing transitivity. ∎

Section 6.2

1. d) $j + 2$

4. a) 12

5. a) 14

12. a) Let $\mathbb{Z}$ be the set of vertices and let the edges be from z to $z + 1$.

13. Proof. Suppose that G has n vertices, $n - 1$ edges, and no circuits. We need to show that G is connected, because then it is a tree. Suppose that G can be split into some number k of nonempty connected subgraphs with, say, $v_1, \ldots, v_k$ vertices. That is, $\sum_{i=1}^{k} v_i = n$. Since the whole graph has no circuits, no subgraph has a circuit and so each subgraph is a tree. Then by Theorem 6.2.1, the ith subgraph has $v_i - 1$ edges, giving a total of $\sum_{i=1}^{k}(v_i - 1) = n - k$ edges. Since the number of edges is $n - 1$, we see that $k = 1$ and so the entire graph is connected. ∎

Section 6.3

2. a) $5^5 = 3125$

3. b) Think of *BAG* as one unit: $6! = 720$.

6. a) $2 \cdot 9 \cdot 10^{11} \cdot 13 = 2.34 \times 10^{13}$ (2 for the choice between $+$ and $-$, 9 for the first digit, and 13 places for the decimal point.)

12. a) $12!/12 = 39916800$ (We can start anywhere since the table is round.)

13. $1 - (\frac{35}{36})^{24} = 0.4914$. 25 rolls give a probability of $1 - (\frac{35}{36})^{25} = 0.5055$.

17. Use Theorem 6.3.2 to get $_{n+1}P_3 = \frac{(n+1)!}{(n+1-3)!} = \frac{(n+1)!}{(n-2)!} = (n + 1)n(n - 1) = n^3 - n$.

Section 6.4

2. b) $\binom{5}{2}/\binom{12}{2} = 5/33 = 0.1515\ldots$

4. a) $\binom{11}{6} = 462$

7. a) $7!/2! = 2520$

9. a) $\binom{5}{2} = 10$: There will be three segments going right and two going up. The two segments going up can come in any two of the five places.

11. a) $\binom{52}{5} = 2{,}598{,}960$

b) Probability is $0.42257 = \frac{1{,}098{,}240}{2{,}598{,}960}$.

15. a) Choosing a subset of size k is equivalent to choosing its complement, which is a subset of size $n - k$.

Chapter 7

Section 7.1

1. a) True b) False c) True d) True
e) False f) True

2. a) commutative and idempotent. Neither a nor b is an identity, since $a * b = c$. Similarly, $c * a = b$ shows that c is not an identity. No identity implies no inverses. $(a * b) * c = c \neq a = a * (b * c)$.

5. a) Yes

7. d) No: $\max(5, 1 + 2) = 5 \neq 10 = \max(5, 1) + \max(5, 2)$.

11. a) No e) Yes

13. a) Identities must be "two-sided." Thus, $x - 0 = x$ isn't sufficient for identity: $0 - x$ does not usually equal x, showing that 0 is not an identity. Without an identity, inverses are meaningless.

Section 7.2

1. a) False (S_n)
b) False ($\mathbb{N}$ is closed in $\mathbb{Z}$ under $+$.)
c) True d) True

2. All operations have closure.
a) Not a group; max is associative and 1 is the identity, but inverses fail to exist, since $\max(2, x)$ is at least 2, not 1.
b) Group

3. Proof. For closure, let 3^x and 3^y be in the set, where $x, y \in \mathbb{Z}$. Then $x + y \in \mathbb{Z}$, so $3^x \cdot 3^y = 3^{x+y}$ is in the set. For associativity, let 3^x, 3^y, and 3^z be in the set. Then $3^x \cdot (3^y \cdot 3^z) = 3^{x+(y+z)} = 3^{(x+y)+z} = (3^x \cdot 3^y) \cdot 3^z$. For the identity, note that $1 = 3^0$ is in the set and $1 \cdot 3^x = 3^x = 3^x \cdot 1$. For inverses, for 3^x in the set, note that 3^{-x} is in the set and $3^x \cdot 3^{-x} = 1 = 3^{-x} \cdot 3^x$. ∎

8. a) Use $b = e$.

12. c) $\{0\}$ $\{0, 6\}$ $\{0, 4, 8\}$ $\{0, 3, 6, 9\}$
   $\{0, 2, 4, 6, 8, 10\}$ $\mathbb{Z}_{12}$

13. a) $\{(1)\}$ $\{(1), (12)\}$ $\{(1), (13)\}$
   $\{(1), (23)\}$ $\{(1), (123), (132)\}$ S_3

19. a) Let $H = \{(1)(12)\}$. $(13)H = \{(13), (123)\}$ $(23)H = \{(23), (132)\}$ H

21. b) This is a proof.

Section 7.3

1. a) False ($\mathbb{Z}$)   b) True   c) True
   d) False ($x^2 + 1 = 0$ in $\mathbb{R}$.)
   e) False (Ex. 8)
   f) False (Consider $\mathbb{Z}_5$.)

3. $(10x + 5)(10x + 5) = 100x^2 + 100x + 25 = 100x(x + 1) + 25$. We use distributivity for each of the equalities. The factor of 100 means that all the terms in $100x(x + 1)$ have place values bigger than the final 25, which therefore will appear on its own at the right of the product.

4. a) Proof. Let $a, b \in R$. Then $b + -b = 0$, so $a \cdot (b + -b) = a \cdot 0 = 0$, from part i. By distributivity, $a \cdot b + a \cdot (-b) = 0$, which shows that $a \cdot (-b)$ is the additive inverse of $a \cdot b$. That is, $a \cdot (-b) = -(a \cdot b)$. For the other equality, start with $(b + -b) \cdot a$. ∎

10. a) For commutativity of multiplication, $(a, b) \cdot (c, d) = (a \cdot c, b \cdot d) = (c \cdot a, d \cdot b) = (c, d) \cdot (a, b)$, using the commutativity in R and S. Associativity of multiplication and distributivity are similar. If 1_R and 1_S are the unities of R and S, respectively, then $(a, b) \cdot (1_R, 1_S) = (a \cdot 1_R, b \cdot 1_S) = (a, b)$, and similarly for $(1_R, 1_S) \cdot (a, b)$. Thus, $(1_R, 1_S)$ is the unity of $R \times S$.

16. Vector addition forms a commutative group on $\mathbb{R}^3$, and $\times$ is distributive over vector addition. $\times$ is not associative, since $((1, 0, 0) \times (0, 1, 0)) \times (1, 1, 0) = (-1, 1, 0) \neq (0, -1, 0) = (1, 0, 0) \times ((0, 1, 0) \times (1, 1, 0))$. $\times$ is not commutative: $(1, 0, 0) \times (0, 1, 0) = (0, 0, 1) \neq (0, 0, -1) = (0, 1, 0) \times (1, 0, 0)$. There is no unity, since $(1, 0, 0) \times (x, y, z) = (0, -z, y)$ is never $(1, 0, 0)$. Hence, inverses fail.

17. a) We can't cancel x in $x^2 - 2x = 0$ because $x = 0$.
   c) This is a proof.

Section 7.4

1. a) False   b) True   c) False   d) False
   e) False (logically equivalent, not equal)

3. Left lattice: not distributive: $B \sqcap (C \sqcup D) = B \sqcap A = B \neq E = E \sqcup E = (B \sqcap C) \sqcup (B \sqcap D)$.

5. a) The properties of idempotency, associativity, commutativity, and absorptivity hold for any subset, so we need only show that the operations are closed. Let $x, y \in L$, and suppose that $x \sqsubseteq a$ and $y \sqsubseteq a$. Then $x \sqcap y \sqsubseteq x$, so by transitivity, $x \sqcap y \sqsubseteq a$, showing closure for $\sqcap$. For $\sqcup$, we use distributivity and our assumptions to get $(x \sqcup y) \sqcap a = (x \sqcap a) \sqcup (y \sqcap a) = x \sqcup y$. Thus, $x \sqcup y \sqsubseteq a$, showing closure for $\sqcup$. ∎

8. a) The identity for gcd is n, since for $d, n \in \mathbb{N}$, if $d|n$, $\gcd(d, n) = d = \gcd(n, d)$. Similarly, 1 is identity for lcm, since for $d \in \mathbb{N}$, $\mathrm{lcm}(1, d) = d = \mathrm{lcm}(d, 1)$.

9. b) Proof. Let $a \in B$, and suppose that both b and c are complements of a in B. Since $a \sqcup b = 1$ and $c \sqcap a = 0$, we see by distributivity that $c = c \sqcap 1 = c \sqcap (a \sqcup b) = (c \sqcap a) \sqcup (c \sqcap b) = 0 \sqcup (c \sqcap b) = c \sqcap b$. Hence, $c \sqsubseteq b$. Switching the roles of b and c gives $b \sqsubseteq c$. By antisymmetry, we have $b = c$. ∎

Section 7.5

2. a) Proof. Note that $f : \mathbb{R}^* \to \mathbb{R}^*$ is a function. Let $x, y \in \mathbb{R}^*$. Then $f(xy) = (xy)^2 = x^2y^2 = f(x)f(y)$, showing that f is a homomorphism. ■ It is not an isomorphism, since f is not 1–1: $f(-1) = 1 = f(1)$.

3. a) $\mathbb{Z}_6$ is commutative, but S_3 is not. $\mathbb{Z}_6$ has only two elements that are their own inverses (0 and 3), while S_3 has four such elements $((1), (12), (13),$ and $(23))$.

4. b) The constant functions $f(x) = c$, where $c \in \mathbb{R}$.

15. Let $R = \mathbb{Z}$ and $S = \mathbb{Z}_6$ and define $f : \mathbb{Z} \to \mathbb{Z}_6$ by $f(x) = y$, provided $x \equiv y \pmod 6$. (Note that $2 \cdot 3 = 6 \neq 0$ in $\mathbb{Z}$, but $2 \cdot 3 \equiv 0 \pmod 6$.)

17. a) Define $f : \mathbb{C} \to \mathbb{R} \times \mathbb{R}$ by $f(x + iy) = (x, y)$. Proof is similar to Problem 16a.

b) $\mathbb{C}$ is a field. However, with $(1, 1)$ the unity of $\mathbb{R} \times \mathbb{R}$ with this multiplication, $(1, 0)$ has no inverse.

22. a) Define $f : D_6 \to D_{10}$ by $f(2^i 3^j) = 2^i 5^j$.

28. a) We can't deduce $x * e = x$ from $f(x * e) = f(x)$. The claim is false.

Chapter 8

Section 8.1

1. a) False b) True c) True d) False
   e) True f) False g) False h) True

3. c) Use $0 < z - y$ and axiom iii to get $0 < x(z - y) = xz - xy$. Add xy to both sides.

5. a) Use cases, axiom i for order, and Problem 3.

8. Proof. Let B be a nonempty bounded subset of $\mathbb{R}$. So there are $a, c \in \mathbb{R}$ such that for all $x \in B$, $a \leq x \leq c$. Define $-B = \{-x : x \in B\}$. Then, for all $-x \in -B$, we have $-c \leq -x \leq -a$, so $-B$ is also bounded and nonempty. By the axiom of completeness, $-B$ has a least upper bound, say, $-u$. That is, for all $-x \in -B$, $-x \leq -u$, and for all upper bounds $-a$ of $-B$, $-u \leq -a$. Then we have for all $x \in B$, $u \leq x$, and for all

lower bounds a of B, $a \leq u$. Thus, u is a greatest lower bound. For the proof of uniqueness, mimic Theorem 8.1.1. ■

9. b) l.u.b. $= \frac{3+\sqrt{17}}{2}$, g.l.b. $= \frac{3-\sqrt{17}}{2}$

14. c) The average of two rationals with denominators of 2^k and 2^{k+i} has a denominator of 2^{k+i+1} or, if it can be reduced, of a smaller power of 2.

16. a) Proof. Let A and B be nonempty bounded sets with l.u.b. u_A and u_B, respectively. Let u be the maximum of u_A and u_B. Let $c \in A \cup B$. Case 1. If $c \in A$, then $c \leq u_A \leq u$. So u is an upper bound. Case 2 is similar. Hence, u is an upper bound of $A \cup B$. Let d be any upper bound of $A \cup B$. Hence, d is an upper bound of both A and B. Thus, $u_A \leq d$ and $u_B \leq d$. Since u is one of the two numbers u_A and u_B, we have $u \leq d$, showing that u is the l.u.b. of $A \cup B$. ■

20. b) In Case 2, we don't know that $y < 0$, so when we multiply by $1/y$ we might not reverse the inequality. The claim is false in general.

Section 8.2

1. a) False b) False c) True d) True
   e) True f) False g) True

2. Any $\delta \leq \epsilon/7$ for the given ϵ. For example, a, $\delta = 0.012$

3. a) Proof. Let $\epsilon > 0$ and pick $\delta = \epsilon/5$. Let $x \in \mathbb{R}$ and suppose that $0 < |x - (-1)| < \delta$. Then $|5x + 3 - (-2)| = |5x + 5| = 5|x + 1| < 5\delta = \epsilon$. ■

6. a) Proof. Let $k : \mathbb{R}^* \to \mathbb{R}$ be $k(x) = \frac{|x|}{x}$. For a contradiction, suppose that $\lim_{x \to 0} k(x) = w$, for $w \in \mathbb{R}$. Case 1. Suppose that $w \geq 0$. Pick $\epsilon = 0.5$. Let $\delta > 0$ and pick $x = -\delta/2$. Then $0 < |x - 0| < \delta$, but $|k(x) - w| = |-1 - w| = 1 + w > \epsilon$. This contradicts the definition of a limit. Case 2 is similar, using $x = \delta/2$. ■

17. a) Proof. Let $B > 0$. Pick $\delta = \frac{1}{\sqrt{B}} > 0$. Let $x \in \mathbb{R}^*$ and suppose that $0 < |x - 0| < \delta$. Then $\frac{1}{x^2} > \frac{1}{\delta^2} = B$. ■

b) If $x < 0$, then $f(x) < 0$, not "big."

20. a) It is inappropriate to replace $|x + 2|$ with $|2 + 2|$, since x can be larger than 2. In fact, x can be up to $2 + \delta$.

Section 8.3

1. a) False   b) True   c) False   d) False
e) True

2. a) Proof. Let $c \in \mathbb{R}$ and $\epsilon > 0$. Case 1. Suppose that $c = 0$. Pick $\delta = \epsilon > 0$. Let $x \in \mathbb{R}$ and suppose that $|x - 0| < \delta$. Then $|f(x) - f(c)| = ||x| - 0| = |x| < \delta = \epsilon$. Case 2. Suppose that $c > 0$. Pick $\delta = \min(\epsilon, c)$. Let $x \in \mathbb{R}$ and suppose that $|x - c| < \delta$. Then $x > 0$, so $|x| = x$ and $|c| = c$. Then $|f(x) - f(c)| = |x - c| < \delta = \epsilon$. Case 3. Suppose that $c < 0$. Pick $\delta = \min(\epsilon, |c|)$. Let $x \in \mathbb{R}$ and suppose that $|x - c| < \delta$. Then $x < 0$, so $|x| = -x$ and $|c| = -c$. Then $|f(x) - f(c)| = |-x - (-c)| = |c - x| < \delta = \epsilon$. ∎

5. a) Proof. Suppose that $f : (a, b) \to \mathbb{R}$ is continuous at c. Let $\epsilon > 0$. From the continuity of f, there is $\delta_f > 0$ such that for all $x \in (a, b)$, if $|x - c| < \delta_f$, then $|f(x) - f(c)| < \epsilon$. Pick $\delta = \delta_f$. Let $x \in (a, b)$ and suppose that $|x - c| < \delta$. Then $|g(x) - g(c)| = ||f(x)| - |f(c)|| \le |f(x) - f(c)| < \epsilon$. ∎

7. a) Proof. For all x in the domain $[-1, 1]$, we have $-|x| \le x^2 \le |x|$. We know from Problem 2a that $|x|$ is continuous at 0. By Problem 13 of Section 8.2, with $m = -1$ and $b = 0$, $\lim_{x \to 0} -|x| = -\lim_{x \to 0} |x| = 0 = -|0|$, so $-|x|$ is continuous at 0. Thus, $\lim_{x \to 0} x^2 = 0 = 0^2$ by the Squeeze Theorem. So, x^2 is continuous at 0. ∎

b) Pick $\delta = \frac{\epsilon}{|x+c|}$. If $|x - c| < \delta$, then $|x^2 - c^2| = |x + c| \cdot |x - c| < |x + c| \cdot \delta = \epsilon$.

12. a) Let $f(x) = \lfloor x \rfloor$ on $[0, 2]$ and $Y = 0.3$.

16. d) We can't assume that $f'(c) \ne 0$ in order to divide by $f'(c)$. Also, even if $f'(c) \ne 0$, we can't guarantee that $|\frac{f(x) - f(c)}{x - c}| < f'(c)$.

Section 8.4

1. a) False   b) True   c) False   d) True
e) True   f) False   g) False

2. a) converges to $1/2$ c) diverges

3. a) Proof. Let $\epsilon > 0$. Pick N to be the smallest integer greater than $1/\epsilon$. Let $k \in \mathbb{N}$ and suppose that $k \ge N$. Then $|\frac{k}{k+1} - 1| = \frac{1}{k+1} < \frac{1}{k} \le \frac{1}{N} \le \epsilon$. ∎

7. a) Proof. Suppose that $\lim_{n \to \infty} x_n = L$. Let $\epsilon > 0$. From the given limit, with $\epsilon_L = \epsilon/2$, there is $N_L \in \mathbb{N}$ such that for all $n \in \mathbb{N}$, if $n \ge N_L$, then $|x_n - L| < \epsilon/2$. Pick $N = N_L \in \mathbb{N}$. Let $n \in \mathbb{N}$ and suppose that $n \ge N$. Then $|(2x_n + 3) - (2L + 3)| = 2|x_n - L| < 2\epsilon/2 = \epsilon$. ∎

12. Proof. Suppose that for all $n \in \mathbb{N}$, $a_n \le c$ and $\lim_{n \to \infty} a_n = L$. To obtain a contradiction, suppose that $L > c$. For $\epsilon = L - c > 0$, the limit gives us $N \in \mathbb{N}$ such that for all $n \in \mathbb{N}$, if $n \ge N$, then $|a_n - L| < \epsilon$. However, $a_n \le c < L$, so $|a_n - L| \le |c - L| = \epsilon$, a contradiction. So, $L \le c$. ∎

21. d) The first step of the argument assumes inappropriately that the infinite sum converges to some number x. It doesn't converge at all. In addition, for a geometric series to converge, the ratio must be between -1 and 1, but this ratio is 2.

Section 8.5

1. a) False   b) True   c) False   d) False
e) False   f) True   g) True   h) False

2. c) $x = \pm\sqrt{2}$.

3. c) $\sqrt{2}$ attracts on $(-\sqrt{2}, 1 + \sqrt{3 + 2\sqrt{2}})$; $-\sqrt{2}$ does not attract at all.

4. a) The fixed point $x = 2$ is not attracting. The derivative at $x = 2$ is 1. All points to the left of $x = 2$ diverge to $-\infty$.

6. b) The period 2 points, $x = 0$ and $x = 1$, attract, since the derivative of p^2 at them is 0.

7. b) The period 2 points, $x = \frac{3 \pm \sqrt{5}}{2}$, do not attract, since the derivative of q^2 at them is -4.

11. b) $N'(x) = 1 - \frac{f'(x)f'(x) - f(x)f''(x)}{(f'(x))^2}$

$$= 1 - 1 + \frac{f(x)f''(x)}{(f'(x))^2} = \frac{f(x)f''(x)}{(f'(x))^2}.$$

13. Given $x = 0.x_1x_2x_3 \ldots$ and $y = 0.y_1y_2y_3 \ldots$ and $\epsilon > 0$, there is $n \in \mathbb{N}$ such that $10^{-n} < \epsilon$. Pick z to be the rational number $0.x_1x_2x_3 \ldots x_n y_1 y_2 y_3 \ldots y_n$ (the first n decimals of x followed by the first n decimals of y). Pick $k = n$. Then $|z - x| \le 10^{-n} < \epsilon$ and $|f^n(z) - y| \le 10^{-n} < \epsilon$. ∎

Chapter 9

Section 9.1

1. a) False b) False c) True d) True
   e) True f) False

6. b)

M	a	b	c
a	a	c	b
b	c	b	a
c	b	a	c

c) No. Axiom ii forces the table to look like the following:

M	a	b
a	a	
b		b

However, no choice for aMb can now satisfy the uniqueness of axiom iv.

7. a)

M	a	b
a	a	b
b	b	a

12. (Many answers are possible.) The sentence Q refers to itself. If we add it to the axioms, Gödel's reasoning, working on the new system of axioms, will give yet another sentence Q', which is unprovable from these new axioms.

Section 9.2

1. a) False   b) True   c) True   d) False
   e) True   f) True   g) False   h) False

INDEX